PHYSICS

TEACHERS' GUIDE 1

UNITS A to G

General editor, Revised Nuffield Advanced Physics
John Harris

Consultant editor
E. J. Wenham

Editors of Units in this *Guide*
Mark Ellse
David Grace
Roger Hackett
Peter Harvey
Paul Jordan
Charles Milward
Trevor Sandford
Maurice Tebbutt
Nigel Wallis

Contributors to this *Guide*
Michael Carrick
Bev Cox
Mark Ellse
John Gardner
David Grace
Tom Gregory
Roger Hackett
Peter Harvey
Paul Jordan
April Bueno de Mesquita
Charles Milward
John Minister
Susan Ross
Trevor Sandford
Richard Spiby
Maurice Tebbutt
Mark Tweedle
Nigel Wallis

The Nuffield–Chelsea Curriculum Trust is grateful to the authors and editors of the first edition:

Organizers: P. J. Black, Jon Ogborn;
Contributors: W. Bolton, R. W. Fairbrother, G. E. Foxcroft, Martin Harrap, John Harris, A. L. Mansell, A. W. Trotter.

PHYSICS
TEACHERS' GUIDE 1

UNITS A to G

REVISED NUFFIELD ADVANCED SCIENCE
Published for the Nuffield–Chelsea Curriculum Trust
by Longman Group Limited

Longman Group UK Limited
Longman House, Burnt Mill, Harlow, Essex CM20 2JE, England
and Associated Companies throughout the world.

First published 1971
Revised edition first published 1985
Fifth impression 1994

Design and art direction by Ivan Dodd
Illustrations by Oxford Illustrators Limited

Filmset in Times Roman and Univers
Produced by Longman Singapore Publishers (Pte) Ltd
Printed in Singapore

ISBN 0 582 35417 X

The publisher's policy is to use paper manufactured from sustainable forests.

CONTENTS

FOREWORD

When the Nuffield Advanced Science series first appeared on the market in 1970, they were rapidly accepted as a notable contribution to the choices for the sixth-form science curriculum. These courses were devised by experienced teachers working in consultation with the universities and examination boards, and subjected to extensive trials in schools before publication and they introduced a new element of intellectual excitement into the work of A-level students. Though the period since publication has seen many debates on the sixth-form curriculum, it is now clear that the Advanced Level framework of education will be with us for some years in its established form. That period saw various proposals for change in structure which were not accepted but the debate to which we contributed encouraged us to start looking at the scope and aims of our A-level courses and at the ways they were being used in schools. Much of value was learned during those investigations and has been extremely useful in the planning of the present revision.

The revision of the physics course under the general editorship of John Harris has been conducted with the help of a committee under the chairmanship of K. F. Smith, Professor of Physics, University of Sussex. We are grateful to him and to the committee. We also owe a considerable debt to the Oxford and Cambridge Schools Examinations Board which for many years has been responsible for the special Nuffield examinations in physics and to the Assistant Secretary of the Board, Mrs B. G. Fraser, who has been an invaluable adviser.

The Nuffield–Chelsea Curriculum Trust is also grateful for the advice and recommendations received from its Advisory Committee, a body containing representatives from the teaching profession, the Association for Science Education, Her Majesty's Inspectorate, universities, and local authority advisers; the committee is under the chairmanship of Professor P. J. Black, educational consultant to the Trust.

Our appreciation also goes to the editors and authors of the first edition of Nuffield Advanced Physics, who worked with Jon Ogborn and P. J. Black, the project organizers. Their team of editors and writers included: W. Bolton, R. W. Fairbrother, G. E. Foxcroft, Martin Harrap, John Harris, A. L. Mansell, and A. W. Trotter. Much of their original work has been preserved in the new edition.

I particularly wish to record our gratitude to the General Editor of the revision, John Harris, Lecturer at the Centre for Science and Mathematics Education, Chelsea College, and a member of the team responsible for the first edition. To him, to E. J. Wenham, Consultant

Editor of the revision, and to the editors of the Units in the revised course – all teachers with a wide experience of the needs of students and of the current state of physics education – Roger Hackett, Nigel Wallis, David Grace, Mark Ellse, Charles Milward, Trevor Sandford, Paul Jordan, Peter Harvey, Maurice Tebbutt, David Chaundy, Wilf Mace, Stephen Borthwick, Peter Bullett, and Jon Ogborn, we offer our most sincere thanks.

I would also like to acknowledge the work of William Anderson, publications manager to the Trust, his colleagues, and our publishers, the Longman Group, for their assistance in the publication of these books. The editorial and publishing skills they contribute are essential to effective curriculum development.

K. W. Keohane,
Chairman, Nuffield–Chelsea Curriculum Trust

INTRODUCTION

THE REVISED NUFFIELD ADVANCED PHYSICS COURSE

Some ten years after the Nuffield Advanced Physics course was published, in 1971–2, the Nuffield–Chelsea Curriculum Trust undertook a revision of the course. The set of materials, of which this book forms a part, is the result of that revision.

Many people have contributed, for the most part teachers with, between them, many years' experience of teaching the original course. In particular, each of the twelve Units of the revised course has been produced by a teacher or group of teachers working together. Their names are given on the title page of each Unit in the *Teachers' guides.*

The project has been to revise an existing course – not to produce a new one – and much of the revised course follows the original one quite closely. In particular we have tried to maintain the carefully-thought-out structure of that course and to emphasize the same general aims. (These are discussed in more detail on pages xii and xvii). There are, however, some significant changes in the course content and in the materials produced.

The Consultative Committee for the revision agreed at an early stage to take the opportunity to introduce some new topics into the course. On the one hand it was agreed to include all the topics in the core A-level syllabus produced by the inter-board working group (see for example C. J. Adkins, *Physics Education*, **16**, 128–135, 1981). This has meant making sure that certain topics, not emphasized in the original course, are given due attention (for example, statics, kinetic theory of gases, circular motion, thermal conduction). At the same time the Committee also agreed to the addition of a few other topics which seem particularly important for the 1980s. Here mention can be made of nuclear and other sources of power, and some increase in the work done in electronics using integrated circuits.

To make room for these additions it was clearly necessary to omit some of the material in the original course. This has proved to be one of the more difficult aspects of the revision, and some who know the original course may be dismayed to find one of their favourite episodes has gone! But this is surely preferable to the alternatives: no change means that the course stagnates; no cuts mean that the course becomes over-full. Among items covered in the original course which are either omitted or treated more briefly now are: X-ray diffraction and unit cells; ionic crystals; the speed of mechanical waves; ionization by electron bombardment; induction motors; special relativity; the propagation of electromagnetic waves. Also some of the more detailed work has been omitted from the last two Units on statistical thermodynamics and quantum mechanics.

Another decision, relating to skill rather than content, was to pay more attention to the treatment of uncertainties in experimental results.

As well as the changes in content outlined here each part of the course has been carefully considered and, where improvements have seemed possible, or changes of emphasis desirable, these have been made. In this connection two major surveys, one carried out by Bill Trotter, and another by Maurice Tebbutt (*Physics Education*, **16**, 228, 1981), have provided valuable information, as have the innumerable suggestions made by teachers individually and at meetings.

The materials produced

The written materials produced for the revised course differ somewhat in format and in function from the original edition.

The teachers' material has been condensed to two volumes. This has necessarily meant writing as concisely as possible, and avoiding the repetition of material. So, for example, if the derivation of a formula or relationship is presented in the form of a structured question in the *Students' guide*, we have generally not reprinted the derivation in the *Teachers' guide* but only given a reference to the relevant *Students' guide* question. In general we have tried to use the *Teachers' guide* to describe and discuss teaching sequence and strategy rather than using it for the exposition of standard pieces of physics.

But a *Teachers' guide* such as this does have other functions to fulfil. Extra background information, which is not part of the course but which we think teachers might find useful for their own information, is indicated by use of smaller type. Technical details necessary about equipment for some experiments, references to questions and reading are treated similarly. We hope that this format will help to emphasize the suggested teaching sequence, while at the same time giving teachers useful support.

The two volumes of *Students' guide* contain a variety of materials: Summaries, Readings, Laboratory notes, and Questions for each Unit. The purpose of each of these is discussed in the introduction to the *Students' guide* itself.

Probably the most important point to repeat here, also emphasized in the *Students' guide*, concerns the Summaries. These are intended to be no more than summaries of the main points that a student should know after studying the Unit. The *Students' guide* is not a text book and these summaries do not contain proof, derivation, exposition, description of experiments of the kind one might generally expect to find in a text book. The marginal notes in this part of the *Students' guide* refer to questions where a particular relationship is derived or to experiments which provide evidence for a statement in the text. We expect the summary of a Unit to be most useful to students after they have studied that Unit in class, and for revision.

The Readings in each Unit are short passages, usually concerned with some application of the physics of the Unit. We have included questions to encourage students to practise the habit of reading to a purpose.

We hope that the students' Laboratory notes will make the organization of practical work easier for teachers, particularly when a number of different experiments are going on simultaneously. These notes are not intended to be precise instructions to follow, but they should contain enough information about the equipment needed and enough leading questions to help students get started on their practical work.

A few suggestions for Home experiments are included in some Units. As the name implies, these are practical activities to be carried out by students outside the school or college laboratory. There is a fuller discussion of this teaching strategy on page xxxv.

The various kinds of Questions (Introductory; Learning; Practice; Essay, estimation, and discussion; Review and revision) are identified by a code letter. The introduction to the *Students' guide* has a brief description of the purposes of the various question types.

It is worth repeating that the *Students' guide* is not a text book for the course. Students should have a text book of their own as well as their *Guide*, and we hope they will have access to more texts and reference works in the library. Nor should having a *Students' guide* prevent them from making their own notes, writing up experiments, and generally making their own record of the course as they go along.

As well as the *Teachers'* and *Students' guides* we have produced a series of background books. Some of these contain supplementary reading for students – articles either reprinted or specially commissioned which are relevant to the physics of the course. One of the teachers' background books deals with the assessment of the course (*Examinations and investigations*), and one is concerned with apparatus. We hope it will be possible to update these books from time to time.

Two packs of computer software (containing teachers' notes as well as discs or tapes of the programs themselves) have also been produced: 'Software for Nuffield Advanced Physics' and 'Dynamic modelling system' (Longman Micro Software).

Units

The course consists of twelve Units. Their titles and the suggested teaching times for each are shown in table 1.

A	Materials and mechanics	5 weeks
B	Currents, circuits, and charge	5 weeks
C	Digital electronic systems	3 weeks
D	Oscillations and waves	4 weeks
E	Field and potential	4 weeks
F	Radioactivity and the nuclear atom	4 weeks
G	Energy sources	3 weeks
H	Magnetic fields and a.c.	5 weeks
I	Linear electronics, feedback and control	3 weeks
J	Electromagnetic waves	4 weeks
K	Energy and entropy	3 weeks
L	Waves, particles, and atoms	3 weeks
	Total for 12 Units	46 weeks
	Investigation 2×2	4 weeks
	Total	**50 weeks**

Table 1
Suggested time allocations for the twelve Units.

Each student also does an independent investigation in each year of the course. The second investigation is assessed and contributes to the A-level grade (see the book *Examinations and investigations* for more details). Each investigation lasts two weeks, giving a total time of 50 weeks.

Not all teachers will want, or be able, to teach the twelve Units in this order, and it is not necessary to do so. The teachers' notes for each Unit includes a plan of the Unit showing what previous work it assumes, and which other parts of the course depend upon it, and also a section 'The place of the Unit in the course'. Teachers making up their own routes through the course should find these useful.

Examination

The course is examined by the Oxford and Cambridge Schools Examination Board, acting on behalf of all Boards. Individual schools and colleges enter candidates for the examination through the Board which they normally use, not directly through the Oxford and Cambridge Board. But teachers thinking of entering candidates for the first time may find it useful to contact the Oxford and Cambridge Board at an early date to obtain information, for example, regarding the system for assessing investigations. Past examination papers can be obtained from the Oxford and Cambridge Board.*

The examination itself consists of a variety of written papers, a practical paper, and an individual investigation. Details are given in the book *Examinations and investigations.*

THE PLAN OF THE COURSE

Clearly the success of a course such as this depends very much on the teachers teaching it and on their ability to transform the written word into an effective and exciting piece of education.

The making of a curriculum begins and ends with judgments of value: judgments about selection of content and judgments about teaching and learning strategies. No doubt different people would have made different judgments, and would have produced different ideas for a course based on those judgments.

For these reasons we think that teachers will be better able to interpret the course and use it more effectively if they have an idea of the principles which have seemed important to those who have contributed to its planning and development.

Principles underlying the course

The general aims of the course are discussed at some length on pages xvii to xxvi. The thinking behind the aims of the course and its plan is summarized here.

Aims

The Nuffield Advanced Physics course is intended to be useful to the student in his or her future life. A substantial proportion of sixth-form students go on to further education, in a wide variety of courses, usually within some pure or applied science.

One cannot teach sixth-form students all the things that they might be expected to know in the future. But we think that it is possible to select a limited number of important ideas, each of rather wide usefulness, and to teach these well. One of our basic decisions has been to sacrifice a wide acquaintance with many ideas for a deeper understanding of fewer. If the right ideas are chosen, and if students do understand them, they should be able to use these ideas in later learning of many sorts. While students are still at school it is impossible to

*Oxford and Cambridge Schools Examination Board, 10 Trumpington Street, Cambridge CB2 1QB.

predict the many different kinds of demands they will face in later life – both because they do not know what they are going to do, and because of continual changes in science, technology, and society. So in planning the course, we have tried to concentrate upon the deepest, most generally useful concepts and modes of thought within physics. We hope that a good grasp of these ideas will enable students to learn new ideas as they need them, in the future.

For the same reason, we have tried to build a course that could reveal the structure of physics: the kinds of argument physicists use and the kinds of problem they tackle. We hope, also, that students will be helped to learn in the future by finding that new problems fit into a recognizable framework.

But understanding fundamental ideas and knowing how they fit together are not enough to enable someone to learn effectively in the future. Some skill in learning and in thinking is also needed; in particular the skill required to enable one to learn from one's own reading, without the immediate assistance of a teacher. So the course attempts to develop this kind of ability, and the independence and maturity that go with it.

We have also tried to produce materials that will encourage students to become more thoughtful. While we see this as a valuable end in itself, we also see it as a necessary part of actually using ideas when one is faced with some problem, whether this arises as a part of further education or in one's later life and career. In physics, the ability to think effectively depends upon having some rather definite skills and knowledge, particularly on having some mathematical understanding. So the course includes some work on mathematical ideas and techniques within the teaching of physics.

It is important that the course should show that physics is useful, and illustrate the kind of impact that discoveries in physics have had on the way people live. We also hope that it will show something of the differences between the work of physicists and of engineers; the first continually probing and analysing, the second having to put things together for a purpose, using inventiveness and intuition, especially where the problem is too complex to be analysed in terms of fundamental principles. Therefore some parts of the course are deliberately designed with this in mind, and nearly all parts are provided with background reading about the uses of the physics they discuss.

We are not the first to hold that 'Physics is not just one damn thing after another'. A course which presents physics as a collection of more or less unconnected topics is much less likely to be effective than one which shows the unity of the subject. Ideas and concepts first encountered in one context can be applied in another, at first sight, unrelated area. We have tried to select valuable pieces of physics which could be used more than once in the course (and in future learning as well), in preference to those which would only appear once and remain isolated.

So, for example, Unit L, 'Waves, particles, and atoms', uses, among other ideas met earlier in the course, kinetic and potential energy, field and potential, and standing waves. Indeed the original course, on which this revision is based, was so well constructed from this point of view of coherence, that one of the most difficult tasks in the revision was to identify parts of the structure which could be removed (to make room for new material) without the whole edifice collapsing.

The course is intended to hang together, to tell a connected story, and to make sense in ways which are a fair reflection of the shape of the subject. It is meant to show students which are the fundamental ideas, and to help them to become better at using these ideas.

Last, but certainly not least, we have tried to provide a course which students, and teachers, will find enjoyable. Enjoyment promotes motivation and enthusiasm which surely means better teaching and more effective learning. These may, in turn, lead to a lifelong interest in the subject, and encourage more students to choose to study more physics, or a related subject, in higher education.

Themes

Another criterion used in the selection of the content of the course has been the need to represent adequately three important kinds of thinking in physics. These different kinds of thinking are themes which run throughout the course, and are visited and revisited in a variety of contexts. These themes, and their relationship to the content of the twelve Units in the course, are shown on the block diagram of the course (table 2).

The theme 'Matter and atoms' concerns the continual shifts physicists make from looking at things on the large scale to looking at them on the small scale: the constant attempt to find a microscopic understanding of macroscopic events, or to create models of what matter or atoms might be like 'inside'.

Another quite different theme is concerned with forces (including action at a distance) and the various field concepts physicists use to help them to understand the interactions of one thing with another.

The theme 'Stability and change' (shown on the right of table 2) is different again. Its concern is with the conditions for change and the description and analysis of problems involving the way things change. For example, it pays special attention to differential equations, treated in a simple way, these being vital tools in scientific thinking.

Forces and fields	Matter and atoms	Stability and change	Unit
	Solids: deformation and structure Gases	Statics: equilibrium Dynamics: collisions	A
Current; potential difference; capacitors; charge	Conduction Electrons	Decay of charge	B
		Digital electronics: gates; memory	C
		Waves; oscillations; resonance; standing waves	D
Electric and gravitational field Inverse-square law Potential		Circular motion	E
	Radioactivity Rutherford atom Ionization	Exponential decay; randomness	F
Nuclear forces	Nuclear binding energy	Fission; fusion	
		Energy supply and demand Thermal conduction Energy conversion	G
Magnetic field; electromagnetic induction			H
		Linear electronics: amplification; oscillation; feedback and control	I
Electromagnetic spectrum Electromagnetic waves		Optics: diffraction and interference	J
		Entropy Boltzmann factor Absolute temperature	K
	Photoelectric effect Spectra; energy levels Wave/particle duality for photons, electrons Atoms as standing wave systems		L

Table 2
Block diagram of the course.

Balance

We have tried to produce a balanced course: balanced in regard to content and in regard to style. It would be perfectly possible, though not very sensible, to produce a physics course at this level which concentrated exclusively on just one area: perhaps electricity, or the atom, or oscillations and waves. Or a course might emphasize one theme, such as the 'Matter and atoms' to the virtual exclusion of the other concerns of the physicist. This course tries to present a balanced view of the subject and give fair representation to the competing specialisms within it. It was largely this concern with balance that led to the decision to include the statistical approach to thermodynamics which is a feature of Unit K, 'Energy and entropy'. Or again a course might deal solely with basic and theoretical ideas of physics as if they existed in isolation, leaving no time for the engineering and technological applications of these ideas and their effect on people's lives.

We have tried to present a balanced view for two reasons. Such a view is a better representation of how the many people who helped produce the course see the subject. We are also fairly certain that a balanced course reflecting a variety of styles of thinking, dealing with a range of topics, and showing something of the usefulness and impact of physics, will appeal to more students (and teachers) than a narrowly focused one.

There is another kind of balance we think important to strive for: balance of style. So, for example, we hope that from Unit L, 'Waves, particles, and atoms', students may begin to understand the excitement of a piece of theoretical physics. But the style of Unit C, 'Digital electronic systems', and Unit I, 'Linear electronics, feedback and control', is closer to engineering: systems are built up rather than taken apart for analysis.

Successful learning involves a variety of activities. We do not believe that any single teaching or learning strategy is so successful that it can be used to the exclusion of all others. So, while following the course, we expect every student to engage in many different activities including independent study (questions, reading); working with one or more fellow students on an experiment; reporting results of experiments, or what has been read, to the class; discussion; a concentrated spell of independent work (the investigation); watching films or videos; using computer programs; and so on.

AIMS

Some of the general aims of the course were touched upon in 'The plan of the course' on page xii. These aims are discussed more explicitly here, with examples of the objectives they imply, and ways in which the course tries to achieve those objectives.

These aims are not offered as ideal aspirations which are impossible to achieve with any but exceptional students. They are intended to imply practical objectives, to be developed in the course and to be tested in the examination.

Teachers should know that there is also a brief discussion of the aims of the course, its methods and strategies in the introduction to the *Students' guide.*

Learning in the future

Few things are more certain than that students will need to learn more science – pure and applied – in the future. Many will go on to higher education, and after that will need to learn yet more for a career. Students will be helped in their future lives in a changing society, if they can learn more and adapt themselves to new careers. It is not possible to predict all their future needs in detail. We hope that, as a result of the course, students will be able more successfully to meet such demands.

Language

One objective that follows from this aim is the knowledge of enough of the language of science, particularly of physics, to enable students to read without undue difficulty, and to take part in discussions. They should be able to recognize some terms when they meet them: examples include stress, strain, elastic modulus, plastic, elastic, and resistivity, insulator, conductor, to take a few from early parts of the course. They should also be able to use terms clearly enough to make up effective arguments. Neither of these objectives will be achieved simply by learning definitions, but by talking and reading about ideas, and by using them often.

Learning new ideas

It seems to us that later learning will be assisted by a first acquaintance with two kinds of ideas. One kind – the basic ideas like charge, field, and potential – have long and rightly formed part of school physics courses, though their difficulty may not always be matched by an adequate amount of time being given to them.

Other kinds of ideas are those like the application of quantum ideas to the structure of the atom, and a simple statistical approach to thermodynamics. Before the publication of the first edition of the Nuffield Advanced Physics course such ideas had not been thought suitable for introduction with any degree of seriousness in A-level courses. It is certainly true that no school course could set an objective of fully understanding such ideas; but we do think it likely to be helpful in the future if a student has at least met the ideas which form the basis of physics as it now is, and which will be used in a wide variety of fields, including chemistry and many applied sciences. Experience of the last ten years suggests that the introduction of these ideas was a good decision, and that the more 'modern' parts of the course have been appreciated by teachers and students alike.

Reading

The general aim of assisting learning in the future implies other sorts of objective, concerned with the skills needed to learn effectively. We hope, as a result of the reading suggested as an integral part of the course, that students will become better at extracting information from books and from papers. We hope they will know that books differ in quality, and will be better able to find the ones that suit them. We hope also that students will have come to expect to find interesting new material in magazines, and will tend to seek more of it for themselves. We know it is not always easy to persuade students to read, but we do think it important enough to be worth the effort needed from the teacher.

Arguing

Learning probably involves being able to sustain a discussion with others, and maybe with oneself. Here the emphasis in the course on students reporting to other students on their reading and experimenting has an important part to play. The teacher's role is decisive: in encouraging the first fumbling attempts to express ideas and arguments, and in discerning and building on what is of value in what students say. It is all too easy unintentionally to stifle discussion by well meant attempts to ensure that every error is dealt with, which, in effect, seems to prove to the student that he or she is incapable of effective argument.

Mathematics

Mathematical skills will often be needed in the future. Some of the most generally useful ones are those connected with changes and rates of change. Simple approaches to first and second order differential equations are developed within the course, and the techniques are used again when opportunities arise. The rapid growth of computing in industry, research, and education is changing the balance of emphasis between analytical and numerical methods. Thus we suggest numerical rather than analytical approaches to several problems involving rates of change. We certainly do not want to see ready-made computer programs used exclusively for these problems. But expressing the physics of the situation in mathematical form, putting the mathematical equations together in an appropriate order, and carrying out a few steps of the iterative calculation either numerically or graphically, should help students to understand the physics of the problem, what it is that the computer can do, and some of the work necessary to obtain a computer solution.

More generally the course should provide plenty of opportunity to develop and practise the mathematical abilities listed by the inter-board working group on the core syllabus for A-level physics (see, for example, C. J. Adkins, *Physics Education*, **16**, 128–135, 1981).

Translating information

Much of the information a student will need to acquire in the future will be presented in a coded form. We have put a special emphasis on translating information between graphical, verbal, and numerical forms.

Independence

An important difference between learning in school and learning afterwards is that students need to become more independent and self-sufficient. Probably no one school course can do very much about this. But we hope that various aspects of this course will help a little to promote independence and a mature approach – as, for example, when students are given separate individual tasks and asked to take on their own individual investigations.

Understanding physics

This aim overlaps a good deal with the first: 'Learning in the future'. The aim is that students understand the important physical concepts and relationships taught in the course. As explained in 'The plan of the course' we have tried to concentrate on a comparatively small number of powerful and generally useful concepts that are of quite wide application, rather than the details of many specific pieces of physics. Take, for example, oscillations: we expect students to know that the condition restoring force $\propto -$(displacement) leads to simple harmonic motion; that for S.H.M. the frequency is independent of amplitude, and the displacement varies sinusoidally with time. But we do not expect students to memorize the particular formulae for the frequencies of a number of different oscillators (pendulum, torsion balance, liquid in U-tube, and so on) – though, given a suggested formula, they should be able to show whether or not it has an appropriate form. In electronics we do not expect students to know the details of the construction or operation of any particular kind of transistor. But we do expect them to appreciate the general concepts of switching, amplification, feedback, and oscillation, and be able to use them in specific contexts.

Time needed for teaching for understanding

Teaching for understanding takes much time; time that is needed for students to try out their ideas in discussion, time to make mistakes and muddles and to get out of them again, and time to reflect and think privately. This is why we have deliberately tried to reduce the number of pieces of physics in the course as a whole, to concentrate on the most generally useful ones, and to limit the time spent on less important byways.

Examination questions

The examination reflects the aim of understanding physics. Formulae, data, and other detailed information are given, so that time is not spent on memorizing information better stored in books. Questions usually ask for knowledge to be used rather than recalled and set down.

Understanding

To devise tests for understanding, one must have some idea of what understanding means. We mean more than recall or recognition, but less than the ability to solve new problems.

The latter serves well enough as a rough guide, but is too strong to be taken literally; every physicist is only too conscious of his or her incapacity in the face of a really novel difficulty. What seems to us to be the mark of competence is the ability to talk sense about a physical problem; to produce relevant, sensible thoughts rather than irrelevant, completely muddled ones. It is this level of understanding which we aim for: students should be able to contribute effective arguments towards the solution of a problem.

Gradual development of concepts

If students are to grasp the basic concepts well enough to use them in argument, plenty of time must be given in the course to the teaching of each concept. For example, the simple but essential idea that electric charge is a quantity of something, not a strength or intensity, needs more than a passing mention. Indeed, Unit B, 'Currents, circuits, and charge' contains a long piece of experimenting with capacitors intended to bring out this idea. Similarly, Unit D, 'Oscillations and waves' devotes time to experiments about what is summed up in saying that waves superpose. Some may feel that we have given these simple, largely qualitative principles too much time. We would prefer to have done that, rather than to give them too little.

We have also planned the course so that it returns again and again to the important concepts, developing each a little further at each meeting, or showing a new use in a fresh field. Students will come to expect to use old ideas in new ways; and the renewed acquaintance should strengthen their grasp of the ideas. At the start of each Unit in the *Teachers' guide* there is a Plan of the Unit. Part of its purpose is to point out how earlier ideas are used in the Unit, and how the new ideas developed in the Unit are used later in the course.

Teaching for understanding

It is the way the course is taught, not what is taught, that will contribute most to the aim 'Understanding physics'. Teachers need to encourage much talk and discussion so that students can rehearse their ideas, make mistakes and correct them, and develop a belief that they can talk sense about physics. In the *Teachers' guides* we have suggested ways of promoting talk and discussion, and ways of approaching pieces of physics as problems deserving argument and experiment.

Understanding the nature of physical inquiry

We hope that students will become better able to talk sense about the various kinds of activity in scientific inquiry: about testing a theory, making a deduction, trying an experiment, or making a guess. They should be able to recognize the differences between these activities – not as a matter of vague philosophizing, but as a matter of practical importance in deciding what is going on in an argument. For example, we would expect students to be able to answer questions like this:

'How good a description of Rutherford's method of argument for a nuclear picture of the atom is each of the following?

A He imagined a model, and compared the consequences with experiment.

B He summarized a collection of experimental evidence in a simple law.
C He chose the simplest model, for lack of experimental evidence.'

Discussion of theories and models
The course, like physics, makes use of models of various kinds, and there should be discussion of their roles, usefulness, and limitations. For example Unit A, 'Materials and mechanics' introduces models of various solids (metals, ionic crystals, polymers) in which the individual atoms or ions are represented by balls arranged in particular patterns. But in a gas the molecules are randomly placed. Such models are certainly useful – to what extent they are 'true' representations should be discussed. Work on fields should draw attention to the value of inventing theoretical concepts (such as field, potential) and their practical usefulness.

At several points, especially in work on oscillations in Unit D, the role of mathematical models can be brought out, and questions asked about whether the real world does, or could, conform to an idealized model, or how far the model should conform to the way things are. Unit L, 'Waves, particles, and atoms' raises even deeper issues about models in the context of the nature of light and of matter.

Guesses and estimates
Some questions in the *Students' guide* ask for guesses or estimates of orders of magnitude, illustrating the value of such methods as well as helping to develop better physical understanding.

The value of understanding what physics is like
We think this important for several reasons. We wish students to see physics as what it is, and not, for example, as magic, absolute truth, or arbitrary theory. In later learning students will often meet models, theories, experiments, and guesses. The conflict of wave and particle ideas is likely to recur for many, while others may have to think about the purpose of building a small-scale test model of a bridge, or to assess a mathematical model of an industrial activity. A beginning in thinking about such situations can be made in school.

This aim is also connected with the aim of understanding physics, for it should be easier to argue effectively about a problem if one is clearer about what kind of problem it is. For example, the distinction between a model and the data it is used to throw light on, is crucial to understanding what is going on in most scientific arguments.

Learning to inquire

Many students will do science of some kind in the future, at various levels ranging from research and industrial application to tackling the practical problems of everyday life. We hope that they will become more successful at pursuing their own personal inquiries in physics.

We do not know how to teach people how to inquire, and the advice of Bruner seems the best:

'Practice in inquiry, in trying to figure out things for oneself is indeed what is needed – but in what form? Of only one thing am I convinced: I have never seen anybody improve in the art and technique of inquiry by any means other than engaging in inquiry.'
BRUNER, J. S. *On knowing*, page 94. Belknap Press, 1964.

Much of the work of the course can be taught in an inquiring, rather than a didactic spirit. For example, the work with puzzle boxes and with capacitors in Unit B, 'Currents, circuits, and charge' is intended to help teach simple ideas in a problem-like way. Several Units begin with pieces of rather open experimenting, whose purpose is partly to give students a concrete idea of what is being talked about next, but partly also to introduce a feeling of openness and inquiry.

Investigations

We hope that experiments in the course will be undertaken in a spirit of investigation. But it is inevitable that, to achieve the ends in mind at each point, carefully designed apparatus has to be provided, which limits the possibility of 'wrong' paths of investigation and steers students towards the ideas which we hope will emerge.

By way of contrast, part of the course consists of some completely open investigations for students to undertake on their own. They are not a marginal item, to be omitted if time is short, but a fundamental part of the design of the course. Students are expected to identify their own problems, invent their own (simple) experiments and deal themselves with the difficulties that arise. Topics will be very simple (or will seem so to teachers), and are not necessarily limited to topics covered in the course. They will be chosen so that there are likely to be some surprises and problems as well as a reasonable chance of some achievement. We have in mind such problems as 'What happens when rubber bands are stretched?', 'How does water emerge from a jet as the speed of flow varies?', or 'How does a ball roll on a spinning turntable?'

These investigations, one in each year, form an integral part of the course. They aim to help students to become better at doing physics and not to teach them more physics, so the choice of topic can be very wide.

In revising the course we have been impressed by the value which teachers place on investigations, by the enthusiasm which students bring to them, and by the value of much that they achieve. This has encouraged us to keep open the modest amount of time they require, rather than give in to the pressure to include extra content. It has also helped to confirm the view that investigations can make a unique contribution towards the important aim of learning to inquire.

The investigation is discussed at greater length in the book *Examinations and investigations.*

Use of instruments

Another necessary part of being able to inquire is the possession of some practical skill. An objective here is that students should come to handle competently the usual ranges of ammeters and voltmeters, and also oscilloscopes, as well as the simpler laboratory instruments. Other instruments, such as the signal generator and the scaler, should also be familiar.

Students should develop a critical awareness of the factors which affect the uncertainty of their measurements. These factors include the limitations of measuring instruments themselves. We hope that by the end of the course students will appreciate some key concepts including range, resolution, sensitivity, response time, loading effects. Understanding such concepts will help them to choose the most appropriate instrument for a particular job, and to appreciate the limitations of the measurements obtained. We do not expect anything at a high level, or a formal treatment here. To take a few examples: we would expect students to know that they would need a picoammeter rather than an ammeter or milliammeter to measure an ionization current or the current in a photocell (sensitivity); that an oscilloscope would be more useful than a moving-coil meter (or digital meter) to follow the rapid discharge of a capacitor (response time); that a low resistance voltmeter would not give a good measurement of the e.m.f. of a cell with high internal resistance (loading effect).

A related aspect is the way in which the uncertainties in individual measurements are combined to give an estimate of the maximum uncertainty in the result. Here we certainly do not expect any statistical treatment; simply the standard techniques for products or quotients, and sums and differences.

There is more detail on these matters in 'Training in experimental work' on page xxxi of this book, and in 'Experimental work' on page xii of *Students' guide 1.* Specific issues should, of course, be brought out in the context of actual experiments and demonstrations. In both *Teachers'* and *Students' guides* we have suggested some such occasions.

Enjoyment – seeking to gain further understanding

Students will not learn more science in the future, nor will they learn much during the course, if they do not enjoy learning it now. The original course team tried to make the work pleasurable, possible, and profitable. Many comments about the course suggest that it has had some success in this aim.

This course is about physics, but the majority of the students will have to put whatever understanding of physics they have gained to use in other fields. Without enjoyment, there is little hope of their trying to apply ideas and methods learned in physics to these other fields. So the aim of enjoyment seems necessary in order to achieve our other aims, especially those that look to the future, where students will more and more have to find their own rewards and do without the external praise or blame of a teacher.

The course involves many pieces of individual work for students, all intended to be tasks at which they can succeed. Successfully solving a problem for oneself can be very rewarding, as well as contributing to other aims. We hope also that students will become sufficiently personally involved in their investigations to want to do more in the future.

Awareness of the social significance of physics

It would be wrong to expect students to acquire one particular attitude towards the relationship of science and society, and it is too much to hope that they will develop a deep and lasting concern with such problems simply as a result of taking this course. But we do hope that at least they become conscious of the existence of questions about the influence of physics on society, about the ways in which physics can be applied to meet human needs, and about the need for people willing to apply their talents in that direction. And we hope that students will leave the course knowing that there are opportunities open to people who want to work in engineering and technology of one kind or another.

Several of the Units have a particularly practical flavour. The two short Units on electronics are concerned with the creation of useful circuits, and challenge students to solve some specific problems: the approach here is intended to reflect the spirit of engineering as much as that of physics. In Unit G, 'Energy sources' the discussion of thermal conduction is based on the important practical problem of energy loss from a home. And we have tried to maintain as practical a flavour as possible in Unit H, 'Magnetic fields and a.c.'.

But our attempt to make students aware of the social significance of physics is not limited to certain Units which lend themselves to a particular approach. Much of the reading which is part of the course – both background articles and the shorter pieces for each Unit in the *Students' guides* – are on applications of the physics in the Unit: the use of radioisotopes; electrostatic photocopiers and printers; the avoidance of oscillations in large structures, and so on. Questions – both in the *Students' guides* and in the examination – often require students to use their knowledge of basic physics in unfamiliar situations involving Man-made or natural artefacts (sparking plugs, windmills, icebergs...).

We hope in these ways to be able to do something to combat the impression, often given at school, that the pure physicist is a more interesting or more worthy sort of a person than the engineer or technologist.

TEACHING THE COURSE

Here we discuss the practical problems teachers may meet, the problems of matching the course to previous courses, mathematics, experimental work, and some of the suggested teaching methods.

Time required

The course is designed to fit into the average sixth-form time allocation, which may be about seven periods per week (each of 40 minutes), or their equivalent. In planning, we assumed that a term contains ten useful weeks, and that there are five useful terms, thus, we hope, leaving time for desirable things like school plays and outings, for revision at the end of the course, and for teachers to develop their own interests at greater length. The suggested time allocations for each Unit are given on page xi, table 1.

Experimental and theoretical work go together in much of the course, and as much of the teaching time as possible should be spent in a laboratory. It would be very difficult to teach the course at all if fewer than four of the seven periods were taught in a laboratory.

It is convenient if much of the work is time-tabled in double periods: a system which sets aside a long session for practical work and so makes practical work more difficult at other times is less convenient.

Laboratory requirements

Gas and water are not often needed; the main need is for plenty of mains electricity outlets. Two mains sockets per pair of students would not be too many.

Low voltage alternating or direct voltage supplies to the benches, permanently wired, need not be provided, if suitable portable sources are provided. Dry cells can be used for many of the experiments. Further suggestions about power supplies appear in the *Apparatus guide*.

A laboratory design which allows very flexible use of space is ideal.

Safety

Since the first edition of the course was published the education service as a whole has become more safety conscious, and in preparing the revision we have borne safety considerations in mind. Although there are only a few experiments in the course where changes have seemed necessary, observant teachers who are familiar with the original course will, we hope, notice them. Thus, for example, the potentially dangerous h.t. power supply is no longer recommended for the demonstration of electrophoresis ('Conduction by coloured salts'); safety spectacles have been added to the apparatus list for several experiments; plutonium is no longer a recommended radioactive source. While we have included notes on safety and suggested precautions where it seems appropriate, it is of course finally the responsibility of the teacher in the laboratory to ensure that due care is taken in any potentially dangerous situation. We draw teachers' attention to the DES booklet *Safety in science laboratories* (3rd edition, 1978), and in particular to the Administrative Memoranda which deal with the use of ionizing radiations (7/76 or later) and lasers (7/70 or later). CLEAPSE School Science Service has made helpful comments on the safety aspects of some of the suggested practical work.

Links with Revised Nuffield Advanced Chemistry

Teachers should be aware of the development of ideas in other science courses, seek opportunities to relate them to their own, to use them, and even, perhaps, to arrange for occasional joint teaching.

We have a particular responsibility to ensure that schools teaching both Revised Nuffield Advanced Physics and Chemistry are aware of where the courses overlap. The main overlaps are shown in table 3. It indicates which Topic in the Chemistry course deals with an idea that is also in the Physics course.

Revised Nuffield Advanced Physics	Revised Nuffield Advanced Chemistry
Unit A, 'Materials and mechanics'	
the mole	Topic 1
kinetic theory of gases	Topic 3
Unit F, 'Radioactivity and the nuclear atom'	
Rutherford atom; neutrons; isotopes; ionization energy	Topic 4
Unit H, 'Magnetic fields and a.c.'	
mass spectrometer	Topic 4
Unit K, 'Energy and entropy'	
diffusion of molecules	Topic 3
arrangement of quanta	Topic 4
vapour pressure	Topic 10
rate of reaction	Topic 14
equilibrium	Topics 10, 12, 15
Unit L, 'Waves, particles, and atoms'	
spectra, energy levels	Topic 4

Table 3
Links with Revised Nuffield Advanced Chemistry.

Previous physics courses

The original Nuffield Advanced Physics course was written at a time when there was considerable debate and uncertainty as to the future pattern of sixth-form studies, and of the A-level examination. (The debates about the alternatives: first Q & F, then N & F, are now almost forgotten!) And the revision has been done against a background of debate and uncertainty about the future pattern of examining at 16+. Perhaps it is inevitable that curriculum development is surrounded by some such uncertainties. But two other things look fairly clear: that the pattern of three-year university and polytechnic courses leading to a first degree will remain; and that, in an attempt to provide (at least to some extent) a common starting point for degree courses, all A-level examinations in physics will be expected to encompass a common core syllabus. Placed, as we are, in a position of tension between more specifically stated requirements from above, and more uncertainty below we have made what seems to be the only possible decision. We have provided a course which takes as its starting point a level of knowledge and understanding that is currently expected for the O-level examination. It is unreasonable to expect students to get to A-level standard in a normal two-year sixth-form course if they start without this basis.

But O-level courses vary, and even if the proposed 16+ examinations based on agreed aims, objectives, and minimum core content are introduced, not all students will be starting their A-level courses from the same position. We have tried to indicate (in the *Teachers' guides* and to the students through introductory questions) what we assume students to know before they start the work of a particular Unit. For some the introductory work on a particular topic may be familiar (but still perhaps useful as revision). For others it may be necessary to fill in some background knowledge which we have assumed.

Although this course is intended to be consistent with the aims and the content of the Revised Nuffield O-level Physics course, we do not assume that course as a starting point. However, there are some specific features which affect the teaching of this course to students who have followed the earlier Nuffield course. These are outlined below.

Dynamics

The Nuffield O-level course has a very full treatment. Students may feel that they already know the material dealing with momentum and collisions in Unit A, 'Materials and mechanics' of this course. Nevertheless some revision will probably do no harm.

Kinetic theory of gases

This topic is part of the core syllabus in physics at A-level. It is also an important part of the Nuffield O-level course. The treatment offered here goes a little further, but students who have followed the O-level work will have a good start.

Conservation of energy

Students who have not met the principle of conservation of energy at all will be in considerable difficulties. The O-level course deals with it thoroughly, and the Advanced course provides no special place for teaching this principle which is used in many different contexts almost from the beginning of the course.

Optics

There is no geometrical optics in this course. We think that the Nuffield O-level Physics course includes just about the right amount, concentrating on the formation of images. We have added no more to the Advanced course. Like any other omission this is sure to be regretted by some, but it is an example of the way in which we have tried to keep down the number of topics studied to allow more time for understanding.

Mathematics

Parts of mathematics which we regard as especially valuable in the education of a scientist are taught within the physics course, along with the physics for which they are used. These are the handling of simple differential equations, and the exponential and the sinusoidal functions. Other pieces of mathematics, mainly various integrations, can be avoided, and ways of avoiding them are suggested. There is advantage in using mathematics when students have it, and many university physics teachers make a plea for this to be done. But there is a disadvantage in insisting on a formal treatment when it is not necessary to achieve a physical understanding of a problem, if students find the formal treatment opaque, difficult, or frightening.

We have included several informal physical arguments, such as that used in question 45 of Unit E, 'Field and potential' to explain, without doing a formal integration, how a flat sheet of point charges, each giving an inverse square field, together produce a uniform field. But if we have to err, this is the side most of us prefer to err on. (Dirac, himself a mathematical physicist, made a good remark about this matter. He said, 'I understand what an equation means if I have a way of figuring out the characteristics of its solution without actually solving it'.)

But stressing this informal kind of argument, which we believe can be an important step in understanding, should not be taken to imply a devaluing of certain more traditional skills. For example, it is important that students are confident in simple algebraic manipulation (changing the subject of an equation, and so on). Teachers may well need to provide more examples than we have been able to in the limited space available.

Attached to the core syllabus for physics at A-level is a list of mathematical abilities which is regarded as part of A-level physics. This list is reprinted in the *Syllabus statement* for the course, available from the Oxford and Cambridge Schools Examination Board. This course should provide ample opportunity for developing and practising these abilities.

Students who also take a sixth-form mathematics course should have no special difficulties. Others may have two sorts of difficulty. They may have trouble with the bread and butter mathematics used in the course: proportion, powers of ten, logarithms, and so on. They may also need more practice with the mathematics taught within the physics course than there is time for.

The teachers' book, *Supplementary mathematics*, prepared for the original version of the course, will help with these problems. It contains revision of mainly O-level mathematics, some introductory calculus, more practice with differential equations and with the functions needed in the course, and a little on statistics. The time needed will vary with the particular students involved. For most students not taking sixth-form mathematics a one-year course of about four periods a week should suffice.

Training in experimental work

There is a great deal of experimental work in the course and, as is appropriate at A-level, one of its functions is to promote a critical approach to accuracy. The experiments themselves range from frankly rough measurements to procedures where an appreciable degree of precision is to be expected. Experience shows that there is a danger here, that pupils may tend to be impressed by the need for maximum care and thought in the 'accurate' experiments, but to regard the 'rough' experiments as not calling for much in the way of critical thinking.

The first thing that needs to be emphasized, therefore, is that *every* experiment needs to be thought about carefully as regards its significance. Indeed, in a sense, the more uncertainty there is in an experiment, the *more* vital is it to assess the uncertainty: only by doing so can we decide whether the outcome is significant or merely indeterminate.

An agreed vocabulary is called for. As a general point, we feel that the phrase 'experimental error' is to be discouraged. A very clear distinction needs to be maintained between a human experimenter's failure to act sufficiently carefully (which is truly 'error'), and the unavoidable limitations imposed by the nature of the instruments. These we refer to as *uncertainties.*

Uncertainty begins with the limitations of measuring instruments. Formal teaching on this is possible, but most teachers seem to agree that it is preferable to proceed informally, making the relevant points at appropriate opportunities as the course proceeds. By the end of the course students will be expected to understand the significance of a number of key concepts. These are elaborated in the passage 'Experimental work' in *Students' guide 1* (page xii), and comprise: range, resolution, sensitivity, response time, accuracy, linearity, repeatability, systematic error, drift, loading effects, validity. Although this may look a fairly long list these concepts are familiar to teachers, and many will seem 'common sense' to students.

From the individual uncertainties of readings, we need to be able to calculate the resulting uncertainty in the final outcome. This uncertainty can be expressed in either of two ways. It can be as a 'Standard error' (standard deviation) which says in effect, 'I believe there is about a

70 per cent chance that the true value lies between these limits', or it can be as a 'maximum uncertainty', which says, 'I am confident that the value is not outside these limits'. In a real-world context, which of these one needs to know depends on the individual case. Standard deviation is appropriate for building bricks, but 'guaranteed minimum contents' is needed for packets of sugar. For A-level purposes, we feel that simple 'maximum uncertainty' is the appropriate criterion. It avoids statistical arguments (which are often in fact rather meaningless where small numbers of readings are concerned), and its meaning is immediately understandable. Should it occur to a student that the chances of all of six readings conspiring to produce a low result are somewhat remote, then all credit to him, but this is not an observation we would necessarily expect at A-level.

The techniques for calculating the uncertainty of the final result from the uncertainties in individual measurements are familiar to teachers. They are explained for students in the passage 'Experimental work', together with some notes on basic graph techniques. The notes on uncertainty calculation cover the treatment of algebraic formulae only, since we do not expect students necessarily to compute uncertainties involving trigonometric or exponential functions. As regards graphs, students will need further guidance, for example, in the transposition of formulae required to give straight line plots, or in the use of logarithmic plots for power law determinations.

For some experiments we have included, in the *Teachers' guides*, specific suggestions on the way in which uncertainty might be discussed. These examples, we hope, will help to make clear the level of treatment we have in mind.

One question frequently raised is: to what extent are students expected to write formal reports on experiments? This is a matter very much for the teacher's discretion. There are so many experiments that the answer may well be: comparatively rarely. Students will want to make notes, and in the main it may be best to leave it to them to decide what to write.

Occasions should certainly be found, however, for each student to write a good coherent account of some experiment, and opportunities for this occur at several points, where simultaneous experiments are exploring different aspects of the same theme, and each experimenting group needs to pass on its conclusions to the rest of the class. The aim, overall, should be that by the end of the course each student can if necessary write an acceptably rigorous account of an experiment. Investigations, of course, provide an ideal opportunity to concentrate on this.

Teaching methods and organization of practical work

At the risk of offending those to whom these ideas are obvious, or to whom they are second nature, we outline below some of the teaching methods suggested together with other information about organization, particularly of practical work. There is of course a vital link between methods and the aims discussed on pages xvii to xxvi.

Discussion

It would be unreasonable to expect a student to be able to talk sense about physics if he or she had not been encouraged to talk a good deal beforehand. What little is known about learning beyond the early years of life suggests that talking about ideas, thinking them through by oneself, trying them out on others and so discovering muddles in one's thinking, and having one's errors corrected by others who are more expert, are important ingredients in learning. It seems to be especially important in 'getting ideas inside oneself'. Most teachers have noticed that it is only when we try to explain something to someone else that we find out how little we understand ourselves.

Another reason for encouraging discussion is that a subject breathes and has its life, not in books, but in conversation.

Therefore, we have tried to find ways of encouraging discussion that goes beyond question and answer exchanges between teacher and class, and that shows some sense of the conjectures and dilemmas which are the essence of a living subject.

Some of the questions in the *Students' guides* (category E – essay, estimation, and discussion) may be useful as starting points. It may be wise to brief one student beforehand to prepare some ideas, or to ask all students to write down something.

Some demonstrations are best as a polished performance by the teacher, but many others can be arranged as the focus of a discussion, with students helping in the performance of the demonstration, and doubtless hindering it with well meant suggestions. It is also useful to ask a student to prepare a demonstration for a later lesson. Students are more likely to be critical and thoughtful about a demonstration done in this way, than they are about one done by the teacher. The teacher is then free to play the role of a critical, questioning physicist, without the double burden of also trying to make a point with the apparatus.

Different experiments shared round the class: reporting back

On a number of occasions in the course, when several different but related experiments are possible, we have suggested that each pair of students does only one, and later reports its findings to the class. In this situation, each student has to be very clear about what he or she has done. Others have to learn from him or her and, with encouragement, soon learn to ask searching questions. The teacher has an important role: that of setting a standard of critical thought and questioning.

The first one or two occasions when this is done may prove to be very painful. Reports are ill prepared, data are presented in a confused way, and the experimenters reporting back blame everything but themselves. But it does seem that students' desire to do well is strong enough to produce quite rapid improvements.

Besides stimulating discussion, this technique helps students to become a little more independent. It seems worth while giving them the responsibility not only of reporting verbally, but also of preparing a written summary of the experiment and of the results, for duplication to the rest of the class if the experiment is important enough. Again, criticism from their colleagues can quite quickly improve the quality of such notes.

There are difficulties in organizing such experiments: apparatus may be unfamiliar, it may not work, or students cannot think what to do. The students' laboratory notes are intended to help them get started by showing how to set up the apparatus and suggesting questions that can be answered with it. We hope they will ease the burden on teachers in these situations.

Individual exploratory experiments

The essence of an experiment is that one decides to try something, not knowing what will happen. The conventional school experiment is, in these terms, barely an experiment at all. Everything is decided beforehand, and what will, or 'should' happen is easy to guess, if it is not already obvious.

We think that sometimes, not always, students ought to have to decide for themselves what to do with some apparatus, and see for themselves what happens. This can only be done if the apparatus is simple, and is best if almost anything a student can do with the apparatus is interesting in some way.

A few experiments in the course can be used in this way. For example, before electric charge is discussed in Unit B, 'Currents, circuits, and charge', students can be given, say, one capacitor, a single dry cell, and a single milliammeter. If they connect them in series, the meter gives a single flick, returns to zero, and stays there. Breaking and making the circuit produce no further change (indeed, some may miss the initial flick). Several things may suggest themselves. Connecting the charged capacitor and meter together without the cell produces a flick in the opposite direction, after which the first experiment 'works' again.

A second meter shows that there is an equal current pulse in each lead to the capacitor. Further cells on offer suggest more experiments.

The teacher can control events by rationing or issuing apparatus, or by offering to provide what is wanted if a reason for wanting it is given.

There will be failures. Just as, if students are given clear instructions, some will not follow them, so, if they are asked to decide for themselves what to do, some will find it hard. Sometimes failure can be turned to advantage. The confusion that results when one has no plan can bring out how an experiment involves deciding what to do.

Home experiments

The suggestion that the course should include some independent practical activities for students to carry out at home came from a teacher, M. R. Moore, who had developed this teaching strategy in his own school. The idea was well received by other teachers and we have, accordingly, included a few examples in several Units under the heading 'Home experiments'. Teachers will, no doubt, be able to think of others.

Although the home experiments are to be done in the students' own time, some class time will be needed when a task is assigned, especially for discussion and comparison of results. It may sometimes be useful to introduce an element of competition into these activities.

Reading and text books

We think it important that students should read books and papers. One reason is that the skills of extracting information from scientific writing are not trivial ones, and need practice.

Another reason concerns authority. Only by looking at several sources can students find out which things are commonly agreed, and which are not. So they should be encouraged to use more than one text book, and also a variety of other reading, from paperbacks to original papers. By reading widely students will discover that scientists communicate in more than one tone of voice, and not only in the level monotone of many a text book.

We know that it is not easy to get students to read. Merely exhorting them to do so will not succeed. We think that reading must be built into the course, and have suggested several ways of doing this.

Early in the course, in Unit A, 'Materials and mechanics', there is an opportunity for students to be sent away to find out about modern, composite materials. Each can have a different task, one to read about reinforced concrete, another about glass fibre, and so on. Later, each can be asked to present a summary report to the others. In Unit F, 'Radioactivity and the nuclear atom' there is opportunity for students to read about and report on various properties of the alpha particle, and the experimental evidence before going on to learn about their use in scattering experiments. Part of Unit G, 'Energy sources' is structured in a way which relies heavily on students reading and reporting back on a variety of energy alternatives. Reading exercises can also be made part of Unit H, 'Magnetic fields and a.c.' and Unit L, 'Waves, particles, and atoms'.

The emphasis on individual reading tasks is a way of making it more likely that they will be done, and also a way of encouraging the growth of mature independence. Reading assignments like the examples given here, together with the extracts reproduced in the *Students' guides*, the background articles, the references and lists of books, are an attempt to assist teachers help students to learn how to read, and to discover how to do it better.

Since the original version of the course was published several new text books, or new editions, covering its content have appeared. Text books are included among the sources suggested for reading tasks. The *Teachers' guides* also contain references to a number of text books, indicating which of them contain good material on the various parts of the course. Sooner or later, students must find out that books differ in their adequacy on various topics, and that their treatments suit different individuals.

The teaching of theory

An American teacher once said that the best audio–visual teaching aid she had ever seen was 'a teacher on her own hind legs, her back to the blackboard, a piece of chalk in her hand, talking with her class about something that mattered to all'.*

When theory is taught at the blackboard, the clarity, precision, and sweep of a scientific argument can be given powerful expression.

But such teaching has its dangers. Too often one finds one person at the front doing all the work, facing a dozen other people whose minds are not engaged with the matter in hand. If questions are asked, answers tend to come from those who answered the time before (and the time before that) unless the teacher adopts the useful strategem of asking everybody to write down an answer.

Some pieces of theory can be dealt with in other ways. Most of the important theoretical arguments in the course are covered in structured learning questions in the *Students' guides*. These are intended to lead a student, step by step from what he or she already knows, to some new

*Quoted by JENNINGS, F. G., in MILES, M. B. (Ed.) (1964) *Innovation in Education*, page 565, Bureau of Publications, Teachers' College, Columbia University.

result. They are not self-contained learning programmes, and it may be necessary to spend time after students have tried them, clearing up difficulties. It will also be important to help students to go beyond the detail, to see the character of the argument as a whole, asking such questions as, 'What physical ideas went into the argument?' and 'What ideas of value came out?'

Sometimes it will be best to use such a question on its own, sometimes to use it before talking through the theory, and sometimes after doing so. It is worth trying all of these methods, if only to introduce some variety.

Making notes

The existence of summaries in the *Students' guides* might seem to make it unnecessary for students to make their own notes. But many teachers will probably feel that it is still important for students to summarize for themselves what has been discussed in a lesson, or learned from a question, from reading, or from an experiment. Although evidence about the importance of notes is inconclusive, we have some sympathy with this view: setting down one's ideas on paper is not a bad way of assessing what one has understood.

But students cannot simultaneously take notes of an argument and contribute to discussion. To do so involves too large an overloading of a person's linguistic capacities. One useful method is to appoint one or two students to take notes of discussion between the teacher and the rest, later producing a summary which is duplicated and circulated. This method can only be used from time to time, as one person's notes will not always please another. But their very inadequacy can be turned to advantage, and lead to an improvement.

Using computers

The number of microcomputers in schools is growing rapidly and will continue to do so during the lifetime of this course. We have tried to take account of this recent and very significant development by suggesting several ways in which a computer can be used in teaching the course.

One of the significant new developments of the revision is the 'Dynamic modelling system'. This is a powerful general purpose program which can be used to obtain graphical or numerical solutions to problems involving change, including (but not exclusively) problems in dynamics. The user defines the problem by typing in (or selecting from a prepared collection) a set of simple equations which the system turns into a computer program. The user must also provide numerical values for any constants, and initial values for the appropriate variables. The system then runs the program and will plot or tabulate values of any chosen variable against any other. Writing or choosing the appropriate equations must entail thought about the physics of the problem and its mathematical representation. So this technique should have great potential for developing students' understanding of a wide range of problems, and perhaps the process of problem solving itself. Also of course it provides a convenient way of displaying the effect on a solution of a change in value of one of the variables. An early version of the system was used in several schools during the revision, and as a result several improvements have been made. More details are given in the booklet *Dynamic modelling system* (Longman MicroSoftware).

As well as the general purpose 'Dynamic modelling system' a number of programs for specific purposes have been developed. Some of these are really essential to the teaching of the course as suggested in the *Teachers' guides* (those dealing with the solution of the Schrödinger equation for a hydrogen atom, the distribution of energy quanta in an Einstein solid, and the relationship between gravitational field strength and potential). Others are not essential but should be useful aids in teaching the course. These programs are described in *Software for Nuffield Advanced Physics.*

As well as these specially developed programs, there is an increasing number of programs from other sources that might usefully supplement the teaching of the course. Where we have known about such programs and felt them to be worth recommending we have referred to them in the same way as to a film, or other teaching aid.

In spite of the growing number of programs available we are sure that many teachers, and students, will want to develop their own. This can be a very valuable activity for the reason mentioned above in the discussion of the 'Dynamic modelling system'. We would not want to discourage it. However, we would warn any teachers who may be unfamiliar with computer programming that the part of the work that involves thinking about the *physics* of the problem is often very much less than the effort needed to deal with peripheral problems like scaling and plotting, tabulating, and so on.

During the period of the revision equipment manufacturers have begun to produce a wide variety of interfaces for a number of different computers. And sensible ideas as to how these might be used in physics experiments are beginning to appear. Although we have not written these techniques into the revised course we hope that those teachers who have the equipment and the necessary expertise will continue to develop and publish their suggestions.

A related development – but one which can be independent of the microcomputer – is the microprocessor based data logger.

This is a tool of great potential which enterprising teachers will surely find many uses for. For an account of some of the possibilities see LAMBERT 'VELA: a microprocessor-based laboratory instrument', *School Science Review* **65**, 38–47, 1983, or DEESON *Physics Education* **18**, 300–301, 1983.

A final word about computers concerns significant figures. Ask a computer to do a simple sum, say 6.2/5.5 and, like a calculator, it will give an answer with as many digits as it can handle. In this example the machine's answer might be 1.127 27 or even 1.127 272 727. But in a calculation involving numbers with two figures the result should only be quoted to two significant figures. We suggest that teachers draw attention to the spurious figures produced by machines to reinforce ideas about significant figures.

Teaching about applications of physics

Increasing students' awareness of the social significance of physics is one of the aims of the course. It is discussed on page xxv and some of the strategies adopted in the course towards achieving that aim are mentioned there. Clearly there is a limit to what can be provided by the project itself and we urge teachers to supplement what we furnished. Local and topical examples are likely to be much more telling than anything which is written into the course. A. A. Bartlett has written convincingly about this. In 'Are we missing something' (*Amer. J. Phys.* **49**, 1105–1119, 1981) he tells how he is able to relate the everyday phenomena he has noticed on his way to class to the physics which his students are learning. We may not all be able to do this but magazines such as *New Scientist*, television or radio programmes (*e.g., Tomorrow's World, Horizon, Science Now*) often provide a topical example that is related to the physics being discussed in class. Happily there seems to be an increasing number of sources to draw on. Among others which have recently appeared are SNIPPETS (Institute of Physics), Hobsons Science Support Series (CRAC Publications, Cambridge), *Physics at work* (a joint ASE/B.P. publication)). Among books which deserve a place in the school, if not in the teacher's library, are Jearl Walker's *The flying circus of physics*, John Wiley, 1977; and *The physics of everyday phenomena*, a collection from *Scientific American* published by W. H. Freeman, 1979.

In the end it is the teacher who counts

It should be clear that those responsible for developing this course (which means P. J. Black and Jon Ogborn, the original Course Organizers, and their project team, as much as anybody else) thought deeply about the aims of an A-level physics course, the plan for such a course, and possible teaching methods.

The purpose of this rather lengthy introduction to the *Teachers' guide* has been to explain some of these things to those who will use the course. We hope it is clear that we do not believe in one simple maxim or recipe for successful teaching. We do believe that it is the business of teachers to think hard about the aims of what they are doing, and to experiment with a variety of methods, in various mixtures, watching carefully to see what happens. Teaching is so personal a matter that each teacher will have to make the aims and methods his or her own, in his or her own way.

ACKNOWLEDGEMENTS

One of the pleasantest aspects of the development of *Revised Nuffield Advanced Physics* has been the willing way in which so many people have contributed and become involved in the work. Above all, teachers have helped in many ways, and the very number who have done so makes it impossible to acknowledge the contribution of each individual. Many have offered suggestions at meetings or have written in with ideas for questions, demonstrations, and so on. We have tried to consider carefully all the suggestions put forward and, inevitably, it is impossible to give proper credit to the source or origin of every idea we have used. One who has made a particularly valuable contribution in this way is Colin Price. To him and the many others whose contributions go unacknowledged, we offer our sincere thanks.

Other teachers have helped by conducting trials of some of the more radically changed parts of the course, and of a major innovation – the 'Dynamic modelling system'. The trial schools are: Aylesbury Grammar School; Beechen Cliff School, Bath; Bexley-Erith Technical High School, Bexley; Bishop Hedley High School, Merthyr Tydfil; Cheltenham College; Esher College; Forest Hill School, London; Godolphin and Latymer School, London; The Grammar School, Batley; The Greenhill School, Tenby; Haverstock School, London; Heathland School, Hounslow; Henbury School, Bristol; Highfield School, Wolverhampton; Howell's School, Llandaff; King Edward VI College, Nuneaton; Kingsbridge School; Lady Margaret High School, Cardiff; Malvern College; Marlborough College; Netherhall School, Cambridge; North London Collegiate School; Northgate High School, Ipswich; Oulder Hill Community School, Rochdale; Richmond-upon-Thames College; Royal Grammar School, High Wycombe; Rugby School; Samuel Ward Upper School, Haverhill; and Sutton Manor High School.

We are grateful to the Inner London Education Authority for trying some of our material on electronics in their 1983 Summer School for sixth-form students at the North London Science Centre.

Mark Ellse has read and commented on much of the draft material, and has made particularly useful suggestions about the up-dating of some experiments and pieces of equipment.

Thanks are due to a group of teachers, convened by Bob Fairbrother, who met several times to discuss assessment. Their suggestions lead to some changes in the structure of the examination.

Others, as well as teachers, have helped, of course. While he was working as a technician at the Centre for Science and Mathematics Education, Chelsea College, Phil Webb found time in a busy schedule to try out ideas for demonstrations and experiments, and to suggest ideas for new apparatus.

CLEAPSE School Science Service reviewed all the suggested experiments and demonstrations and made useful suggestions on the safety aspects of some of them.

Industry has helped too, and, among others, we are indebted to Rank Xerox, Amersham International P.L.C. and the CEGB for technical help and information.

Examination questions in the *Students' guide* are reprinted by permission of the Oxford and Cambridge Schools Examination Board. With one exception all are taken from Oxford and Cambridge Nuffield A-level Physics papers. The exception is one question taken from an Oxford and Cambridge O-level Nuffield Physics paper. Where guide lines for answers to examination questions are provided it must be understood that these are not the Examination Board's responsibility.

The Consultative Committee have, I believe, been asked to work harder and contribute more than is usually expected of such a group. As well as attending many meetings they have read and commented in detail on draft manuscripts – sometimes in a far from ideal state – and they have done all this most willingly.

It is a pleasure to acknowledge E. J. Wenham's help and sound advice. Much of what is written in these books has benefited from his knowledge and experience as teacher and author.

All of us who have contributed to these books owe a great debt of gratitude to Nina Konrad and her colleagues in the Publications office of the Nuffield–Chelsea Curriculum Trust for their thorough and painstaking work in preparing our manuscripts for the printers and our sometimes quite inadequate drawings for the artists.

Finally, I would like to express my sincere thanks to Paul Black and Jon Ogborn. Their help and support has been invaluable. During a period when both have been particularly busy, they have still found time to give advice both on general matters and on points of detail. They were, of course, the chief architects of the original Nuffield Advanced Physics course. Their willingness to be involved with what must at times have seemed like a severe distortion of their original plans, says much about their generosity of spirit.

John Harris

Unit A
MATERIALS AND MECHANICS

Roger Hackett
Christ's Hospital, Horsham

Suggested time allocation: five weeks

PLAN OF THE UNIT

	Section A1 The behaviour of materials	
	Stretching and breaking	
	Stress, strain, and breaking stress	
	The Young modulus	
kinetic energy; potential energy ▶	**Energy stored in a spring**	▶ energy stored in a capacitor: Section B3
	Section A2 The structure of solid materials	
oil film experiment ▶	**Size of an atom**	▶ Unit F, 'Radioactivity and the nuclear atom' Unit L, 'Waves, particles, and atoms'
diffraction ▶	**Structure of metals**	
GCSE chemistry ▶	**The Avogadro constant and the mole**	▶ Unit K, 'Energy and entropy'

Prior knowledge		Topic		Links
force, work, and energy	▶	**Intermolecular forces and energies**	▶	forces between ions: Section E3
		Chain molecules, slip, and crystal imperfections		
		Models		

Section A3
Statics, structures, and composite materials

Prior knowledge		Topic		Links
centre of gravity; moments	▶	**Addition and resolution of forces**		
		Composite materials		

Section A4
Momentum and the simple kinetic theory of gases

Prior knowledge		Topic		Links
GCSE	▶	**Momentum, collisions**	▶	mass of α-particles: Section F1
GCSE	▶	**Kinetic theory of gases**		ionization of gases: Section F3
GCSE	▶	**Temperature**	▶	thermodynamic temperature: Section K3

INTRODUCTION

In the order suggested here, simple tests on the stretching and breaking of solid materials lead quickly to a selection of more detailed experiments emphasizing different aspects of the application of stress, the resulting strain, and the breaking stress for a given material. Terms used to describe the behaviour of materials are defined and students make a careful measurement of the Young modulus of a metal. The idea of dimensional analysis to check formulae and the estimation of both random and systematic uncertainties in measurement are introduced. Extensometers as examples of commercial testing devices are mentioned briefly.

The storage of energy in a stretched spring or elastic cord and its retrieval are then examined. Situations in which force is not proportional to extension are included where it is necessary to determine energy changes by estimating the area under a force–extension graph. This is tested experimentally. The design of shock absorbers and of car seat belts provides examples of the application of this knowledge.

After offering some evidence for the existence and the size of atoms and molecules, Section A2 deals with models of the structure of simple solids, with particular emphasis on metals. The Avogadro constant is calculated using measurements of atomic diameters from X-ray diffraction experiments. Two coupled vehicles on a linear air track provide a model to illustrate simple ideas about intermolecular forces and energies. The shapes of the graphs of force and of potential energy against separation for a pair of molecules are introduced. The stretching, the Young modulus, and the breaking stress of materials are then related to these graphs. Defects in a crystal lattice and their effects on the properties of the material are considered briefly. This is followed by an examination of the properties of metals, glasses, and rubbers with reference to the differences between crystalline, amorphous, and polymeric substances. The section also provides opportunities for considering the use of models in physics and their validity.

Section A3 is about statics. Forces between bodies in contact are due to their deformation and a link is made with earlier work on tension and compression. Students who are familiar with the basic facts about forces in equilibrium can approach the subject through a model bridge building session. Alternatively, a series of experiments can be used to develop the triangle of forces and the principle of moments, and so lead to a consideration of the requirements for building a stable, static structure. The bridge building contest may then be used to reinforce these ideas. The section ends with a passage on the materials used by engineers. The limitations of such simple materials as metals and woods indicate the need for the composite materials including glass fibre, reinforced concrete, and new metal alloys.

Section A4 deals with gases. Some preliminary work (restricted to two dimensions) establishes that, whilst momentum is always conserved, kinetic energy is conserved only in elastic collisions. A conventional treatment of a collection of ideal gas molecules, which may be largely revision for some students, leads to the results $pV = \frac{1}{3}Nm\overline{c^2}$, $p = \frac{1}{3}\rho\overline{c^2}$ and to the kinetic theory interpretation of temperature. Orders of magnitude for intermolecular spacing, mean free path, and molecular speed are reviewed.

THE PLACE OF THE UNIT IN THE COURSE

'Materials and mechanics' has been designed as an introductory unit. Several of the important themes which run through the course are met for the first time, and the methods of work suggested in this *Teachers' guide* will give students a first taste of some of the kinds of activity that will be expected of them in the rest of the course.

One of the major concerns of physics, reflected in the course and introduced in this Unit, is the explanation of behaviour on a large scale in terms of models based on the very small. Thus the properties of metals, plastics, and rubber are better understood if one has models of how atoms and molecules are diffcrently arranged in these materials. The Young modulus, E, of a metal can be related to the force constant, k, of its interatomic bonds. At a more general level, the use of models, of various kinds, as tools in physics, is a recurrent theme in the course.

Another characteristic of the course which is introduced here is an emphasis on the usefulness of physics. The Unit is concerned with everyday materials (metals, rubber, plywood, reinforced concrete...) and with real structures (cranes, bridges...) and their importance in the world, as well as with the simplified, abstract, and powerful laws of physics.

In contrast with their studies at O-level or at GCSE, the students are introduced to an atmosphere of more independent learning. The emphasis on the practical approach gives the students opportunities to become familiar with the laboratory and its apparatus; to make observations and measurements in simple experiments and to extend these to more complex and realistic situations; to take notes and write up experiments concisely and accurately; to draw graphs; to use the method of dimensions to check formulae; and to consider the effects of uncertainties in measurement on their results.

There are opportunities for students to read independently and to report back to the class (composite materials) and to undertake a mini-project (bridge building).

Many of the specific topics introduced in this Unit are used or developed further later in the course. The plan of the Unit (page 2) shows these links. Teachers who prefer to start the course with Unit B, 'Currents, circuits, and charge' will be unable to use the analogy between $\frac{1}{2}CV^2$ for a charged capacitor and $\frac{1}{2}kx^2$ for a stretched spring unless some of the work from Section A2 is transferred to Section B3.

LIST OF SUGGESTED EXPERIMENTS AND DEMONSTRATIONS

A1a	Experiment	The stretching and breaking of metals to compare strength, ductility, and hardness *page 8*
A1b		A preliminary study of the force–extension relationship for lengths of various materials *10*
A2a		A study of how the length of a rubber band varies with the applied force *13*
A2b		The stretching of a nylon fishing line or a strip of polythene *13*
A3	Group of experiments (A2a–A6)	Measurement and prediction of the breaking force for aluminium samples *15*
A4		Measuring the breaking strength of a glass fibre *17*
A5		Measuring the strength of paper *17*
A6		A short investigation of spring behaviour *19*
A7	Demonstration	The effect of cracks *21*
A8	Experiment	Measurement of the Young modulus and breaking stress of a wire *24*
A9		Testing the formula for translational kinetic energy *27*
A10	Group of experiments (A9–A11)	Measuring the elastic strain energy stored in a spring *29*
A11		Changing elastic strain energy into gravitational potential energy or translational kinetic energy *31*

SECTION A1
THE BEHAVIOUR OF MATERIALS

STRETCHING AND BREAKING

An immediate start can be made on experiment A1 to focus the students' attention on to their *Guides*, both the 'Laboratory notes' and the 'Summary of the Unit'.

The two parts of the experiment can be done together or separately. Widely differing properties of materials emerge, which can be classified in standard terms. Students can then think about uses and suitability.

Time: not more than a double period.

EXPERIMENT

A1a The stretching and breaking of metals to compare strength, ductility, and hardness

ITEM NO.	ITEM
	Selection of materials:
1501	stainless steel wire, 1 m lengths, 0.08 mm diameter
1501	copper wire, 1 m lengths, 0.28 mm diameter
1153	other metal wires, *e.g.*, iron, nichrome, fuse wire, etc., as available
1153	selection of metal strips about 100 mm × 10 mm × 1mm *e.g.*, copper, steel, aluminium, etc.
44/2	G-clamp, small
522	Hoffman clip
1153	2 wooden blocks, about 20 mm × 20 mm × 10 mm and 2 strips, about 60 mm × 10 mm × 5 mm
1153	2 dowel rods, 1–2 cm diameter; 10–15 cm long
503–6	retort stand base, rod, boss, and clamp
1153	centre punch
1153	plastic, card, or metal tube (about 30 cm long, to take centre punch)
24	hand lens
1502	safety spectacles

Safety note: Accidents have occurred when nylon line under tension suddenly broke and flew up into a student's face. Safety spectacles are recommended for all experiments involving stretching wires or filaments, or breaking brittle materials.

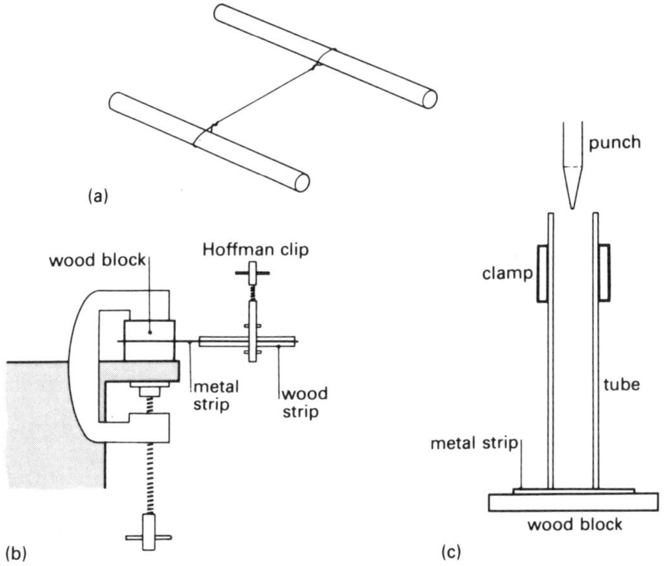

Figure A1
Simple testing of metals: (a) stretching; (b) bending; (c) hardness.

Samples of different metals should be pulled using two dowels until the wire snaps – figure A1(a). The steel wire is very thin, so not much energy is stored in it when it fractures and danger from a broken end whipping back is minimized. Warn students that the broken end or edges may be sharp; they should wear safety spectacles.

The ends of the wires can be observed with a hand lens for evidence of brittle fracture (clean ends) or plastic yielding (figure A2).

Figure A2
Necking at point of fracture of copper wire.

The strips can be flexed until broken. The wood sandwich held by a Hoffman clip gives a handle and also ensures that the specimen is only flexed at one point – figure A1(b). A centre punch is used to test impact hardness, dropping it down a narrow vertical tube onto the metal sheet placed on a piece of wood – figure A1(c). Another method is to place a 1 kg mass on the punch to press it into the sheet's surface. A hand lens will be needed when estimating the diameter of the indentation.

See also Revised Nuffield Chemistry *Teachers' guide II*, Experiments A14.6 and B15.8.

EXPERIMENT

A1b A preliminary study of the force–extension relationship for lengths of various materials

ITEM NO.	ITEM
1153	rubber bands
1153	thin rod of Plasticine
	paper
1153	polythene (*e.g.*, from a food bag)
1153	nylon line
1153	strip of expanded polystyrene
1501	steel wire
1501	copper wire } as in A1a
1153	2 dowel rods } as in A1a
1502	safety spectacles
1153	plastic ruler

Safety note: Students should wear eye protection.

Students are asked to sketch force–extension curves and to estimate the force needed to break the specimen. Emphasize the importance of a preliminary trial – a quick experimental test – before proceeding to a more elaborate investigation. Comparison of estimates of breaking forces will bring out the need for careful measurement rather than human estimation. The importance of the cross-section of the material may also be appreciated. The small amount of springy stretching of a copper wire may be missed here but should become evident when measuring the Young modulus (experiment A8).

Discussion: the choice of material

Before or after experiment A1 discuss, for example: what materials are used in cars? Why is each chosen? What different materials have been used for frying pans? (Cast iron, steel, copper, glass, ceramics.) What advantages or disadvantages does each have? (Strength, cost, thermal conductivity, corrosion, density, etc.) What materials are used for wrapping and tying things? (Paper, cord, polythene, string, steel bands, etc.) What determines the choice? How are things fixed together? (Glue, nuts and bolts, screws, rivets.) The questions in the students' laboratory notes are intended to stimulate thinking along these lines. These questions will be examined in more detail later in the Unit.

Terms to describe materials

A table of properties may help students to describe the materials simply and accurately using the terms in the *Students' guide* (page 2).

A 'stiff' material has a high Young modulus, whereas a 'strong' material needs a large stress to break it (see experiment A8). Tough and brittle also have precise meanings: tough materials absorb energy by plastic deformation before breaking; a brittle material breaks without any plastic deformation. Encourage students to use such terms more precisely as their experience with materials increases.

	Tough	**Stiff**	**Strong**
Biscuit	no	yes	no
Steel	yes	yes	yes
Nylon	yes	no	fairly
Jelly	no (it cracks)	no	no

Table A1

Add 'dense' and most worthwhile information about the material is represented.

Questions

Question 3 is about the use of terms to describe materials. Questions 1, 2, and 4 are about graphical description and prediction of behaviour.

Reading

GORDON *The new science of strong materials.* This discusses terms informally in Chapter 2. Why materials break can be found in Chapter 4.
AKRILL, BENNET, and MILLAR *Physics.* Chapter 5 distinguishes terms clearly.
Schools Council Engineering Science Project *The use of materials.* Chapter 1 is about the strength and properties of materials.
The Institution of Metallurgists publishes various booklets about materials. In *Which materials?* one of the sections is devoted to the metal can.
There is a short passage in the *Students' guide* (page 23) on the mechanical testing of materials, with a few questions.

History

The broader social significance of the availability of materials with various properties can be shown from history:
Stone age (up to about 2500 BC): hard, brittle material.
Bronze age (from 2500 to about 500 BC): hard, tough material, somewhat ductile.
Iron age (from about 500 BC): hard, tough, ductile material, cheap.
More recently (19th century): steel – less brittle and stronger than iron.
Present (20th century): plastics and composites – new possibilities.
(The dates given are only approximate; they refer to Western Europe.)

STRESS, STRAIN, AND BREAKING STRESS

The importance of cross-section and extension per unit length of a specimen being stretched should be discussed if it has not already been noted.

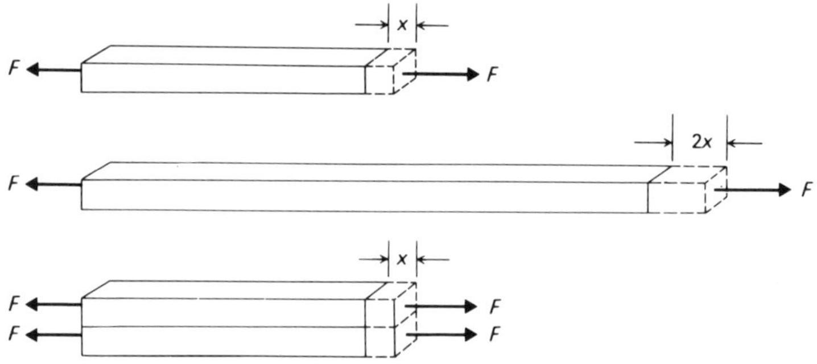

Figure A3
Effect of length and cross-section.

To introduce the terms *stress* and *strain* ask students to imagine extending a piece of rubber of double the length or double the thickness of the original, with a fixed force (figure A3).

The question 'Is steel really the strongest of the materials investigated so far?' can bring out the need to calculate stresses rather than simple stretching forces.

Mention the breaking stress of a material – the stress at which it breaks. Designers and engineers must have knowledge of this before using a material safely for any purpose.

Organization of a group of experiments

Experiments A2 to A6 form a group on a common theme: stress, strain, and breaking stress. Each experiment is short, and working on a circus basis, students can do as many as time permits. Two double periods should suffice, after which pairs of students should briefly describe the different experiments to the class. Experiment A3 raises an important general issue: the treatment of uncertainties.

Experiment A2 takes a precise look at materials that can withstand a large strain. It can be linked later with experiments A9 and A10. Remind students that two strands of material are being stretched. The polythene example could be kept as part of demonstration A7 for the teacher when the group of the experiments A2–A6 are discussed together.

EXPERIMENT

A2a A study of how the length of a rubber band varies with the applied force

ITEM NO.	ITEM
1153	size 32 rubber band, 75 mm × 3 mm × 1 mm unstretched
31/2	hanger and slotted masses, 0.1 kg, up to 2 kg
503–6	retort stand base, rod, boss, and clamp
44/1	G-clamp
501	metre rule

EXPERIMENT

A2b The stretching of a nylon fishing line or a strip of polythene

ITEM NO.	ITEM
1153	nylon fishing line (breaking load about 10 N), 1.5 m long
503–6	retort stand base, rod, boss, and clamp
31/2	hanger with slotted masses, 0.1 kg
1153	polythene strip, 200 mm × 10 mm (*e.g.*, 500 gauge, about 0.1 mm thick)
1183	2 polarizing filters, 50 mm × 50 mm
1153	adhesive tape
501	metre rule
1502	safety spectacles

Safety note: Accidents have been reported in which the end of a nylon filament has whipped up into a student's face. Eye protection should be worn.

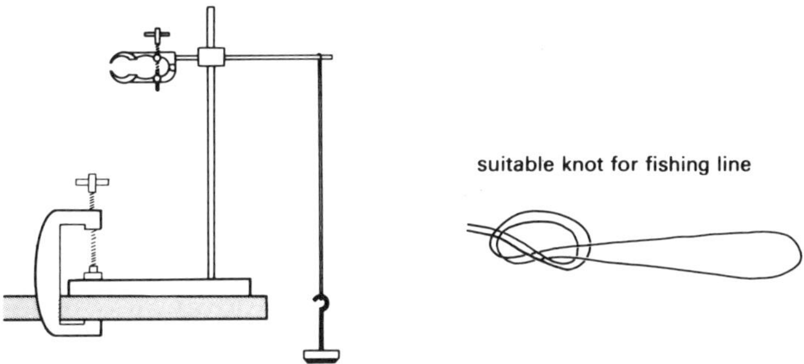

Figure A4
Stretching of rubber or nylon.

The rubber band or nylon can simply be looped over a clamp and the masses suspended from the loop (figure A4).

Students may notice the temperature changes when rubber is rapidly stretched and released (the lips are particularly sensitive). There may be an opportunity to explain this effect later in Unit K, ‘Energy and entropy’.

Students should observe the unusual properties of rubber: it can undergo very large strains; there is a large hysteresis effect between loading and unloading. Students draw a force–extension graph rather than a stress–strain curve. The rubber band may have a permanent extension after a large strain (loads up to 15 N).

Rubber becomes stiffer with large loads. This will be explained later in molecular terms (demonstration A16).

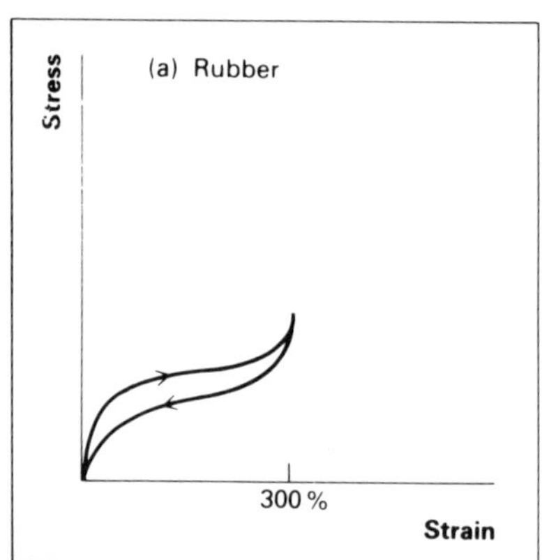

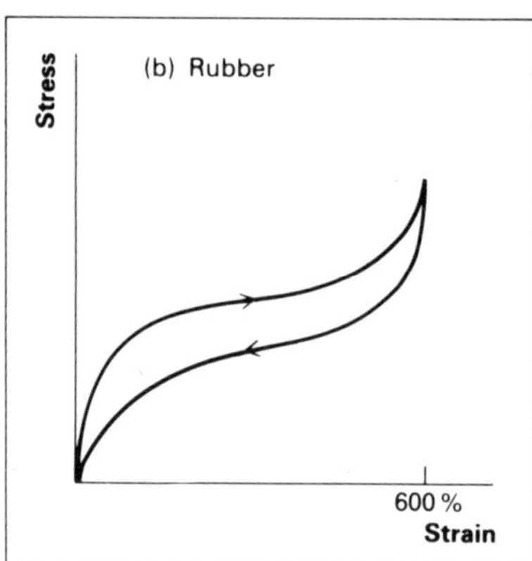

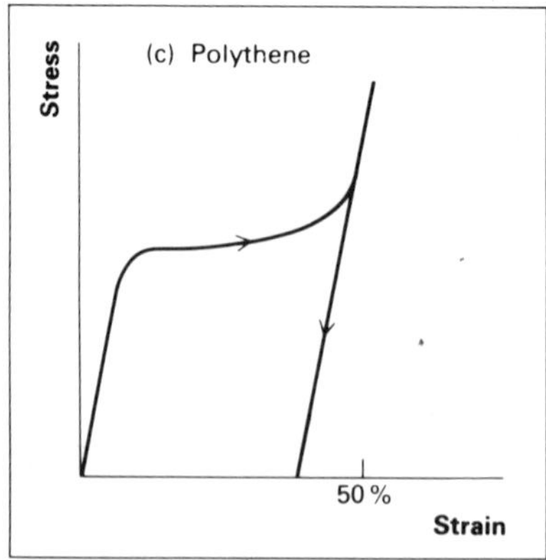

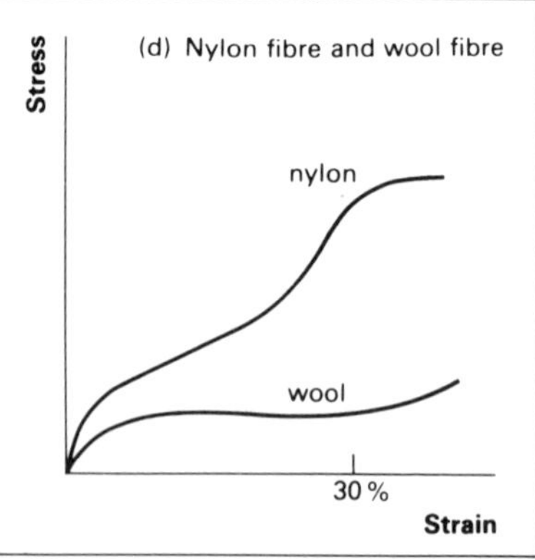

Figure A5
Stress–strain curves for rubber, polythene, and nylon and wool.

Rubber is stiffer than it appears. The decrease in cross-section means that the true stress is greater than the nominal stress and so the shape of the true stress–strain graph is steeper than one drawn assuming constant cross-section. (The volume of the rubber remains nearly constant.)

Nylon is a fairly strong material. Contrast its use for clothing with garments made of wool.

Polythene strip must be cut with clean, not ragged edges. The stiffness is constant at first as polythene behaves elastically. The strip suddenly necks and there is a large increase in strain for little increase in stress. The narrowing does not continue. The material becomes about as stiff or slightly stiffer than before and is again fairly elastic. When released it has a large permanent set. If the narrowed strip is boiled in water, it returns more or less to its original length, though the striations which have developed remain. (Demonstration A16a will suggest an explanation in molecular terms.)

Questions

Questions 6 and 7, about stress and strain, refer to rubber and steel samples.

Reading

BOLTON *Materials.* Chapter 1 has an appendix on nylon socks.

Apart from emphasizing the term 'stress', experiment A3 gives a good opportunity to introduce and discuss random and systematic uncertainties, indicating how to obtain quantitative estimates where possible.

EXPERIMENT
A3 Measurement and prediction of the breaking force for aluminium samples

ITEM NO.	ITEM
1153	strip of aluminium (kitchen) foil, 100 mm × 10 mm
1155	micrometer screw gauge or Vernier calipers
1153	ruler
1153	strip of aluminium sheet, 100 mm × 10 mm × 1 mm
1153	adhesive tape
	either
81	newton spring balance, 10 N
	or
31/2	hanger with slotted masses, 0.1 kg

Students make 'handles' of paper and adhesive tape to attach to the aluminium foil (figure A6) and find a value for the force required to break the foil by pulling along its length.

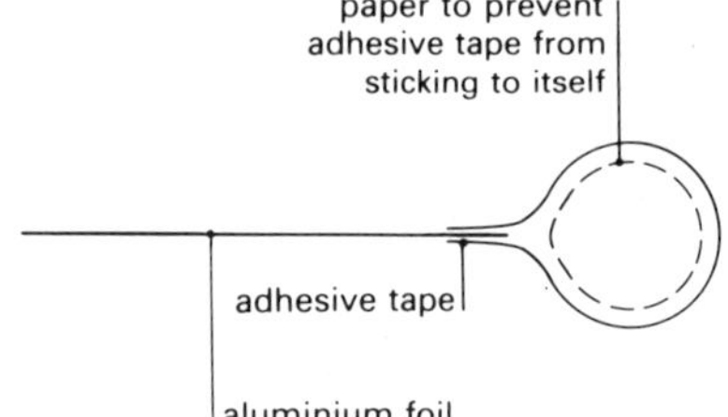

Figure A6
Adhesive tape 'handles' for pulling aluminium foil.

The aluminium foil is broken by a force of several newtons. Students use a micrometer to measure the thickness of the foil and of the strip, and hence calculate the force needed to break the thick strip. They are asked to estimate the uncertainty in their result.

Treatment of uncertainties

The uncertainty in any measurement should be implied in the number of significant figures quoted. The results of any calculations using these measurements must not then be quoted to more significant figures, even if the pocket calculator provides extra meaningless figures.

This is a good opportunity to explain what a percentage uncertainty is and take a few simple numerical examples using two quantities to show the rule:

Σ (maximum percentage uncertainties in individual measurements)
= maximum percentage uncertainty in result, whether the quantities are multiplied or divided.

The pocket calculator makes this a very quick exercise.

An algebraic treatment is too advanced – a rule of 'thumb' is all that is needed. Include powers as well. For example, percentage uncertainty in x^3 is 3 $\times$ percentage uncertainty in x.

When measurements are added or subtracted the absolute, not the percentage uncertainties are added.

The difference between random and systematic errors should be explained. (See also *Students' guide*, page xiii, and *Teachers' guide*, page xxxi.)

Reading

Two suitable references about uncertainty in measurements and results are:

AKRILL and MILLAR *Mechanics, vibrations, and waves.* Chapter 1.

HOCKEY and MILLS *Physics by experiment.* Appendix 1.

EXPERIMENT
A4 Measuring the breaking strength of a glass fibre

ITEM NO.	ITEM
1155	soda glass rod, 0.2 m long, about 3 mm diameter
503–6	retort stand base, rod, boss, and clamp
508	Bunsen burner
1155	micrometer screw gauge
31/2	hanger with slotted masses, 0.1 kg
24	hand lens
1153	string
1502	safety spectacles

Safety note: Students should wear safety spectacles.

Students may need help in making a straight glass thread. Instructions are given in the laboratory notes for this experiment (*Students' guide* page 40–41). A warning should be given about hot glass not appearing so.

Glass softens gradually as it is heated. It can be pulled into very long threads. Such threads do not stretch appreciably, but can carry surprisingly large loads. As in experiment A3, students estimate the strength of the glass rod. The estimate of strength is unreliable and the questions in the laboratory notes are meant to lead the student towards the weakening effect of scratches and cracks. Glass suffers brittle fracture which can be demonstrated to the class in demonstration A7.

Home experiment

Home experiment AH1 is about a bundle of fibres – human hair – which also suffer brittle fracture (*Students' guide* page 61).

Experiment A5 emphasizes stress through compression rather than tension. Time of action also affects deformation or strength of material.

EXPERIMENT
A5 Measuring the strength of paper

ITEM NO.	ITEM
	sheet of thin paper, A4
	sheet of corrugated paper at least 30 mm × 200 mm
1153	hardboard square about 80 mm × 80 mm
32	2 masses, 1 kg
31/2	10 slotted masses, 0.1 kg
1153	adhesive tape
1153	raw potato
1153	paper drinking straw
501	metre rule

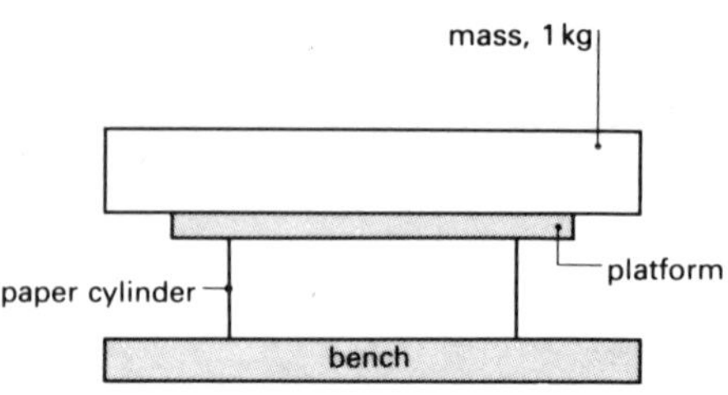

Figure A7
Measuring the strength of paper.

Students make thin paper cylinders of various diameters and load them until they collapse. They should find that if the paper cylinder is vertical, the 'collapsing' load is proportional to the length of the paper (that is, the circumference of the cylinder). A cylinder made of typist's copy paper collapses easily. The use of corrugated paper, or plain paper folded concertina fashion, illustrates a simple design improvement which greatly increases the load-bearing ability of the cylinder. Corrugated paper resists shear forces in one direction. It makes a useful model bridge building material. Repeat measurements of the crushing strength for paper will give a wide range of values.

A raw potato can be stabbed successfully with a paper straw (instructions are given in the students' laboratory notes), illustrating that time of interaction affects material properties.

Design problem or competition

Experiment A5 can be used to provide the basis for a short contest at some stage in the Unit or as an end of term 'entertainment' in place of model bridge building (page 60), for example. It could also be used as a basis for an investigation.

The aim of the contest is to support the greatest static load on a paper tower. The load is applied to a 20 cm × 20 cm × 1 cm plywood or blockboard platform by piling on 1 kg masses. Each contestant is provided with a 50 cm × 30 cm sheet of cartridge paper or *very thin* card (well designed card structures will support a student), PVA adhesive, elastic bands, and pins to hold the paper while the glue sets.

One double period is required for design and construction. The glue dries overnight so the test and discussion requires another double period. Rules can be varied, for example, the tower must be 30 cm high, but can be of any diameter – the remaining paper being used as stiffening diaphragms or corrugations, etc. Another possibility is to allow any height of tower but look for the maximum (weight supported) × $(\text{height})^2$, or some similar expression.

The last experiment in this group illustrates stress and strain by combining springs in series or parallel, effectively doubling the length, l, or area, A, of the spring pulled by the same force. When the Young modulus, E, is introduced later, it will be related to the spring constant, k, by

$$E = kl/A$$

Revised Nuffield Physics *Pupils' text Years 1 and 2.* Chapter 5 'Investigating springs' covers much of the next experiment in an elementary form.

EXPERIMENT
A6 A short investigation of spring behaviour (using a mass hanging from a spring)

ITEM NO.	ITEM
2A	4 expendable steel springs
31/2	hanger with slotted masses, 0.1 kg
503–6	retort stand base, rod, boss, and clamp
1501	short length of stiff wire (*e.g.*, 0.90 mm diameter copper, or coat hanger wire)
507	stopwatch

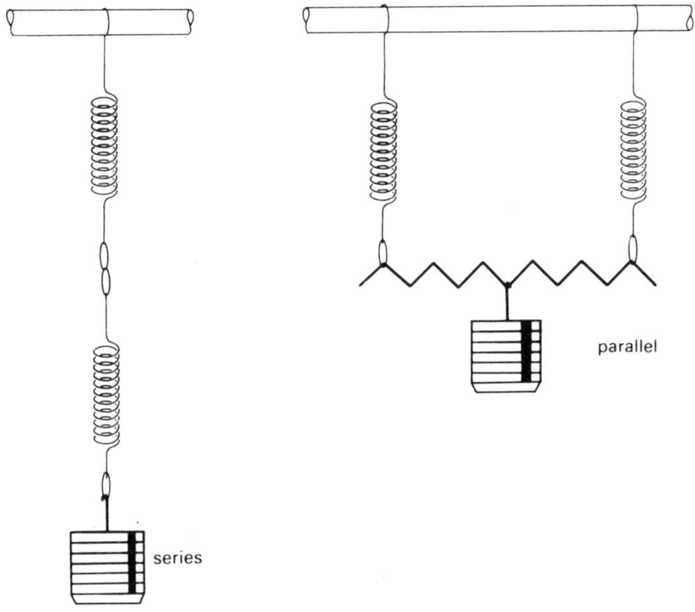

Figure A8
Springs in 'series' and in 'parallel'.

Students are asked to find the spring constant for a single spring and then for a combination of springs in 'series' and 'parallel' (figure A8).

A rule relating the stiffness of a combination of n identical springs to that of a single spring, k, can be stated:

in 'series' $k_s = k/n$ (weaker)

in 'parallel' $k_p = nk$ (stronger)

Contrast the examples of a mass tethered on a frictionless surface by a compression spring to make 'series' and 'parallel' clear (figure A9).

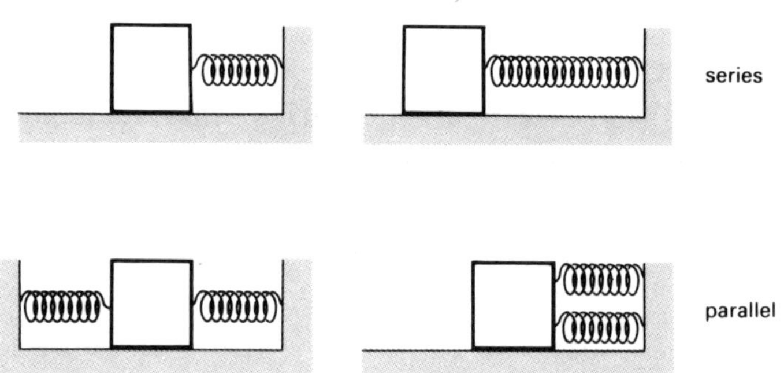

Figure A9

The approach will be helpful when making an analogy between a capacitor and a spring in Unit B, 'Currents, circuits, and charge'. The oscillation of a mass and spring system is examined in detail in Unit D, 'Oscillations and waves'; fast students may extend this experiment to investigate the dependence of period (T) on mass (m) and spring constant (k).

Question 5 repeats much of the experiment.

Home experiment

Home experiment AH2, An accurate newtonmeter, challenges students to construct a sensitive device for measuring forces up to 1 N.

Discussion of a group of experiments

A discussion session about the preceding experiments can be linked with demonstration A7. Students can give a brief (2 minute) account, if possible with the apparatus in front of them. The main point which should emerge is the importance of relating extension to the length of the specimen (strain), and force to the cross-sectional area (stress).

Students have measured the breaking strength of various specimens and have been asked to estimate the force needed to break thicker specimens of the same material. The next demonstration shows the importance of cracks on the strength of materials.

DEMONSTRATION
A7 The effect of cracks

ITEM NO.	ITEM
508	Bunsen burner
1155	3 lengths of soda glass, 0.1 m, about 3 mm diameter
1155	file, *e.g.*, triangular
	matchstick
1153	2 strips of polythene, 100 mm × 10 mm (*e.g.*, 500 gauge, about 0.1 mm)
529	scissors
1155	safety screen
1153	Perspex strip, 250 mm × 10 mm × 6 mm (clear faces 6mm wide)
	slide or overhead projector
1502	safety spectacles
	either
1183	2 polarizing filters, about 50 mm × 50 mm
	or
	liquid crystal display (LCD) filters, 105 mm × 35 mm

Safety note: Demonstrators should wear eye protection.

A7a File a nick across a glass rod, place the rod over a matchstick with the nick uppermost and above the match, and press the ends gently. The rod should snap easily and cleanly. To emphasize the crack-opening action, try two rods together, one with the filed nick on its underside, and one with the nick upwards.

A7b Put one polarizing filter in the carriage of the slide projector and place the other over the lens, crossing them. Hold the strip of polythene in the projector and focus the strip on the screen. (Alternatively, place one polarizing filter on the OHP and clamp the other, crossed, above it. The strip of polythene between them can then be focused on the screen.)

Pull the polythene strip. Colours develop over the image as the strip is stressed.

Make a cut halfway across a second strip, and pull that. Colours develop at the tip of the crack. A more permanent demonstration can be made using acetate or Perspex strip with a notch cut in it.

LCD filters consist of a plane polarizer bonded to a quarter wave plate. Using these in place of polarizing filters enables strain in all directions to be displayed equally. With simple plane polarizers only the components of strains in some chosen directions are shown.

Concentration of strain can also be shown by ruling a grid on a sheet of rubber, cutting a nick, and stretching it (see figure A10).

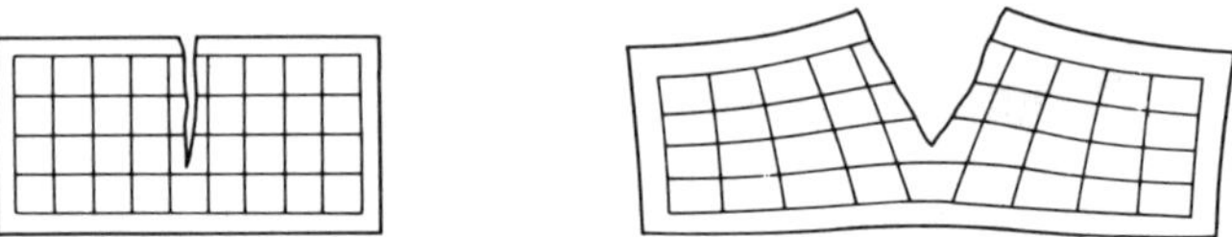

Figure A10
Strain concentrations due to a nick in a rubber sheet.

A7c Repeat experiment A7b, flexing the Perspex strip. Equally spaced lines will appear across the strip as one edge becomes compressed and the other goes into tension.

A7d Make a new glass fibre (as in experiment A4) about 0.5 m in length. It should be possible to bend it into an arc (figure A11).

Use a safety screen to remove any possibility of students being hit by flying glass fragments.

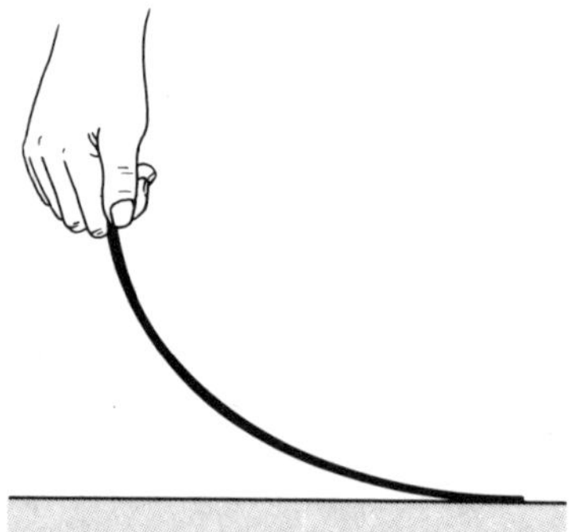

Figure A11
Unscratched glass fibre bent into an arc.

Scratch the fibre by running it between your fingers and again bend the glass. It should snap. Repeat, with a *new* fibre, touching first the inside and then the outside of the bent glass with the edge of another broken glass fibre. When touched on the outside, with practice, the glass snaps at once.

The stress concentrations at a crack were shown using the polythene and polarizing filters. Glass does not deform plastically, and so the high local stress at the top of the crack causes brittle fracture, making the crack deeper, increasing the stress, until the sample breaks. Stress concentrations can sometimes be relieved by drilling a hole at the end of a crack to widen it, so lowering the local stress. Copper can withstand cracks because it is ductile and yields, reducing the localized stress. Ductile and brittle fractures were observed with a hand lens in experiments A1 and A4.

Cracks are always a source of weakness because they increase stresses. The same argument applies to holes – hatchways in ships, for example. Stress multiplication by 3 mm rivet holes is said to have contributed to the disastrous failures in one of the first commercial jet aircraft, the Comet. Think also of the perforations between postage stamps. Ask 'What would be the best shape for the holes?'.

All materials can and do crack. 'Tough' is the word used for materials not susceptible to cracks, for example, nylon, rubber. Knowledge of the structure of each material leads to an understanding of why some materials can withstand deeper cracks than others without failing (see Section A2).

Reading

GORDON *The new science of strong materials.* Chapter 4 gives a clear account of the problem of cracks.

GORDON *Structures.* Chapter 3 is entitled 'The invention of stress and strain', and Chapter 5 considers energy changes in determining the propagation of cracks.

The purity and the previous history of a specimen, for example, how much it has been flexed (fatigue), also affect the stress at which it is likely to fracture (see Section A2).

Creep effects (a continuing extension of the material with time for a constant load) may have been noticed. The effect is quite small for metals at normal temperatures, and only becomes serious at higher temperatures where the effect can be quite dramatic. For materials such as polythene and rubber, creep is more easily observed. It will also be noticed in the next experiment when the stretched copper wire is near its breaking stress. This may be taken up again in Unit K, 'Energy and entropy'.

THE YOUNG MODULUS AND BREAKING STRESS

Students are now asked to measure the Young modulus and breaking stress (properties independent of the shape and size of the specimen) for a length of wire stretched horizontally across the bench (experiment A8). Many textbooks describe this experiment, but the students should be encouraged to design as accurate an extension measuring system as they can, using simple apparatus. The experiment and account should not take more than a double period.

EXPERIMENT

A8 Measurement of the Young modulus and breaking stress of a wire

ITEM NO.	ITEM
44/1	G-clamp
1153	2 wooden blocks
40	single pulley on clamp
501	metre rule
1153	adhesive tape or gummed paper tape
31/2	hangers with slotted masses, 0.1 kg, up to 2 kg
1155	micrometer screw gauge
1502	safety spectacles
1501	lengths of iron wire, 0.20 mm diameter
1501	lengths of copper wire, 0.28 mm diameter
1501	lengths of stainless steel wire, 0.08 mm diameter

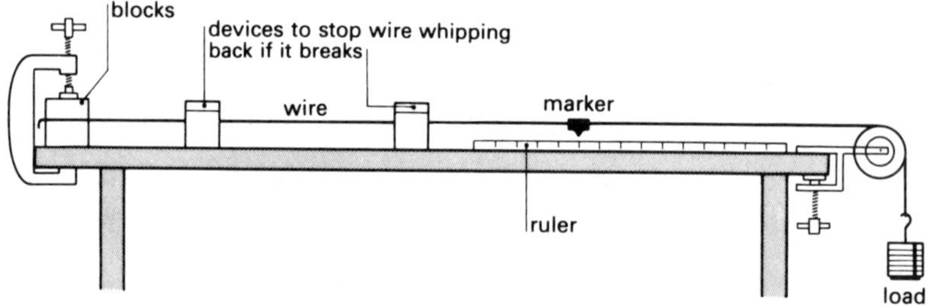

Figure A12
Measurement of the Young modulus and breaking stress.

Safety note: Some simple device such as two or three shapes (⊓) of timber should straddle the wire to avoid any broken end whipping back into a student's face. They should wear safety spectacles.

A scale marked in millimetres will serve if the wires are about 2 m long. An adhesive tape flag on the wire can be used as an indicator. Alternatively a Vernier could be used or even a lever system to magnify the extensions for the steel wire (for example, see Revised Nuffield Physics *Teachers' guide Years 1 and 2*, Demonstration 33).

A cardboard carton containing, perhaps, scraps of plastic foam placed beneath the load will protect the masses and the floor when the wire breaks (and keep students' feet away!).

Students measure force and extension, plot a stress–strain graph, estimate values for the Young modulus and breaking stress of the metal. They are asked to give an estimate of the uncertainty of their results. The Young modulus is measured for the Hooke's Law region of the graph, that is, where the material has a constant stiffness. Beyond this the stiffness decreases. (For substances like rubber, the Young modulus value quoted is an average value over the region of strain being considered.)

Graph plotting

It is conventional to plot the dependent variable on the y-axis (ordinate) and the independent variable on the x-axis (abscissa). In the laboratory students vary the tension in the specimen and measure the resulting extension. So strict adherence to the convention would require extension to be plotted as ordinate and tension as abscissa. However, it will be much more convenient for future work, especially when calculating energy changes, to plot tension as ordinate and extension as abscissa. And this is how such graphs are usually presented.

Questions

Question 8 is about the Young modulus. Question 9 is about breaking stress and yield stress.

Reading

The stretching of the copper or iron wire leads to a graph which has a number of important features. Students should be encouraged to look in textbooks where these curves are described in detail and to make their own notes accordingly.

Suitable references are:

AKRILL, BENNET and MILLAR *Physics.* Chapter 5.

COLLIEU and POWNEY *The mechanical and thermal properties of materials.* Chapter 3.

There is also the short passage on the mechanical testing of materials in the *Students' guide* (page 23) which may have been used already with experiment A1.

Industrial measurements

Industrial extensometers are usually automatic and measure the force which must be applied to the ends of a rod specimen to produce a constantly increasing strain. Thus some features in the stress–strain curve appear which would not otherwise be easy to observe, for example, at the yield point of a low-carbon steel wire where the nominal stress falls but the strain continues to increase. Beyond the maximum load the true stress continues to rise until fracture. The specimen usually necks and as a result the force that must be applied falls as the cross-section of the specimen is smaller. It is usual engineering practice to calculate the stress using the original area of cross-section (nominal stress) so the stress appears to fall from the maximum load to the fracture point with increasing strain.

The method of dimensions

The method of dimensions is part of the core A-level syllabus and might well be introduced here for the first time. Students should appreciate that while the method can serve as a useful check on the validity of a relationship it cannot prove that the relationship is true – though it may of course prove that it *cannot* be true. Nor can the method tell us anything about numerical (that is, dimensionless) constants.

The currently accepted convention is to use square brackets around a physical quantity to mean the dimensions of that quantity. For example,

$$[\text{velocity}] = \text{LT}^{-1}$$

or

$$[v] = \text{LT}^{-1}$$

means 'the dimensions of velocity (or v) are those of length divided by time'.

In mechanics only the base dimensions mass, length, and time (symbols M, L, T) are needed. The dimensions of current and temperature are I, Θ. (See ASE, 1981, *SI Units, signs, symbols and abbreviations.*)

Reading

AKRILL and MILLAR *Mechanics, vibrations and waves.* This has a section on dimensions in Chapter 1.

COLLIEU and POWNEY *The mechanical and thermal properties of materials.* Chapter 3, section 13.

Question 10 uses the method of dimensions, and involves spring constant, stress, strain, and the Young modulus.

Organization of the energy experiments

In the next part of the work the relationship energy stored $= \frac{1}{2}Fx$ (if Hooke's Law is obeyed) is derived theoretically, and tested experimentally. How much revision of energy ideas is needed will have to be judged. Students need to be able to use $\frac{1}{2}mv^2$ and mgh confidently, as in experiments A10 and A11 stored elastic energy is measured by transforming it to kinetic energy or gravitational potential energy. For a class with good understanding only one of the following experiments need be attempted. Weak students may need to do all three. Revised Nuffield Physics *Pupils' text Year 4*, Chapter 5 on kinetic energy, and especially experiments 55 to 57, cover much of this work.

For students who have already done a lot of work with dynamics trolleys, alternatives are given using an air track. A maximum of two double periods is suggested for experiments A9 to A11.

If students are familiar with experiment A9, each pair should be asked to use one method for experiment A10, or possibly two parts of experiment A11, as these are shorter. Results can be reported back to the class with brief demonstrations where appropriate.

EXPERIMENT

A9 Testing the formula for translational kinetic energy

A9a Catapulting an air track vehicle

ITEM NO.	ITEM
1019	air track and accessories
1020	air blower
1503	timer, resolution 1 ms
130/2	photodiode assembly with light source
501	metre rule
106/2	3 elastic cords for accelerating trolleys
1153	thin card
1504	balance, 1 kg

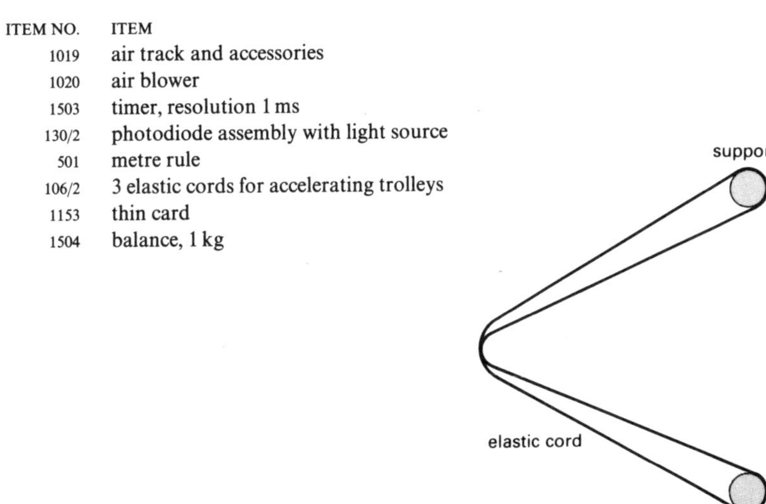

Figure A13

Students catapult the air track vehicle with one, two, or three elastic bands or threads all at the same stretch, so that the vehicle is given one, two, or three units of kinetic energy.

They measure the speed of the vehicle in each case and repeat, having changed its mass, to test the expression for kinetic energy: $\frac{1}{2}mv^2$.

A9b Measurements with potential energy changing to kinetic energy

ITEM NO.	ITEM
	either
	Apparatus as Experiment A9a (air track, etc.) with
31/1	hanger and slotted masses, 0.01 kg
1153	string
38	single pulley
	or
107	runway
106/1	2 dynamics trolleys
40	single pulley on clamp
1153	string
31/2	hanger and slotted masses, 0.1 kg
108/1	ticker-tape vibrator
108/2	carbon paper disc
108/3	ticker-tape
27	transformer
1504	balance, 2 kg

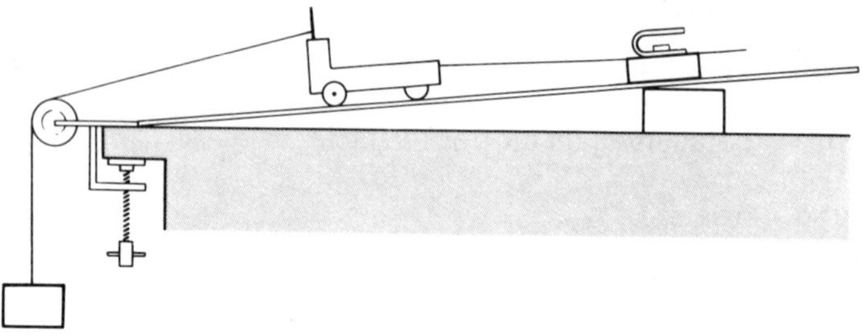

Figure A14

i Air track: the track must be carefully levelled. Students use a falling mass to accelerate the vehicle whose speed, v, is found for various distances, h, fallen by the mass, using the photodiode assembly.
ii Trolley and runway: the runway must be friction compensated. Students use a falling 0.4 kg mass to accelerate the trolley (figure A14). The trolley's speed, v, for various distances, h, fallen by the mass can be found from a single ticker-tape record.

Students then change the value of the falling mass and/or use more trolleys stacked together (or a train of air track vehicles) to show how the kinetic energy depends on mass. (The total mass being accelerated is of course the falling mass plus the mass of the trolley or air track vehicle.)

ENERGY STORED IN A SPRING

If a spring obeys Hooke's Law, then a graph of force against extension will be linear and pass through the origin. At some extension of the spring, x, the tension in the spring will be F. To extend the spring by a further small distance Δx will require work $F\Delta x$ to be done. This will be stored as elastic strain energy, equal to the shaded pillar in figure A15.

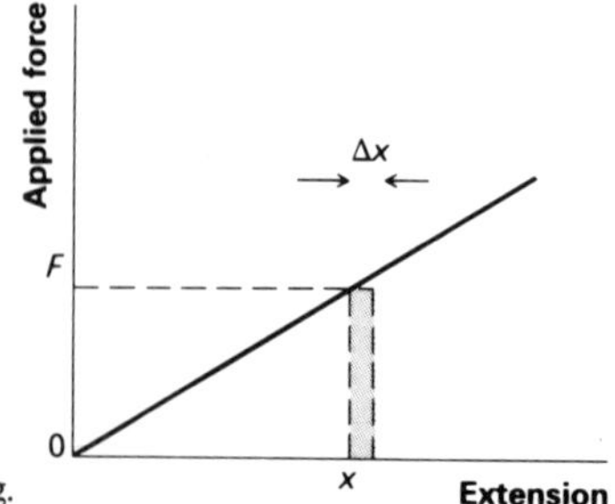

Figure A15
Energy stored in a spring.

From the beginning of the stretching process to its final extension, the total strain energy stored will be equal to the sum of the terms (tension at each extension multiplied by small change in extension). In the limit for tiny extensions, this will be equal to the area under the graph, that is, for a final tension, F, and extension x, the stored elastic strain energy $= \frac{1}{2}Fx$. Since $F = kx$, the stored energy is $\frac{1}{2}kx^2$.

In many situations, for example, using rubber bands, the graph of force against extension will not be a straight line. The same small strips are drawn under the graph to find the total strain energy at the maximum force. The energy stored in any situation is then found by counting the squares on the graph paper, knowing the 'energy value' of a single square.

Reference should be made to experiment A2a where the force–extension curve formed a loop. The hysteresis effect is significant for rubber and is a major cause of heating in motor car tyres.

The next two experiments examine the interchange between elastic strain energy and gravitational potential or kinetic energy. There will always be a mixture of forms of energy, and the questions in the laboratory notes are meant to point out that the moving spring or rubber band will retain some of its elastic strain energy as kinetic energy. This effect is more significant in certain cases than others.

In experiment A10 the force–extension graph should be linear but this is not the case in experiment A11 and the questions emphasize this point.

The alternative method of recording results photographically is described in Revised Nuffield Physics *Teachers' guide Year 4*, Chapter 1.

EXPERIMENT
A10 Measuring the elastic strain energy stored in a spring

Method 1

ITEM NO.	ITEM
107	runway
106/1	dynamics trolley
2A	expendable steel spring
81	spring balance, 10 N
503–6	retort stand base, rod, boss, and clamp
44/1	G-clamp
	either
108/1	ticker-tape vibrator
108/2	carbon paper disc
108/3	ticker-tape
27	transformer
	or
1503	timer, resolution 1 ms
130/2	photodiode assembly with light source
501	metre rule

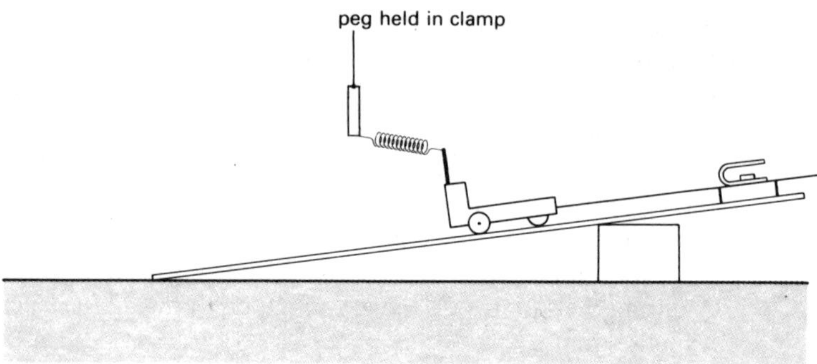

Figure A16

The trolley runway must be carefully compensated for friction. The retort stand (G-clamped to the bench) is used to hold a peg vertically over the middle of the runway. The lower end of this peg should be a few millimetres higher than the peg in the trolley. The spring is hooked over the two pegs, the trolley pulled back and released. As the spring becomes slack it should fall from the fixed peg (figure A16).

Students measure the extension of the spring, x, and the final speed of the trolley, v, for extensions of the spring up to 120 mm. They calculate the kinetic energy given to the trolley in each case. The spring balance is used to obtain a force–extension graph for the spring over the range of x used.

Method 2

ITEM NO.	ITEM
	Apparatus as for method 1 except for:
106/1	dynamics trolley with spring plunger
1504 or 32	balance, 2 kg or masses, 1 kg
1153	wooden block at least 8 cm high
1153	adhesive tape
	graph paper

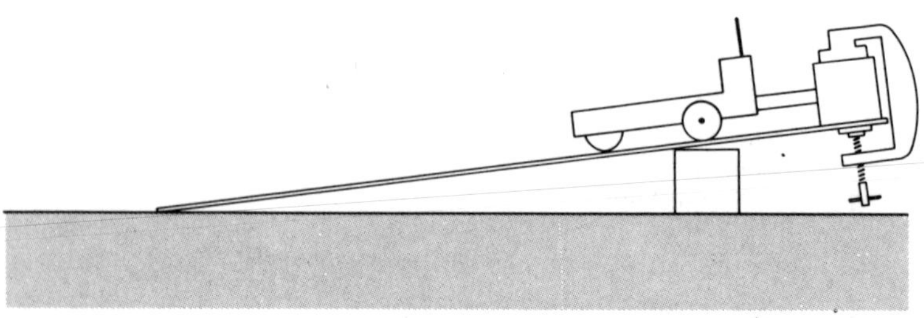

Figure A17

Students calibrate the spring plunger using the compression balance or by loading the plunger with masses. They measure the maximum speed attained by the trolley on a friction compensated runway for various compressions of the plunger (figure A17).

For both methods students compare $\frac{1}{2}mv^2$ for the trolley with energy stored in the spring (from area under force–extension/compression graph). For a Hooke's Law spring, the trolley's speed, v, should be proportional to x, the extension or compression of the spring.

Not all the elastic energy originally stored in the spring will be transformed to kinetic energy of the trolley.

EXPERIMENT

A11 Changing elastic strain energy into gravitational potential energy or translational kinetic energy

A11a

ITEM NO.	ITEM
	Apparatus as for experiment A9a, plus:
	either
81	spring balance, 10 N
	or
31/2	hanger and slotted masses, 0.1 kg

Students plot force–extension curves for a single elastic cord using the arrangement of experiment A9. They find the energy stored for a given extension of the cord (from area under the force–extension curve) and compare this with the final kinetic energy of the air track vehicle.

A11b

ITEM NO.	ITEM
1153	steel nail, 100 mm or equivalent mass of wire, 10–15 g, bent into shape of horseshoe staple
503–6	retort stand base, rod, boss, and clamp
1153	card or board
501	metre rule
31/2	hanger and slotted masses, 0.1 kg
1153	rubber band (size 32, *i.e.*, 75 mm × 3 mm × 1 mm unstretched)
1502	safety spectacles
1504	balance, 0.1 kg

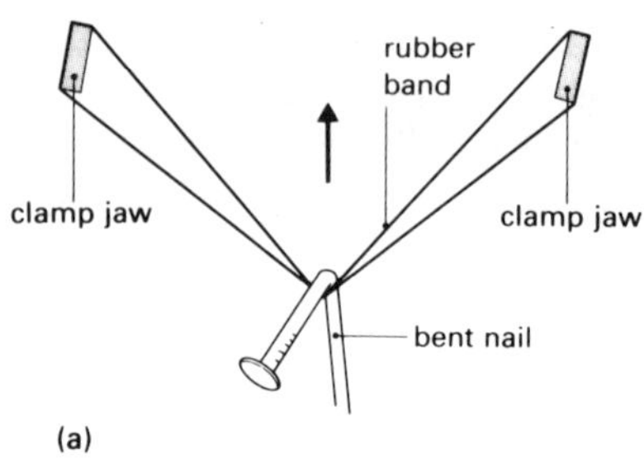

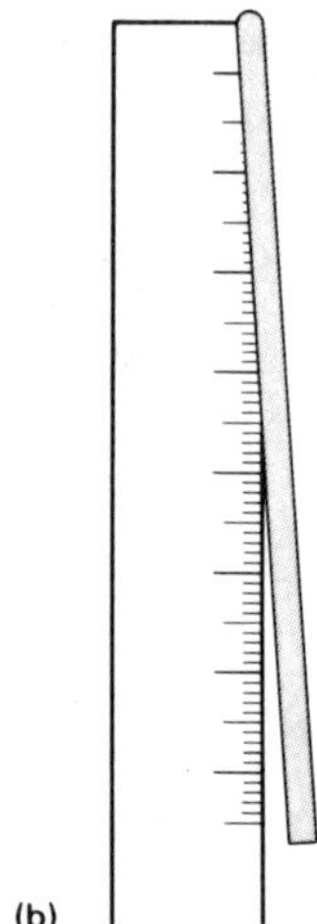

Figure A18

Safety note: Students should wear safety spectacles.

i Students stretch the rubber band between the jaws of a clamp to make a narrow horizontal loop – figure A18(a). They place the bent nail across both strands and fire it vertically upwards about 0.5 m to 1.0 m. They compare the maximum potential energy of the nail with the elastic strain energy stored in the stretched rubber band. (They use the hanger and masses to obtain a force–distance graph for the rubber band and hence the stored energy in the band from the area under the curve.)
ii Instead of firing the nail into the air, the rubber band is stretched over the end of the ruler to make a narrow vertical loop, which, when

released, rises about 1.0 m – see figure A18(b). Students compare the maximum gravitational potential energy of the band with the elastic strain energy stored when it is stretched, again found from the area under the curve.

The force–distance graphs will be different for the two different experimental arrangements – but the area under the curve is always a measure of work done in stretching the rubber.

The students' laboratory notes hint that in these experiments not all the stored elastic strain energy is transformed to kinetic or gravitational potential energy. The transfer is more complete for method **b***ii* than for method **b***i*.

Energy to deform material

The preceding experiments have directed attention to the energy stored in a spring which can be recovered later. But not all deformation is elastic and not all strain energy can be recovered. The last experiment in this section deals with strain energy which cannot be recovered. There are obvious applications to climbing ropes, car seat belts, crash barriers, and so on, which are required to absorb energy and not return it.

EXPERIMENT
A12 Energy absorbed in deformation

ITEM NO.	ITEM
1501	iron wire, 1.0 m long, 0.71 mm diameter
1153	wooden rod, plastic tube, or similar, *e.g.* cylindrical former about 35 mm diameter and at least 10 cm long
31/1	hanger with slotted masses, 0.01 kg
503–6	retort stand base, rod, boss, and clamp
501	metre rule

The choice of wire diameter and spring diameter are quite critical for good results.

Use the rod or tube as a former to make a coil of iron wire – about 8 turns and about 3.5 cm diameter. Hang it from a clamp. Hang a load of 0.1 kg from the lower end and support it by hand so that the 'spring' remains unstretched (figure A19).

When the load is released the coil should unwind almost completely, without oscillation.

The experiment could be extended. Rewind the coil and plot a force–extension curve for loads up to 0.2 kg. From the area under the curve up to the extension caused by the original 0.1 kg load, calculate the work done in uncoiling the spring. It should be within 20 % of the gravitational potential energy lost by the falling mass.

(An attempt to perform this experiment with cored solder wire may turn into a nice demonstration of creep!)

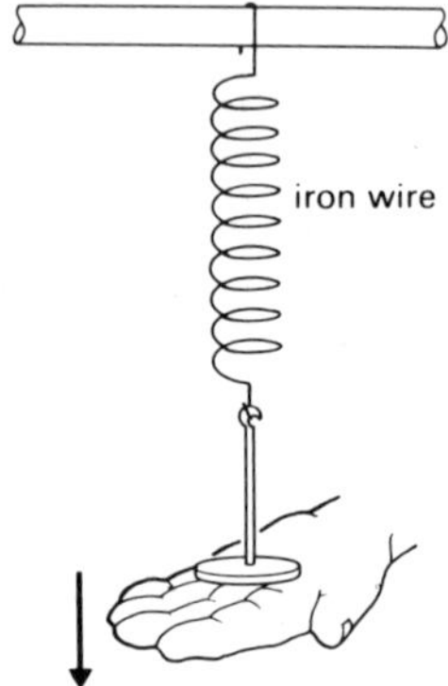

Figure A19

Questions

Question 11 leads to the expression $\frac{1}{2}Fx$.
Question 12 is about a catapulted vehicle which requires a photographic record to be analysed. Question 13 is a problem about energy transfer in an oscillating trolley-and-spring system.
Question 14 leads to the formula for the energy stored per unit volume. Question 15 asks for an estimate based on that formula. Question 16 asks students to apply the method of dimensions to the formula. Question 17 is a revision question on stretching and storing energy.

Reading

GORDON *Structures.* Chapter 5 is about strain energy and modern fracture mechanics.

All of the main textbooks have chapters which deal with mechanical energy. These range from a very mathematical to an almost qualitative treatment.
P.S.S.C. *College Physics,* Chapters 17, 18 or P.S.S.C. *Physics* (2nd edition), Chapters 23, 24 give a suitable treatment of stored mechanical energy for this early stage in the course.
CRAC, *Stress, strain and strength* mentions many interesting applications.

Preparation of students' reports for the last part of Section A3

Students can be asked now to begin to collect information on a composite material, for example, in note form or as a summary on about one side of paper, including diagrams and references where appropriate. Later the collected information can be duplicated for the whole class and/or selected students asked to give a brief report on their findings.

Some suitable materials include:

concrete; glass fibre; titanium; nylon; carbon fibres; ceramics; rubber; adhesives; aluminium and alloys; diamond and graphite; thermoplastics; types of steel; road surface materials; plywood; toughened glass; materials for nuclear reactor vessels.

Important factors to consider include:

stiffness/strength/toughness	cost per unit strength/volume
density/weight	structure
durability/corrosion	cost/ease of manufacture/machining
electrical/magnetic/thermal conductivity	cost of installation
	availability
melting point	uses

Film

The film 'Materials for the engineer' (five examples of technical advances in the use of new materials in the building, metallurgy, and communications industries) will help to suggest ideas.

Some suitable sources of information for students

Schools Council Engineering Science Project *The use of materials.*
MCSHEA *Materials technology.*
Nuffield Advanced Physics *Physics and the engineer.*
Materials. SCIENTIFIC AMERICAN.
LOFTAS and GWYNNE *Advances in materials science.*
MARTIN 'Strength of materials'.
GORDON *The new science of strong materials.*
Pamphlets from the Institution of Metallurgists, *e.g., Which materials? Materials in action: Concorde; Materials in action: Materials and the human body.*
Magazines, *e.g., Project* (1975–1980), *New Scientist, Scientific American, Atom,* etc.

SECTION A2
THE STRUCTURE OF SOLID MATERIALS

INTRODUCTION

This section includes evidence for the size of an atom, the arrangement of atoms in crystalline and amorphous substances, and the interatomic force. Properties of materials observed in Section A1 are related to these. The interatomic force is discussed in more detail in Unit E, 'Field and potential'; diffraction in Unit J, 'Electromagnetic waves'.

Lord Rayleigh's oil film experiment appears in Revised Nuffield Physics *Teachers' guide Years 1 and 2*, Chapter 7. The Avogadro constant appears in the Nuffield O-level Physics and Chemistry courses (Physics *Year 4*, Chapter 8; Chemistry Topics 14 and 17). The optical analogue for von Laue X-ray diffraction patterns is experiment A14.4b Revised Nuffield Chemistry *Teachers' guide II*. Crystal structure, heat treatment of steel, and bubble rafts also appear under Topic A14. There is a brief mention of the intermolecular force curve in Revised Nuffield Physics *Teachers' guide Year 4* page 89, after demonstration 49. The approach here is very similar. Thus for students who have studied earlier Nuffield courses in physics and chemistry much of the work of this section is revision or extension of concepts already introduced, albeit briefly.

Revised Nuffield Advanced Chemistry *Students' book I* deals with crystal structure in Topic 7.

Timing

The whole Section should not occupy much more than one week. The temptation to go into detail should be avoided.

SIZE OF AN ATOM

Experiment A13 gives evidence about the size of grains of lycopodium powder. If the oil film experiment, which gives information about the size of molecules, has not been done before, it can replace the experiment described here. From it students find that oil molecules are about 2 nm long. For comparison, an air molecule is about 0.4 nm across (some students will have met this earlier in chemistry or as a deduction from kinetic theory). The purpose of experiment A13 is to show by analogy how the size of atoms is found from X-ray diffraction measurements, and to justify the use of X-ray diffraction photographs as evidence for the amorphous (disordered) or crystalline (ordered) structure of a material. Single crystals produce spots on a photograph, polycrystalline samples sharp rings, and amorphous substances fuzzy rings.

Experiment A13 is not essential and could be omitted if time is short. The electron microscope photographs in the *Students' guide* may be considered sufficient evidence for the size of atoms and their regular arrangement in crystals. It might then be worth using experiment A13 at the end of this section, to link with the brief discussion of disordered structures.

EXPERIMENT/DEMONSTRATION
A13 An optical analogue of X-ray diffraction

(Experiment)

Each student requires:

ITEM NO.	ITEM
	either
92R, 92T	m.e.s. bulb, 2.5 V, 0.3 A, in holder
1033	cell holder with 2 cells
	or
72, 74	lamp, 12 V, 24 W, in holder
27	transformer
1153	card
1153	pin
1167/3R	colour filters
1000	leads
	(Demonstration)
1505	laser
1501	several short lengths of wire of diameter between 0.28 mm and 0.08 mm
1167/2M	adjustable slit
1153	plastic ruler
	(Experiment and Demonstration)
24	hand lens
1153	piece of cloth with regular structure, *e.g.*, cotton handkerchief, or piece of fine mesh nylon, or Terylene net or umbrella fabric *and/or* set of diffraction grids
1155	microscope slide dusted with
7L	lycopodium powder

Experiment

A point source of light is required. An m.e.s. bulb at a distance of about two metres in a darkened room, or a 12 V, 24 W lamp with a card containing a pinhole are both suitable.

When students hold the diffraction grid or piece of material close to the eye and look through at the point light source they see a pattern of diffraction spots. The pattern will rotate if the grid or material is rotated, or change shape if the material is stretched. The random grid or slide dusted with lycopodium powder will give at least two fuzzy rings centred on the light source.

Red, green, and blue filters can be used to show that the diameter of the pattern depends on the wavelength of the radiation. The shorter the wavelength of the radiation the less it is spread out. This is essential information for the scaling down to X-ray diffraction.

Demonstration

Safety note: Teachers should observe the recommendations on the use of lasers issued by the D.E.S. (Administrative memorandum 7/70 'Use of lasers in schools and other educational establishments'.)

The experiment can be repeated, placing the grid, material, or slide close to the laser and projecting the resulting diffraction pattern on a screen at least 1 m away.

An adjustable slit in the laser beam shows that the narrower the gap the light passes through the wider the beam becomes.

Short lengths of copper wire ranging from 0.28 mm to 0.08 mm also give good diffraction patterns and enable one to show that the width of the pattern (measured with a plastic ruler) is inversely related to the thickness of the wire. Direct comparison with the diameter of the two dark rings caused by the lycopodium powder, suggests that the particle size is close to 25 μm (any change from linear to circular geometry having been ignored).

(The particles' size can be estimated more directly by observing them under the microscope.)

Some students will already be familiar with the effects of change of wavelength, and gap or barrier width on the diameter of the diffraction pattern (for example, from Revised Nuffield Physics *Pupils' guide Year 5* Chapters 8 and 9). Students should also understand that light and X-rays are electromagnetic radiations differing in wavelength.

The wavelength of the He–Ne laser light is about 600 nm. In experiments similar to von Laue's, X-rays of the order of 0.1 nm wavelength are used, about 6000 times smaller than the wavelength of laser light. For an X-ray pattern of similar size to the optical one (diffraction angle identical), simple proportion suggests that the particles must also be 6000 times smaller, that is, $\frac{25}{6000}$ μm, about 4 nm. Wider patterns are caused by smaller particles still.

Videotape or broadcast

Several minutes at the beginning of 'Electron diffraction' in the Granada television series *Experiment: Physics* show the diffraction of visible light passing through a regular network pattern of tiny holes in an opaque screen. This grating is then replaced by one made up of many small pieces which are aranged at random, showing how the spots become sharp rings, that is, transition from a single crystal to a polycrystalline sample. The main part of the film is more useful for Unit L, 'Waves, particles, and atoms'.

Questions

Question 18 revises the oil film experiment.
Question 19 is about the scaling of the analogue experiment.
Question 21 calculates the size of an atom using the Avogadro constant.

Experiment A13 is a simple analogue in two dimensions. The Braggs were able to develop von Laue's experiments on single three-dimensional crystals and polycrystalline specimens and produced a simple mathematical analysis from which the positions of the atoms in simple crystals could be calculated.

The question to answer next is, 'How are atoms arranged?' The Avogadro constant can then be calculated knowing the size of an atom and the regular structure of the solid.

STRUCTURE OF METALS (SIMPLE CRYSTALS)

Before discussing the close-packed arrangement of atoms in a metal crystal, it may be helpful to review briefly bonding between atoms and molecules. Atoms are held together by electrical forces. There are three types of strong bond (ionic, covalent, and metallic) between atoms and ions, and also the weak van der Waals bond between molecules. The covalent bond is directional. The ionic bond is between ions of different charge. The metallic bond between positive ions in a sea of electrons exerts equal forces in all directions. Metal atoms are therefore likely to be packed together as closely as possible. Identical polystyrene spheres, marbles, or coins can be used to see how identical objects are arranged when packed together as closely as possible. Soap bubbles are like metal atoms in that they are attracted to each other by radially directed forces. A quick look at the bubble raft model of experiment A16b might be useful now.

Reading

Textbooks with good diagrams of packing arrangements include:
DUNCAN *Physics: a textbook for Advanced level students.* Chapter 1.
BOLTON *Patterns in physics.*

WENHAM, DORLING, SNELL, and TAYLOR *Physics: concepts and models.* Chapter 14.
NUFFIELD REVISED CHEMISTRY *Handbook for pupils.* Chapter 9.

Most of the next experiment is a repeat of Revised Nuffield Chemistry experiment A14.5d (see *Teachers' guide II*). In Revised Nuffield Physics *Teachers' guide Years 1 and 2*, Experiments 7 and 8, spheres are stacked to illustrate crystal structure.

Students should know that a layer of close-packed balls (or atoms) forms a hexagonal array, and see how these hexagonal layers can form three-dimensional close-packed structures. The details of the hexagonal close-packed (h.c.p.) and face-centred cubic (f.c.c.) structures and their unit cells are not important.

DEMONSTRATION/EXPERIMENT
A14 How atoms are arranged in metals

ITEM NO.	ITEM
1016/2	40 expanded polystyrene spheres, 50 mm diameter
	4 books (or wooden battens 0.25 m long)

Glass marbles or small polystyrene spheres may be used if preferred. This can be done as a demonstration or class experiment in small groups.

Lay out a square 5×4 ball array inside a $0.25\,m \times 0.2\,m$ boundary fence, and pile a second, third, and fourth layer of balls on top of the first. The close-packed hexagonal array appears on the sides of the pyramid.

Repeat the experiment with a wooden triangle so that the base layer is hexagonal (close-packed) rather than square. As successive layers are added students may notice that two alternative arrangements are possible.

The ABABAB arrangement in which each ball in the third layer is immediately above a ball in the first layer, and the ABCABC arrangement, form the hexagonal close-packed (h.c.p.) and face-centred cubic (f.c.c.) structures respectively. In both arrangements the balls are as close together as possible. Zinc and magnesium are examples of h.c.p.; copper, aluminium, and iron between 910 and 1400 °C are f.c.c.

These details are not important for our students, but they should notice that when hexagonally packed layers are stacked together each ball has 12 nearest neighbours. This is the closest possible packing of equal sized balls in three dimensions.

The more open body-centred cubic (b.c.c.) structure cannot be built up simply by stacking balls: they have to be held in place or glued together. In the b.c.c. structure each ball has only 8 nearest neighbours. But b.c.c. metals (for example, sodium, chromium, and iron below 910 °C) are not necessarily less dense or softer than the others. The transition of iron from b.c.c. to f.c.c. at 910 °C is one of the factors which makes iron (and steel) such a versatile metal (see page 48).

To balance the emphasis on metal bonding, mention briefly an ionic material like NaCl with its simple cubic array of large Cl^- ions separated by the much smaller Na^+ ions. Each Na^+ has six nearest neighbours: all are Cl^-; and vice versa.

Question 23 is about hexagonal layers.

THE AVOGADRO CONSTANT AND THE MOLE

The *Students' guide* (page 8) mentions the unit for amount of substance, the mole. Much confusion is probably caused by students linking the word in their minds with molecule. In fact mole means 'heap' or 'pile' – in this case a heap consisting of 6.02×10^{23} particles. The particles may be atoms, ions, molecules, electrons... but must be specified.

The Avogadro constant, *L*, can be estimated in a number of very different ways, for example *a* from radioactive decay, *b* from electrolysis, *c* by Perrin's method considering the distribution of particles in a column of fluid, and *d* by X-ray crystallography.

The electrolysis methods (using the Faraday constant and the charge on the electron) and the size of the atom method (from X-ray crystallography) are both reasonably accurate methods of historical importance. The measurement of the size of the atom depends on how accurately the wavelength of the X-rays can be determined. This is done using a grating.

At present X-ray crystallography provides the most accurate method for finding the Avogadro constant. Two alternative approaches can be used:

a Find the number of atoms in and the length of the side of a unit cell of copper. From the size of the cell find the number of unit cells in the volume of a mole of copper atoms and hence the Avogadro constant.

b Use the geometry of stacking spheres (experiment A14) to calculate the number of atoms in a rectangular box of volume of a mole of copper atoms.

Questions

Question 20 hints at the electrolysis method of finding *L*. Question 24 is an estimation of *L*.

Reference

Revised Nuffield Advanced Chemistry *Students' book I* and *Teachers' guide I* deal with the mole in Topic 1, 'The Periodic Table'.

There is an opportunity here for students to practise the use of indices in the form of powers of ten, and to think about orders of magnitude. It is important to be able to recognize the order of magnitude of the quantities involved in order to think sensibly about problems on the microscopic scale, for example, question 22.

INTERMOLECULAR FORCES AND ENERGIES

Stress that the interatomic and intermolecular forces responsible for holding materials together are electrical and not gravitational. This point will be emphasized quantitatively in Unit E, 'Field and potential'.

Point out that we know that there are attractive forces between molecules in a solid – if there were not, a solid rod would be unable to resist tension. But there must also be a repulsive force between them which prevents them getting too close.

A simple model of the intermolecular force is constructed to enable discussion of and simple justification for the shapes of the force–separation and potential energy–separation curves for two isolated spherical molecules.

Reference

TABOR *Gases, liquids and solids.* Chapter 2 and page 138 contain more detailed information for teachers.

DEMONSTRATION
A15 An intermolecular force model using a linear air track

ITEM NO.	ITEM
1019	air track and accessories
1020	air blower
106/2	elastic cord
92A	2 Ticonal magnets

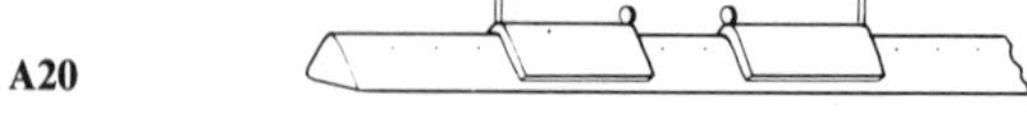

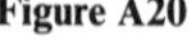

Figure A20

Connect the two vehicles by the rubber cord, pull them apart and release. (Attractive force.)

Remove the rubber cord, and attach small magnets arranged to repel when facing each other, to the inner ends of the two air track vehicles. Hold the vehicles together and release. (Repulsive force.)

Using both the magnets and the cord together produces an equilibrium separation (see figure A20). The vehicles will then move together or apart when displaced from this position and oscillate about it. A suitable length of rubber cord to balance the repulsive force between the magnets must be found in advance.

Evidently the equilibrium position is the one in which the stored energy is a minimum.

In this model the elastic force becomes stronger as the vehicles move apart, unlike the electrical attraction it represents. A better model would have the magnets attracting and compression spring buffers to supply the repulsive force.

The two vehicles oscillating about this equilibrium position can be used as a crude model for a pair of atoms or molecules with thermal energy.

Force–separation and energy–separation curves

The *Students' guide* (page 9) has sketches of the attractive and repulsive forces between molecules, and also the net force, as a function of separation. It would be useful to discuss the general shapes of these graphs while demonstration A15 is on the bench.

Emphasize that the repulsive force (between molecules) is only significant at small separations, but that it varies much more rapidly than the attractive force; and that the equilibrium position is the separation at which these two forces balance.

Note that repulsive forces are shown as positive; attractive forces as negative. The same convention will be used when gravitational and electrical forces are discussed in Unit E, 'Field and potential'.

The relationship between force–separation and energy–separation curves is important. Some difficult ideas which will be met later in the course (especially in Unit E) can be introduced here.

At the equilibrium position the net force is zero and the potential energy has its minimum value. (The notion of a potential 'well' is a useful one: energy is needed to compress the molecules against the repulsive force or to separate them from the attractive force.) The potential energy is taken to be zero when the molecules are 'infinitely far apart', and is negative at the equilibrium position. Students will have other opportunities later in the course to come to terms with the idea of negative potential energy.

Questions

Questions 25 and 26 help to construct the shape of the force–separation curve. Question 27 relates the force–separation and potential energy–separation curves together.

Microscopic and bulk properties

Picturing the atoms in a solid as being held together by tiny springs obeying Hooke's Law enables us to relate the force constant between atoms, k, to the Young modulus, E, a bulk property.

Imagine a pair of atoms, r_0 apart, being pulled a distance Δr by a force F (figure A21(a)).

$F = k\Delta r$

Each of these atoms is part of a layer of atoms (figure A21(b)).

Area occupied by each atom is $r_0{}^2$

Number of atoms per unit area $= 1/r_0{}^2$

The stress between layers $=$ (force between pair of atoms) $\times$ (number of atoms per unit area) $= F/r_0{}^2 = k\Delta r/r_0{}^2$

The strain $= \Delta r/r_0$

$\Rightarrow$ The Young modulus, $E = k/r_0$

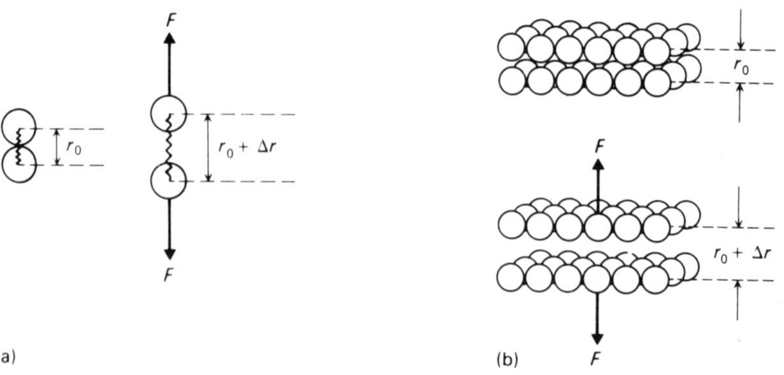

Figure A21
The interatomic force constant and the Young modulus.

Thus the interatomic force constant (the slope of the force–separation curve close to $r = r_0$) can be calculated from the Young modulus and the atomic separation. For steel $E \approx 2 \times 10^{11}$ N m^{-2} and $r_0 \approx 3 \times 10^{-10}$ m giving $k \approx 60$ N m^{-1}.

Questions

Question 29 is useful preparation for the calculation of the interatomic force constant.
Question 30 goes through the above argument step by step.

CHAIN MOLECULES, SLIP, AND CRYSTAL IMPERFECTIONS

Only simple regular structures have been considered so far. Real materials are not perfect; most are not simple structures.

To complete the section, the arrangement of the long chain molecules of rubber or polythene weakly linked together is used to explain the unusually large strains possible in these materials (experiment A2). This is unlike the stretching of the interatomic bond which allows a metal to stretch elastically by a small fraction only.

A simple model predicts that metals will require very large stresses to deform plastically by layers of atoms slipping over each other. However, in the most nearly perfect metal crystal that can be obtained, there are some defects which enable slip to occur at stresses well below the predicted value. Observed features of the stress–strain curves can be explained in terms of dislocations, other crystal imperfections, and changing crystal structure.

Any discussion on dislocations and the effects of other crystal imperfections must be kept short, providing no more than a brief introduction and a glimpse at the immense subject of metallurgy and materials science. It is important to emphasize that this picture of the effects of dislocations and positioning of foreign atoms within a structure is very over-simplified.

Reading

If students are interested in looking into the subject further they can be referred to books, for example:

AKRILL, BENNET, and MILLAR *Physics.* Chapter 4.
FARRAR *The mechanical properties of materials.*
COLLIEU and POWNEY *The mechanical and thermal properties of materials.* Chapter 3.
MARTIN and HULL *Elementary science of metals.*
SCIENTIFIC AMERICAN *Materials.* This has a section 'The nature of metals', by A. H. Cottrell.
ROSENBERG *The solid state.* Chapters 3 and 4.
GORDON *The new science of strong materials.* Chapter 4.
MOFFATT, PEARSALL, and WULFF *The structure and properties of materials.* Volume 1 Chapter 4.
Schools Council Engineering Science Project *The use of materials.*

EXPERIMENT/DEMONSTRATION

A16 Chain molecules, dislocations, and alloys

a Splitting of stretched rubber/polythene
b Bubble raft model: dislocations, grain boundaries, and foreign atoms
c Heat treatment of steel

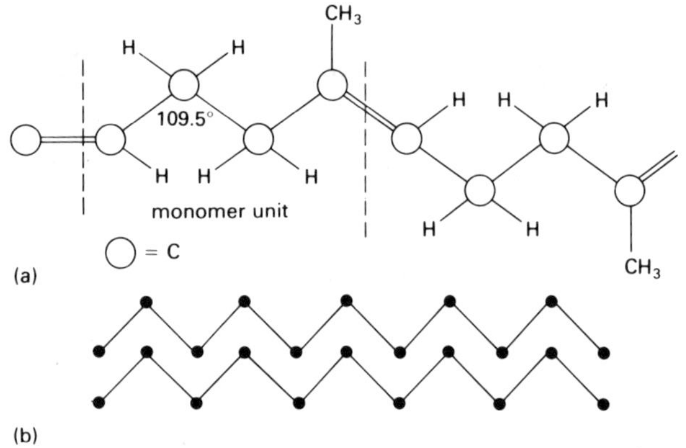

Figure A22
(a) Part of the molecular chain of rubber.
(b) Crude picture of stretched rubber.

A16a Splitting of stretched rubber/polythene

ITEM NO.	ITEM
1016/1	12 expanded polystyrene spheres, 25 mm diameter
1153	11 cocktail sticks
	Each pair requires:
1153	2 squares of balloon rubber, 50 mm × 50 mm
1153	polythene strip, 100 mm × 10 mm, 250 gauge (about 0.05 mm, *e.g.*, from a food bag)
	pin

A rough model of a rubber molecule can be made from polystyrene balls and cocktail sticks pushed in at about 110° to each other – figure A22(a). Aligned at random, the chain curls up on itself. It unwinds if pulled out carefully.

The bond angles between the carbon atoms are fixed but rotation can occur about these bonds, so that the chain bends and wriggles randomly. When the rubber is stretched, the chains are straightened, and also tend to line up – figure A22(b).

One student holds a square of rubber, cut from a toy balloon, whilst another pierces it, stirring the pin about gently. A second square of rubber is then stretched as hard as possible in one direction and pierced near one corner. Next, this square is stretched at right angles to the first direction and pierced again. The size and direction of all three holes should then be compared. (The rubber splits along the stretched direction, suggesting that the molecules are lined up.)

It is useful to split stretched polythene too.

Not all materials made from long chain molecules will split like this. Plastics can be flexible or brittle depending on how strong the cross-linking bonds between molecules are.

A16b Bubble raft model: dislocations, grain boundaries, and foreign atoms

ITEM NO.	ITEM
1155	Petri dish
1155	length of rubber tubing to fit gas tap
1155	hypodermic needle, 25 gauge
1501	L-shaped pieces of wire, about 2 mm diameter and about 10 cm long
1156	bubble solution: 1 Teepol, 8 glycerol (propane-1,2,3-triol), 32 water, parts by volume
522	Hoffman clip
508	Bunsen burner

Bubbles are blown by connecting the needle to the gas tap. The size of the bubbles depends on both the rate of flow of the gas and how deep the needle is in the solution. The clip is used for fine adjustment to the rate of flow of gas. The bubbles need to be about one to two millimetres in diameter for good results. It takes several minutes to make a raft big enough for experimental purposes. Wrongly sized bubbles can be burst with a hot wire and the whole raft cleared by playing a lit Bunsen burner over it quickly.

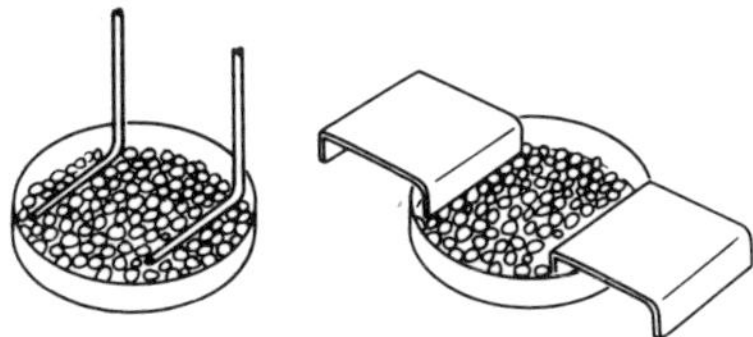

Figure A23
Means of compressing a bubble raft.

Soap bubbles make a good model of a two-dimensional layer of atoms with surface tension acting as the attractive force and the pressure inside the bubbles which stops them collapsing providing the repulsive force.

Very rarely will a perfect raft be formed. Grain boundaries are clearly observed. The dislocations are most easily seen if the raft is viewed almost in line with the surface of the bubbles. A sheet of white paper underneath the dish and perhaps some side illumination may make the effects easier to see.

Dislocations can be made to move backwards and forwards by carefully compressing and stretching the raft with the L-shaped pieces of wire.

The best way to demonstrate the movement of the bubbles is to project them onto a screen using an overhead projector. Small-sized bubbles are just distinguishable and form a dark surface in which the movement of dislocations is strikingly obvious.

A perfect lattice would be strong, but a moderate number of dislocations reduces the strength by a large factor.

Work hardening functions by generating so many dislocations that they pin each other down (that is they prevent each other moving). Foreign atoms in the lattice (as in steel) may make for greater strength, sometimes by pinning down dislocations as shown in experiment A16c.

Students should know that the properties of a material can be changed dramatically by altering the number of foreign atoms, by heat treatment, or by work hardening, but the mechanisms of these changes should at most only be briefly hinted at.

A16c Heat treatment of steel

ITEM NO.	ITEM
	either
1153	4 steel sewing needles
	or
2A	5 cm wire from expendable steel spring
1155	2 pairs of pliers
508	Bunsen burner
1502	safety spectacles

When a needle or piece of steel wire is bent with pliers some bending occurs before it snaps. If made red hot and allowed to cool very slowly in air it becomes very soft and is easily bent (annealing). If it is heated red hot and then rapidly cooled in water it becomes brittle and will fracture if bent (quench hardening). If quenched steel is heated to an intermediate temperature (for example, just enough to cause a blue oxide colour to appear) and then cooled, a useful balance of properties is obtained (tempering).

Steel is iron with the addition of small amounts of carbon (typically between 0.02 and about 2 %) and other elements. It is important and has many uses because it is relatively cheap and because its properties can be altered over a wide range by varying the composition and by heat treatment. Iron itself exists as a body-centred cubic structure up to 910 °C and as close-packed face-centred cubic structure between 910 °C and 1400 °C. The metallurgy of steel is quite a complex subject.

Tempering is important for other alloys too – for example modern lightweight aluminium armour plating.

Reading

BRONOWSKI *The ascent of Man.* Chapter 4 'The hidden structure' describes the ritual of the making of the Japanese sword.

NUFFIELD REVISED ADVANCED CHEMISTRY Special Study, *Metals as materials.*

A reading passage in the *Students' guide,* 'Looking at the structure of materials' contains a collection of illustrations of crystal structure, grain boundaries, dislocations, X-ray and electron diffraction patterns.

Questions

Question 31 is about rubber chains and uses as background the random walk result from the kinetic theory of gases in Section A4 (see also Revised Nuffield Physics *Pupils' text Year 4*, Chapter 8).

Questions 32 and 33 distinguish amorphous and crystalline states.

Questions 34 to 36 distinguish between the effects of cracks, dislocations, and the introduction of foreign atoms in both amorphous and ductile materials.

MODELS

Much of the work of this section uses models. An atom has been thought of as a hard sphere similar to a glass marble or a polystyrene sphere. Does the atom have a definite boundary as the model implies? Is it connected to a neighbouring atom by a spring? What is the purpose of describing the observed behaviour of atoms in this naive way? A brief discussion of the purpose and use of such models is called for at this point. Models are used to help one to visualize what things may be like and as tools for thinking about the behaviour of the thing modelled. A model of the arrangement of atoms in a crystal makes it possible to predict behaviour. A model may take the form of a mathematical equation or perhaps a set of equations – there will be many examples in the course; the solution of the equation may suggest new experiments, predicting behaviour as yet not observed. This distinguishes between a useful and a less useful model and suggests limits to the use of any particular model for prediction.

Sometimes scientists imagine a model on very slender evidence and then test it. How far did our use of a model based on polystyrene balls for a metal structure follow this pattern? How useful was this model? How much of the behaviour of copper can we predict by applying it? (Stacking arrangement, strength, slip, etc.) Is a model which uses a bubble raft more or less useful?

The session might end with questions about the propriety of using the imagination in this way in a science.

Questions

Questions 23, 25, 26 and 29 refer to models explicitly. The reading passages 'Models' and 'Theories – true or not?' in the *Students' guide* can be used to promote discussion on models, their purpose, and their validity.

Reading

Anyone wishing to read further about the use of models in scientific inquiry could be referred to:

ROGERS *Physics for the inquiring mind.* Chapter 24.

WATSON *The double helix.* This gives a description of how the model of DNA was conceived with only minimal access to experimental facts.

Further ideas about scientific inquiry as well as the use of models can be found in:

BRONOWSKI *The common sense of science* and *The ascent of Man.*

BONDI *The Universe at large.*

SECTION A3

STATICS, STRUCTURES, AND COMPOSITE MATERIALS

TREATMENT OF STATICS

The aims of this section are modest: students should understand the conditions required for equilibrium of rigid bodies; they should know how to resolve forces and take moments to see if these conditions are satisfied. The stresses within a structure can then be appreciated. All the forces considered lie in a single plane.

Limited time is available and the work is done near the beginning of the course, so we do not expect students to be able to solve complicated problems. Nevertheless it is important that they see the relevance of these methods to real problems involving cranes, bridges, buildings, and so on. Some of the suggested experiments and questions in the *Students' guide* try to make this connection.

The summary in the *Students' guide* points out that the forces involved in statics problems are usually the weights of objects, and the contact forces between them which arise from deformation (tension, compression). It is often convenient to resolve the contact force into two components, the normal force (perpendicular to the surface), and a frictional force tangential to the surface.

We advise against using the term 'reaction' which can lead to a misunderstanding in several ways. (For example, it has suggestions of cause and effect. See J. W. Warren's *Understanding force* for a useful discussion of some of the conceptual difficulties that may arise.)

Experiments A17 to A20 introduce the resolution of forces into components, the triangle of forces, and the need to take moments to solve problems involving non-concurrent forces. Students are reminded of the concept of centre of gravity. Finally the forces within a simple structure, a roof truss, are analysed.

A model bridge building competition can make a good climax, reinforcing the ideas of the experiments. Finally, students present their reports on composite materials (see page 34).

ADDITION AND RESOLUTION OF FORCES

Vector addition, by resolution or a graphical method, is examined experimentally in experiment A17a. In preparation for the test of

Coulomb's Law in Unit E, 'Field and potential', this experiment is also used to show that if an object (weight W) hung on a string is pulled sideways a small distance d, the horizontal force on it is proportional to Wd.

Experiment A17b emphasizes the need for the three non-parallel forces to pass through a point for equilibrium. Moments can be used but are not necessary. The difference between a strut (compression) and a tie (tension) is also examined.

A brief reminder that friction exists in all real situations also gives another opportunity for a resolution of forces example.

EXPERIMENT
A17 Measuring the forces on systems in equilibrium

ITEM NO.	ITEM
503–6	2 retort stand bases, rods, bosses, and clamps
44/2	2 G-clamps
81	2 spring balances, 10 N
32	mass, 1 kg
550	protractor
1153	0.5 m lath or rule
1153	string

A17a Triangle of forces and resolution

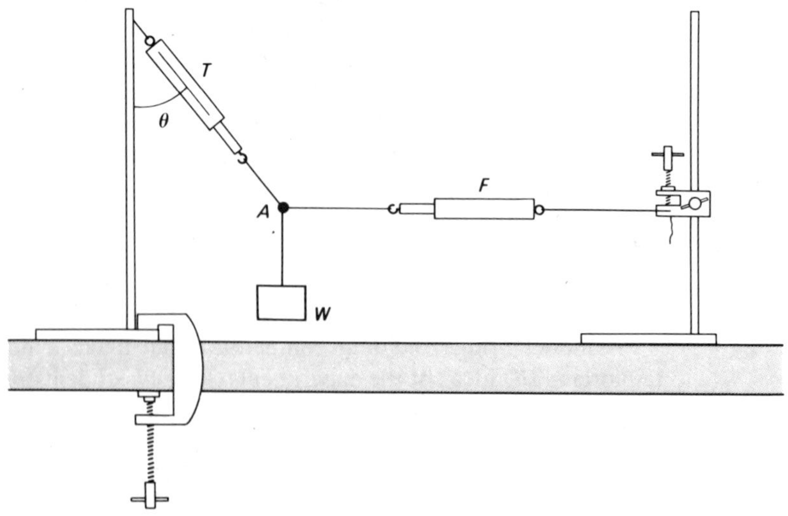

Figure A24

The mass in figure A24 is tied in the middle of a length of string with a loop at either end, to which spring balances are attached. Students measure T, F, and θ, for various θ and draw space and vector diagrams to verify that the three forces make a closed triangle in equilibrium. For small angles θ, F is proportional to d, the horizontal displacement of the weight, W.

A17b Strut or tie?

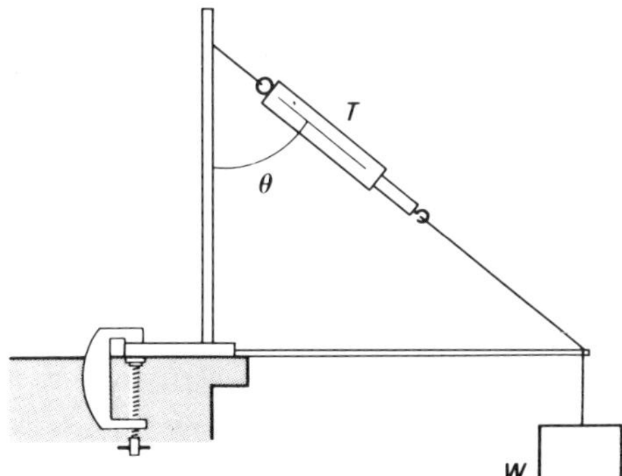

Figure A25

The mass is tied to the end of the rule. The retort stand is clamped close to the edge of the bench so that the rule can be pivoted freely at the edge as in figure A25. Students use the spring balance to measure the tension in the string when the rule is horizontal.

They are asked to infer the magnitude and direction of the force in the rule and to compare this situation with experiment A17a. Here the rule is in compression; in A17a the string was deflected from the vertical by a tension force in the horizontal part of the string.

In structures, members in compression are known as struts, those in tension are called ties.

If three non-parallel forces are to be in equilibrium then their lines of action must pass through a point. The students' laboratory notes ask them to use this fact to decide which of the arrangements sketched in figure A26 are stable.

If we neglect the weight of the strut then there are only three forces acting: T, W, and the contact force, F, at the lefthand end of the strut. (F can be thought of as made up of two components: one horizontal, the other vertical. Each of these components will be due partly to the normal force and partly to friction at the bench and at the retort stand.)

For arrangement (a) to be stable F must pass through the point of intersection of T and W. The dotted line shows the required direction for F.

For arrangement (b) to be stable F must be purely horizontal.

For arrangement (c) to be stable F must have an appreciable downward component (due to friction between the end of the strut and the retort stand base). A practical attempt to set up a stable arrangement like this will probably fail.

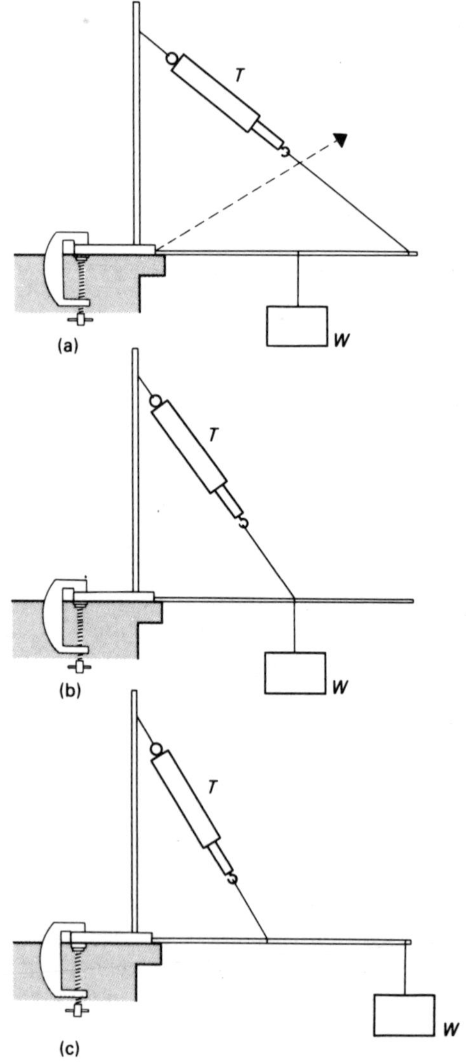

Figure A26

A17c Determination of the coefficient of static friction for two wooden surfaces

ITEM NO.	ITEM
107	trolley runway
1153	wooden block, about 20 cm square, with screw eye
32	several 1 kg masses
31/2	hanger and slotted masses, 0.1 kg
40	single pulley on clamp
1153	string

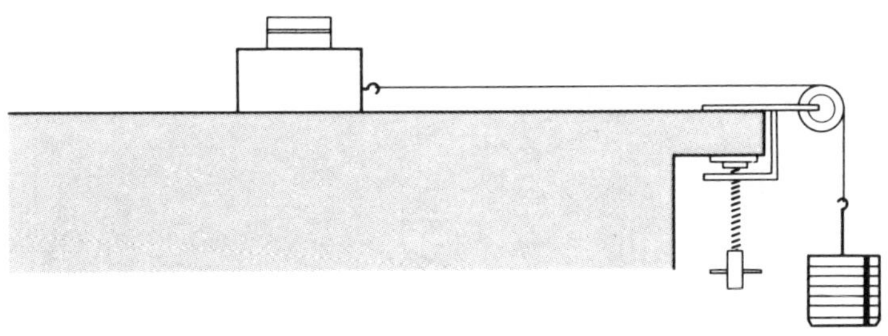

Figure A27

A cardboard carton containing perhaps scraps of plastic foam placed beneath the load will protect the masses and the floor when the block slips (and keep students' feet away).

The laboratory notes for this experiment introduce the terms limiting friction and coefficient of friction, μ. Students vary the load on the wooden block to check whether the frictional force is proportional to the normal force, and they obtain a value for μ.

Students can also obtain a value for μ from $\mu = \tan\theta$, where θ is the angle of the plane at which the block just begins to move.

Questions

Questions 37 to 40 all refer to experiment A17.
Question 39 emphasizes that the contact force is perpendicular to the surface if there is no friction.

The next experiment concentrates on moments and may be revision in a practical context for some. Revised Nuffield Physics *Pupils' text Years 1 and 2*, Chapter 4, refers briefly to the principle of moments.

EXPERIMENT

A18 Investigating the forces at the supports of a loaded bridge

ITEM NO.	ITEM
81	2 spring balances, 10 N
503–6	2 retort stand bases, rods, bosses, and clamps
501	metre rule
31/2	2 hangers and slotted masses, 0.1 kg
1153	6 string loops

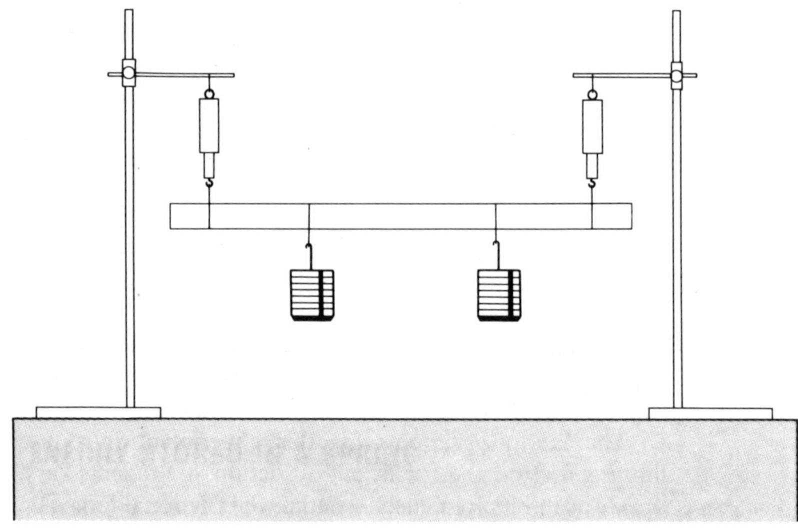

Figure A28

Students investigate the way the forces at the bridge supports (spring balance readings) change with the positions of the two loads. They are asked to compare this very simple model with a bridge.

The need to take the weight of the bridge or structure into consideration and a reminder of the concept of centre of gravity are the main reasons for the inclusion of the next simple experiment which has been left as a problem for the student to solve.

Students need to know that the weight of a body may be taken to act at its centre of gravity; they will not be required to find the centre of gravity of any except the simplest, symmetrical shapes. (We are concerned with the force on a body in a uniform gravitational field. Therefore the term *centre of gravity*, rather than *centre of mass*, is appropriate.)

EXPERIMENT
A19 'Weighing' a retort stand

ITEM NO.	ITEM
503, 504	2 retort stand bases and rods
505, 506	boss and clamp
32	mass, 0.5 kg (or 1 kg)
1153	string

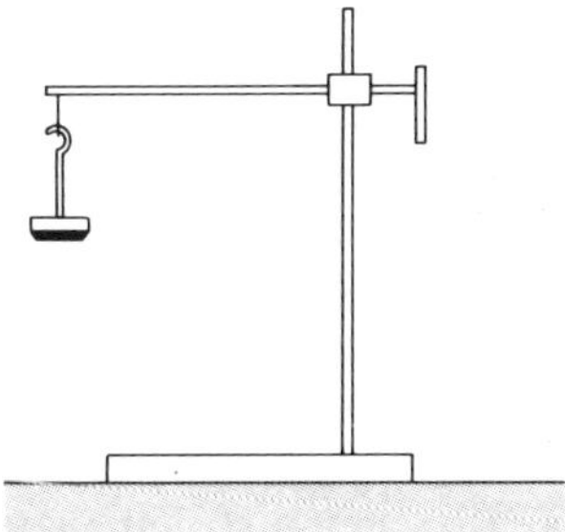

Figure A29

The students are told, 'with only the apparatus given, find the mass of the retort stand'.

The students first find the centre of gravity of one retort stand by balancing it on the clamp as a fulcrum. Next the stand is moved to a new position and rebalanced using the mass provided as a counter-weight. The stand's mass can then be evaluated.

Questions

Question 40 uses the principle of moments and centre of gravity concept.
In question 41 the weight of the bridge has been ignored and only the turning forces caused by the extra load are to be calculated.

Conditions for equilibrium

The conditions for equilibrium (of coplanar forces) are summarized in the *Students' guide*, page 15.

a The sum of the resolved components of all the forces in any two perpendicular directions is zero.
b The resultant moment about any point in the plane is zero.

Forces in a plane can be resolved in any two convenient directions which may be perpendicular to each other, most frequently parallel and perpendicular to a slope, or horizontally and vertically.

Note that condition *a* could also be stated as 'The vector sum of all the forces is zero', and is equivalent to saying that the triangle of forces (or more generally the polygon) must close.

The most convenient point to choose when taking moments is one that gives the least number of unknown quantities in the resulting equation.

The conditions for equilibrium look rather formidable written in a short mathematical form. The simplest rule to help solve any problem is probably to 'resolve, resolve and take moments'; alternatively 'take moments, three times'. There are often several ways of solving a particular problem and confusion arises when different methods are mixed.

It is tempting to go on here to develop the polygon law for addition of forces by using Bow's notation to analyse the stresses in a bridge structure and to introduce the important concepts of shear force and bending moment. But there is no time for more. In the reading recommended below on structures these ideas are explained fully and simply for those who are interested.

Questions

Questions 43 to 45 give opportunity to practise application of the conditions for equilibrium in examples concerning simple structures. For some students practice with more standard elementary problems may be more appropriate.

Reading

BLUNDELL, HAWKINS, and LUDDINGTON *Structures*. This covers the work of the whole of this section at the level required.

Schools Council Engineering Science Project *Structures* is an excellent A-level reader giving both theory and practical applications for the whole chapter.

GORDON *Structures*. Part 3 Compression and bending structures.

AKRILL and MILLAR *Mechanics, vibrations and waves*. Chapter 11, 'Rigid bodies in equilibrium' is a formal exposition of statics with examples.

BOLTON *Materials*. Chapter 2 'Structures', is a good short account of equilibrium with application to buildings.

DUNCAN *Physics: a textbook for Advanced level students*. Chapter 6 covers all the statics required.

WILLIAMS *Force, matter, and energy*. Chapter 10, 'Forces in equilibrium' is another good short account of statics.

The last statics experiment looks at the design of the forces in one simple rigid framed structure.

EXPERIMENT
A20 Investigating the forces in a roof truss

ITEM NO.	ITEM
106/1	2 dynamics trolleys
1153	2 laths about 0.5 m long (or 0.5 m rules) with holes at 5 cm intervals
32	mass, 1 kg
81	spring balance, 10 N
1153	bolt with nuts and washers (about 5 mm diameter, 40 mm long)
1153	string
	Plasticine

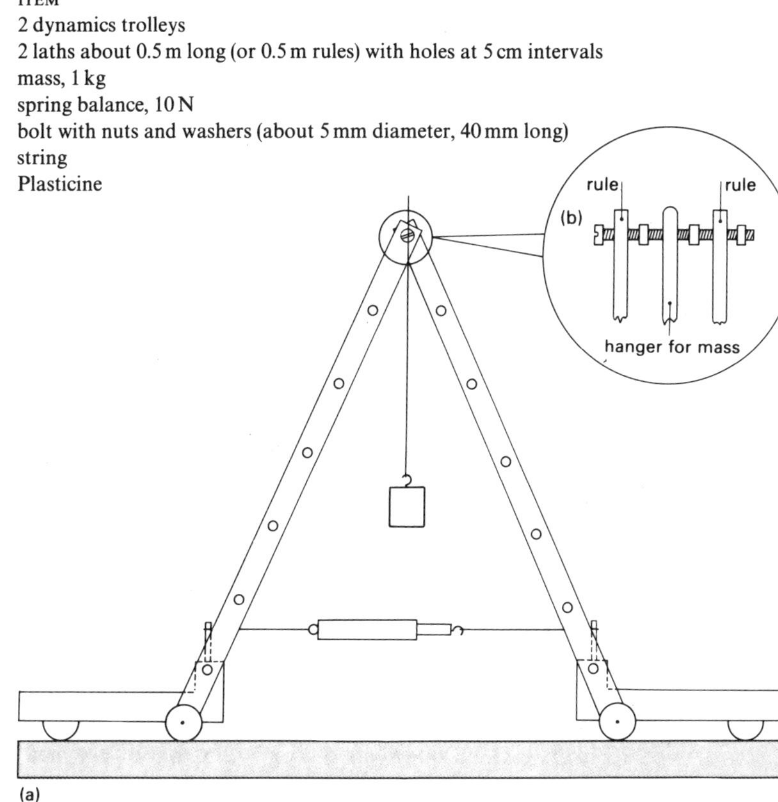

Figure A30
Investigating the forces in a roof truss.

The two rules are bolted together as shown in figure A30(b), so that they can pivot freely and there is space between them to hang the 1 kg mass. The lower end of each rule is fixed to the trolley as follows: a trolley wheel is removed and the screw which holds it to the trolley is passed through a hole (about 7 mm diameter) near the end of the rule; the wheel is then loosely screwed to the trolley again thus holding the rule between the wheel and the body of the trolley. The rules must pivot freely at both ends.

A less satisfactory alternative is to embed the lower end of each rule in a lump of Plasticine on the top of each trolley.

The spring balance acts as a tie bar between the two dynamics trolleys. Students measure the force in the 'tie' bar and at the apex angle.

The struts support the load hung from the apex, and are in compression. Students draw force diagrams for each corner of the structure to determine the force in each strut and the normal force on each trolley.

They are asked to predict the effect of lengthening the tie bar (less steeply pitched roof) on the tension in it, and the force in the struts.

The students are to draw space and force diagrams and decide whether it makes any difference where the tie bar is placed.

Finally, they are asked to contrast this model with a real house in which the walls are fixed to the ground, rather than being mounted on 'frictionless' trolleys.

Films

The film 'Let's make a model' emphasizes the use of models showing scale effects. Three school laboratory experiments and five different engineering situations using models are included illustrating both dimensional and time effects. The film should provide discussion and develop interest. It can be used as an introduction to a model bridge building session or as an entity on its own. Other films which might be of interest at this time or for Unit D, 'Oscillations and waves', include the 'The slender span', 'Dynamic behaviour of tall buildings', 'River to cross', 'We build for the world'.

Building a bridge

To complete this part of the work students can be given a double period to build a model bridge from corrugated cardboard. Empty boxes from the local supermarket are a good source of material.

There can either be a contest to build the strongest bridge, or each pair of students can be given a different type of bridge to make to scale. Although a contest is more enjoyable, the second approach has the advantages that the principles of the construction of the different types of bridge will be more clearly seen, and the results are likely to be more elegant than in a contest where students will make maximum use of all the materials, including glue and thread, to produce a strong structure.

Each group (2 students) should be given the same materials:

1 piece of corrugated card, 50 cm × 20 cm
not more than 30 cm of adhesive tape
a measured amount of PVA adhesive
two wooden blocks, 5 cm × 5 cm × 5 cm
two G-clamps
75 cm of thread
a single-edged razor blade
scissors
pencil and ruler

The roadway of the bridge is to be 20 cm long and 3 to 5 cm wide. It should be taped to the two blocks which may be clamped to the bench, if the model is to be loaded to destruction. The glue should be allowed to dry overnight.

Information about the design and construction of bridges can be found from many references including those below. Models of various types can be made, including suspension, arch, Warren girder, cantilever, stayed girder, bowstring girder, box girder, etc. (See BLUNDELL, HAWKINS, and LUDDINGTON *Structures.*) The model should be correctly scaled in the vertical plane, that is, ratio of height to length of bridge correct, even if the 3 to 5 cm width of the bridge is excessive on this scale. 40 cm long bridges can be built if a central tower is included, for example, for a cantilever bridge, or for a suspension or stayed girder bridge with spans of 10 cm, 20 cm, and 10 cm.

Factors such as use and loading, environment, type of foundations possible, materials, cost, etc., are important in the design of bridges, and can be linked to the final part of the section, about composite materials.

Reading

FISHER (1982) 'Design and construction of the Humber Bridge' *Physics Education*, **17**, 198–203 (reprinted in the Reader *Physics in engineering and technology*).
OVERMAN *Roads, bridges, and tunnels.* Chapters 4 and 5 give an excellent account of the design and construction of great bridges.

COMPOSITE MATERIALS

Linked with the bridge building project, or as an alternative to it, the information on various materials collected over the last week or so can now be used as suggested on page 34. The *Students' guide* has a very brief summary (page 16) which students must supplement by wider reading.

Questions

Questions 46 to 49 are about materials.
The reading passage 'Materials used in architecture' (*Students' guide* page 33) is also useful here. Motor cars, plastics used at home and in school, space vehicles, nuclear reactor vessels, etc. all make suitable themes on which to build a discussion on the use and development of materials, to make a change from the largely structural approach developed so far.

Home experiments

Home experiments AH3, The jelly column, and AH4, Cement (*Students' guide* page 61) are relevant here, and students could be asked to bring the results of their efforts to the discussion on composite materials.

Further information

The Institution of Metallurgists, Northway House, High Road, Whetstone, London N20 9LW produces and promotes much information about materials science for schools.

SECTION A4
MOMENTUM AND THE SIMPLE KINETIC THEORY OF GASES

INTRODUCTION

The first three sections of this Unit considered matter in its solid form where atoms have no more than vibrational motion. In this last section we consider the freely moving independent particles of a monatomic gas. The analysis requires some knowledge of dynamics and of momentum conservation. Most of the work is covered in Revised Nuffield Physics *Year 4* Chapters 4, 6, 7, and 8. The work on momentum in this section goes beyond what is required for the simple kinetic theory of gases. It covers the other atomic collision ideas required in Unit F, 'Radioactivity and the nuclear atom'.

Timing

The material within this section could be covered in about one week, if most of it is revision work. However, if the ideas are new, more time will be needed.

MOMENTUM AND ITS CONSERVATION

Revised Nuffield Physics *Year 4* Chapter 4 contains a number of simple momentum experiments involving trolleys (experiments 39, 40, 44, 45, 49). A circus of these could be used for any students who have not yet met the concept of momentum. For revision, or as an introduction from a different view point, the following alternative experiments and demonstrations are suggested. The purpose here is to understand momentum, to see evidence for its conservation and the conditions in which the principle of the conservation of linear momentum can be applied. Experiment A21 is to re-introduce the idea of momentum and to contrast it with kinetic energy.

Newton's three laws of motion can be revised to bring out the ideas:
that the motion of the centre of mass of a system of bodies is unchanged unless an external force acts on the system (Newton 1);
that any internal forces (or impulses) add to zero (Newton 3);
and that the total momentum of the system is unchanged unless an external force acts (Newton 2).

Many problems are more easily tackled by considering motion of the system's centre of mass, rather than by considering the changes of motion of its individual parts.

The linear air track can be used quickly and effectively to perform a number of simple collision or explosion experiments using light-operated gates and clocks, or by coupling the gates directly to a microcomputer, demonstration A22. The ideas can then be reinforced by discussing a line of colliding balls, Newton's cradle, demonstration A23. The course does not deal with rotational motion, and students must realize that they are considering situations where only translational motion occurs; once rotation is introduced, the dynamics is more complicated. One-dimensional momentum experiments can be concluded by measuring the speed of a rifle bullet using a ballistic pendulum (demonstration A24). This requires use of both momentum and energy conservation in different parts of the experiment. The demonstration can also be compared to the method in Revised Nuffield Physics *Year 4* Demonstrations 50 and 51.

Finally, momentum conservation in two dimensions can be observed using air or ice pucks, experiment A25.

Textbooks

Most textbooks have a suitable chapter on momentum.

EXPERIMENT

A21 Measurement of momentum change and impulse

The experiment is a repeat of A9b, page 27.

After the emphasis on energy in Section A1, the concept of momentum is introduced by the contrast:

force × time = change in momentum

force × distance = change in kinetic energy

The questions in the students' laboratory notes should lead them to consider the significance of the areas under the force–distance and force–time curves, and to realize that changes in momentum or energy can be measured by this method whether the force is constant or not.

The meaning of the term 'impulse' should also be clearly defined.

Questions

Question 50 is about the changes in momentum and energy which occur at the impact of a ball with a surface. Question 51 emphasizes the constant motion of the centre of mass during interaction processes where no external forces exist. Question 52 concerns a car hitting a wall and is a direct follow up to experiment A21.

DEMONSTRATION

A22 Collisions on an air track

ITEM NO.	ITEM
1019	air track and accessories
1020	air blower
130/2	2 photodiode assemblies and light sources
1153	card
1503	2 timers, resolution 1 ms

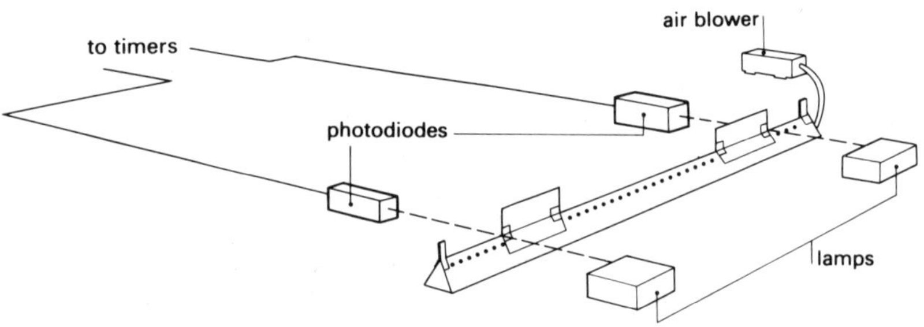

Figure A31
Collisions on an air track.

The track and vehicles need to be kept clean, and vehicles should be stored and handled in such a way that they do not become deformed. The loads on a vehicle should balance, or it will lift at the less loaded end and be driven along by air from the track. Use one vehicle first on the air track to check that it is level.

The speed of each vehicle can be measured by the card breaking the light beam to a photodiode as the vehicle passes (figure A31). Two light gates and timers are required and four time measurements may be needed. When a light beam is interrupted more than once it is of course necessary to note the intermediate reading of the timer, giving a maximum of four time measurements.

Alternatively, the four time measurements could be achieved by the use of a microcomputer or perhaps a data logger. It is possible that the use of sophisticated equipment may divert attention from the original aim of the demonstration – a study of momentum. But it may be worthwhile as it provides an illustration of the use of a modern technique which does have important applications where the simpler methods which can be used here would be inadequate. It can also stand as an example of a digital electronics system – the subject of Unit C of this course.

Students should see elastic and inelastic collisions and the explosive separation of two vehicles of the same and different masses, and should measure the velocities of the vehicles before and after the interaction.

In testing conservation of linear momentum the experimental uncertainties involved should be discussed.

The different roles of kinetic energy and momentum are again emphasized by calculating the kinetic energies before and after each collision as well as the momenta. Students are asked to explain where the missing kinetic energy has gone.

Questions

Question 53 deals with the completely inelastic impact of two air track vehicles. Question 54 is about Newton's Third Law.

Conservation laws

This is a good moment to point out the importance of conservation laws in physics. Because the quantity mv is conserved in a closed system, it is useful and worth naming.

Collision examples in the course

Examples of elastic and inelastic collisions involving atomic and nuclear particles will appear in Unit F, 'Radioactivity and the nuclear atom'. The term 'elastic' here means an event in which no kinetic energy is lost, whereas 'inelastic' is used where some kinetic energy is not recovered. (Some textbooks use the term 'inelastic' to indicate that the objects adhere on collision and then distinguish between the other two situations as 'perfectly elastic' and 'elastic'.)

Collisions between electrons and gas atoms may be elastic (in which case very little kinetic energy is transferred to the much more massive atom), or inelastic leading to ionization (or excitation).

Collision between an α-particle and a nucleus gives evidence for an upper limit to nuclear size. The collision can be regarded as elastic, and the α-particle loses negligible energy to the 'fixed' nucleus.

Collision between an α-particle and an air molecule is inelastic: ionization leads to the formation of a track in a cloud chamber, and the energy of an α-particle can be estimated from the length of the track.

Collision between an α-particle and a helium atom can be regarded as elastic, and gives evidence that an α-particle and a helium atom have (very nearly) the same mass.

Collisions between neutrons and gas atoms, and interaction of neutrons in proton-rich material provide means of detecting neutrons and estimating their mass ($\approx$proton mass).

Energy transferred in collisions

There is some value in showing the more mathematically able students the analysis of a head-on elastic collision between two masses, M and m, where mass m is initially stationary.

The resultant velocities are given by:

$$v_m = \frac{2Mu}{(M+m)}$$

$$v_M = \frac{(M-m)}{(M+m)} u$$

where u is the initial velocity of mass M.

The amount of kinetic energy transferred from M to m can also be deduced:

$$\text{Kinetic energy of } m = \frac{4\,mM}{(m+M)^2} \times \text{(initial kinetic energy of } M\text{)}$$

The three cases $M \ll m$, $M = m$, and $M \gg m$ can be considered. If not dealt with quantitatively, the results for these three cases should be discussed qualitatively. Little kinetic energy is transferred if m and M have significantly different masses. Maximum energy transfer occurs for $m = M$. Elastic collisions between electrons and atoms will leave the electrons with almost the same kinetic energy. Loss of kinetic energy indicates the presence of inelastic collisions. The fable of the 'elephant-in-a-fog' (Revised Nuffield Physics *Teachers' guide Year 4*, pp. 93–4) can be told if students have not heard it. Some brief reference can be made to each of the collision examples listed above.

Questions

Questions 55 and 56 give examples of the redistribution of momentum and kinetic energy in an elastic collision. Question 57 concerns two 'elephant-in-a-fog' type collisions. Question 58 deals with an elastic and a completely inelastic collision between objects of very different mass.

Impulse

Newton's cradle can be used to reinforce the idea of equal and opposite impulses on the two objects in an isolated collision. It is of course an example of the special case of a collision between two objects of equal mass with one initially stationary. The interaction between any pair of balls must be considered independently of all of the others in a definite time sequence. The collision of two balls, pulled aside, with the other three, to cause the subsequent motion of two at the far end, rather than of all three, can then be explained simply.

DEMONSTRATION
A23 Newton's cradle: investigation of a line of colliding balls

ITEM NO.	ITEM
1506	Newton's cradle

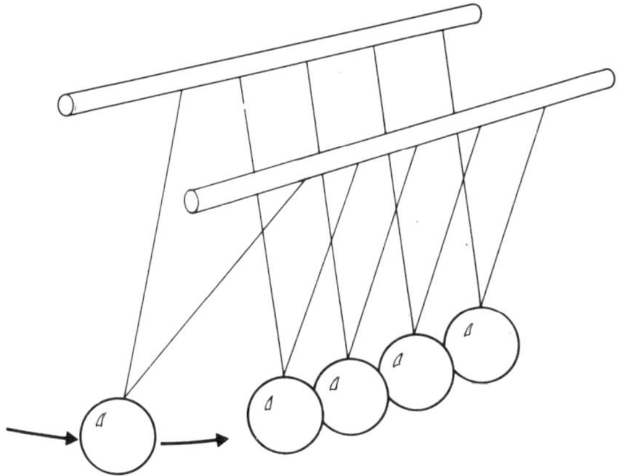

Figure A32
Newton's cradle.

It is important that the centres of mass of the balls lie in a straight line and that all balls just touch.

Start by showing what happens when one ball collides with the others, which are stationary.

By considering a succession of elastic collisions between a moving ball and a stationary one of the same mass, students should be able to show that the observed result is the only way in which both momentum and kinetic energy can be conserved. The laboratory notes ask students to consider why the line of balls must be horizontal, and why they should just touch, and also to consider what will happen if several of the balls are stuck together or if one ball is of greater density than the others. The idea of a mis-match can be used again when considering the fraction of a mechanical wave pulse transmitted or reflected at a boundary in Unit D, 'Oscillations and waves'.

An alternative demonstration using steel balls or glass marbles rolling on a piece of horizontal curtain track or on the edges of two adjacent metre rules would be less satisfactory for two reasons. First, there is no easy way to judge whether or not the ball which leaves the stationary group has the same speed as the incident one. And second, because the balls are rolling some of their momentum and kinetic energy is rotational. Rotational dynamics is *not* part of the course. This may be a suitable moment to emphasize that only translational motion is being considered.

As a final example of momentum conservation in one dimension, the speed of an air rifle pellet can be found using a ballistic pendulum. This experiment differs from the method of Revised Nuffield Physics *Year 4* Demonstration 50, in that the energy of the target object rather than its momentum is measured to find its initial velocity. The problem therefore involves both kinetic energy and momentum.

DEMONSTRATION

A24 Measurement of the speed of an air rifle pellet using a ballistic pendulum

ITEM NO.	ITEM
159	air rifle and pellets
1153	Plasticine, about 150 g
1153	cotton or thin string
503–506	2 retort stand bases, rods, bosses, and clamps
501	metre rule
1504	balance (resolution 1 g)

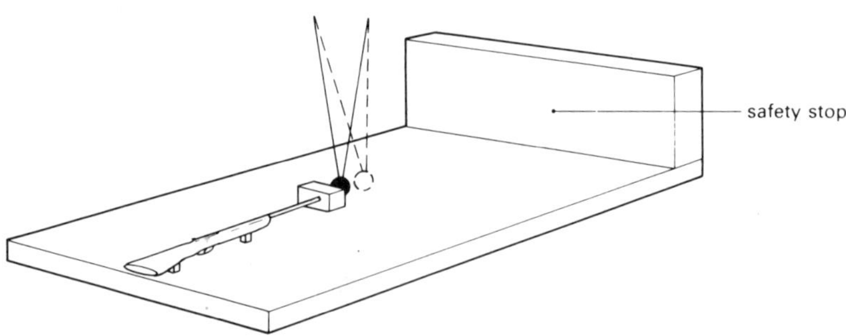

Figure A33
Measuring the speed of an air rifle pellet.

The rifle must be bolted on its side to a board near one end as shown in figure A33. *Safety note:* Ensure that there is a safety stop for pellets which miss or pass through the Plasticine. The safety stop should be soft enough to prevent pellets ricochetting off it.

Suspend the lump of Plasticine from a strong cotton or a fine string bifilar suspension to form a pendulum about 0.7 m to 1.0 m long. Clamp a paper or card scale to measure the vertical swing, h, of the pendulum. Keep the board clear of stands, bases, rules, etc. which could cause a pellet to ricochet.

Conservation of energy gives the initial speed of the Plasticine-plus-pellet as $u=\sqrt{2gh}$. Conservation of linear momentum on impact of a pellet of mass m and pendulum bob of mass M, gives $mv=(m+M)u$, where v is the velocity of the pellet. Hence

$$v=(1+M/m)\sqrt{2gh}$$

A typical result is in the range 120–150 m s^{-1}. The muzzle speed of the pellets is not constant and may depend on the length of time between loading and firing the rifle.

A slight modification of this experiment can be used as an analogue to explain how Chadwick deduced the mass of the neutron. The pellet is invisible to the eye but its presence is noted by the recoil of the suspended lump of Plasticine. Assuming that the pellet's velocity is always the same, the recoil of different masses of Plasticine can be measured and hence the mass of the pellet found.

In Chadwick's experiment, the neutron replaces the pellet and the nucleus of a gas atom replaces the lump of Plasticine. But the collision between the neutron and the nucleus is *elastic*, so the Plasticine of the analogue would have to be replaced by a hard material, such as steel, which the pellet would bounce off. Chadwick used different gases in a cloud chamber, and calculated the kinetic energy of the nuclei from their visible tracks. Hence he was able to calculate the mass of the invisible neutrons.

EXPERIMENT
A25 Collisions in two dimensions using pucks

This is a repeat of Revised Nuffield Physics *Year 4* Demonstration 41 using magnetic pucks, either CO_2 or on an air table. The photographic techniques required for making accurate measurements are given in the *Teachers' guide Year 4* Appendices 1 and 2.

EM NO.	ITEM
95	Edinburgh CO_2 pucks kit
19/1	CO_2 cylinder
19/2	dry ice attachment
133	camera
134/1	motor-driven stroboscope
161	gantry for CO_2 pucks kit
	cleansing materials for glass
1502	safety spectacles

Safety note: Lumps of solid carbon dioxide should always be handled with tongs and/or thick gloves, 'snow' with a spatula or spoon. Eye protection should be worn. Before crushing, lumps should be covered with a cloth.

If CO_2 pucks are being used, the glass plate must be carefully cleaned. The plate or air table must be carefully levelled.

All students should be shown the elastic collision between two pucks of equal mass, one initially at rest, with the resulting 90° angle between the directions of separation of the pucks. If the mass of the incident puck is increased, the collision will then produce an angle of less than 90°, and vice versa.

Photographs take time to produce so teachers may prefer to keep suitable ones for class analysis and allow this experiment to be done as a longer experiment during a circus by a few interested students.

Cloud chamber photographs of α–H, α–He and α–N collisions (*Students' guide* page 19) show the relevant angled tracks indicating that alpha particles have the same mass as helium nuclei. (Of course all three tracks must be in the same plane for the analysis to be simple, and such events are rare.)

For two pucks of the same mass, with one initially at rest, see figure A34.

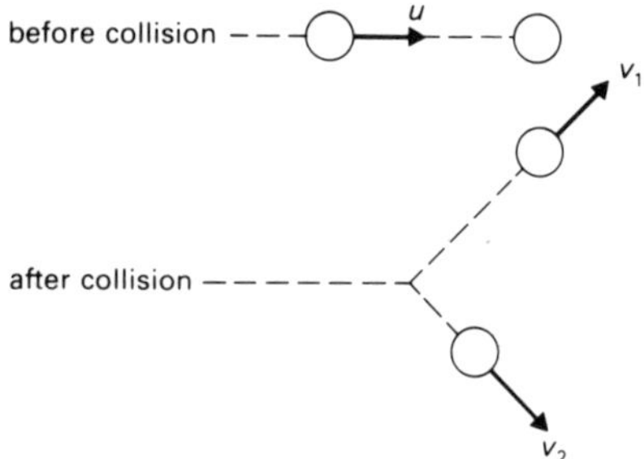

Figure A34

momentum conservation requires that, vectorially,

$$\boldsymbol{u} = \boldsymbol{v}_1 + \boldsymbol{v}_2$$

and kinetic energy conservation requires that

$$u^2 = v_1{}^2 + v_2{}^2$$

Both equations can only be satisfied by a vector triangle where the angle between v_1 and v_2 is 90° (figure A35). (This is Pythagoras' theorem.)

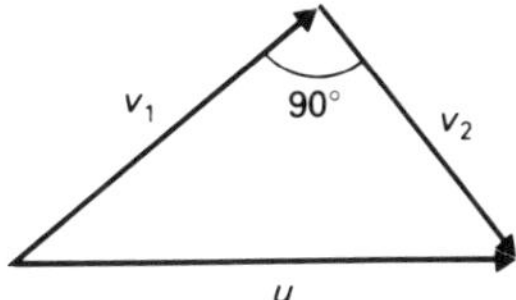

Figure A35

Question 59 is about the conservation of momentum in two dimensions and may be solved by calculation or drawing vector triangles.

THE SIMPLE KINETIC THEORY OF GASES

The purpose of this part of the work is to make some simplifying assumptions about the nature and behaviour of molecules and to apply the laws of mechanics to the motion of these molecules in order to explain and make predictions about the macroscopic behaviour of an ideal gas. This work is covered in Revised Nuffield Physics *Year 4* Chapters 6, 7, and 8. Thus for many it will be revision. Revised Nuffield Advanced Chemistry *Students book I* and *Teachers' guide I* deals with the kinetic theory of gases in Topic 3.

All students, whether they have covered the earlier Nuffield course or not, are expected to be familiar with the following:

Evidence for the kinetic theory of matter from Brownian motion and diffusion experiments.

Pressure is caused by the molecules bombarding the walls of the container; hence a possible molecular explanation of Boyle's law.

The three gas laws relating pressure, volume, and absolute temperature combined in the equation

$$\frac{p_1V_1}{T_1} = \frac{p_2V_2}{T_2}$$

Real gases at low pressures and ordinary temperatures (that is, far from conditions which would cause them to liquefy) behave like an ideal gas; but at lower temperatures they will liquefy.

Evaporation causes cooling of a liquid by removal of the fastest moving molecules, suggesting that temperature is related to the kinetic energy of the molecules.

Plan of section

The ideal gas equation $pV=nRT$ is introduced first. Assumptions of the kinetic theory are stated and the resulting model used to derive the formulae $pV=\frac{1}{3}Nm\overline{c^2}$ and $p=\frac{1}{3}\rho\overline{c^2}$.

The ideal gas equation and the pressure formula are combined to show that the kinetic energy of a mole of the gas is $\frac{3}{2}RT$. The formula can be tested by measuring the molar heat capacity at constant volume of a real monatomic gas under 'near ideal' conditions.

This is followed by the kinetic theory explanation of Avogadro's Law. Finally, the concepts of mean free path and random walk of a molecule in the gas are discussed.

No class experiments or demonstrations are included in this section, but references are given to suitable ones in Revised Nuffield Physics *Teachers' guide Year 4*.

The ideal gas equation

An ideal gas is one to which the equation of state $pV = nRT$ applies. In this equation n is the number of moles of gas particles in the sample and R is the gas constant, 8.31 J mol^{-1}K^{-1}. It follows that pV/T is constant for a fixed quantity of gas.

The equation of state summarizes Boyle's Law, Charles's Law, and the pressure law. Boyle's Law ($p \propto 1/V$ at constant temperature) is an idealization which holds well enough for air at normal temperatures and pressures – indeed for all gases far from liquefaction. The 'pressure law' ($p \propto T$ at constant volume) provides the definition of temperature on the ideal gas scale. $V \propto T$ at constant pressure ('Charles's Law') follows automatically from $p \propto 1/V$ and $p \propto T$.

Questions

Question 60 defines the molar volume of a gas and introduces R and k. Question 61 gives some practice with the ideal gas equation.

Derivation of $pV = \frac{1}{3}Nm\overline{c^2}$ and $p = \frac{1}{3}\rho\overline{c^2}$

The assumptions of the simple kinetic theory of gases are printed in the *Students' guide* (page 20).

The derivation of the pressure formula suggested here is one of several simple methods. It gives the correct answer to this particular problem but is not suitable for some other kinetic theory problems.

Reading for teachers

ADKINS *Thermal physics.* The Appendix gives a more rigorous approach to the kinetic theory.

TABOR *Gases, liquids and solids.* Chapter 4 gives the simple theory suggested here and Chapter 5 a more rigorous version.

Background for teachers

There are two choices. Molecules may be considered as point objects making no collisions with each other. Or one can assume that the number of collisions per unit time is so large that at all times there will be the same number of molecules with the same velocity travelling in the same direction, so that the effect of collisions can be ignored on a macroscopic scale. This implies that each molecule can be considered to travel from container wall to container wall at a constant speed. Again we can assume that the distribution of speeds between the molecules remains constant. From earlier in the section, students should be aware of the interchange of kinetic energy between objects in elastic conditions.

Real molecules stick to real surfaces on collision and leave randomly some time after collision. However, since there are so many molecules per unit volume we can assume that there is one molecule leaving the wall at the instant of impact of another, behaving as if it was the incident molecule making an elastic collision.

A molecule colliding elastically with a wall causes an impulse on the wall which recoils. The returning molecule can only return at its initial speed if the collision is superelastic. This problem can be avoided in many ways, for example by arguing that at this same instant a molecule of the same speed gave an equal and opposite impulse to either the other side of the wall or to the opposite wall of the box.

Remind students that the sum of many impulses in a very short time averages out to give a constant force, for example, pouring sand slowly and steadily onto a top-pan balance so that the sand then slides off the pan (Revised Nuffield Physics *Year 4* Demonstration 67). Students can be asked about the juggler who stands on a weighing machine and juggles with Indian clubs. What do the scales register?

Reading

The derivation below appears in many textbooks including:

P.S.S.C. *Physics* (2nd edition). Chapter 25.

COLLIEU and POWNEY *The mechanical and thermal properties of materials.* Chapter 10.

TABOR *Gases, liquids and solids.* Chapter 4.

AKRILL, BENNET, and MILLAR *Physics.* Chapter 10.

DUNCAN *Physics: a textbook for Advanced level students.* Chapter 19.

WENHAM, DORLING, SNELL, and TAYLOR *Physics: concepts and models.*

OGBORN *Molecules and motion.* This is a useful background reader for the whole of this section.

CRAC *Gases and gas laws.* This deals with a variety of applications.

A rectangular box of dimensions $l_x \times l_y \times l_z$ contains N molecules (figure A36(a)).

A molecule, mass m, moves in an arbitrary direction with velocity c and components c_x, c_y, and c_z along the axes where

$c^2 = c_x{}^2 + c_y{}^2 + c_z{}^2$ (figure A36(b)).

The molecule making an elastic collision with face $l_y l_z$ (perpendicular to the x direction) experiences a change in momentum $= mc_x - (-mc_x)$.

The impulse exerted on the face $= 2mc_x$.

The molecule next collides with this face after a time $2l_x/c_x$. The rate of transfer of momentum to the face $l_y l_z$ by this molecule is:

$$\frac{2mc_x}{2l_x/c_x} = mc_x{}^2/l_x$$

This is also the force exerted by the molecule on the face $l_y l_z$.

The total force from N molecules on face $l_y l_z = m(c_{x_1}{}^2 + c_{x_2}{}^2 + \ldots + c_{x_N}{}^2)/l_x$.

Let $\overline{c_x{}^2} = \sum(c_{x_i}{}^2)/N$. The total force is thus $N m\overline{c_x{}^2}/l_x$.

The pressure on the face is $p = \dfrac{Nm\overline{c_x{}^2}}{l_x l_y l_z} = \dfrac{Nm\overline{c_x{}^2}}{V}$,

where V is the volume of the box.

So $pV = Nm\overline{c_x{}^2}$.

The molecules have random motion, there are many of them, and there is no preferred direction of motion, so

$\overline{c_x{}^2} = \overline{c_y{}^2} = \overline{c_z{}^2} = \frac{1}{3}\overline{c^2}$ is a reasonable assumption.

Thus $pV = \frac{1}{3}Nm\overline{c^2}$.

Since the density, $\rho = Nm/V$, we also have $p = \frac{1}{3}\rho\overline{c^2}$.

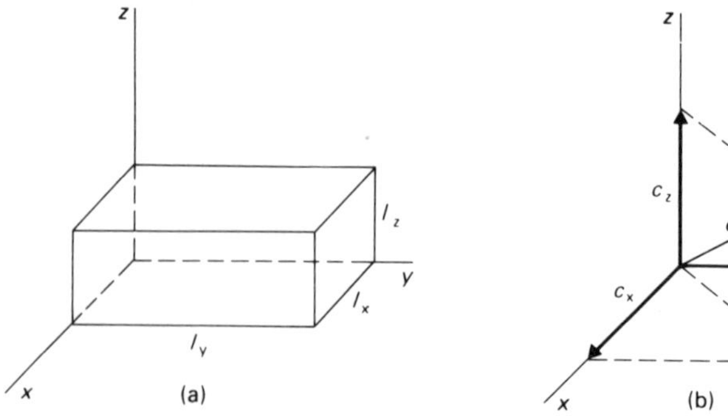

Figure A36

Questions

Question 62 goes through this derivation as a series of short questions. Question 64 asks students to think about the simplifications and justification for this derivation.

Velocities in kinetic theory

Experimental evidence for the thermal speeds of molecules comes from diffusion of bromine gas in a vacuum (Revised Nuffield Physics *Year 4* Demonstration 77) and the speed of sound in air (Demonstrations 82, 83).

Students should be aware that there is a difference between the mean speed $\bar{c}$ and the r.m.s. speed $c_{\mathrm{r.m.s.}} = \sqrt{(\overline{c^2})}$. (For reference, $\bar{c} = \sqrt{\dfrac{8RT}{\pi M}}$ and $c_{\mathrm{r.m.s.}} = \sqrt{\dfrac{3RT}{M}}$.)

Question

Question 63 gives a simple numerical example for six molecules to illustrate the difference between $\bar{c}$ and $c_{\mathrm{r.m.s.}}$.

Interpretation of temperature

The assumptions of kinetic theory lead to the result $p = \frac{1}{3}\rho\overline{c^2}$ (page 74). For one mole of particles $\rho = Lm/V_{\mathrm{m}}$ (where L is the Avogadro constant, m the mass of one particle, and V_{m} the molar volume). Hence $pV_{\mathrm{m}} = \frac{1}{3}Lm\overline{c^2}$. This result is combined with the equation of state ($pV_{\mathrm{m}} = RT$, for one mole), to give the kinetic theory interpretation of temperature.

$$RT = \tfrac{1}{3}Lm\overline{c^2}$$

$$RT = \tfrac{2}{3}\left(\tfrac{1}{2}Lm\overline{c^2}\right) = \tfrac{2}{3}\left(\tfrac{1}{2}M\overline{c^2}\right)$$

$\frac{3}{2}RT = \frac{1}{2}M\overline{c^2}$ where M is the molar mass.

The total translational kinetic energy of one mole of gas is $\frac{3}{2}RT$. Or, for one molecule

$$\tfrac{1}{2}M\overline{c^2}/L = \tfrac{3}{2}RT/L$$

$$\tfrac{1}{2}m\overline{c^2} = \tfrac{3}{2}kT$$

where $k = R/L$ is called the Boltzmann constant.

The mean kinetic energy of translation of a molecule at any given absolute temperature is the same for all gases behaving as ideal gases.

Hence absolute temperature could be defined in terms of the kinetic energy of an ideal gas. For two gases, A and B, at the same temperature

$$\tfrac{1}{2}m_{\mathrm{A}}\overline{c_{\mathrm{A}}^2} = \tfrac{1}{2}m_{\mathrm{B}}\overline{c_{\mathrm{B}}^2}$$

If the particles of smoke in Brownian motion are considered as 'large molecules', that is, gas B, then a study of their motion gives us information about the energy of the air, gas A, in which they are mixed.

The Brownian movement sets a lower limit to the use of measuring instruments. For example the suspended mirror of a sensitive galvanometer vibrates randomly from collisions with the air. The random motion of electrons in electronic circuits causes background noise, which is amplified with any tiny signals and can swamp them. Brownian motion can be reduced by working at lower temperatures.

Questions

Question 65 deals with the equation $\frac{1}{2}m\overline{c^2}=\frac{3}{2}kT$.
Question 66 is about Brownian motion.

Molar heat capacity at constant volume: C_V

The internal energy, U, of a mole of an ideal monatomic gas is the translational kinetic energy of its molecules. There are negligible forces between molecules or between the molecules and the container, so there is no potential energy. At constant volume any thermal energy change is a change in internal energy. Since $U=\frac{3}{2}RT$,

$$\Delta U=\frac{3}{2}R\Delta T \quad \text{so} \quad C_V=\frac{\Delta U}{\Delta T}=\frac{3}{2}R$$

Real monatomic gases, for example, He, Ar, have C_V values of $12.5\ \mathrm{J\ mol^{-1}\ K^{-1}}$ at low pressure and room temperature, that is, under ideal conditions.

Suppose a gas in a cylinder is compressed by a moving piston. Question 57b shows that the molecules move faster after collision with the piston. Work is done on the gas. Its mean kinetic energy increases. The temperature rises.

It is important that students appreciate the power of the kinetic theory in explaining and predicting the behaviour of gases. Some examples are suggested.

Avogadro's Law and Dalton's Law of partial pressures

Using $pV=\frac{1}{3}Nm\overline{c^2}$ and $\frac{1}{2}m_A\overline{c_A^2}=\frac{1}{2}m_B\overline{c_B^2}$, both laws follow immediately.

Graham's Law of diffusion through a small hole or a porous wall

From $\frac{1}{2}M\overline{c^2}=\frac{3}{2}RT$ it follows that $c_{\mathrm{r.m.s.}}=\sqrt{\frac{3RT}{M}}=\sqrt{\frac{3p}{\rho}}$ and as $\bar{c}\propto c_{\mathrm{r.m.s.}}$, Graham's Law follows (rate of diffusion $\propto 1/\sqrt{M}$). The agreement of experiment and theory indirectly confirms the assumptions of the kinetic theory.

No use of Graham's Law is made in the course. It is mentioned because it is a good piece of experimental evidence to justify and test the assumptions of kinetic theory. It is important that it should not be confused with the diffusion of two gases as in the bromine experiments mentioned below.

There is no time in the course for discussion of deviations from the ideal gas laws and the behaviour of real gases. It should be noted that the experimental laws have been formulated for near ideal gas conditions, that is, at low pressures and at temperatures significantly above those at which the gas liquefies.

Finally, the effects of intermolecular collisions are considered. The molecules have been treated so far as point masses. This final section leads to further information about atomic size and distinguishes between thermal and drift velocities. If the diffusion of bromine gas into a vacuum and through air have not been seen before, these demonstrations should be shown here. (Revised Nuffield Physics *Year 4* Demonstrations 76, 77, and 86.)

Mean free path

The average distance a molecule travels between collisions can be calculated using the following arguments.

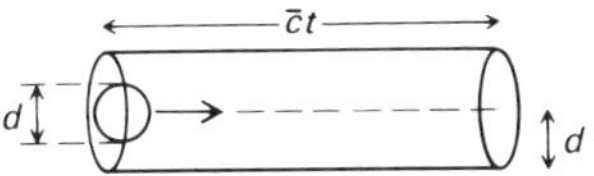

Figure A37
Mean free path of a molecule.

A molecule of diameter d will collide with any other molecule of the same gas whose centre is within a distance d of the centre of the chosen molecule. In time t it sweeps out a collision volume $\pi d^2 \bar{c} t$ (figure A36).

For a gas with n molecules per unit volume, the chosen molecule makes $\pi d^2 \bar{c} t n$ collisions. The mean free path, $\lambda = \frac{\bar{c}t}{\pi d^2 \bar{c} t n} = \frac{1}{\pi d^2 n}$. Note that this expression is independent of the molecule's speed.

This simple derivation has assumed that all molecules but the chosen one are stationary. Using the Maxwell–Boltzmann distribution of velocities for the molecules introduces a factor of $\frac{1}{\sqrt{2}}$, *i.e.*, $\lambda = \frac{1}{\sqrt{2}\pi d^2 n}$.

Random walk in two dimensions

A molecule makes N collisions in one second in travelling from A to B. The total distance travelled will be $N\lambda$. In terms of two arbitrarily defined perpendicular axes, the direct distance travelled, L, is given by $L^2 = X^2 + Y^2$, where X and Y are the resultant distances travelled in the x and y directions.

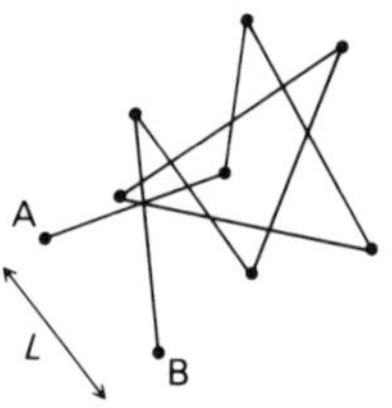

Figure A38
Molecular motion from A to B — a random walk.

$$X = \Sigma X_i$$

$$Y = \Sigma Y_i$$

Now $X^2 = \Sigma X_i^2$ as the sum of the cross terms (for example $2X_1X_2$) is zero, because these terms are as often positive as negative, ranging in magnitude between 0 and $2\lambda^2$. (See, for example, Revised Nuffield Physics *Teachers' guide Year 4*, page 164.)
Thus $L^2 = \Sigma X_i^2 + \Sigma Y_i^2$

$$= \Sigma(X_i^2 + Y_i^2) = N\lambda^2$$

i.e., $L = \sqrt{N}\lambda$

A random walk can be simulated easily on most microcomputers.

Questions

Question 67 uses the theoretical expressions for the mean free path and the distance travelled in random walk to obtain an estimate of the diameter of a molecule. This result can be contrasted with those from other methods in Section A2. Question 68 is about the diffusion process and drift velocity caused in other ways. Question 31 uses the idea of random walk to explain the properties of rubber.

Work done in expansion

The *Students' guide* simply quotes the expression $W = p\Delta V$ for the work done by a gas expanding reversibly at constant pressure. The derivation is of course a simple application of the idea work = force × distance, and is treated simply in nearly all textbooks. The meaning of 'reversible' will need to be discussed. The expression $p\Delta V$ is used later in Unit K, 'Energy and entropy'.

Question 69 takes students through the derivation of $W = p\Delta V$.

SUMMARY

The theoretical results above and experimental measurements on bromine diffusion through air at room temperature and pressure give a consistent picture of molecules of about 3×10^{-10} m in diameter, moving at about $500\,\mathrm{m\,s^{-1}}$, and travelling a distance of about 10^{-7} m between collisions.

Thus each molecule makes about 5×10^9 collisions per second so that it diffuses randomly and slowly. A molecule would take more than a week to cross a large room by this process alone.

Quantity	Experiments	Theory	Results
Average speed of molecules, $c_{\mathrm{r.m.s.}}$	Speed of sound in air Bromine diffusion in vacuum	$c_{\mathrm{r.m.s.}} = \sqrt{\dfrac{3RT}{M}} = \sqrt{\dfrac{3p}{\rho}}$ using density and pressure measurements	$500\,\mathrm{m\,s^{-1}}$
Mean free path, λ	Bromine diffusion in air	Random walk gives $\lambda = L/\sqrt{N}$, where N is the number of collisions in travelling distance L	10^{-7} m
Molecular diameter, d	Change of volume from liquid to gas, X-ray diffraction, etc.	$\lambda \approx \dfrac{1}{\pi d^2 n}$	3×10^{-10} m

Table A2

Bromine is about 5.5 times denser than air. Revised Nuffield Physics *Year 4* Chapter 8 makes many useful comments about the roughness of the calculations and what corrections must be made between results from bromine and air. The results depend on the estimate of the 'half brown' distance after 500 s.

Students' guide: randomness; liquids

The *Students' guide* points out the importance of the theme of randomness in physics, and mentions some of the other places in the course where it will appear. It also has a short paragraph on liquids to complete the picture at the end of a Unit which has dealt in much more detail with solids and gases.

Unit B
CURRENTS, CIRCUITS, AND CHARGE

Nigel Wallis
Archbishop Holgate's School, York

Mark Tweedle
The Grammar School, Batley

Suggested time allocation: five weeks

PLAN OF THE UNIT

Section B1
Things which conduct

Prerequisite		Topic		Link
		Evidence for a flow		
(GCSE) definition of a volt	►	**Conversion of energy while current flows**	►	$E = V/d$: Unit E, 'Field and potential'
		Current–voltage characteristics		
		Ohm's Law, resistance, series and parallel combinations		
		$R = \dfrac{\rho l}{A}$	►	bridge circuit: Section B2
		Change of ρ with temperature	►	Unit I, 'Linear electronics, feedback and control'

Section B2
Currents in circuits

Prerequisite		Topic		Link
energy conservation	►	**Definition of e.m.f.**	►	e.m. induction: Section H2
		Internal resistance		

$\mathscr{E} = IR + Ir$ and its application to loading of sources of e.m.f.

Characteristics of voltmeters

Maximum power theorem

Use of potentiometer for control and supply ► Unit C, 'Digital electronic systems'; Unit I, 'Linear electronics, feedback and control'

Balancing a potentiometer – null methods

Wheatstone bridge circuit, particularly in 'out of balance' mode ► control and measurement: Unit I, 'Linear electronics, feedback and control'

Kirchhoff's Second Law, $\sum_{\text{loop}} IR = \mathscr{E}$

Section B3
Electric charge

(GCSE) $Q = It$ ► **Measurement of Q by constant current charging of a capacitor**

Q/V is constant ► $Q/V = \varepsilon_0 A/d$: Unit E, 'Field and potential'

continued

$C = C_1 + C_2$ **and** $\frac{1}{C} = \frac{1}{C_1} + \frac{1}{C_2}$

$Q = It$
$\Delta Q = I\Delta t$ **for varying current**

compare with elastic stretching Unit A, 'Materials and mechanics' ► **Area under V–Q graph → energy**

Decay of charge

$$\left.\begin{array}{l} V = Q/C \\ V = IR \\ \Delta Q = -I\Delta t \end{array}\right\} \frac{\Delta Q}{Q} = \frac{-\Delta t}{RC}$$

Exponential changes and linear–lg plot ► radioactive decay: Unit F, 'Radioactivity and the nuclear atom'

Section B1 ► **Production of an electron beam**
$eV = \frac{1}{2}mv^2$

Sign of charge on beam

Particulate nature of charge – Millikan's experiment

INTRODUCTION

This Unit provides the fundamental electrical ideas that will be needed for the rest of the course. The suggested order starts with the idea of flow of charge through individual components, moves on to consider the behaviour of complete (d.c.) circuits, and in Section B3 the measurement of charge and the behaviour of capacitors is investigated. Finally, some properties of the electron, including its particulate nature, are discussed, using electron beams and Millikan's oil drop experiment.

The Unit starts by discussing flow in general terms, and looks at the direct and indirect evidence for considering electric current as a flow. The fact that a supply of energy is needed to drive the flow leads on to potential difference, V–I characteristics, and resistance. $R=\rho\left(\frac{l}{A}\right)$ is developed and related to other flow equations.

Section B2 returns to the idea of energy exchanges in a completed circuit, deriving $\mathscr{E}=I(R+r)$. This equation is used as the basis for the analysis of such problems as the drop in p.d. of a supply when current is drawn, and the behaviour of potentiometers. The latter are considered both as devices for control and supply of p.d. and in their role as null devices for balancing and comparing p.d.s. Finally, a simple bridge circuit is taken as an extension of the potentiometer, with a brief look at its applications to measurement and control when it is used off balance.

Section B3 establishes the fundamental properties of capacitors. Q is measured by charging at constant current, and later (once the principles of charge-sharing have been established) using a high-impedance voltmeter. The ratio Q/V is then seen to be an important constant. The ability of a capacitor to introduce a *time* element into the behaviour of a d.c. circuit is investigated, and curves for current and charge against time are obtained experimentally. This is the first time students meet rates of change in the course, and provides a first opportunity for numerical integration and introduction to the properties of an exponential change. There is a brief discussion of analogies, taking the spring versus capacitor as an example.

The Unit concludes with electron beams and the energy needed to accelerate them. $eV=\frac{1}{2}mv^2$ is discussed and leads to the electronvolt as a useful 'atomic-sized' unit of energy. The information that beams give us on the nature of charge is assessed, and the evidence from Millikan's oil drop experiments is considered.

THE PLACE OF THE UNIT IN THE COURSE

This Unit sets the scene for the substantial electrical content of the course as a whole, and must necessarily precede Units C, 'Digital electronic systems', E, 'Field and potential', and F, 'Radioactivity and the nuclear atom'. The work on circuits in Section B2 is perhaps fairly self-contained, but even here the treatment of the potentiometer is taken up in Unit C, 'Digital electronic systems' and Unit I, 'Linear electronics, feedback and control', and in the latter Unit the out-of-balance bridge is used. The treatment of capacitors in Section B3 is taken up in Unit E, 'Field and potential' in considering the uniform electric field between parallel plates, and later in Unit J, 'Electromagnetic waves' in discussing the propagation of an electromagnetic pulse along a wave guide.

The work on electrons returns to the microscopic world. It is extended in Unit F, 'Radioactivity and the nuclear atom' where other atomic particles are also considered, and in Unit L, 'Waves, particles, and atoms' where the wave nature of the electron is discussed.

LIST OF SUGGESTED EXPERIMENTS AND DEMONSTRATIONS

B13	Experiment	Detection of small resistance changes *118*
B14	Experiment	Capacitors and charge *124*
B15	Demonstration	Charging a capacitor at a constant rate *127*
B16	Demonstration	Spooning charge *128*
B17	Demonstration	The energy stored in a charged capacitor (motor) *130*
B18	Demonstration	Energy proportional to V^2 (lighting lamps) *133*
B19	Demonstration	Energy proportional to V^2 (heating a coil) *134*
B20	Demonstration	Decay of charge *136*
B21	Demonstration	Electron streams *141*
B22	Optional experiment	The Millikan experiment (charge on an electron) *143*

SECTION B1
THINGS WHICH CONDUCT

IDEAS OF FLOW

The kinetic theory of gases (Unit A, 'Materials and mechanics') presented students with a model of the microscopic world which could be tested by considering the consequences of the model in the large-scale, macroscopic world of everyday experience and simple laboratory measurements. Since much of electricity is concerned with a particle which is several orders of magnitude smaller than gas molecules, and since it is a major job of physics to probe beyond the perception of our senses, it seems appropriate to start this Unit with two questions:

What is actually happening in a wire (or other conductor) when a current flows?

How do we actually know that this picture is correct? (Or, perhaps more correctly, how can we have reasonable confidence that it is a good picture?)

The evidence for a flow of something is not clear-cut and many students will be happy to take it on trust. There will probably be some, though, who even at this stage need convincing that the current really is the same at all points in a series circuit and is not 'used up'. This fact can lead to a discussion of Kirchhoff's First Law for branches and loops in a circuit, and is our main evidence for flow.

Teachers could pursue with more perceptive students the justification for taking a meter deflection (or digital display) as a measure of the *flow*. The important point is that Kirchhoff's First Law holds whatever device (electromagnetic, chemical, etc.) is used to indicate the current.

Students may have seen demonstrations of electron beams in earlier courses, or at least be familiar with the operation of the cathode ray tube (or 'television picture tube'). These can be used to point out that it is not always necessary for the flow of charge to be trapped in a conductor.

DEMONSTRATION
B1 Conduction by coloured salts

This demonstrates the migration of coloured ions in an electrolyte when a p.d. is applied, the movement being either in the direction of conventional current or opposite to it depending on the charge on the ion. In any event, reversal of the p.d. reverses the direction of ion flow.

ITEM NO.	ITEM
1156	copper(II) sulphate crystals
1156	potassium manganate(VII) crystals
1156	aqueous ammonia (see below)
1155	filter paper
1155	microscope slide
1155	2 large pins
52K	2 crocodile clips
59	power supply, 25 V d.c.
1000	leads

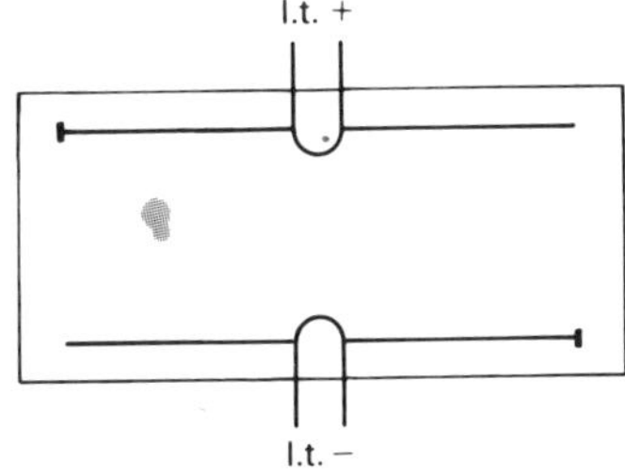

Figure B1
Movement of coloured ions (using 25 V d.c. supply).

Safety note: The h.t. power supply (which can supply up to 60 mA at 300 V) should not be used for this demonstration.

The aqueous ammonia (traditional name 'ammonium hydroxide solution') should be strong enough to dissolve the precipitate which it would form with copper(II) sulphate solution, giving an intense blue cuprammonium sulphate (systematic name 'tetraamminecopper(II) sulphate') solution. One part of concentrated aqueous ammonia with one part of water is satisfactory.

Eye protection should be worn when diluting the concentrated aqueous ammonia.

Put a piece of filter paper about 50 mm by 20 mm on the microscope slide and wet it with aqueous ammonia. Let excess run off if it will. Lay pins across each side of the paper and clip them to the slide with crocodile clips. Put one or two very tiny crystals of each sort on the paper, and connect the clips to the supply and switch on. The colour of the manganate(VII) (traditional name 'permanganate') will move appreciably in a few seconds. The copper is slower, and can be spoiled by diffusion.

The paper should be horizontal so that the solution stays still upon it. If a p.d. of 25 V is applied across the *width* of the slide, as in figure B1, the colours move fairly quickly.

This is one of the few convenient direct demonstrations that current and movement are linked; students could be invited to say how conclusive they found it. The very slow speed of the colour spread, compared with the speed at which electrical information is known to travel, should certainly be brought out.

Speed of conducting particles

Question 1 in the *Students' guide* derives the relation $I = AvnQ$ for the current carried by n charge carriers per unit volume, each of charge Q, moving at velocity v through a conductor of cross-sectional area A. Students make an order of magnitude estimate for v in a liquid. The small speeds ($\approx 10^{-6}\ \mathrm{m\,s^{-1}}$ for a liquid) tend to confirm the observation of demonstration B1.

The calculation is important for a number of reasons:

the actual theoretical result is used later in the course; it is a good example of the way rates of change can be handled ('consider a time interval t ...', rather than just taking 1 second);

the intermediate result that volume flow rate is Av can be applied to any fluid flow (and perhaps the extension to mass flow rate);

lastly, the possibility of extending the argument to making predictions about other kinds of flow (for example, transport: buses versus cars in cities).

Finally, if the cathode ray tube has already been introduced and students have covered e/m for electrons in an earlier course, the velocity which free electrons have acquired at the anode in a cathode ray tube may be calculated and compared with the speeds considered above. The calculation uses the energy equation $\frac{1}{2}mv^2 = eV$ which looks ahead to the next part of this Section and the meaning of potential difference. Teachers will use their judgment whether they can use it here.

Reading

There are few references to specific books or other sources in this Unit. Most textbooks deal adequately with the material, and students should be encouraged to read the relevant chapters.

ENERGY CONVERSION AND POTENTIAL DIFFERENCE

The central idea of this passage recurs frequently throughout the course and it seems appropriate to raise it at this early stage. However, teachers with a class which has been well grounded in an earlier course may choose to pass it by or treat it cursorily; others may prefer to postpone it and move directly to circuit measurements, continuing to treat p.d. intuitively as 'what a voltmeter reads'. Note that the treatment in Section B2 of electromotive force and the complete circuit does require the formal definition of potential difference.

The basic fact of energy conversion in electrical circuits should be familiar enough to students. The important thing is to be able to quantify this, perhaps leading students to suggest 'joules per coulomb' as a useful quantity. (On the point of nomenclature and language used – it is a good idea, while a concept is being fixed in the students' minds, to use almost synonymously the name of the physical quantity and its unit of measure. This practice is open to purist criticism, but at least in speech using 'joules per coulomb' interchangeably with 'energy transformed per unit charge passed', particularly if the two forms are used together, seems useful teaching practice and helps to identify the SI unit with the quantity it measures. In formal written work students should, of course, be encouraged to use the more formal term.)

The following points might usefully form the basis of a discussion to revise and extend earlier ideas of potential difference:

i On an electricity bill what are we actually paying for? (Approximately $\frac{2}{3}$ of the amount is for coal burned.) From the previous work on flow it cannot be for 'electricity consumed' (whether for a.c. or d.c., all that flows in flows out).

ii We are somehow paying for the electricity passing through the house wiring. What would be the consequences (for the consumer and Electricity Board) of replacing the conventional 'meter' (kWh or 3.6 MJ) with a coulombmeter?

iii As an example, a torch bulb and a 60 W lamp both take the same current (about 0.3 A) – they could be demonstrated connected in series. How does their behaviour differ? '... more joules for the same number of coulombs'. (The coulomb can be introduced here quite informally as the unit of charge, current × time. Most students are likely to have met it already, and the more detailed connection between charge, current, and time is dealt with in Section B3.)

In a complete circuit we are concerned with *two* energy transformations: that in the power supply (battery, dynamo, etc.) from chemical, mechanical, etc. *into* electrical energy, and that in the external circuit *from* electrical energy into other forms (heat, kinetic, etc.). The term 'electromotive force' (e.m.f.) applies to the former and is followed up in Section B2. The external circuit typically consists of a series of components – lamps, resistors, etc. – in each of which an energy transformation takes place. The term 'potential difference' (p.d.) can then be applied to each of these components and relates to all energy transformation happening between the boundary points of the components as the current flows. Since in most cases the transformation will be to internal energy ('heat'), the term 'dissipation' can be used (the energy change is irreversible).

The description 'electrical energy' is conventionally used in discussions of this kind, but can be criticized as being meaningless and vague. Where is this energy located? How is it related to the properties of the flow – is it to do with the kinetic energy of the charge carriers, or perhaps with the electromagnetic field around the conductor? Teachers should be aware of the inadequacy of the term, but it is useful to retain it since it acts as a useful intermediate stage between the *measurable* absorption of energy in the power supply (particularly if it is a dynamo) and the *measurable* output of energy in parts of the external circuit, using non-electrical methods. And it is perhaps no more vague than 'chemical energy' or even 'nuclear energy'.

Perceptive students might like to explore the semantics of whether a current flow causes a potential difference, or vice versa.

The end point of such discussion on energy transformation is the formal definition of potential difference as energy transformed per unit charge passed, and the volt as the shorthand unit for joule per coulomb. Since the strict SI base electrical unit is the ampere, the rigorous definition of p.d. is power converted per unit current flowing ($1\,\mathrm{V} = 1\,\mathrm{W\,A^{-1}}$) and students should be familiar with this even though the charge–energy relationship is perhaps more intuitive. It does of course lead directly to the power–p.d.–current relation that most students will be familiar with from introductory courses.

Questions

Questions 3 to 6 are on electrical energy and power.

It is perhaps worth now and again enquiring how meters and other instruments acquired their calibrations. This is tedious if taken to excess, but basic physics can be revised and useful points made about laboratory and national standards by chasing a line of reasoning back to definition and fundamental measurement. The next experiment makes this point with p.d. and the voltmeter.

DEMONSTRATION

B2 Measuring p.d. without a voltmeter

This demonstration could be used in either of two ways. It may be 'talked through' as part of the discussion above, almost as a commentary, with the obvious sacrifice of experimental technique and discussion of accuracy; alternatively it may be approached as a careful experimental calibration of a voltmeter with opportunity for a full consideration of experimental uncertainty from heat losses and other factors. Performed this way it need not, of course, be a demonstration.

ITEM NO.	ITEM
1011	apparatus for measuring joules per coulomb
1508	demonstration meter, 1 A
507	clock
542	thermometer, 0–50 °C in 0.2 °C
31/2	slotted masses, 1 kg
501	metre rule
e.g., 59	power supply, 12 V
1000	leads

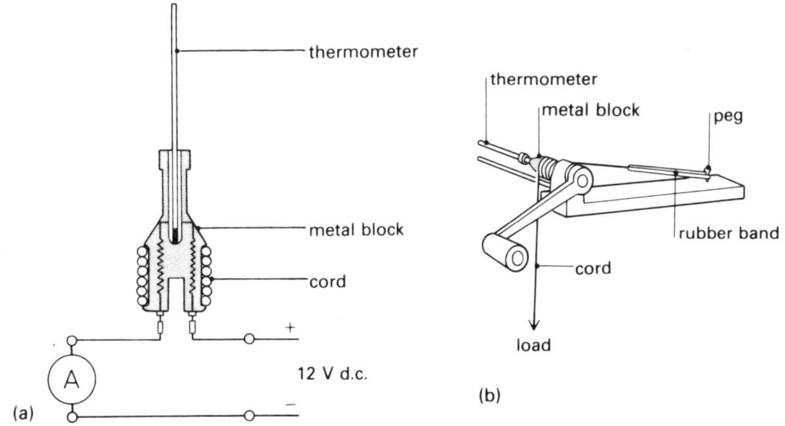

Figure B2
Apparatus for measuring p.d. without a voltmeter.

Warm the metal block (item 1011) electrically, from room temperature through about 10 K, stopping the current after this rise but continuing to record the temperature until it begins to fall again. A 12 V supply is needed. It is best to wind 5 or 6 turns of cord around the block so that conditions are as far as possible the same as in the second part of the demonstration.

Before doing the second part of the demonstration, the block must be returned to room temperature – an aerosol freezer is useful. The block is then turned against the cord which carries a load of about 8 kg and acts as a friction band.

As far as possible, the temperature should be made to rise at the same rate as previously, stopping after the same temperature rise. The temperature rises will not be exactly the same, but the losses will be nearly enough equal to be ignored.

The energy supplied mechanically in the second stage is given by $T \times 2\pi r \times n$, where $T(=mg)$ is the tension in the hanging cord, r is the block radius, and n the number of revolutions. The energies supplied in the two stages are in the ratio of the temperature rises, enabling the electrical energy to be calculated. The p.d. may then be found from the ratio of the energy delivered in joules to the number of coulombs that passed (current × time). It might be worth pointing out that this must have been, at least in principle, how the 'first voltmeter' got its calibration!

Voltmeters

Voltmeters are used in practice for a variety of purposes, not all of which relate directly to energy. However, it might be appropriate to refer to them more often as 'energy conversion meters', particularly in applications which stress the conversion. One demonstration voltmeter could perhaps have a permanent 'joules/coulomb' marking stuck on. Teachers who discussed the cathode ray tube at the end of the work on flow might like briefly to show it and discuss the energy transformations in more detail and the versatility of the voltmeter: '... a moment ago it was used with a hot metal block, now it helps us find electron velocities'.

RESISTANCE

Most of the time for this part can be devoted to measuring, either in small groups or in demonstrations in which members of the class participate. Necessary learning and practice questions are in the *Students' guide*, questions 7 to 11.

At the end of the work students should:

i have developed a familiarity with ammeters and voltmeters and be able to make sensible judgments about what meters to use in any situation;

ii appreciate that for any conductor there is some particular relationship between the current flowing through it and the p.d. across it. The relationship is often linear but in many important cases is not. The I–V relationships should be sketched and thought of as 'fingerprints' which characterize a device. (Hence the term 'characteristic'.)

EXPERIMENT
B3 Two-terminal boxes

ITEM NO.	ITEM
1507	milliammeter, 100 mA, occasional access to other meter ranges
	either
1033	2 cell holders with four cells
	or
59	l.t. variable voltage d.c. supply
1147	two-terminal boxes
1000	leads
	graph paper

Suggested contents for the boxes are:
1 MΩ resistor
tungsten filament lamp
germanium diode with safety resistor in series
three boxes containing marked resistors:
500 Ω
200 Ω and 300 Ω in series
820 Ω and 1200 Ω in parallel
cadmium sulphide cell with a hole in the box so that light can enter it
100 μF, 50 V, electrolytic capacitor

The two terminals of each box should be different colours so as to be distinguishable, but not red and black, which might imply a direction for the current. The boxes should conceal what is inside when they are placed bottom downwards on the bench (for example, the lamp should not visibly emit light). There is no reason to prevent students looking inside if they want to, but the experiment will be duller for them if they do.

A kit of boxes from a manufacturer may not contain all the things listed. Teachers should add whatever they think suitable. Students could make up puzzles for one another if they want to.

Students may be told, 'Here is a set of boxes, each with two terminals, most containing something simple which you have met before. The bottom of the box is open, so you could look in it, but try to find out what is in it by electrical measurements as if the bottom were sealed up. When you think you know what is in your box, tell me. You may put as much as 12 volts across any box and the current will not exceed one-tenth of an ampere. Assume dry cells give 1.5 volts each unless you think they need checking.'

The object of the puzzles is not to test knowledge but to increase experience and confidence, so help should often be given and whatever a student discovers should be treated as an achievement. Ideally students would apply different p.d.s in both directions and plot graphs. If the p.d. is not varied, a lamp cannot be distinguished from a resistor. If the p.d. is not reversed, a diode cannot be distinguished from a resistor. A 1 MΩ resistor requires a sensitive meter, and so does the diode in its reverse direction – if a delicate galvanometer is used with the diode it is wise to check that students avoid using it for the forward direction. There is no way of distinguishing the three boxes containing 500 ohms. A student who knows that his box contains a 200 ohm and a 300 ohm resistor in series must have found out by looking. The cadmium sulphide (photoconductive) cell should give inconsistent results, and finally show that there may be factors (such as illumination) other than temperature which affect the resistance. The capacitor is to be used later and practical acquaintance with it will be useful. If students find it baffling, agree with them.

Students should have a chance to use most of the boxes, so as to get experience of different meters, and a brief discussion at the end should elicit from them the characteristics by which they recognized the devices in the boxes.

In discussing the graphical results of the experiments, teachers could bring out the following points:

i the difference between a linear relationship and direct proportionality (latter passes through origin);

ii the ratio p.d. to current is *called* the resistance (a measure of 'difficulty' of passing current) *irrespective* of the device's characteristic;

iii Ohm's Law ($I \propto V$) relates to a particular class of materials (mainly metals) and there is no immediately obvious reason why the proportionality should hold so well. The relationship between gradient and resistance should be brought out.

In discussing Ohm's Law, and giving a formal statement of it, the specific condition of constant temperature should be brought out.

Questions

Questions 7, 8, and 9 are about I–V relationships, resistance, and two-terminal boxes.

EXPERIMENT
B4 Comparison of rheostat and potentiometer as controllers

In experiment B3 students probably changed the applied p.d. in steps. If asked how the change could be made continuous they will almost certainly suggest connecting a rheostat in series, as in figure B3(a), and this suggestion should be tried first.

ITEM NO.	ITEM
1507	ammeter, 1 A
1507	voltmeter, 10 V
541/1	rheostat, 10–15 Ω
1151	diode
1040	clip component holder
	and/or
92R	m.e.s. lamp, 2.5 V, 0.3 A
92T	holder
1033	cell holder with two cells
1000	leads

In the first part students should set up the circuit, as in figure B3(a), to vary and measure the current through the diode or lamp and the p.d. across it. They should observe the limited range of variation possible using the rheostat in this way. Teachers may need to make local variations to the component values suggested above to ensure that the proportional variation in the total resistance produced by the rheostat is not too great.

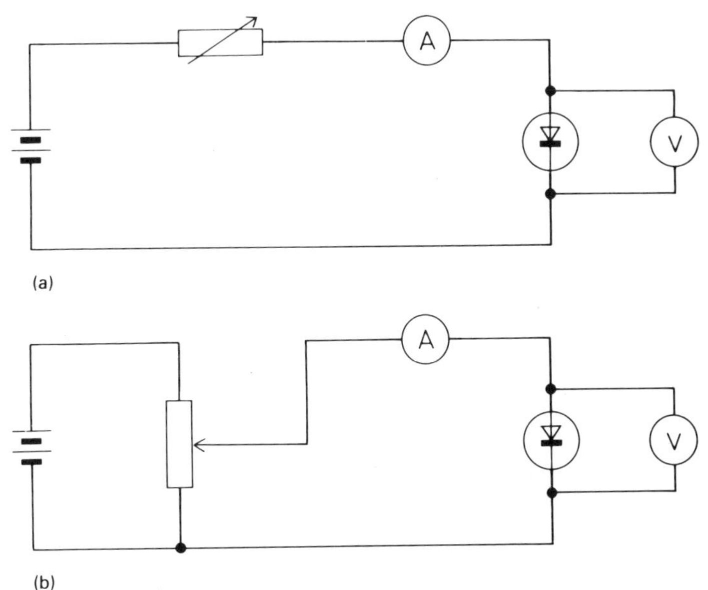

Figure B3
Rheostat and potentiometer as controllers.

In the second part, shown in figure B3(b), the same rheostat is now used as a potentiometer (potential divider is perhaps a better description, but a more cumbersome term). It is obviously essential that all three terminals are accessible. Also if small rheostats – such as those in the circuit board kits – are used, their power handling capacity should be checked: the full supply p.d. is of course across the potentiometer, and there can be problems when a large current flows through the end few turns.

The arrangement with the potentiometer could now be used to plot a diode or lamp characteristic with fine control over the whole range of the applied p.d., but this is by no means essential. No formal treatment of the potentiometer is needed at this stage. The level of explanation can be along the lines of, '... here is something which gives us a "continuously variable battery". When the wiper is at the top, the full p.d. is applied; when at the bottom no p.d. is applied to the rest of the circuit. It seems reasonable to guess that with the wiper halfway down we shall be applying half the battery voltage'. This last statement could be checked by connecting the voltmeter directly to the potentiometer's output. The diode or bulb should be disconnected to remove the complications of loading effects.

This use of the potentiometer (or potential divider) to provide continuous control of an applied p.d. is very common and extremely important (and it will often be used in the course in this way).

Extension

Question 12 is an *experimental* question based on experiment B3. It requires the same apparatus as this experiment and could well be set immediately after the experiment as a 10-minute extension to it. The rheostat (and any concealed resistance if necessary) should be such that the p.d. across the lamp does not get as low as 0.2 V nor as high as 2.8 V.

The following further points can usefully be made from this experiment:

i The distinction between dependent and independent variables can be brought out. Arguably in the rheostat version the current is the independent variable, while with the potentiometer it is the applied p.d.

ii Following from *i*, students should certainly be aware that if they are primarily interested in controlling the *current* over a specific range then they will need a rheostat; if they need an 'adjustable battery' then a potentiometer is needed. They should further appreciate that it is not of course *necessary* for a potentiometer to deliver current at all whereas a rheostat is useless unless current flows through it (try connecting a battery through a rheostat to an oscilloscope).

iii Finally, looking forward to electronics, the terms input and output can be introduced, together with the convention of 'signal flow' from left to right.

Question 20 explores the control use of rheostat and potentiometer.

Resistance combinations

The formula for a parallel combination is developed theoretically in question 10. Some students may need convincing that p.d.s add up in a series circuit. A quick demonstration using a voltmeter (it is best to use the same meter in different places to avoid discrepancies of up to 10 per cent between meters), and the theoretical point that we are really adding up different contributions to the total energy transformed, may be helpful.

Factors determining resistance

Shape and material of conductor

Students are quite happy to accept that for a uniform wire $R \propto l$, and possibly showing the markings on reels of wire (ohms per metre) is sufficient. A very simple line of reasoning is to take the series combination formula and consider a wire of length $2l$ as two wires, each of length l, in series.

The effect of cross-sectional area may not be obvious to some students. It is useful to show some high current capacity cables (presumably therefore low resistance) compared with low rating ones: domestic 30 or 60 A cable and lighting flex; car starter-motor rotor (copper bars) and the generator coil; different thicknesses of fuse wire. Two simple lines of reasoning to show $R \propto l/A$ are offered; while not leading to a very startling conclusion they are good examples of a physicist's argument.

i We can take the equation $I = AvnQ$

If we were to assume that a fixed p.d. over a fixed length implies a fixed drift velocity v (a big step at this stage) then $I \propto A$. Since for constant V, $R \propto 1/I$ the relation $R \propto 1/A$ follows. (There is 'more room' for the charges to flow – wide and narrow water pipes.)

ii We can imagine a wire of cross-section $2A$ made up of two separate wires of area A each in parallel. Applying the parallel resistance formula shows the combination (thicker wire) to have half the resistance of either thinner wire.

The theoretical reasoning above, together with $R \propto l$, leads to the result that for *wires of a given material:*

$$R \propto \frac{l}{A}$$

or

$$R = \rho\left(\frac{l}{A}\right)$$

which defines a constant ρ for the material, the *resistivity*.

Demonstration B5 is in two parts. First different lengths and thicknesses of the same material (constantan) are used to confirm experimentally that $R \propto l/A$. Teachers can decide whether they need both experiment and theory to make this point. In the second part (which should *not* be omitted) similar wires of different materials are used. A comparison of resistances shows that resistivity is a property of the material (the value for nickel–chromium is about twice that for constantan).

DEMONSTRATION

B5 Effect of size and material on resistance

ITEM NO.	ITEM
1507	ammeter, 1 A
59	l.t. variable voltage supply
1507	voltmeter, 10 V
1501	constantan wire, 0.56 mm and 0.28 mm diameter
1501	nickel–chromium wire, 0.28 mm diameter
1155	micrometer
1000	leads

(Covered wire can be used if the insulation is removed at appropriate points.)

Initially use constantan only. Apply, say, 2 V across 0.2 m of 0.28 mm diameter wire and record the current. Increase the p.d. in steps of 2 V, increasing the length of wire to keep the current the same. The applied p.d. is a measure of the resistance (I is fixed), and its proportionality to length is clear.

Repeat the readings with 0.56 mm diameter constantan (which has a diameter twice that of 0.28 mm diameter), bringing out the dependence of resistance on cross-section.

Now repeat the first part, using 0.28 mm diameter nickel–chromium. Comparing resistances of similar wires of the two materials will show that nickel–chromium has approximately twice the resistivity of constantan. Table B1 may be useful.

Diameter/mm	Constantan/$\Omega\,m^{-1}$	Nickel–chromium/$\Omega\,m^{-1}$	Copper/$\Omega\,m^{-1}$
0.56 (24 s.w.g.)	1.96	4.20	0.07
0.28 (32 s.w.g.)	8.12	17.40	0.29

Table B1
Resistance per unit length for various wires.

A version of this demonstration can be devised using 'conducting putty', but it is not of course possible to change the medium, only to show the effect of changes in l and A on R.

Questions about why the thick constantan conducts four times as well as the thin constantan; how much better copper conducts than constantan; how to calculate the resistance of a copper bar (say 10 mm × 20 mm cross-section) carrying current in a power station; or how fine a copper wire would need to be to have a resistance of, say, 2 ohms per metre, are all useful simple applications of $R=\rho(l/A)$.

Like the Young modulus, a value of the resistivity of a material is a compact way of summarizing useful information of practical value, although tables of resistance per metre, while less compact (why?), are quicker for the engineer to use. It has already been hinted that the resistivity equation relates to that for flow. Combining $R = \rho(l/A)$ with $R = V/I$ leads to:

$$I = \left(\frac{1}{\rho}\right) A \left(\frac{V}{l}\right)$$

The electrical conductivity $(=1/\rho)$ could usefully be introduced here. The structure of this equation is common to many flow-type situations (fluid, thermal flow of energy, magnetic flux). Teachers can use their discretion on how far to push this analogy into unfamiliar situations at this stage. The significance of (V/l) can be pointed out though: since the current in a conductor is everywhere the same, then for a uniform conductor the potential *gradient* is constant. This is important for more formal work on the potentiometer, and, of course, links with Unit E, 'Field and potential'.

Students who can take it could argue from dimensions that (V/l) has the units of force per unit charge. This could help to justify the assumption made earlier that conductors with the same p.d. over the same length (of the same material) have the same drift velocity for their charge carriers.

Questions 13 to *17* are on resistivity.

Home experiment

Home experiment BH1 challenges students to construct a 25 Ω resistor from household materials.

Effects of temperature

The proviso of constant temperature in Ohm's Law will already have arisen, perhaps in the context of the increasing resistance of the lamp filament with current. The effect of temperature change is briefly explored here for the following reasons:

i Students are made aware that resistance is only a 'local constant' under prescribed conditions.

ii Changes in resistance with temperature have important practical applications in resistance thermometers.

iii The way particular materials respond to temperature changes gives valuable information on the microscopic processes in the material.

Although *ii* and *iii* are not explored in the course in detail they provide a good example of the two-edged nature of many physical phenomena – the technical application and the opportunity to probe further into the physical world.

DEMONSTRATION

B6 Effect of temperature on resistance

ITEM NO.	ITEM
59	l.t. variable voltage supply
1507	ammeter, 1 A
1507	milliammeter, 10 mA
1501	enamelled copper wire, 3 metres, 0.2 mm diameter
512/2	beaker
1021	aerosol freezer
1151	carbon resistor, 0.125 W, 150 Ω
1040	clip component holder
1033	cell holder with one cell
1151	thermistor
	hot and cold water

Connect the wire, the ammeter, and the l.t. variable voltage supply (about 2 volts) in series. Measure the current when the wire lies in a tight bundle and also when it is laid out more loosely. Also measure the current when it is put in hot water, in cold water, and when cooled by the aerosol freezer.

Show the opposite effect on the carbon resistor when subjected to the same temperature changes, using the milliammeter and the single cell. The effect is small.

The thermistor, again with a single cell, shows a larger effect.

Summary and suggestions for further work

The enormously wide range of resistivities found in materials is shown in a logarithmic plot as figure B65 in the *Students' guide*. The direction of change of resistivity with temperature is also shown. Apart from pointing out the usefulness of such a scale, the diagram could form the basis of a speculative discussion on the processes involved at the microscopic level in different types of material, and in particular the role of semiconductors.

Resistance change is the principle of many types of transducer, in particular temperature sensors, light-dependent resistors, and strain gauges. In the last case the principle follows directly from Unit A, 'Materials and mechanics' – stretching a wire makes it longer and thinner.

If students have measured the reduction in diameter of a wire when it is stretched (or they could be given Poisson's ratio from a reference book) they are in a position to calculate the proportional change in resistance for a given strain. If teachers feel their students need something of a challenge at this stage to round off this work they could consider tasks from the following suggestions:

i Take the strain gauge idea as an example of how to deal with small changes. With a slight change of context this would also be an opportunity to discuss the accumulation of small individual uncertainties to an overall uncertainty in an experimental result. The treatment could be informal using simple algebra or even concrete arithmetic if necessary, but the mathematically able might enjoy a calculus approach.

ii Actually calculate the resistance change for say a 1 per cent strain in a given wire. (Excellent practice for using a reference book to find appropriate numerical values.)

iii How can you detect or measure such a change in resistance, using the ideas of the Unit so far? (A more professional approach to this task is considered in the bridge circuit of experiment B13.) Set up the wire in a suitable circuit, load it and try to measure ΔR.

A starting point for *i–iii* is question 16 in the *Students' guide.*

iv Finally, returning to temperature change, try to calibrate a thermistor, say, as a thermometer over a definite range. Is the response linear? Possibly linear over a small temperature range?

These ideas are fruitful areas for investigation topics, and teachers should seize any opportunity for putting suggestions for likely titles to the class.

Further reading

CRAC *Stress, strain and strength.*
CRAC *Instrumentation systems.*

SECTION B2
CURRENTS IN CIRCUITS

SIMPLE NETWORKS

The work of the previous Section was chiefly concerned with the current in, p.d. across, and resistance of individual simple components such as resistors, wires, lamps, diodes, and so on. The next stage is to build up experience of circuit behaviour as a whole, in particular consideration of the effects of the power supply and any meters connected to the circuit. In principle this only involves application, possibly repeated, of $I = V/R$ to the whole or part of the circuit. Teachers should not, however, underestimate the time needed for students to proceed from competence in dealing with a single component to the ability to handle complete circuit calculations with confidence. The first experiment starts this process by considering networks of resistors. It is also important because it gives more practice in connecting meters to circuits, and in choosing meters.

EXPERIMENT
B7 Four-terminal boxes

Students are asked to find out what they can about a set of boxes, each of which has four terminals. Two kits are available and boxes **A–D** in kit 1 are easier than those in kit 2. Experience suggests that all students should investigate kit 1 and at least one box from **E–G**. Boxes **I** and **J** may be kept in reserve for a minority of faster students.

ITEM NO.	ITEM
1033	cell holder with four cells
1507	milliammeter, 100 mA
1507	voltmeter, 10 V
1148	four-terminal boxes
1000	leads

The circuits are shown in figures B4(a) and (b). Students may be told that each box contains a simple circuit and that the green terminals are connected together. Some connections to get students started are shown in figure B5. They may also be told that there is at least 60 Ω between any other pair of terminals so that a 100 mA meter can be used with 6 V.

This work is not aimed primarily at circuit analysis, though better students might get as far as that by themselves. It aims to improve their ability to think out simple electrical experiments, choose and use apparatus, and relate what they observe to what they might expect to observe. We also hope that students will afterwards use electrical instruments with greater confidence, and expect to meet and tackle similar simple puzzles at other times. Students need not be expected to solve every, or even any, box they try. Any correct partial deduction about the contents is a success, as is any thought-out series of observations which reveal how a box behaves.

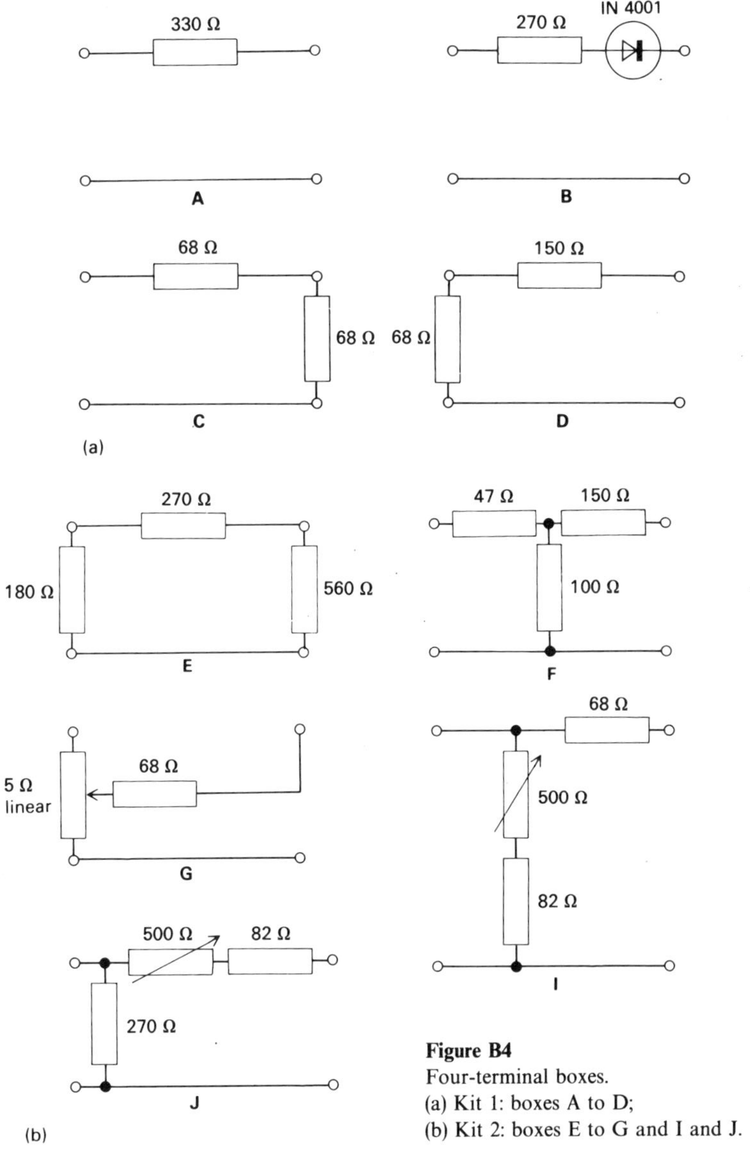

Figure B4
Four-terminal boxes.
(a) Kit 1: boxes A to D;
(b) Kit 2: boxes E to G and I and J.

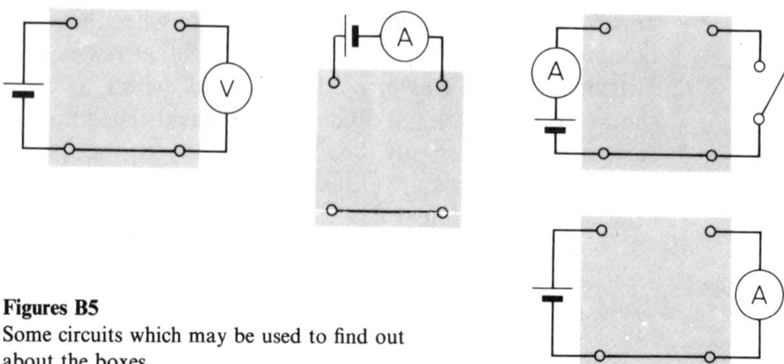

Figures B5
Some circuits which may be used to find out about the boxes.

Boxes **E** and **F** are indistinguishable except that the tolerance of the resistor values will make slight differences. Most students could probably show (knowing the answer) that the two are equivalent, but the best mathematicians could be invited to consider how to construct a T- (or star-) network (box **F**) which is equivalent to any particular π- (or Δ-) network (box **E**).

As an introduction to the next work on internal resistance and effects of loading a source of e.m.f., students who have used box **F** could be invited to compare a given voltage reading with that given by a lower resistance voltmeter (item 80 or 179). There is a drop in p.d. when a low resistance voltmeter is connected (whether or not the high resistance meter is also connected). This observation could be presented as a puzzle to be argued over. No detailed explanation is required at this stage.

Questions

Question 18 deals with conclusions from a four-terminal box experiment; question 19 with measurements in such an experiment.

ELECTROMOTIVE FORCE AND INTERNAL RESISTANCE

The term e.m.f. has effectively been defined already. It is best presented as a quantity which is a property of a power source and characterizes the energy transformation in it. As a formal definition the e.m.f.

$$\mathscr{E} = \frac{\text{power transformed in the source}}{\text{current flowing}}$$

(The symbol $\mathscr{E}$ is used instead of E to avoid confusion with electric field strength, and possibly energy.)

Different types of power supply with the same e.m.f. differ widely in their ability to drive current into an external circuit. This difference is caused by wide variations in their *internal resistance*. For a commercial mains operated power supply the internal resistance may have been set deliberately by the manufacturer; for a cell it will depend on the state of, and physical processes in the chemical contents.

As an introduction to the idea, a student could be invited to short-circuit a 5 kV supply across a milliammeter. Full details are provided in the last part of demonstration B8. There may be some initial reluctance to do anything so drastic, but the outcome is quite dramatic if the scene has been well set. The reading of the voltmeter on the supply falls virtually to zero and the milliammeter shows perhaps 2 mA. 'Why cannot the supply drive the expected very high current?' Presumably because it has some (very high) resistance of its own.

An informal discussion at this stage can lead to a calculation that the internal resistance

$$r \approx \frac{5 \times 10^{3}\,\mathrm{V}}{2 \times 10^{-3}\,\mathrm{A}} \approx 2\,\mathrm{M\Omega}$$

Real power supplies can be modelled as an ideal source of e.m.f., $\mathscr{E}$ (with no resistance) in series with a resistance, r (the internal resistance). Such a source connected to an external resistive load, R, would therefore have the equivalent circuit shown in figure B6.

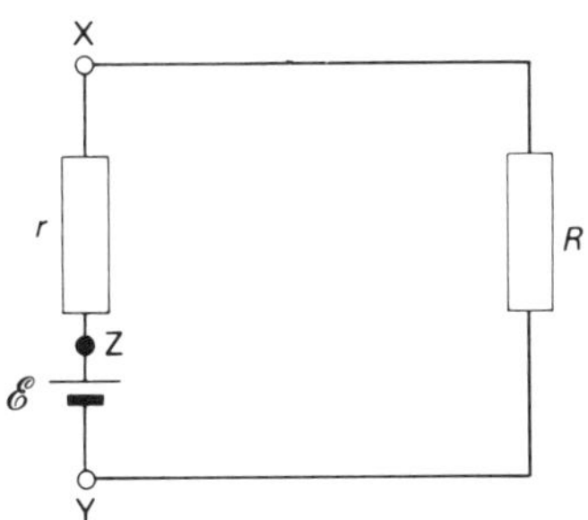

Figure B6
Internal and external resistances.

This is, of course, the conventional representation but it is worth pointing out that it is a simple example of modelling: the relatively complicated processes inside a real cell affect an external circuit connected to its terminals in an identical way to a connection to XY in the circuit in figure B6. Point Z is, of course, an inaccessible virtual point and it is not worth speculating where it 'really' is in a cell.

Applying the Law of Conservation of Energy,

power transformed in source into electrical energy
= power transformed from electrical energy in the circuit

$\mathscr{E}I = I^2r + I^2R$ where I is current flowing

$\mathscr{E} = I(r + R)$

The importance of this relation is that provided r is known it enables the current to be found – students should be encouraged always to regard this as the first stage in any calculation, at least until they get confidence in any short-cut tricks.

The equation may be used in two ways to make important statements about circuits:

i If R is put equal to zero we have a short-circuit of the power supply, with short-circuit current $\mathscr{E}/r$. This shows why it might be permissible briefly to short-circuit a laboratory l.t. supply (under supervision!) but highly dangerous to short-circuit a car battery where r may be of the order of hundredths of an ohm. (In an accidental short circuit in a car the steel brake cable has been known to act as a fuse!) Students could be invited to apply $V = IR$ to the zero external resistance to make a surprising prediction about the p.d. between the battery terminals.

ii Rearranging the equation

$IR = \mathscr{E} - Ir$

or

$V_{\text{load}} = \mathscr{E} - Ir$

where V_{load} is the p.d. across the external load R.

This expresses the very important point that, in practice, the p.d. across the load (or the terminal p.d. of the supply) is always less than the e.m.f., provided some current flows.

The Ir term is frequently referred to as the 'lost volts', and $\mathscr{E}$ thought of as the 'total available voltage'. These may well be intuitively useful terms for those well used to handling circuits but they are not recommended as a starting point. Here is a case where insistence on some rigour and appeal to fundamental ideas (energy transformation and conservation) are physically more illuminating.

This result for the drop in p.d. across a source delivering power summarizes a family of circuit effects which may be termed 'loading behaviour'. Students may have come across some already and commented on the discrepancy between a laboratory l.t. supply nominal voltage and the actual measured p.d. (Some manufacturers provide a 'load line' graph on the instrument itself.) The short-circuit behaviour of a 5 kV e.h.t. supply has already been seen and its paradoxical safety compared with the lethal 240 V mains may well have arisen.

A very important example is where a voltmeter itself is the load – clearly low resistance voltmeters will draw more current and produce a bigger voltage drop Ir. Some algebraic manipulation of the circuit formula produces:

$$V_{\text{load}} = \mathscr{E}\left(\frac{1}{1+r/R}\right)$$

Only as $R \to \infty$ will V_{load} approach $\mathscr{E}$ and this leads to the alternative view of e.m.f. as being the open circuit (zero current) p.d. across the supply. It also indicates how we may in practice measure e.m.f. (the actual definition not being particularly helpful here) – we seek a voltmeter which draws negligible or zero current. The oscilloscope and digital voltmeter (and electrometer) fall into the 'negligible current' category; a balanced potentiometer and electrostatic meter draw truly zero current. Note, though, that it is the ratio r/R, not the absolute value of R, which determines how useful a given voltmeter is for measuring e.m.f. Students should be aware that there can be occasions when even high resistance moving coil meters can effectively short-circuit the terminals.

Questions

Questions 20 and 21 are on various aspects of e.m.f., internal resistance, and use of voltmeters.

Reading

The useful concept of *loading* is extended to other areas in the article 'Systems' in the Reader *Physics in engineering and technology*.

Demonstrations B8 and B10, and experiment B9, explore the predictions of the circuit equation. In a sense they are not true experiments but provide reassuring confirmation of the theory. Arguably, though, they *are* experiments in that they test the validity of the original circuit model, figure B6.

DEMONSTRATION

B8 Drop in terminal p.d. of source on load

This demonstration uses three power supplies – a cell, an l.t. variable, and an e.h.t. – and students observe the drop in p.d. when current is drawn and, where appropriate, the effect of increasing the current. Teachers may already have demonstrated the effect with the e.h.t. supply. They may well wish to cover it again more formally now. If they omit it here, this could become a class experiment.

ITEM NO.	ITEM
52	Worcester circuit board kit (3 lamps and 1 cell)
59	l.t. variable voltage supply
73	lamp, 12 V, 36 W
74	lampholder (s.b.c.) on base
14	e.h.t. power supply
1508	demonstration meter, 10 V d.c., 15 V a.c., and 10 mA d.c.
1000	leads

Connect the voltmeter across one cell. Then connect first one, then two, and then three 1.25 V lamps on the circuit board in parallel across the cell, noting the falls in p.d. An ammeter may be added if desired, and students may prefer other ways of changing the current, such as a variable resistance.

Set the l.t. supply at 12 V, and connect the lamp (item 73) across the alternating output. Note the drop in p.d. The experiment can be repeated with a d.c. supply.

Set the e.h.t. supply to give 1000 V and connect the milliammeter across the output. The voltmeter reading should fall to zero, the current being a few milliamperes. Any very high resistance, shown as a possible connection on the front panel of the power pack to limit current even further, should not be used.

This experiment should not be laboured; its main aim is to help students build up an instinctive feeling of how real power supplies behave. It can, of course, reinforce the theoretical work in a more formal way if precise current and p.d. values are measured and used to calculate internal resistance.

Home experiment

In home experiment BH2 students make a 'voltaic pile' to be no larger than a matchbox, but to produce as large as possible an e.m.f.

Comparing voltmeters

To support the theoretical points made about the actual measurement of e.m.f. and to bring together earlier observations which may have been made about the loading effect of voltmeters themselves, we suggest a quick experiment and then a demonstration to compare different voltmeters on the same source of e.m.f. Teachers may choose to combine them.

EXPERIMENT
B9 Comparison of voltmeters

ITEM NO.	ITEM
1033	cell holder with 3 cells
	either
1151	resistor, 10 kΩ
	or
1017	resistance substitution box
1507	voltmeter, 5 or 10 V
	voltmeter, same range as above, low resistance – see note below
	optional
1511	oscilloscope
1000	leads

Note: There is no need to obtain a low resistance voltmeter specially for this experiment. Proceed without it – or improvise one by putting a 1 kΩ resistor across the terminals of item 1507.

The internal resistance of a cell is increased artificially by placing a high resistance (say 10 kΩ) in series with the cell. Many variations are possible: the appropriate high resistance could be concealed with the cell in a box, with just the terminals presented to the student; or the arrangement could be completely open and either a fixed resistor or the resistance substitution box used as the additional 'internal' resistance. The arrangement adopted will obviously be decided by the way the teacher chooses to present the experiment to the class – as an additional puzzle box (this time active, not passive), or as a more open model of a power supply with very high resistance (say a cell with polarization).

The voltmeters suggested are low and high resistance moving coil meters, and an oscilloscope. The low resistance voltmeter has a resistance of 1 kΩ; item 1507 with a 10 V multiplier has a resistance of 100 kΩ; and the input resistance of an oscilloscope set to d.c. is typically 1 MΩ.

The oscilloscope should be used with the time base off, and set to d.c. If it does not have a calibrated vertical gain it should be calibrated before the experiment against a voltmeter.

Students use each voltmeter in turn, starting with the lowest resistance one, to measure the p.d. across the cell/high resistance combination. Typically they should get readings of 0.4 V, 4.0 V, and 4.5 V respectively.

The essential point is that the series of p.d. readings tends towards a limit which is the cell e.m.f. Students occasionally think that the meters are somehow showing a 'wrong' reading: it is worth emphasizing that they *are* accurately showing the p.d. which happens to be across their terminals. An illuminating extension to the experiment involves starting with the oscilloscope and adding in turn the other meters in descending order of resistance, leaving the previous meters connected. Each additional meter pulls the voltage down to the value it showed by itself, and all meters now show the same.

Later in the course a very high resistance voltmeter (possibly an electrometer/d.c. amplifier) will be used – for example, to measure the p.d. across a capacitor and hence the charge on it, or to measure the p.d. across a very high resistance, and hence the small current in it. This instrument without the added complication of accompanying capacitors or resistors is introduced in the next demonstration which can be seen as an extension of experiment B9.

DEMONSTRATION
B10 High resistance voltmeter

ITEM NO.	ITEM
1033	cell holder with one cell
1151	2 resistors, 220 kΩ
1040	2 clip component holders
1507	voltmeter, 1 V, moving coil
1509	high impedance voltmeter (electrometer/d.c. amplifier would do)
1511	oscilloscope
1000	leads

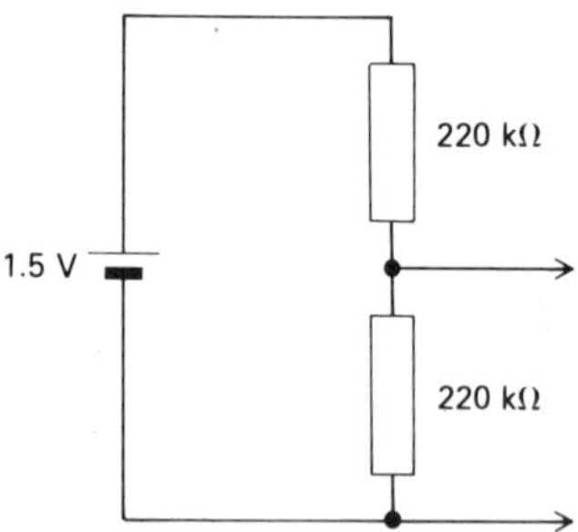

Figure B7

The high impedance voltmeter should have a resistance of at least 10 MΩ (a typical value for a digital meter); the electrometer's input resistance may be as high as 10^{11} Ω; an oscilloscope typically has an input resistance of 1 MΩ on its d.c. setting.

If the electrometer has to be calibrated before use, this should be done, according to the manufacturer's instructions, before the demonstration, so that the electrometer can be presented simply as a very high resistance voltmeter.

Use the circuit of figure B7 to measure the p.d. across one of the 220 kΩ resistors, using each of the instruments in turn: first the moving coil voltmeter, then the oscilloscope, then the digital meter and/or electrometer. Again it is instructive to start with the highest resistance device and to show the increasing load presented as the others are added in turn.

If a modern microprocessor-based instrument (*e.g.*, VELA, GiPSI) is available and will be used later in the course, it too could be used in this demonstration. Students will now have been introduced to the full range of voltmeters they are likely to meet in the course.

POWER IN A COMPLETE CIRCUIT

Since a source of e.m.f. will dissipate power within itself, because of its internal resistance, the efficiency of energy transfer to the external circuit is an important question. The condition that for the useful power transferred to be a maximum, the internal and external resistances should be equal, seems sufficiently important and simple to merit a mention. It is an example of a more general condition, known as the maximum power theorem, that impedances should be matched to attain maximum power transfer. (Another example is the acoustic coupling of sound sources to a transmission medium.)

It is not suggested that a formal treatment should be given and question 22 approaches it through a numerical example. Teachers could, though, consider the general shape of the power–external resistance graph by considering two extremes and relating them to the earlier complete circuit equation:

'... suppose the load resistance R was very big; we have nearly the full e.m.f. applied but very small current. Hence power ($= V \times I$) is very small. If R were made very small the current would be large (nearly the short-circuit current) but the p.d. would be nearly zero. Hence the power is very small again. It seems likely that between these two very small values the power will rise to a maximum for a particular value of R.'

This is a good example of the prediction of a not immediately obvious result from qualitative reasoning.

A minority of students may, of course, relish the calculus approach of differentiating the power with respect to R and equating to zero. We think it is more important, though, for all students to be able to follow (and sometimes construct for themselves) arguments of the kind outlined above, and to do a check of their numerical validity.

An interested student could write a short computer program to calculate the power in the external circuit for various values of R (for a given r).

Question

Question 22 deals with power in a complete circuit.

POTENTIOMETER AND BRIDGE CIRCUITS

The end point of this short sequence of experiments is the Wheatstone bridge circuit, and the appreciation of its sensitivity and versatility as a detector of small resistance changes when used out of balance. It is not presented as a way of measuring an unknown resistance. The circuit is approached through the potentiometer, which has already been met as a control device: here it is used in a null method to balance an e.m.f. and to detect small changes in e.m.f.

It is not part of the course that students should be familiar with the details of the multitude of ways potentiometer and bridge circuits can be used; modern instrumentation in any event makes much of the accurate potentiometric work unnecessary. On the other hand, the out-of-balance bridge does seem useful. The ability to generate a p.d. proportional to a change in resistance (if small) finds many applications in instrumentation and control systems, as will be seen in Unit I, 'Linear electronics, feedback and control'. The p.d. can, for example, be the error signal in the feedback loop of a servo.

EXPERIMENT

B11 Potentiometer balancing an e.m.f.

It is suggested that the traditional form of slide wire potentiometer is used here. It reinforces the idea of constant potential gradient in a uniform wire carrying a current. (Alternatively it shows that such a wire behaves like the sealed rotary potentiometer already used.) It can be used to measure small differences in e.m.f. The alternative, using calibrated vernier potentiometers (a very small fraction of the total rotation can be read accurately), is very expensive and unnecessary.

Schools or colleges which already possess one of the various commercial forms of metre wire apparatus will, of course, use it. However, it is not suggested that special purchase is made for courses being set up: one metre of wire stretched over a rule is quite adequate, having a resistance of about 4 Ω. Constantan or manganin (0.4 mm diameter) would be satisfactory. Likewise the traditional form of slide 'jockey' contact can be improvised by, say, a small electrical screwdriver with crocodile clip connection.

ITEM NO.	ITEM
1501	constantan wire, 0.4 mm diameter
501	metre rule
59	l.t. supply (or 2 V accumulator)
1033	cell holder with one cell
1507	voltmeters, 1 V and 10 V
1507	microammeter, 100 μA
1000	leads

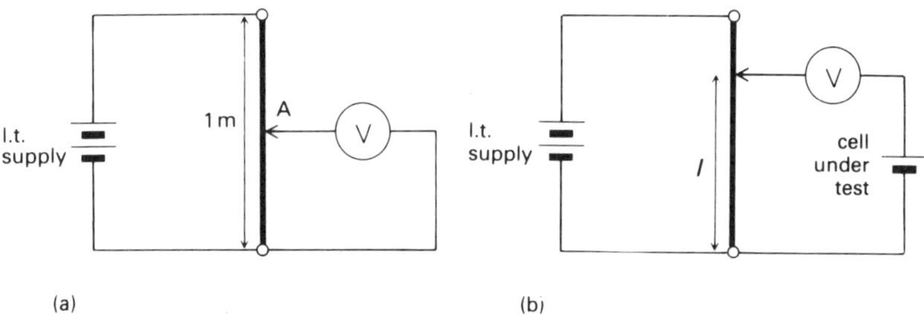

Figure B8
Metre wire potentiometer.

The circuit is initially set up as in figure B8(a). The l.t. supply is set to 2–3 V d.c. and kept constant. Dry cells are not a satisfactory source for the low resistance potentiometer: a 2 V accumulator can be used instead. If necessary, more than one group can operate from the same power supply.

The voltmeter is connected directly across varying lengths of the wire, tapped off by the sliding contact, and students confirm that the p.d. across the meter drops uniformly to zero as the length is reduced. It could be suggested that it is sensible to think in terms of the voltage drop represented by, say, 10 mm of wire.

A very useful teaching point can be made here which helps subsequent circuit analysis, and will be met again in the more formal work on potential (Unit E, 'Field and potential'). This is to label one point in the circuit arbitrarily as 'zero potential', as is commonly done in describing electronic circuits. In later Units this point on the circuit will be labelled 0 V. It is a common convention to take this point as the negative terminal of the supply, but there is nothing special about the choice. With this convention, the voltmeter can be regarded as showing *the* potential at A. The experiment then shows that the *potential* at A gets smaller as A is moved round the external circuit in the direction of the current.

In the second part of the experiment a 1.5 V cell is connected in the voltmeter circuit as in figure B8(b). Students should appreciate that provided the cell's polarity is correct, a point on the wire can be found such that each side of the voltmeter is at the same potential and hence no current will flow through it. The meter can be replaced by or converted to a sensitive galvanometer for the precise determination of the balance point. The length of wire tapped off, l, is a measure of the cell's e.m.f., and if a potential gradient had been calculated in the first part the e.m.f. could be calculated.

An interesting quick extension to the experiment involves replacing the 1.5 V cell with another apparently similar one. The balance point will probably shift a few millimetres, and the direction and magnitude of shift indicates whether the second cell has a larger or smaller e.m.f. and by how much. If the l.t. supply is 2 V, a shift of 5 mm corresponds to 10 mV – would this be detectable on a 0–10 V meter?

Teachers could make two final points about this experiment:

i Since this is a null method, the only requirement of the meter is that it should indicate zero.

ii Since at balance the potentiometer is drawing no current from the cell under test (and hence effectively presents an infinite resistance to it), it is fulfilling the requirement of an 'ideal' voltmeter and is truly indicating the e.m.f. of the cell.

EXPERIMENT

B12 Using two potentiometers to make a bridge circuit

The potential on the righthand side of the voltmeter could equally be derived from a second potentiometer, as in figure B9.

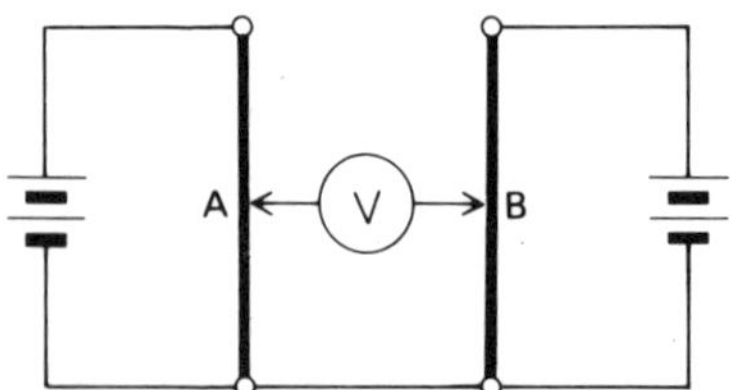

Figure B9
Two potentiometers balanced together.

The lower terminals of the potentiometers are connected to provide a common reference point for both circuits.

Apparatus
As for experiment B11, but with a second metre wire potentiometer and without cell holder and cell.

Although cumbersome on the bench, it is suggested that initially pairs of groups doing experiment B11 are brought together to combine their circuits as in figure B9. The meter should revert to a 0–1 V voltmeter to allow a range of off-balance settings, and the l.t. supply should be approximately the same for both potentiometers. Students should see that for any fixed position of A there is only one balance point B, and for any fixed B only one position of A will give balance. Thus a range of A–B pairs up and down the wires will give balance.

Invite students to state explicitly a condition for balance ($V_A = V_B$). What could upset the balance? (Either power supply could drift away from its set value.)

(The teacher could perhaps demonstrate this point on one set of apparatus as part of the discussion by arranging balance and then deliberately altering one of the voltages slightly.) How could this disadvantage be removed? Draw from the class the suggestion that one power supply is really redundant, and both potentiometers could work off the same source. This gives the circuit of figure B10.

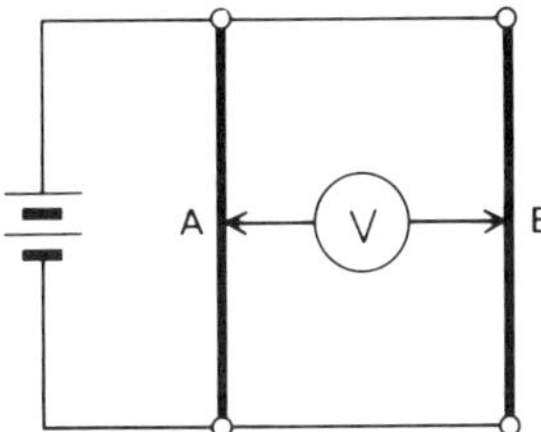

Figure B10
Two potentiometers on a single power supply.

Students should now make this change (which gives the arrangement of the Wheatstone bridge). They should check that when balance is reached variations in the supply voltage have no effect (they affect both sides equally). They should see that the only way to move off-balance is to fix A, say, and make small movements of B.

This last point can be observed by replacing the voltmeter by the microammeter and, keeping A fixed, making small adjustments to B while keeping the current within the range of the meter. Students should observe proportionality between the out-of-balance current and the movement of B.

Bringing experiments B11 and B12 together in summary, students should understand that a single potentiometer is useful for detecting small changes in potential (particularly e.m.f.) and that the bridge circuit will indicate small changes in *resistance* in at least one of the 'arms' of the bridge (see below). It is not necessary to derive the resistance ratio condition for balance: it seems more important to realize that balance is a useful starting point in using such a circuit.

Questions

Questions 23 to 26 are on potentiometers and bridge circuits.

Detection of small resistance changes

In experiment B6 the change of resistance with temperature was observed. It has also been suggested that straining a wire will produce small resistance changes. These changes might be detected by noting the change in current in a series circuit, but since the *proportional* change in resistance may be small they are not necessarily easy to observe. Experiment B12 shows how the out-of-balance bridge current may detect these small changes, with the added advantage of linearity provided the resistance change is small. This experiment sets up a balanced bridge using a thermistor in one of the arms and follows the resistance change produced by small changes in temperature.

EXPERIMENT
B13 Detection of small resistance changes

ITEM NO.	ITEM
1033	cell holder with four cells
1151	thermistor
1510	potentiometer, 1 kΩ or 5 kΩ
1017	resistance substitution box
1507	microammeter, 100 μA
1000	leads

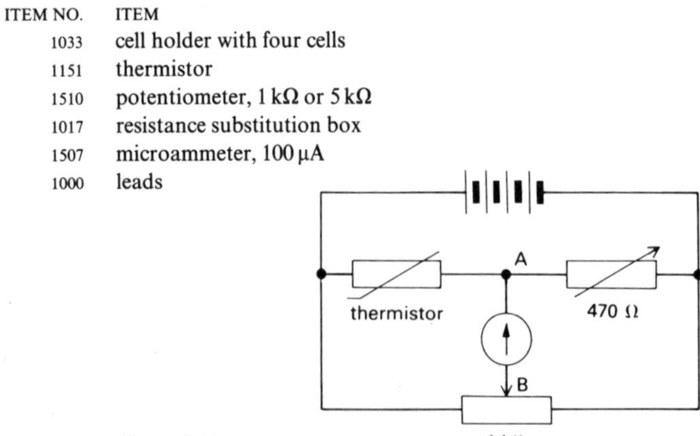

Figure B11
Detection of small resistance changes.

The circuit is shown in figure B11. The thermistor resistance at room temperature is about 400 Ω. The two arms at the top of the bridge contain the thermistor and the resistance substitution box set to 470 Ω. The bottom two arms are provided by the potentiometer which is adjusted for balance. The sensitivity is such that warm breath or a small drop of volatile liquid on the thermistor produces appreciable deflection either way on the microammeter.

This experiment is best regarded as a starting point to indicate to students the power of the technique and to provide them with a useful instrumentation tool. Teachers may well wish to make local variations on the method: a strain gauge (commercial or improvised) is an obvious candidate and could provide a useful way of instrumenting an investigation of mechanical oscillations in Unit D, 'Oscillations and waves'. As it stands, the potential difference between A and B is used to drive current through the meter: it could equally be displayed on an oscilloscope or used to drive a chart recorder or a computer-based data logger. Suppliers' data sheets show a variety of simple and cheap integrated bridge and amplifier circuits.

Reading

The article 'Applications of Wheatstone bridge circuits' (*Students' guide*, page 108) gives three different industrial applications.

KIRCHHOFF'S SECOND LAW

Here we come full circle and consider briefly a more formal and theoretical method for analysing circuit networks. It enables relationships between currents in and p.d.s across various branches of a complex circuit to be written down directly. Its theoretical basis is the Law of Conservation of Energy. In practice, the solution of the resulting set of simultaneous equations is often an exercise in algebraic skill and the physics becomes obscured. Apparently innocuous circuits can yield fearsome equations and teachers should be wary here. Question 27 in the *Students' guide* gives an indication of the level of treatment expected. (However, such sets of equations can legitimately be presented to the maths department as practice in matrix methods of solution.)

The following treatment is suggested:

Consider the loop circuit of figure B12. The p.d. between A and C is the same for either branch (the energy transformed per unit charge must be independent of route – a further look forward to Unit E, 'Field and potential').

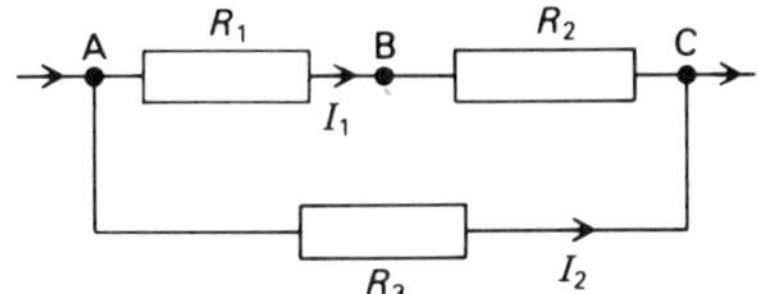

Figure B12
A circuit loop.

$V_{AC} = V_{AB} + V_{BC}$

$I_2R_3 = I_1R_1 + I_1R_2$

If we imagine a journey clockwise round the loop in figure B12 from A, then potential *drops* through R_1 and R_2, and it *rises* through R_3, since the sense is in opposition to the current flow here. If the equation is written $I_1R_1 + I_1R_2 + (-I_2R_3) = 0$ we can say that if we sum round a complete loop in which there is no source of e.m.f., the total potential *drop* is zero. $\sum_{\text{loop}} IR = 0$, remembering that for some of the resistors in the loop the current will be negative.

If the circuit is modified to include an e.m.f. $\mathscr{E}$ (for simplicity with no internal resistance) as in figure B13, we can again start from A and say:

$$\underbrace{(I'_1R_1 + I'_1R_2)}_{\text{drop from A} \longrightarrow \text{C}} - \underbrace{(I'_2R_3 + \mathscr{E})}_{\text{rise from C} \longrightarrow \text{D} \longrightarrow \text{A}} = 0$$

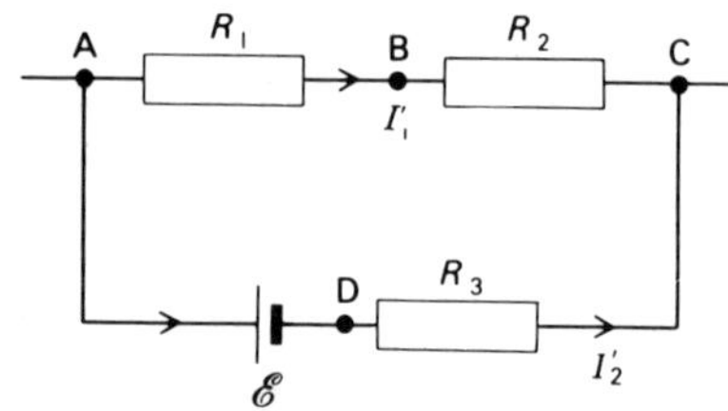

Figure B13
A loop containing a source of e.m.f.

It is a matter of taste whether the second term is regarded as the subtraction of a potential rise or the addition of a (negative) potential drop. There are traps for the unwary here and teachers are advised to be clear which approach they are using! Careful attention to detail on signs here will be repaid in further work on potential.

In any event it is worth pointing out that such a route through the cell in the direction shown, from negative to positive terminal, results in a *rise* of potential. This rise is equal to the e.m.f., whether or not there is any current and regardless of the direction of any current.

The equation becomes

$$I'_1R_1 + I'_1R_2 + (-I'_2R_3) = \mathscr{E}$$

or in general

$$\sum_{\text{loop}} IR = \sum \mathscr{E}$$

where the righthand side is the total e.m.f. in the loop, having regard to sign. This form is known as Kirchhoff's Second Law.

Finally, a numerical example may help to fix these ideas for any who feel on slightly shaky ground. In figure B14 a current of 10 A flows into the loop at A. The circuit equation has already been solved to find the current in each branch, which has been rounded to 1 decimal place.

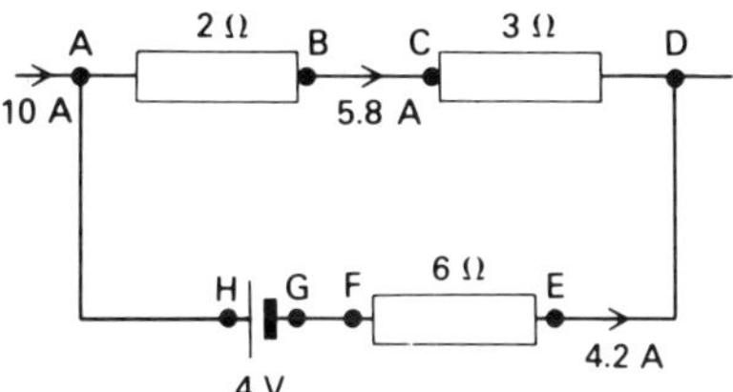

Figure B14

If we regard the negative cell terminal, point G, to be at zero potential (a decision of convenience not principle), then we can plot out how the potential changes as we go round the loop in a clockwise direction from A (figure B15).

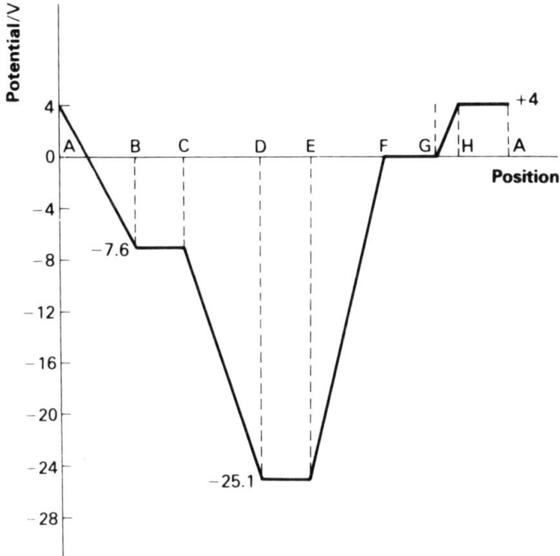

Figure B15
Variation in potential round a loop.

The connecting wires are assumed to have zero resistance (no potential drop). A sketch like this showing variation of potential with position brings out the concept of potential gradient and also shows clearly what would happen if some other point had been labelled zero – the whole pattern would simply be shifted vertically.

SECTION B3

ELECTRIC CHARGE

AN APPROACH THROUGH CAPACITORS

The idea of electric charge is approached through class experiments with capacitors, making use especially of high value electrolytic capacitors which can store large charges for long periods of time. These charges are large enough to give measurable currents for appreciable times when passed through an ammeter, so that students can be clear that charge is measured in ampere seconds or coulombs. It is important for students to realize that charge is a quantity of something, like litres or bucketsful, not a force or intensity. This last confusion can arise if the idea of charge is introduced with ideas about fields and the forces between the charges, and so this is deferred to Unit E, 'Field and potential'.

Question 28 is an introductory question on the relationship between current, time, and charge.

Aims

i Students should become able to use and recall the idea of electric charge as a quantity of electricity measured in coulombs (ampere seconds) in a variety of situations. They should recall that charge is proportional to potential difference, and be able to calculate capacitances from $C = Q/V$.

ii They should understand that when a capacitor discharges through a resistor the rate at which charge leaks is proportional to the p.d., and so to the charge on the capacitor. Later they should come to see this as one among many instances of exponential decay.

iii They should recall and be able to use arguments leading to the energy stored, $\frac{1}{2}QV$, and be ready to compare this with the energy stored in a spring, $\frac{1}{2}Fx$.

Experience with capacitors

Many of the basic ideas concerning charge and the properties of capacitors can be introduced in a series of simple class experiments (B14), which are supported by questions 29 and 30 in the *Students' guide.* A quick demonstration may be the best way of introducing these experiments (figure B16).

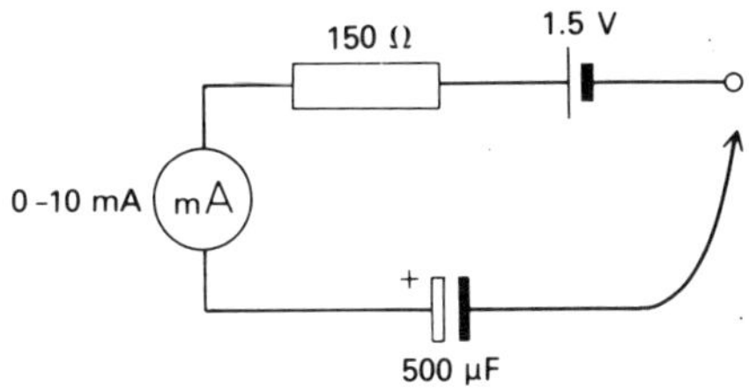

Figure B16
Capacitor charge/discharge circuit.

Observation of the meter readings during charge and discharge might prompt speculation about the construction of the capacitor (conducting plates separated by an insulating layer) and the circuit symbol (+) could now be introduced together with the words charge and discharge.

Students should be warned about the polarity of capacitors. Teachers may well feel it worth while showing the damage to an electrolytic capacitor by connecting it the wrong way round to a supply voltage greater than its rated value. After such treatment it will conduct a small current in either direction (figure B17).

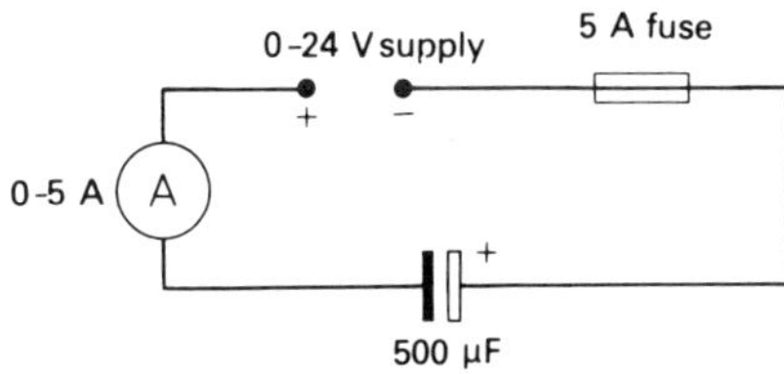

Figure B17
Reverse capacitor connection.

It is worth destroying one by over-running for 2–3 minutes to show its construction. *N.B.* A safety screen is vital in case the capacitor explodes.

EXPERIMENT
B14 Capacitors and charge
(a series of 8 short experiments)

ITEM NO.	ITEM
1033	cell holder with four cells
1040	3 clip component holders
1151	two 500 μF electrolytic capacitors, 10 V
1151	two 1000 μF electrolytic capacitors, 10 V
1151	two 2200 μF electrolytic capacitors, 10 V
1511	oscilloscope
507	stopwatch
1017	resistance substitution box
1507	2 milliammeters, 10 mA
1151	resistor, 150 Ω
1000	leads

Details of the circuits used in experiment B14a–h are given in the students' laboratory notes.

B14a This uses the same circuit as figure B16. Besides investigating charging and discharging for themselves students should find that the meter movement is not repeated if a connection is made a second time.

B14b This adds a second meter to the circuit to investigate the currents flowing in both leads to the capacitor. Students must pay attention to meter connections or incorrect conclusions may be drawn. The results show that as much electric charge leaves the capacitor as enters it (figure B18).

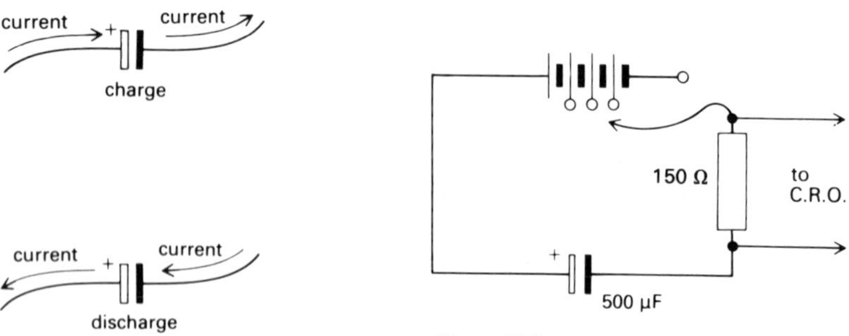

Figure B18

Figure B19
Use of oscilloscope.

Many of the following experiments work well using a moving coil meter to estimate charge. The oscilloscope is introduced because of the increasing use of digital meters which are unsuitable for this work, and the need to forge stronger links between this work and that in electronics, where the oscilloscope, connected across a resistor, is used to display current.

B14c This introduces the oscilloscope in order to find out exactly how the current in the circuit is changing (figure B19).

The oscilloscope must be set to d.c. Initial experiments might best take place with the time base switched off. The time constants of the suggested circuits are approximately 0.1 second, and the time base, when eventually used, should be at one of the slowest settings.

The following points are worth discussing:

i The large difference between the inertia of the electron beam and the moving coil of a conventional meter, and the effect this has on the response time of the two instruments.

ii Even though the oscilloscope is a voltmeter the vertical displacement of the trace is proportional to the current in the 150 Ω resistor.

iii The area under the trace represents the charge flowing around the circuit. It can be estimated if the time base is calibrated (using a stopwatch), and the maximum current is calculated from $I = V/R$.

B14d In this experiment extra cells are used producing larger pulses. Some students may find that if an extra cell is added without previously discharging the capacitor, each extra cell gives an equal extra pulse. The moral 'Equal extra voltage gives equal extra charge' can be drawn, this being one of the simplest experiments to show that charge is proportional to potential difference.

B14e Different values of capacitors can be substituted and the quantity of charge flowing can be estimated from the oscilloscope trace. If students are told that the capacitor rating is the charge stored on each plate for a 1 V p.d., they can check the accuracy of their estimates.

B14f Two capacitors may be tried in parallel. Charging and discharging the capacitors as a pair (figure B20) or singly (figure B21) provides evidence that the charges on the plates add up. A similar investigation with the capacitors in series shows that the combination stores less charge for the same p.d. than does either capacitor separately.

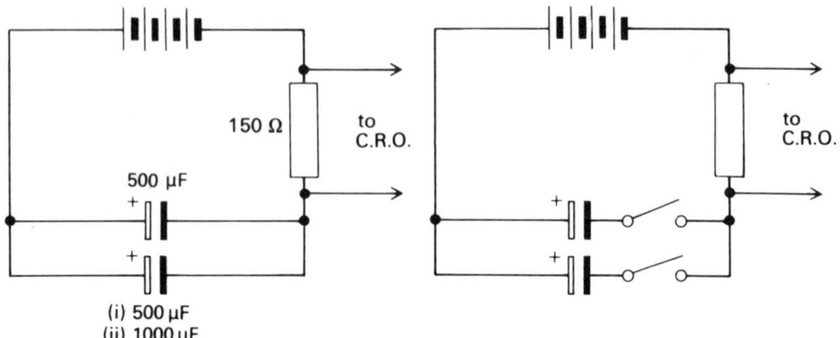

Figure B20
Connections for experiment B14f.

Figure B21
Connections for experiment B14f.

B14g The effect of adding extra resistance to the circuit can be investigated using a resistance substitution box. The 150 Ω oscilloscope resistor must not be changed otherwise the constant of proportionality linking beam deflection and current would alter. From studying the changing shape of the oscilloscope traces students might be encouraged to speculate on the effect of changing the resistance on the total flow of charge around the circuit.

B14h Fast students may want to use the oscilloscope to investigate the p.d. across the capacitor. This opens up the opportunity of applying Kirchhoff's Laws to the circuit. Why does the p.d. across the resistor fall as the p.d. across the capacitor rises? What about the p.d. across both the resistor and capacitor? During discharge why is the p.d. across the resistor negative?

These experiments will provide a valuable link to the later work on electronics. Experiments B14a to h provide for more work than can be fitted into one laboratory practical session. It is well worth while taking up another practical session, but if time is restricted B14f, g, and h may be omitted.

Summary

The following points need to be raised, possibly in group discussion.

i A pulse of current carries charge off one plate and on to the other, leaving the plates with equal but opposite charges.

ii There is evidence that the charge on a capacitor plate is proportional to the p.d. across the capacitor.

iii Adding capacitors in parallel increases the charge which can flow around the circuit. There is evidence that the charges on the plates of the capacitors add up.

iv Reducing the rate at which charge flows on to the capacitor increases the length of time for which the charge flows. The total amount of charge flowing remains the same.

v Capacitors differ in the amount of charge they store for a given potential difference. This is investigated further in the following demonstration which introduces the unit of capacitance.

DEMONSTRATION

B15 Charging a capacitor at a constant rate

ITEM NO.	ITEM
1033	cell holder with four cells
1507	microammeter, 100 μA
1510	potentiometer, 100 kΩ
1151	capacitor, 500 μF, 10 V
507	stopclock
1511	oscilloscope
1040	clip component holder
1000	leads

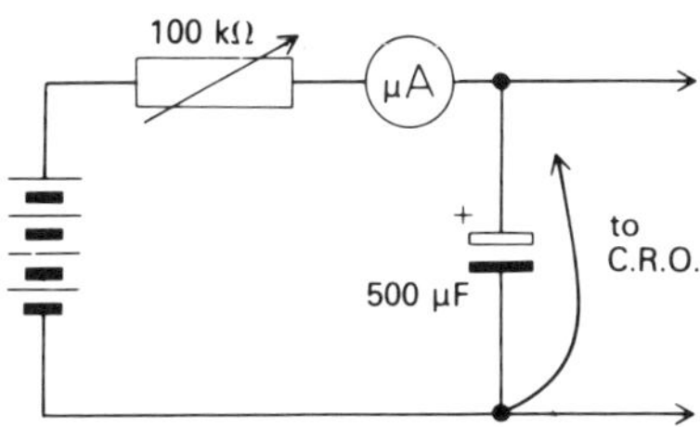

Figure B22
Charging a capacitor at a constant rate.

Set the oscilloscope to d.c. with the time base switched off. Adjust the sensitivity to $1\ \mathrm{V\,cm^{-1}}$. Short out the capacitor and adjust the rheostat for a suitable current (say 80 μA). Remove the shorting lead and rotate the rheostat control to keep the current constant as the capacitor charges up.

Record times for the p.d. across the capacitor to reach 1 V, 2 V, 3 V, etc. The constant current enables the students to 'count the coulombs' say every five seconds; this reinforces the proportionality between the charge and p.d. If the proportionality, $Q \propto V$, is written

$$Q = CV$$

then C is the number of coulombs needed to raise the p.d. by one volt. The capacitance can be estimated from the values of Q and V and compared with the marked rating.

The unit of capacitance (the farad) together with the sub-units microfarad, μF (10^{-6} F), and picofarad, pF (10^{-12} F) should now be mentioned.

Questions

Questions 31 to 33 use $C = Q/V$.

Combinations of capacitors

The formulae for adding capacitors in series and parallel are derived in questions 34 and 35 and used in question 36. Students should know these formulae, especially as the measurement of charge in later experiments relies on charge sharing.

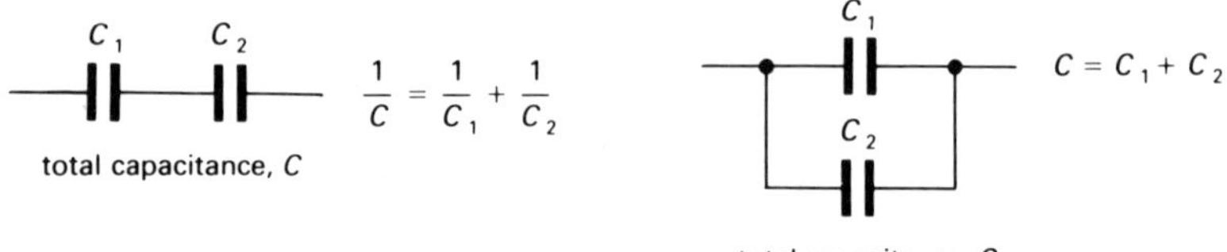

Figure B23
Capacitors in series and parallel.

DEMONSTRATION
B16 Spooning charge

ITEM NO.	ITEM
1512	coulombmeter (see note below)
14	e.h.t. power supply
51K	electrophorus plate
1000	leads

The coulombmeter uses a high impedance voltmeter to measure the p.d. across a capacitor, and so gives a reading proportional to charge.

An electrometer/d.c. amplifier with a 0.01 μF capacitor connected across the input gives a range of 0–10 nC with a resolution of about 0.2 nC. If a digital meter is used, set it to the 200 mV range and connect a 10 μF non-polarized capacitor (or 2 × 4.7 μF in parallel) across the input. This gives a range of 0–2000 nC and a resolution of 1 nC.

Small conductors with 4 mm plugs may be provided as accessories for the e.h.t. supply and electrometer, or they can be improvised. Note that the 50 MΩ limiting resistor on the e.h.t. supply is used.

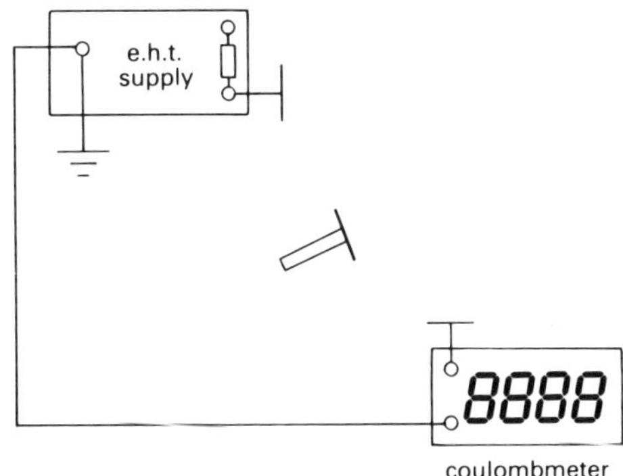

Figure B24
Spooning charge.

Note that expensive damage can result from connecting the e.h.t. directly to the coulombmeter input.

Use the electrophorus plate to transfer charge from the e.h.t. supply (set at a suitable value between 1 and 5 kV), to the coulombmeter. Record the meter reading after every transfer. A graph of meter reading against number of transfers should be a straight line.

The p.d. of the supply can be varied, and a larger 'spoon' used (*e.g.*, item 65 – metal plate with insulating handle – or even a soup ladle with an insulating handle).

Whether a special-purpose coulombmeter, a high impedance voltmeter, or an electrometer is used, it is worth pointing out that the instrument is in fact a voltmeter measuring the p.d. across a capacitor, and so giving a reading proportional to charge. (Just as to measure current one connects a voltmeter, perhaps an oscilloscope, across a known resistance.)

The input resistance of the instrument is high and some calculations on its loading effect might be worth while. For example a 0.01 μF capacitor charged to 1 V carries a charge of 0.01 μC; if the input resistance of the electrometer is $10^{12}\,\Omega$ this quantity of charge would take in excess of 10^5 seconds to leak off the capacitor through the input resistance. If a digital meter with a resistance of $10^7\,\Omega$ is used with a 10 μF capacitor, the time constant will be 10^2 s.

The demonstration should emphasize that charge is a quantity of something that can be measured out and passed from one place to another without being lost (unlike pressure, temperature, or p.d.). It also emphasizes that capacitance is not just a property of capacitors, but that all conductors, when raised to a potential V, store charge Q and have capacitance $C=Q/V$. The capacitance appears to be a function of the size of the conductor, and for faster groups calculations of capacitance are possible. For example, a single transfer of charge raises the p.d. across a 0.01 μF capacitor by 0.2 V.

Charge carried per transfer

$=0.01 \times 10^{-6}\,\text{F} \times 0.2\,\text{V}$

$=2 \times 10^{-9}\,\text{C}$

Potential of plate $= 1000$ V

Capacitance of plate

$=Q/V$

$$=\frac{2 \times 10^{-9}\,\text{C}}{1000\,\text{V}}$$

$=2 \times 10^{-12}\,\text{F}$ or 2 pF

The problem of sharing charges between capacitors is worth mentioning, with the warning that although no significant charge is left on a small plate, this might not be so for a larger object.

ENERGY STORED IN A CHARGED CAPACITOR

Capacitors are not useful ways of storing energy, except perhaps in some research applications. However, these first ideas about electrical energy will be useful later in the course, especially in Unit E, 'Field and potential', where ideas about energy link electrical forces and fields.

DEMONSTRATION

B17 The energy stored in a charged capacitor (motor)

A small electric motor which will run on a very small current shows that the energy stored in a capacitor can be transformed into mechanical energy. The motor efficiency is low ($\approx 10\,\%$) and although the motor can be used to raise a small load, quantitative work is best avoided.

ITEM NO.	ITEM
1034	large electrolytic capacitor, 10 000 μF, 30 V
1033	cell holder with four cells
	either
	solar motor (if available)
	or
9B	small motor/generator unit (energy conversion kit)
1153	length of thin cotton thread
1153	Plasticine
1000	leads

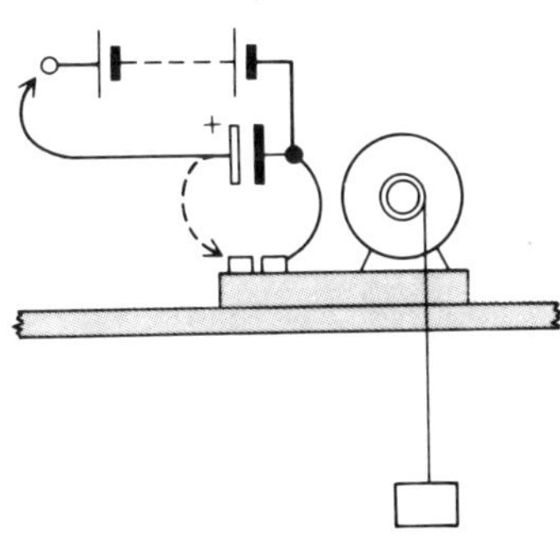

Figure B25
Discharge of capacitor through a motor.

The experiment works much better at higher voltages and if the small motor/generator unit is used, the capacitor should be charged up to 9 V or 10.5 V. (The solar motor may be damaged by any voltage over 6 V.)

Discussion: energy stored is $\frac{1}{2}QV$

If the capacitor starts discharging at a p.d. of V_0, and charge ΔQ flows, the energy transformed is $V_0 \Delta Q$. This follows from the definition of the volt as a J C^{-1}. However, as the charge flows V will drop by an amount $\Delta V = \Delta Q / C$ to a new value V_1 (figure B26).

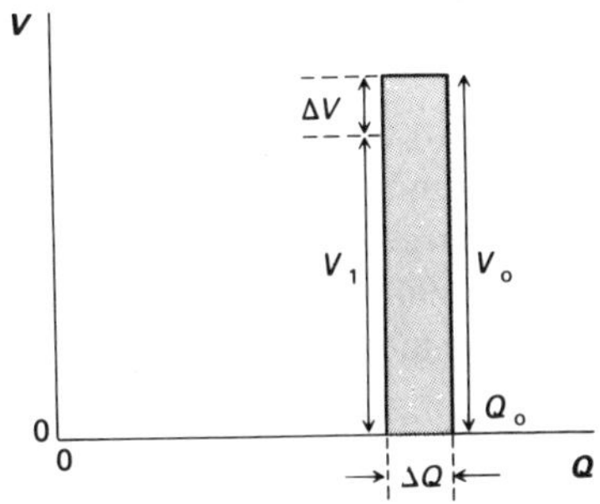

Figure B26
Drop in p.d. across a discharging capacitor.

If the capacitor goes on discharging and further charge ΔQ flows, the new bit of energy transformed is $V_1\Delta Q$. This process continues with V becoming smaller and smaller until eventually $V=0$ ($V=Q/C$). The total energy transformed is the area of all these strips, each $V\Delta Q$, added together, which is the area $\frac{1}{2}Q_0V_0$ (figure B27).

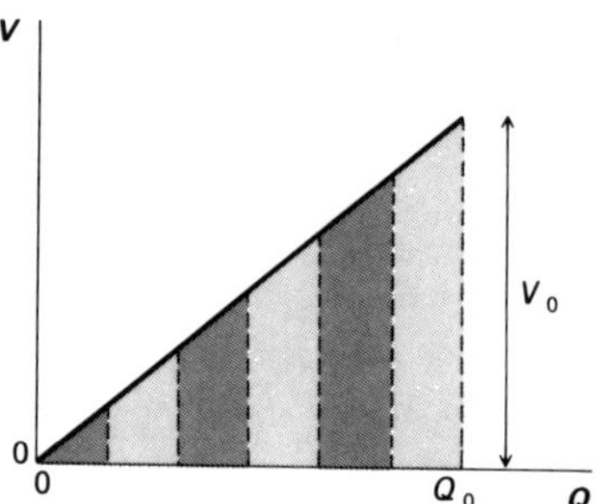

Figure B27
$V\Delta Q$ strips.

Many students will be familiar with the strips and triangle method from earlier courses or from the example of a stretched spring in Unit A, 'Materials and mechanics'. Those who are confident with calculus may be profitably shown or asked how to complete the argument by integration.

See also question 37.

Students should do the algebra

$$\left.\begin{array}{l} W=\frac{1}{2}QV \\ \\ C=\dfrac{Q}{V} \end{array}\right\} \Rightarrow W=\tfrac{1}{2}CV^2;\ W=\tfrac{1}{2}\frac{Q^2}{C}$$

Energy proportional to V^2

The first demonstration, B18 – discharging the capacitor through torch bulbs – is short and effective, and should occupy little time. Heating by capacitor discharge (demonstration B19) is more involved but could well include useful quantitative work and an appreciation of uncertainties in the measurement.

DEMONSTRATION

B18 Energy proportional to V^2 (lighting lamps)

ITEM NO.	ITEM
52	Worcester circuit board kit with three cells
1033	cell holder with three cells
92R	9 m.e.s. bulbs, 2.5 V, 0.3 A
1034	large electrolytic capacitor, 10 000 μF, 30 V

One circuit board and one cell holder are required to give a 9 V supply. The bulbs should be tested beforehand so as to be as nearly identical as possible.

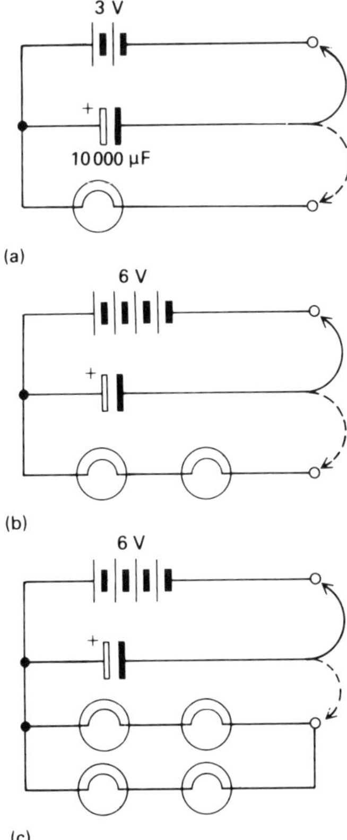

Figure B28
Discharge of capacitor through lamps.

Using circuit (a), charge the capacitor to 3 V and discharge it through the lamp. Note the brightness and length of flash. Doubling the battery voltage and adding another lamp in series (to avoid burning out) as in circuit (b) produces brighter, longer flashes than in circuit (a). If the capacitor is charged to 6 V and discharged through four lamps, as in circuit (c), each will flash with the same brightness as the single lamp in circuit (a), showing that four times the energy is stored by doubling the voltage. At 9 V the capacitor will light three parallel banks of three lamps in series; nine times as much energy is stored when the voltage is trebled.

DEMONSTRATION

B19 Energy proportional to V^2 (heating a coil)

Students should calculate that a 10 000 μF capacitor stores 4.5 joules at 30 volts, enough to warm a small object by a measurable amount. The capacitor is discharged through a small bundle of wire and temperature changes are indicated by a thermocouple. The galvanometer rise after one discharge can be plotted against V or against V^2. A quick economical demonstration is to show four discharges in quick succession at, say, 10 V, giving the same temperature rise as one discharge at 20 V.

ITEM NO.	ITEM
1034	large electrolytic capacitor, 10 000 μF, 30 V
59	l.t. variable voltage supply
1064	smoothing unit
1501	3 m of insulated constantan wire, 0.4 mm diameter
1501	1 m of copper wire, 0.28 mm diameter
1101	sensitive galvanometer
77	aluminium block
1508	demonstration meter, 30 V d.c.
1000	leads

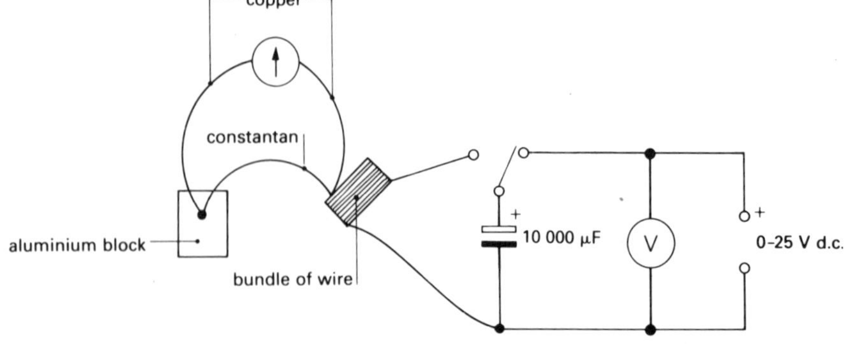

Figure B29
Heating by capacitor discharge.

Use one metre of the constantan to make up the thermocouple (twisted junctions are good enough) (figure B29).

Wind two metres of the insulated constantan wire into a ball about 10 mm in diameter, leaving a small hole for one junction of the thermocouple. The wire bundle has a resistance of about 10 Ω, giving a time constant of approximately 0.1 seconds.

The galvanometer deflection is measurable for p.d.s over 10 V and considerable for 30 V. But the temperature rises are small, and so unless the coil has been heated by several discharges in rapid succession cooling effects are not noticeable.

The smoothing unit is needed to ensure that the voltmeter reading is the same whether or not the capacitor is connected to the supply.

It might be useful to discuss some of the problems involved in obtaining valid reliable quantitative results, possibly justifying energy conservation, from the apparatus. One might raise these points:

i The need to calibrate the thermocouple, and the assumptions that lie behind the making of a scale.

ii Cooling effects, and the uniformity of the temperature rise throughout the coil of wire.

iii The heat capacity of the thermocouple junction: is it significant? – and under what conditions would it be more significant? (Refer to the loading effects of measuring instruments in general.)

iv Having obtained a reliable figure for the average temperature rise of the coil, how could the energy dissipated in the coil be calculated?

Questions 38 and *39* are about energy stored by capacitors; *question 40* invites students to discuss an analogy.

THE EXPONENTIAL DECAY OF CHARGE ON A CAPACITOR

Exponential decay is a common phenomenon in much of science and engineering. The mathematical ideas for handling rates of change (differential equations) are perhaps the most important mathematical tools that a physicist needs. For these reasons this first example is discussed at greater length than it may seem to merit. (Exponential decay will appear again in the course, especially in Unit F, 'Radioactivity and the nuclear atom'.)

In demonstration B20 the charge on a capacitor is allowed to leak away through a resistor and meter, and a decay curve is plotted from the results of the experiment. Then from a discussion of the flow of charge an algorithm is developed which enables a theoretical prediction of the decay curve to be made. Theory and experiment can then be compared. Calculations are made easier if a 500 μF capacitor discharges through a 100 kΩ resistor, giving a time constant of 50 seconds.

DEMONSTRATION

B20 Decay of charge

ITEM NO.	ITEM
59	l.t. variable voltage supply
1507	voltmeter, 10 V
1151	capacitor, 500 μF, 10 V
1040	clip component holder
1508	demonstration meter (if possible), 100 μA
1151	resistor, 100 kΩ
507	stopclock
1000	leads

A 100 μA display meter might not be available. Suitable alternatives are listed below:

Meter (0–1 mA)	$R = 5\ \text{k}\Omega$	$C = 10\,000\ \mu\text{F}$
Meter (0–5 mA)	$R = 2.2\ \text{k}\Omega$	$C = 25\,000\ \mu\text{F}$

Electrolytic capacitors have wide tolerances (typically −20 per cent to +50 per cent) and an untested capacitor of the correct nominal value can give poor agreement between experimental and predicted results. A capacitor should be selected by prior trial and error in the circuit, or by using a capacitance meter. It should then be reserved for this demonstration.

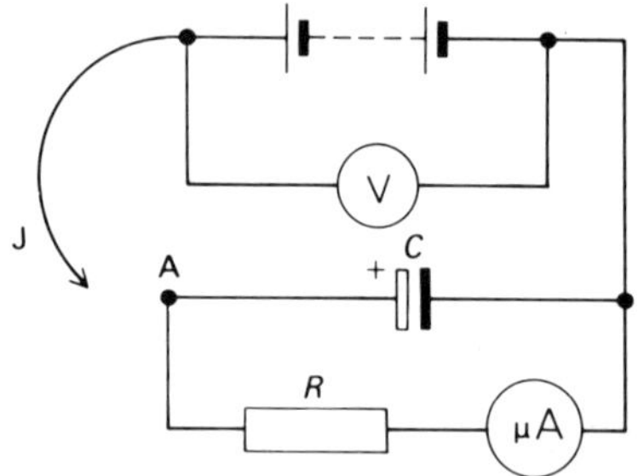

Figure B30
Circuit for investigating decay of charge.

Charge the capacitor by connecting the flying lead J to A. Start the clock and disconnect J from A at the moment the second hand passes zero. Take a current reading every ten seconds.

Alternatively a microprocessor-based data logger may be used. A class discussion might proceed as follows:

'With a 10 volt supply the meter reads 100 μA when the flying lead is connected to A. Would there by any difference if the capacitor were removed? What is the resistance *R*? When J is disconnected a current still flows – why?'

The p.d. across the capacitor and the charge left on the plates can be written down for each current reading. (The calculations are the same every time.)

$V=10^5 I$ (If $R=10^5\,\Omega$)

$Q=5\times 10^{-4} V$ (If $C=500\,\mu F$)

Graphs of current and of charge against time should be plotted using large simple scales. This will be essential later on when the theoretical curve is plotted to the same scale.

The graphs of current and of charge against time will raise the question, 'Why do they have the same shape?' Mathematically there is a constant multiplier linking the two quantities; however, some thought about the physical changes taking place in the circuit will provide a link to the theoretical argument which follows: 'Why does the current get less and less?' (Because the p.d. driving it falls, and that falls because the previous current flow removed some charge. The current falls because of the current itself.)

Explaining the decay curve

Delta notation is introduced in preparation for later developments. The Δ symbol is used because all the changes are finite, ΔQ representing the change in Q (strictly the increase) and Δt the corresponding interval of time. If the initial charge is Q_0 the p.d., V, across the resistor is

$$V=\frac{Q_0}{C}$$

The current I flowing through the resistor is:

$$I=\frac{V}{R}=\frac{Q_0}{RC}$$

During a short interval of time Δt, the current will remain fairly steady and the charge changes by ΔQ:

$$\Delta Q=-I\Delta t$$

$$\Delta Q=-\frac{Q_0}{RC}\,\Delta t$$

The negative sign indicates a discharge, Q drops as the current flows. After time Δt the new value for the charge on the capacitor, Q_1, is given by

$$Q_1=Q_0+\Delta Q$$

$$Q_1=Q_0-\frac{Q_0\,\Delta t}{RC}$$

In general after n intervals of time Δt

$$Q_n = Q_{n-1} - \frac{Q_{n-1}\,\Delta t}{RC}$$

If RC is 50 seconds, and Δt is 5 seconds, then $\Delta t/RC = 0.1$ giving

$$Q_n = 0.9\,Q_{n-1}$$

This equation tells us how Q changes and is a recipe for finding what charge remains on the plates after any time. The fact that with these values for R, C, Δt the charge is always nine-tenths of the previous charge makes the graph easy to draw (figure B31). The two graphs from theory and experiment drawn on the same scale can now be compared. The theoretical graph will run down a little too steeply as the current in each interval was assumed to stay constant at its value at the beginning of the interval. If $\Delta t = RC/10$ (so that there are 10 steps in the time RC), the curve will have fallen to 1/2.9 of its initial value instead of 1/e of this value (e = 2.718...), an error of some −10 per cent.

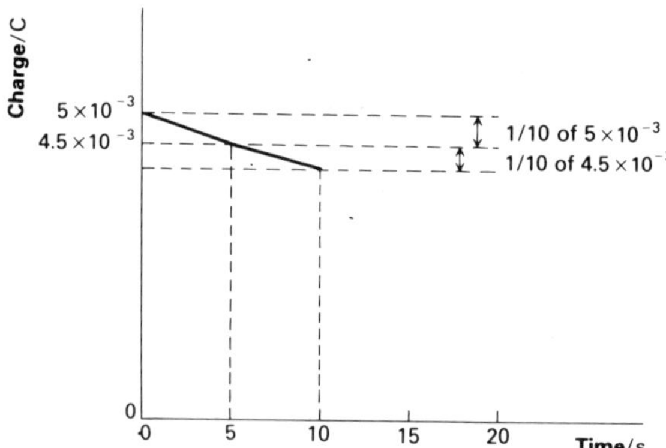

Figure B31
Decay of charge: theoretical curve.

The constant ratio property

Students used a constant ratio rule to plot the theoretical graph. After each five-second interval the charge remaining is nine-tenths of the previous charge. By looking at either curve it can readily be seen that a similar result holds for any time interval: in a given time interval the charge always falls by the same ratio. For example, it takes approximately 34 seconds for the charge to be reduced by one half. The similarity with radioactive decay met perhaps in Revised Nuffield Physics *Year 5* should be noted.

The quantity RC was shown to control the rate of discharge both in the initial class experiments on capacitors, and in this theoretical treatment of the decay. The term 'time constant' should now be introduced and the units of RC shown to be seconds

$$\frac{\mathrm{V}}{\mathrm{A}} \times \frac{\mathrm{As}}{\mathrm{V}} = \mathrm{s}$$

The ratio of charge at one moment to charge at a moment RC seconds later can now be found. Values will vary by 10 per cent or more but will be in the region of 2.7. This number, given the symbol e, should be introduced as a natural constant and given no more significance than, for example, π. The time constant gives a 'rule of thumb' measure for the time taken by a capacitor to charge or discharge, although as table B2 shows, five time constants are required before the charge remaining drops below the 1 per cent level.

Time	**0**	***RC***	***2RC***	***3RC***	***4RC***	***5RC***
Percentage of charge left on plates	100	37	13.5	5.0	1.8	0.7

Table B2

Logarithmic graphs

The series: 10^7, 10^6, 10^5, 10^4, 10^3 ... has the constant ratio property. However the logarithms of that series: 7, 6, 5, 4, 3, etc. decrease in equal steps. Students may be reminded about the use of logarithmic scales earlier in this Unit when displaying the range of resistivities in histogram form (question 13).

A graph of lg Q against t for either the experimental or theoretical values of demonstration B20 is a straight line. The use of logarithmic graphs to test for exponential changes (perhaps in Investigations) is worth emphasizing.

Further mathematical study of the exponential function

For many students a calculus treatment of exponential change is best postponed until radioactive decay is considered (Unit F, 'Radioactivity and the nuclear atom'). However, the solution of the differential equation may be a useful challenge to the mathematical skill of a few students. Point out that the solution in equation form

$$Q = Q_0 \, \mathrm{e}^{-t/RC}$$

is really no different from the graphical solution obtained by numerical integration. This can be shown in a few moments by using a scientific calculator to plot values of Q and t from the equation and comparing the result with the theoretical curve.

The dynamic modelling system

Discussion of the approximations made during the numerical integration (I constant for time Δt) will probably have raised the possibility of making Δt smaller and using the high speed of a computer to produce the decay curve. Many programs which simulate exponential decay are available, and all use basically the same algorithm to generate the solution.

This is a good opportunity to introduce the 'Dynamic modelling system', a general-purpose program which will find many applications in this course. The system allows the user to type in (or to call up from disk) a set of equations describing a dynamic (*i.e.*, changing) situation. The modelling system turns these equations into a new program. This program loops through the equations calculating new values for the variables each time, and tabulating, or plotting the results.

Writing the equations which describe the situation to be modelled should be a class activity. It focuses attention on the physical aspects of the situation, and setting down the equations in a logical order encourages understanding of the physical situation.

The following equations describe the decay of charge on a capacitor:

$$V = Q/C$$

$$I = V/R$$

$$DQ = -I * DT$$

$$Q = Q + DQ$$

$$T = T + DT$$

The computer must be given the values of any constants (R, C, DT in this example), and the initial values of any variables which have to be specified to start the calculation (here V and T).

Various graphs (*e.g.*, current–time, charge–time) can be obtained. It is easy to investigate the effect of changing, for example R or C, or the time increment DT. Progressively more accurate values of e could be determined by decreasing DT.

For more detailed information, see the booklet *Dynamic modelling system.*

Questions

Questions 41 to 44 deal with exponential changes.

ELECTRONS

The particulate nature of electric charge has been implicit in the earlier work of this Unit (*e.g.*, in deriving $I = AvnQ$) and students have necessarily had to take it on trust. The free use without justification of the word 'electron' will probably also have arisen. The last part of the Unit explores these ideas in more detail, qualitatively using electron beam demonstrations and quantitatively by the Millikan experiment using charged oil drops.

This work continues the aspect of a physicist's thought started in Unit A, 'Materials and mechanics' and raised at the start of this Unit: 'What does the evidence suggest about what can't be seen directly?' Asking the question can introduce a discussion of the evidence for electrons which should include a demonstration of the properties of electron beams. This may be revision for students who have covered it in an earlier course: in this case students themselves could present the demonstrations.

DEMONSTRATION
B21 Electron streams

'EM NO.	ITEM
27	transformer
14	e.h.t. power supply
140	stand for tubes
136	Maltese cross tube
137	Perrin tube
138	deflection tube
139	pair of coils
51A/B	gold leaf electroscope
15	h.t. power supply
176	battery, 12 V
61	fine beam tube
62	fine beam tube base
92B	Magnadur magnet
541/1	rheostat, 10–15 Ω
1508	demonstration meter, 100 mA d.c.

Follow manufacturer's instructions for the detailed setting up of the tubes. If an old valve (*e.g.*, pentode EF80 or triode) is available, it is also possible to show the heating of the anode as fast electrons strike it. This shows directly the energy transfer in the beam. The anode potential is raised as high as possible: increasing the grid potential to +50 V causes the anode to glow. This demonstration will ruin the valve.

Safety note: Since the h.t. supply may provide up to 60 mA at 300 V it is recommended that leads with *shrouded* 4 mm plugs (*e.g.*, RS Components stock no. 489–037) be used.

A selection of the following demonstrations may be made:

—stream of electrons making a shadow of an obstruction (Maltese cross);
—stream of electrons coming through a slit to make a splash across a screen;
—deflection of stream of electrons by electric fields;
—fine beam tube, raising and lowering gun voltage;
—deflection of beam (fine beam tube) by electric field;
—deflection of beam (fine beam tube) by magnetic field;
—electrons collected to determine the sign of their charge (Perrin tube).

The demonstrations can use tubes in which electron streams strike an obstacle, make a streak, are bent by fields, and are shown to carry negative charge. The extent to which the experiments do *not* give evidence for some points is worth attention, and a sceptical attitude towards the interpretations offered is to be encouraged. Some points worth noting are:

i None of the experiments points directly to particles of charge.

ii There is evidence for the negative sign of the charge carried by the electron stream.

iii The stream bends in electric and magnetic fields, and the question, 'Why does it bend just the amount it does?', can lead to discussion suggesting that the stream may carry charge and mass.

If necessary, the construction of the cathode ray tube should be explained at this stage.

Later uses of magnetic deflection
In Unit F, 'Radioactivity and the nuclear atom' the deflection of beta radiation by a magnetic field will be used as evidence that the radiation carries negative charge. The deflection of a charged particle moving in a magnetic field is treated formally in Unit H, 'Magnetic fields and a.c.'.

THE ELEMENTARY UNIT OF CHARGE

Historically the idea that electric charge is carried in discrete lumps (the term 'quanta' could be introduced here, with its meaning of discreteness – 'quantum of charge') came from the chemical evidence of electrolysis. The amount of charge passing in a circuit which deposits one mole of atoms of an element at an electrode is either 96 500 C, or exactly two or three times this value. Since one mole is 6×10^{23} atoms, one might imagine that the current is carried in the electrolyte on *ions* which carry a charge of e, $2e$, or $3e$ where

$$e = 96\,500/(6 \times 10^{23}) = 1.6 \times 10^{-19}\,\mathrm{C}$$

This value was well known in the early days of atomic physics. The importance of Millikan's experiments is that they provide *independent* evidence confirming both the discrete nature of charge and the numerical value of the unit.

A modern version of the Millikan experiment using oil drops should be introduced here. It is a lengthy experiment and it is not suggested that all students should do it. A small group with a flair for patient detailed investigation could perform it and present their results to the rest of the class. However everyone should see the apparatus, the outline procedure being 'talked through' and the principle understood.

The film 'Are there electrons?' assumes that students have seen the apparatus and understood the point of the experiment. It shows first the evidence for the 'grainy' nature of the charge and the last part offers a calculation of the charge on an electron from the data obtained.

Question

The theory is developed in question 48.

OPTIONAL EXPERIMENT
B22 The Millikan experiment (charge on an electron)

ITEM NO.	ITEM	
1043	Millikan apparatus	
	either	
14	e.h.t. power supply	according to manufacture
	or	
15	h.t. power supply	
1005	multirange meter	
1000	leads	

Consult the manufacturer's instructions for precise details of adjustment, appropriate power supply, and method of introducing the droplets and ensuring that they are charged. To avoid the complications of a theoretical argument involving viscous forces on falling drops, a graph is shown (*Students' guide*, figure B55) of the weight of a drop against the time taken to fall freely a standard distance (1 mm).

Safety note: Since the h.t. supply may provide up to 60 mA at 300 V connections should be made with leads having *shrouded* 4 mm plugs (*e.g.*, RS Components stock no. 489–037).

An energy argument is used to establish that the electrical force on a drop carrying a charge Q is QV/d, where V is the p.d. needed to hold the drop stationary (*Students' guide*, figure B28). This force is equated to the weight of the drop, obtained from the graph, enabling Q to be calculated.

ENERGY AND SPEED OF THE ELECTRONS

The speed which an electron, charge e, acquires while it is being accelerated through a p.d. V can be calculated directly if it is assumed that the electron is a particle to which the large-scale laws of mechanics apply. The electrical work done is equal to the gain in kinetic energy, and if the initial speed is effectively zero,

$$eV = \tfrac{1}{2}mv^2$$

This result is used in Unit H, 'Magnetic fields and a.c.' Teachers could point out that in using this energy equation it is not necessary to know separately the values of e and m but only the specific charge e/m. Students could be told that this ratio ($1.78 \times 10^{11}\ \mathrm{C\,kg^{-1}}$) can be found from beam deflection experiments such as those in demonstration B21 and that the mass of the electron can then be found by combining the values of e/m and e. The details of the determination of e/m are left until Unit H, 'Magnetic fields and a.c.'

The *electronvolt* as a convenient 'atomic-sized' unit of energy should be introduced here as the energy acquired by one electronic charge (electron, proton, or any singly-charged ion) when accelerated through a p.d. of 1 V. The equivalence $1\ \mathrm{eV} = 1.60 \times 10^{-19}\ \mathrm{J}$ should be brought out explicitly.

Question 46 gives practice in calculating electron velocities for given energies. Able students could be invited to find the velocity for, say, 300 kV accelerating voltage and to comment on the result. (The simple theory here gives a velocity greater than that of light.)

It is not easy to measure electron speed directly (this would involve timing their flight across a cathode ray tube) but it is possible to show that a beam of electrons carries *momentum*, from which the velocity of the electrons can be found (assuming constant mass). The relationship between velocity and accelerating p.d. can then be investigated, confirming that $v \propto \sqrt{V}$.

Such an experiment, in which bursts of electrons strike a suspended vane and deflect it, is shown in the film 'The momentum of electrons' (although it is now difficult to obtain). The film is worth seeing, but in any event calculations can be done using data drawn from it.

Figure B32 shows the arrangement of electron gun and vane. The gun bursts are fired so as to hit the vane each time it swings away from the gun, building up the amplitude of swing. The increase in angle of swing per burst is used as a measure of the impulse delivered by a burst of electrons (and hence the momentum transferred by the beam).

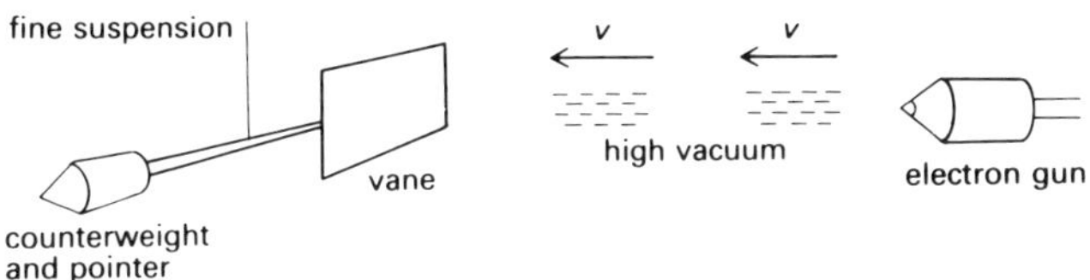

Figure B32
Electron gun and vane used in film 'The momentum of electrons'.

	Increase of swing in each oscillation/ radians per oscillation	**Gun voltage/V**	**Beam current/μA**
1	1.5×10^{-3}	2500	200
2	0.86×10^{-3}	2500	100
3	0.67×10^{-3}	1250	120

Table B3
Data from the film 'The momentum of electrons'.

Results of such an experiment are shown in table B3. Since force is rate of change of momentum the value in the first column is a measure of the force exerted by the beam. Comparison of readings **1** and **2** confirms that if the beam current is halved at the same accelerating voltage (half the number of electrons per burst at the same speed) then the force exerted is approximately halved.

This preliminary result enables the force exerted in reading **3** to be converted to the value to be expected if the beam current were 100 μA (multiply by 100/120). If reading **2** is compared with this revised force, the forces are seen to be in the ratio 1.5:1. Since the energy per electron for the last reading is half that for reading **2** (the p.d. is halved) the ratio for the electron velocities (and hence momentum transferred, since equal numbers of electrons are now being compared) is expected to be $\sqrt{2:1}$:1, *i.e.* 1.41:1.

Teachers will use their judgement as to how much of the detailed numerical work to present here: if time allows, it provides a useful revision of the work on force and momentum started in Unit A, 'Materials and mechanics'.

Questions

Questions 45 to 48 deal with electrons, *e*, the electronvolt, the Millikan experiment.

Unit C
DIGITAL ELECTRONIC SYSTEMS

Mark Ellse
Emanuel School, London

David Grace
Eaglesfield School, London

Suggested time allocation: three weeks

PLAN OF THE UNIT

Section C1
Combinational logic

Systems

Digital signals

The inverter

The NOR gate

The NAND gate

Making more gates

► non-inverting gate as a buffer: Unit I, ‘Linear electronics, feedback and control’

Logic functions on a microcomputer

Prerequisite		Topic		Leads to
		Transfer characteristics		
potential divider: Section B2	►	**Using the transfer characteristics**	►	designing digital systems: Section C3
binary arithmetic	►	**Half and full adder**	►	designing digital systems: Section C3
		Section C2 Sequential logic		
RC circuits: Section B3	►	**Making pulses**		
		The astable circuit	►	designing digital systems: Section C3
		The bistable circuit; memory	►	feedback: Section I2
		Section C3 Designing digital systems		
		A selection of practical challenges		

INTRODUCTION

The importance of including at least some electronics in a sixth form physics course is recognized by the core syllabus for physics at A-level. We include electronics for two reasons:

1 It will be useful in the future. Students will find themselves using electronics in many courses in further education, and in a wide variety of future careers, for instance medicine and engineering, as well as in pure science. Electronics already plays an important part in our lives; this part is increasing rapidly.

2 It gives an opportunity for a new style of work, reflecting the tastes and problems of the engineer rather than those of the physicist. In much of the course students analyse things, taking them apart in the manner of the pure scientist. Here we ask them to synthesize, to put things together to serve a purpose, in the manner of the engineer. We hope that they can be 'engineers for a day'.

This Unit is concerned with digital electronics, an increasingly important part of the subject, using for the most part simple gates and combinations of gates. Later in the course Unit I, 'Linear electronics, feedback and control' focuses on a variety of linear circuits built around an operational amplifier.

After an introduction to the notion of a system of digital signals, students examine the behaviour of a few simple gates (inverter, NOR, NAND) after which they are challenged to use these to build up other gates such as OR, AND, Exclusive OR, three-input gates, and so on. The transfer characteristics of a gate are established, after which it can be used, together with the potential divider, to make, for example, a light-operated switch.

Students are expected to know enough about binary arithmetic to follow the construction of a half and full adder.

Knowledge of *RC* circuits from Unit B, 'Currents, circuits, and charge' is needed in the explanation of pulse-producing circuits. A circuit which produces a single pulse leads on to the astable and bistable circuits, and the use of the bistable circuit as a memory unit.

Throughout the Unit we have tried to point out the ways in which gates and logic circuits find application, for example in microcomputers and in communication.

The final section of the Unit is in a sense the most important. It is here that students are challenged to use what they have learned earlier to solve a selection of practical problems. While some should be simple enough to allow beginners in electronics to achieve success, we hope that all students will find some that tax their ingenuity.

In electronics perhaps more than any other part of the course, students (and teachers) will have widely differing backgrounds and previous experience. We believe that what is presented here and in Unit I, 'Linear electronics, feedback and control' adequately covers the essential ideas that all students should know. The teaching of electronics in school is developing rapidly, and it is certainly possible that some or even much of the material in this Unit will be familiar to some students by the time they reach the sixth form. Teachers who find themselves in this situation will have to decide whether to extend the teaching beyond what is suggested here, or to take advantage of any time that can be saved for other parts of the course.

THE PLACE OF THE UNIT IN THE COURSE

It is suggested that this Unit should come early in the course. It uses ideas from Unit B, 'Currents, circuits, and charge', especially the potential divider and the behaviour of *RC* circuits, and provides a good opportunity to use these ideas in a new context. It also requires students to use the voltmeter and oscilloscope to monitor the output of circuits, and the experience of using two- and especially four-terminal boxes should be useful preparation. But if teachers bear these points in mind and adjust the pace of the teaching accordingly, it would no doubt be possible to teach this Unit first.

LIST OF SUGGESTED EXPERIMENTS AND DEMONSTRATIONS

SECTION C1
COMBINATIONAL LOGIC

A SYSTEMS APPROACH

The designers of electronic circuits and the inventors of electronic devices need more than we shall offer in this course. Those needs must be met later in more specialized courses. We do not approach electronics through the physics of electrons in solids, not because this is uninteresting, but because it would leave less time to show electronics in action. We believe that the user of electronics has a greater need to grasp what electronics can do than to understand the working of particular devices. People buying video and audio equipment or home computers, engineers monitoring the vibrations in a tower or the flow of chemicals in a pipe line, chemists following reactions electrically, physicists counting particles from an accelerator, secretaries choosing a word processor, and many others all need to be able to choose systems that will do the necessary job if combined in the right way. Designing the working parts is a different, more specialized profession. The components available to the engineer have changed rapidly in the past and may continue to change in the future. What changes less quickly is the range of jobs for which electronic devices are useful. Electronics will go on being about combining switches, gates, counters, registers, amplifiers, pulse formers, and oscillators to perform desired functions. In so far as electronics affects and will affect the daily lives of men and women, they will need to understand the system with which they are dealing rather than the working parts of the system. When the central heating fails to come on, finding the fault is a matter of knowing how the system works, of understanding the function of the parts in the whole. Computerized bank accounts, automatic telephone exchanges, and thermostatically controlled electric ovens are all systems; to 'understand' them is primarily to know how they work as systems.

Modern electronic systems are either digital or linear or a combination of both. Industrial control systems are a good example of this mixture: a microcomputer might monitor signals from an industrial process. If these signals are continuously variable, they may need to be changed into digital form so that the computer can handle them, controlling the process on the basis of the data obtained, and storing for future use the data collected and computed. A process may need linear information to control it, as may the flow of a fluid in a pipe. In this case the digital output from the computer has to be changed into linear form. (*Note*: The word 'linear' is used in electronics to describe circuits that handle information that is not digital, that is, information that is continuously variable.) Another example is the transmission of analogue signals without degradation by noise. For instance, an audio or video signal picks up electrical noise while being transmitted. However, if the signal is converted to a series of numbers representing its amplitude at successive intervals of time, these numbers can be transmitted as digital signals, usually in binary form as a series of on/off pulses representing 1 and 0. These are then converted back to analogue form at the far end. Provided that the noise level is not so high that '0's and '1's cannot be distinguished from each other, the transmission will be error free and the signal will be faithfully reproduced. This technique is known as pulse-code modulation. It is particularly useful when signals have to be transmitted over very long distances, such as in transcontinental telephone systems, where the signal has to pass through a series of regenerators. The regenerators are situated at intervals along the cable to amplify the signal so that '0's and '1's can always be clearly distinguished. Both the examples above illustrate the need for digital and linear circuits as well as circuits that can change signals from one form to another.

Electronics appears twice in this course. In this Unit we concentrate on digital circuits; Unit I, 'Linear electronics, feedback and control' deals with linear circuits, and includes mention of the conversion of digital signals to linear (analogue) form.

We hope that the electronics will provide a useful tool throughout the course, and that the experience with relatively simple systems in this Unit will make it possible for students to understand, in principle at least, how some of the more complicated tools they meet later (scalers, timers, frequency meters, etc.) work.

In this first electronics Unit we shall see how simple logic gates can be used to build up more complex systems: digital systems which can count, perform mathematical and logical operations; circuits which can remember and control.

Apparatus

A wide range of apparatus is available for teaching digital electronics. In preparing this Unit, we have tried to find an approach that will work with most types. We expect that some teachers will continue to use single transistor 'basic units', perhaps alongside more sophisticated systems based on integrated circuits. Others will be using kits based on integrated circuits entirely. We have endeavoured to supply enough information for the most commonly available apparatus, but the diversity already present means that both students and teachers will need to become used to dealing with a variety of equipment.

Electronic systems

Students will already be familiar with many electronic systems. Most will have used calculators and microcomputers. The *Students' guide* examines two complicated systems: a hi-fi system and a digital clock. The aim is to show how an apparently complex device can be represented by relatively simple diagrams: the block diagram simplifies the complexity of a detailed circuit diagram without losing the essential nature of the system.

Often the block diagram is more useful than the circuit diagram. In the electronics industry this has become increasingly true with the growing use of integrated circuits. New devices and circuit techniques are continually being developed; but all fit in similar ways into similar systems – the block diagram remains unchanged.

Other systems may be discussed to illustrate the 'state of the art' of modern electronics. Readily available may be:

Calculators – A cheap calculator can easily be opened and the circuit revealed. It is probably advisable to retain a broken calculator for this purpose.

Laboratory instruments – Consult the manufacturer's notes for details.

Microcomputers – With care the circuit boards are accessible for visual inspection.

These examples should illustrate the use of integrated circuits in modern devices; even though other components will be seen.

Questions

Questions 1 to 5 deal with systems.

Electronic systems have a number of common features. They all have inputs (a keyboard on a calculator, the internal oscillator in a watch); outputs (the display on a calculator or watch); some form of processing system; and a power supply. Students should be able to identify these parts of the systems which they examine and compare their relative sizes, perhaps making an estimate of the cost. (Some idea of the relative costs of, say, a keyboard, a calculator display, and a complex integrated circuit can be found in a catalogue, for example, the one produced by RS Components.) This is a good time at which to look at the inside of an integrated circuit. With careful use of a hammer and a fine chisel an integrated circuit such as the 4011 can be cracked open and examined with a low-power microscope (around $\times 30$).

DIGITAL SIGNALS

We start to examine digital electronics by establishing what we mean by a digital signal. Digital circuits respond to, process, or produce such signals. We must first show examples.

If students are at all familiar with 'signals' it will probably be with oscilloscope traces of alternating waveforms and, maybe, voice 'prints' from a microphone connected to the oscilloscope. These are examples of 'linear' or 'analogue' signals. The first demonstration introduces a different concept: that of a signal that has only two values or levels.

The list below gives some sources of digital signals. The two levels present can represent a variety of information: whether a switch is open or closed, whether a signal is present or not, whether a signal is above or below a set level, whether an event has occurred or not, and so on.

We think that all students should see the first three parts of demonstration C1 and perhaps one other example. Many teachers will, of course, wish to add to the list. Circuit details are not important at this stage and the students' laboratory notes do not show them. The students' attention should be focused on the fact that each circuit has only two states. These can be represented by the binary digits 0 and 1.

DEMONSTRATION

C1 Digital signals

C1a Simple circuit having two states

ITEM NO.	ITEM
52L	mounted bell push
92R	lamp, *e.g.*, 2.5 V m.e.s.
92T	holder
1033	cell holder with two cells
1000	leads

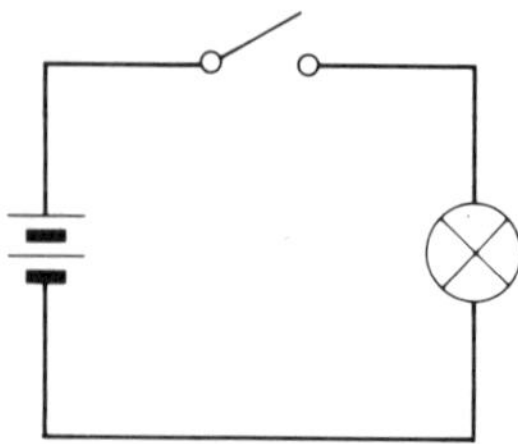

Figure C1
A simple circuit having two states.

The lamp will either be off (low = 0) or on (high = 1).

C1b Slow astable circuit controlling a lamp

ITEM NO.	ITEM
1514	digital electronics kit with power supply
1000	leads

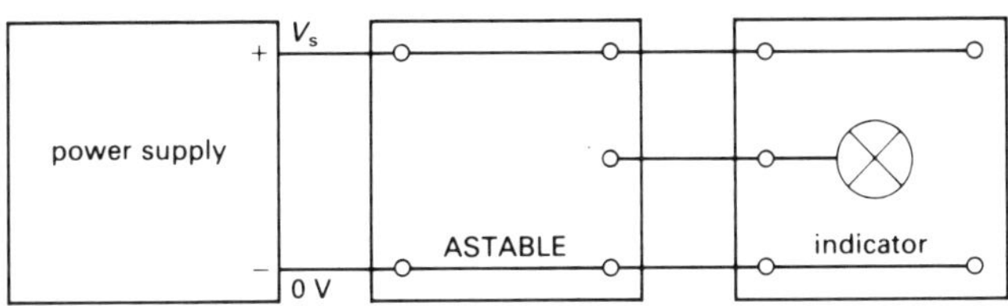

Figure C2
A slow astable circuit controlling a lamp.

Adjust the astable module so that the lamp can be clearly seen to be in one of two states: either on (high) or off (low).

C1c Fast astable circuit

ITEM NO.	ITEM
1514	digital electronics kit with power supply
1511	oscilloscope
1000	leads

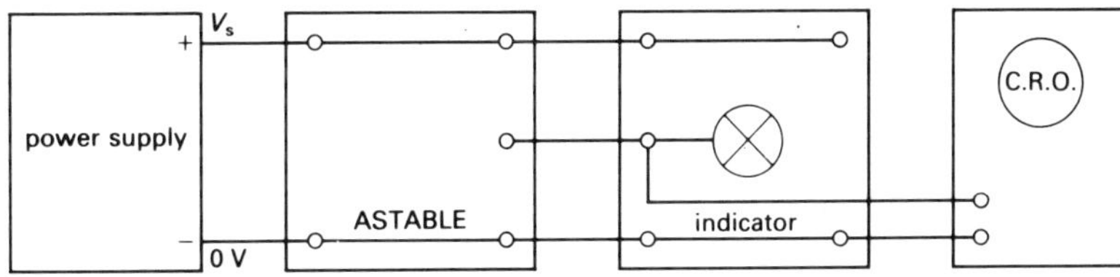

Figure C3
A fast astable circuit controlling a lamp.

Adjust the frequency of the astable circuit so that the lamp cannot be seen to be flashing. It appears to have a brightness somewhere between off and on. The C.R.O. shows that this is because the astable circuit alternates rapidly between its two states.

C1d Other sources of digital signals

Some of the following may be available:
i A microcomputer running a program has digital signals at many parts of the circuit. Enter a program into the computer having first

exposed the main circuit boards. Identify that part of the circuit you wish to monitor. Connect the earth lead from the oscilloscope to the negative supply, 0 V, or earth on the board. Set the time-base to $10\,\mu s\,cm^{-1}$. Run the program and carefully use a fine-tipped probe connected to the input of the oscilloscope to explore either at the legs of an integrated circuit or at a conductor, being careful to touch only one conductor at a time. (Some oscilloscope probes have a special tip for use with integrated circuits.)

ii Most school scaler/timers have sockets at which their internal clock signals are available. Consult the manufacturer's manual.

iii A mechanical telephone dial has a number of switch contacts. The most useful for this demonstration is a pair that is normally closed but is opened by a cam a number of times depending on the number dialled. This may be identified by examining the mechanism at the rear of the dial. This is connected in place of the switch in experiment C1a above. Some manufacturers sell telephone dials ready mounted with connections brought out to sockets.

The nervous system is an example of a digital system and is discussed briefly in the *Students' guide*. Teachers may be interested in the following extract from *The conduction of the nervous impulse* by A. L. Hodgkin, Liverpool University Press, 1964.

'The nervous impulse in one fibre is of constant amplitude and shape and its characteristics cannot be altered by changing the strength or quality of the stimulus. The intensity of a sensation or a movement is controlled by varying the frequency of impulses and the number of fibres in action. In other words, in response to external stimuli nerve fibres produce pulses. A pulse is either there or it isn't. A stronger stimulus produces more pulses. The nerve fibres appear to be biological analogue-to-digital converters.'

The *Students' guide* relates light intensity to electrical impulses in the fibre from a 'rod' on the retina. In fact the smallest amount of light that will produce one pulse is one photon. But as pointed out in the *Students' guide*, a single impulse does not produce the sensation of seeing.

Further reading

HODGKIN *The conduction of the nervous impulse.*
ADRIAN *The nerve impulse.*
The article 'Digital sound' in the *Students' guide* deals with one recent application of digital techniques.

Questions

Questions 6 and 7 examine digital signals.

INTRODUCING THE DIGITAL ELECTRONICS KIT

We expect that a range of digital electronics kits will be available in schools, and probably more than one type of kit in any one school. The picture is likely to get more complicated as time progresses, and as teachers and manufacturers find better ways of doing the same job. At present most of the available apparatus falls into one of three categories. Some information about the three categories is given here.

These most common types are based on

a discrete components, for example, the single transistor 'basic unit'

b TTL (Transistor Transistor Logic) integrated circuits – the 7400 series

c CMOS (Complementary Metal Oxide Semiconductor) integrated circuits – the 4000 series.

Their important characteristics are summarized in table C1.

	Supply voltage (V_s)	**Logic level low (0)**	**Logic level high (1)**
basic unit	+5 to +6 V	less than 0.5 V	more than 2 V
TTL	+4.75 to +5.25 V	less than 0.8 V	more than 2 V
CMOS	+3 to +15 V	less than 0.2 V_s	more than 0.8 V_s

Table C1

Question 19 aims to make students familiar with tables such as these and contains this table with some further information.

References

For more information see, for example:

LANCASTER *Transistor-Transistor Logic cookbook.*

LANCASTER *Complementary Metal-Oxide Semiconductor cookbook.*

Texas Instruments *Designing with TTL integrated circuits.*

MULLARD *CMOS digital integrated circuits.*

From the teaching point of view, the similarities are more important than the differences. All three systems use a supply polarity that is positive with respect to the 'zero volts' line (0 V) and are 'guaranteed' to work on +5 V. In practice, most will work from three fresh 1.5 V cells in series. The supply voltage that is used will depend on the power source and the system used. We shall in this Unit refer to the potential difference of the power supply as the supply voltage (V_s).

For all three systems an input or output that is near 0 V corresponds to binary digit 0, and an input that is near V_s to binary digit 1.

Electronics kits are available based on both the NOR gate and the NAND gate. We think it important that students become acquainted with both types of gate. However, most schools will already have large numbers of NOR gates in the form of the basic unit and for this reason, as well as the fact that the truth table for NOR is perhaps slightly more obvious, we assume that most teachers will want to approach the other gates from the NOR gate. (See Appendix I for advice on using the basic unit kit.)

The following four experiments can follow rapidly after one another, taking rather less than a double period altogether.

EXPERIMENT
C2 Introducing the digital electronics kit

ITEM NO.	ITEM
1514	digital electronics kit with power supply
1507	voltmeter appropriate to the power supply
1000	leads

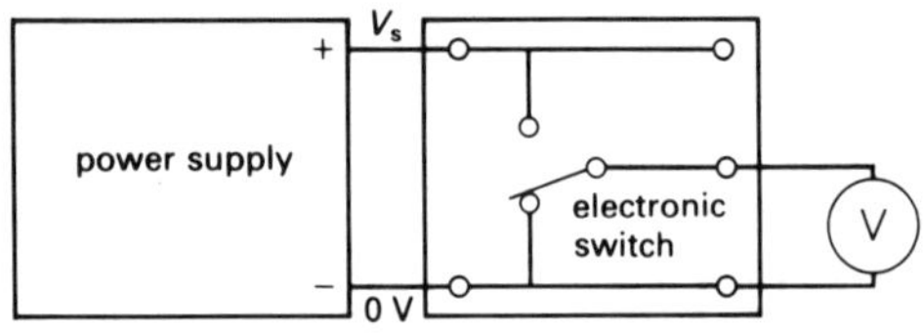

Figure C4
Introducing the digital electronics kit.

Students should be able to determine either from the apparatus itself or from the manufacturer's instructions the supply voltage required and connect it up. The input to this circuit is a 'press' applied to the button of the switch, and the output is an electrical signal going to the voltmeter (figure C4). Again the students' attention should be drawn to the fact that the output as well as the input has only two states and that the output represents these by a potential that is near 0 V for 'logic level 0' (or 'logic 0') and near V_s for 'logic level 1' (or 'logic 1').

The voltmeter is an unnecessarily complicated piece of equipment to use to indicate the state of the output – an indicator lamp or light emitting diode (L.E.D.) would be adequate. If a separate indicator is available in the electronics kit, students should try this instead of the voltmeter.

EXPERIMENT
C3 Investigating a single-input gate

ITEM NO.	ITEM
1514	digital electronics kit with power supply
1000	leads

The two inputs to the gate are connected together so that both may either be made low (0 V) or high (V_s). The indicator monitors the output. Students are asked to produce a table to describe the behaviour of the gate, and from this it is easy to introduce the concept of a 'truth table' describing the behaviour of any gate.

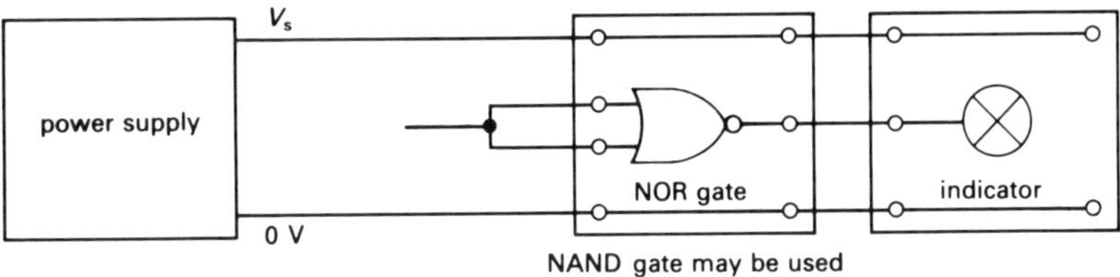

Figure C5
Investigating a single-input gate.

input	output
low	high
high	low

input	output
0	1
1	0

Figure C6

The gate in figure C5 is called an inverter or NOT gate (because the output is NOT the input).

At this stage, students should investigate and remember for future use what happens when the input is not connected either to high or low. On some systems the inputs will drift low when not connected (basic unit), on other systems it will drift high (TTL). Kits based on CMOS circuits usually incorporate high-value resistors to 'tie' the inputs either high or low.

EXPERIMENT
C4 Investigating a NOR gate

ITEM NO.	ITEM
1514	digital electronics kit with power supply
1000	leads

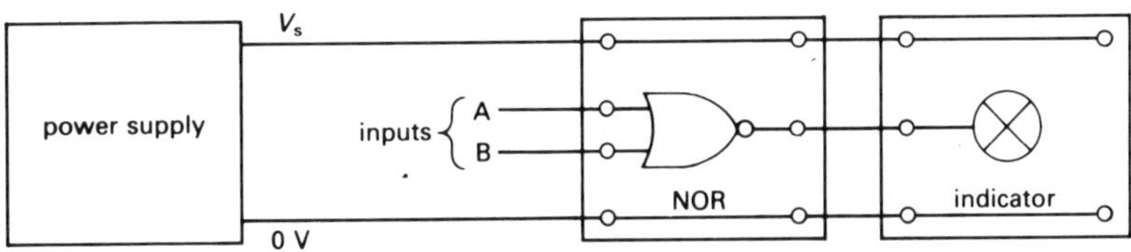

Figure C7
Investigating a NOR gate.

Most teachers will probably want students to investigate the behaviour of the gate and let them establish the truth table for themselves. Some may wish to demonstrate it first. Each input, A and B, can be either high or low. There are four possible combinations for two inputs and the output corresponding to each of these is given in the truth table in figure C8.

inputs		
A	B	output
0	0	1
0	1	0
1	0	0
1	1	0

Figure C8
Truth table for NOR gate.

The output is high when neither input A, NOR input B, NOR both are high.

So far, we have included in the circuit diagrams information that is common to all electrical circuits – the power supply connections. An alternative representation is shown in figure C9, where the power supply leads and connection of lamp to 0 V are not shown. This makes circuit diagrams simpler to draw and to follow. From now on we shall use this method exclusively, including power supply lines only when necessary.

Figure C9
An alternative representation of figure C7.

The symbols used for the gates (shown on pages 165 and 166 of the *Students' guide*) are standard, and students should be able to recognize them and use them with ease.

EXPERIMENT
C5 The behaviour of the NAND gate

ITEM NO.	ITEM
1514	digital electronics kit with power supply
1000	leads

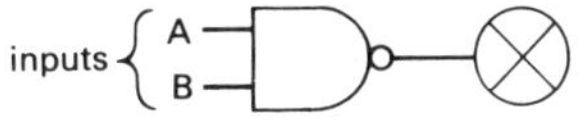

inputs		
A	B	output
0	0	1
0	1	1
1	0	1
1	1	0

Figure C10
Investigating the NAND gate.

This experiment follows the same pattern as the previous one, and should take only a few minutes.

MAKING MORE GATES

We tackle this by setting students carefully specified problems. They should know where they are going and that there may be more than one way of getting there. All the problems can be solved using either NOR or NAND gates although some choice is indicated in case only one of these gates is available. Some problems are easier with NOR, some with NAND.

Questions

Questions 8 to 11 pose similar problems to those that students are asked to tackle practically in the next experiment.

Few scientists or engineers work in isolation. Ideas and solutions to problems grow out of discussion with colleagues. We should encourage students to discuss the problems that they meet. All students should tackle parts **a** to **d** of experiment C6; parts **e** and **f** may be given to faster groups. The time taken should be about a double period.

Solving these problems requires students to produce designs. We are giving them a chance to tackle a problem in the same way as an engineer might. To begin with, students may be unsure of what they are doing and some will need more guidance than others. They will learn by doing. We can help by getting them to look at the problem from different angles. There are some hints in the students' laboratory notes. More may be needed. Encourage students to think of solutions using both NOR and NAND gates, and to produce their solutions on paper before trying them out.

EXPERIMENT
C6 Designing more gates

ITEM NO.	ITEM
1514	digital electronics kit with power supply
1000	leads

Some tasks and suggested solutions are given below. Remind students that NOR and NAND gates can be used as inverters.

C6a Design a two-input AND gate.

Using NAND gates the solution is quite simple (figure C11).

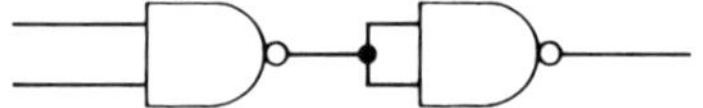

Figure C11
A two-input AND gate made from NAND gates.

Using NOR gates is a little harder (figure C12).

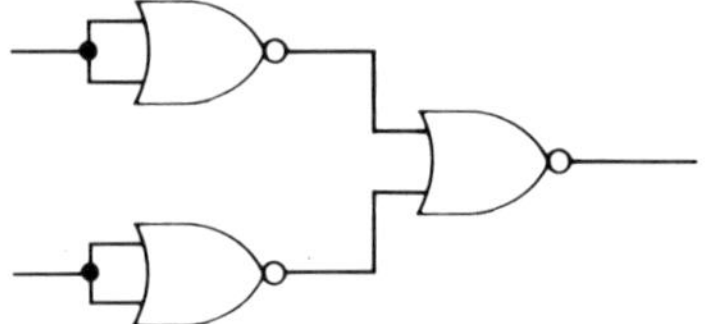

Figure C12
A two-input AND gate made from NOR gates.

The electronics kit may contain an AND gate. Students should check that it behaves in the same way as their circuit.

C6b Design a two-input OR gate.

This is easier using NOR gates (figure C13).

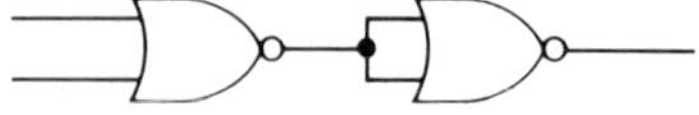

Figure C13
A two-input OR gate made from NOR gates.

But it is harder with NAND gates (figure C14).

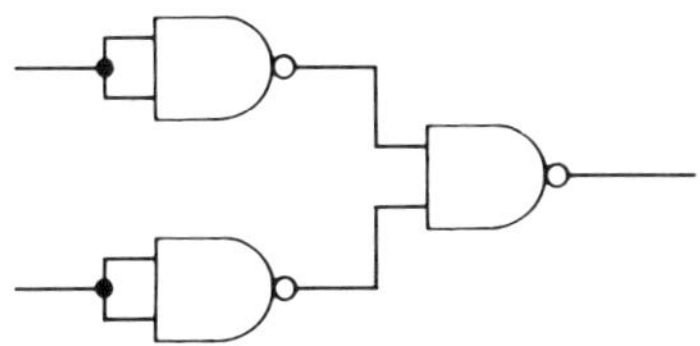

Figure C14
A two-input OR gate made from NAND gates.

C6c Design a non-inverting gate whose output is the same as the input.

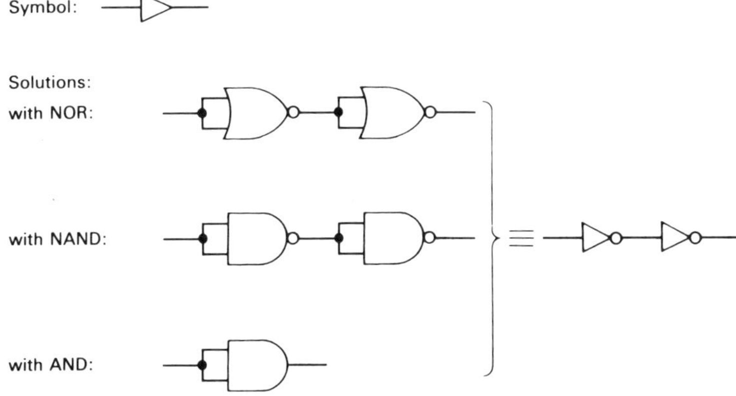

Figure C15
A non-inverting gate.

This gate is used when we want to insulate a system from changes in its load. For example, connecting an astable circuit directly to an earpiece may cause the frequency of the astable circuit to change. Interposing a 'buffer' prevents the astable circuit being affected by the earpiece.

At this stage it is worth while consolidating the information that has been acquired about six basic gates. This is summarized in figure C16. Some students may find the names of the gates difficult to understand. The name describes the input conditions that will cause the output to be high. The best example is the AND gate: calling the inputs A and B, the output is only high when input A AND input B are high.

From AND to NAND is then fairly simple. It is the opposite or inverse of the AND gate or 'NOT AND'; that is NAND for short. The output of a NOR gate is high when neither input A NOR input B are high. It is the inverse of the OR gate. The output of this gate is high when either input A OR input B is high. Note that the output of an OR gate is high when either input is high or when both are high, unlike the Exclusive OR gate.

Gate type	Symbol	Truth table		
Non-inverting gate		input		output
		0		0
		1		1
Inverter or NOT		input		output
		0		1
		1		0
OR		inputs		output
		0	0	0
		1	0	1
		0	1	1
		1	1	1
NOR		inputs		output
		0	0	1
		1	0	0
		0	1	0
		1	1	0
AND		inputs		output
		0	0	0
		1	0	0
		0	1	0
		1	1	1
NAND		inputs		output
		0	0	1
		1	0	1
		0	1	1
		1	1	0

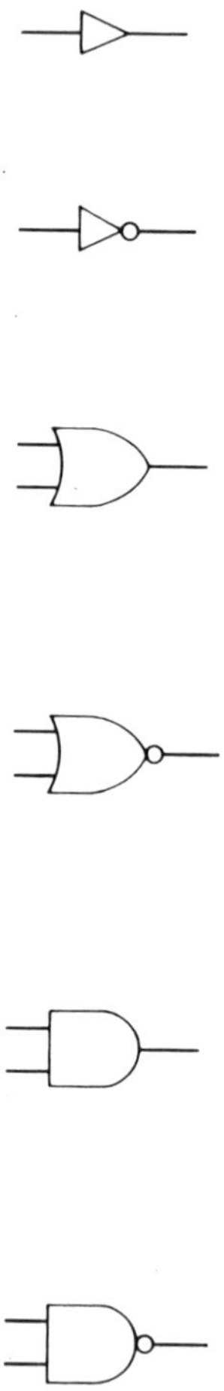

Figure C16
Symbols and truth tables for logic gates.

The inverter and inverse function gates (NOT, NOR, and NAND) are generally more useful than other gates. You can make an AND out of two NAND gates, but you cannot make a NAND out of two AND gates. Students will appreciate this point when they have attempted to solve a few problems.

C6d Design a two-input NOR gate using NAND gates.

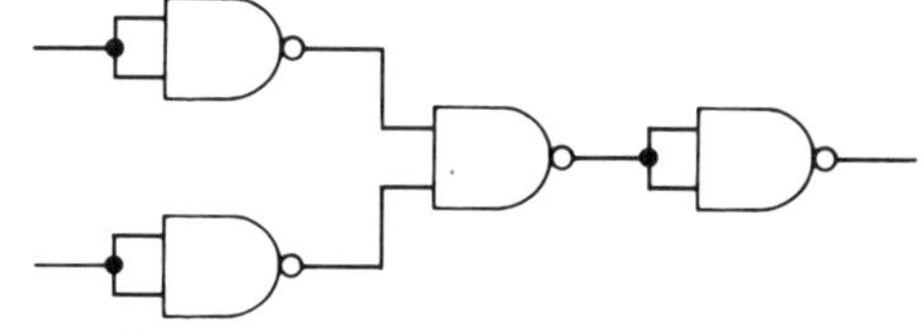

Figure C17
A two-input NOR gate made from NAND gates.

Design a two-input NAND gate using NOR gates.

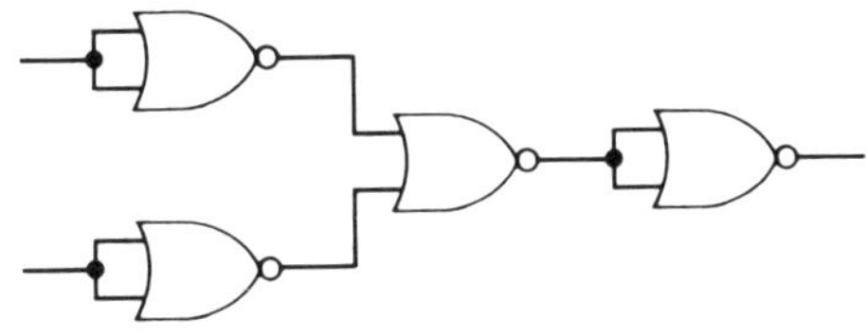

Figure C18
A two-input NAND gate made from NOR gates.

C6e Design a three-input NOR gate using two-input NOR gates. The truth table and symbol are shown in figure C19, and the solution in figure C20.

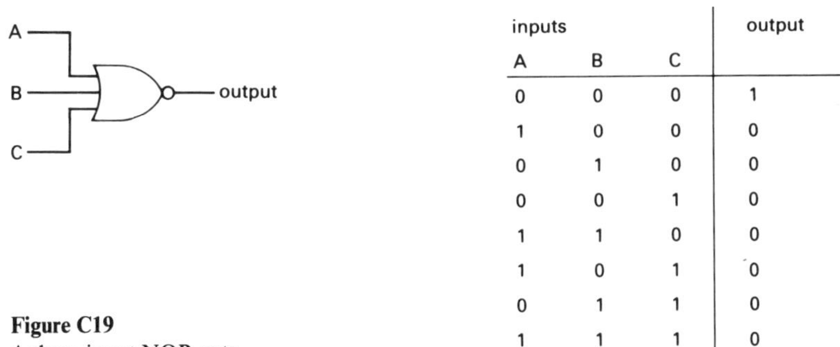

inputs			output
A	B	C	
0	0	0	1
1	0	0	0
0	1	0	0
0	0	1	0
1	1	0	0
1	0	1	0
0	1	1	0
1	1	1	0

Figure C19
A three-input NOR gate.

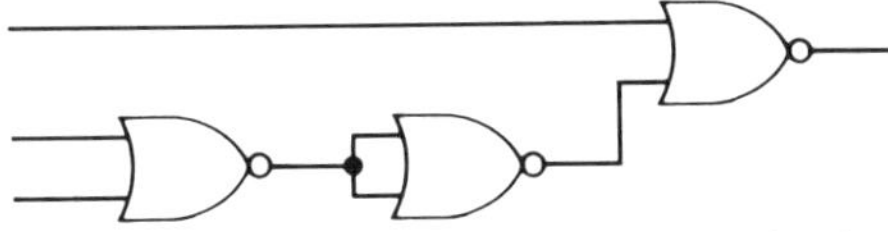

Figure C20
A three-input NOR gate made from two-input NOR gates.

Design a three-input NAND gate using two-input NAND gates. The truth table and symbol are shown in figure C21, and the solution in figure C22.

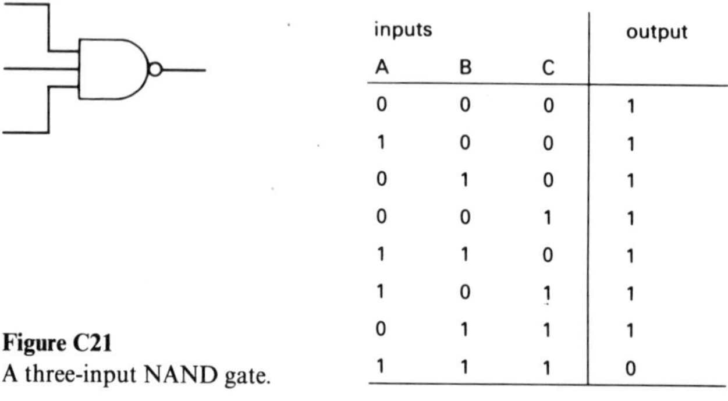

inputs			output
A	B	C	
0	0	0	1
1	0	0	1
0	1	0	1
0	0	1	1
1	1	0	1
1	0	1	1
0	1	1	1
1	1	1	0

Figure C21
A three-input NAND gate.

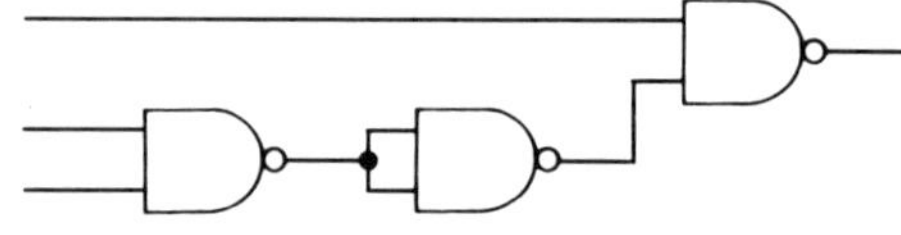

Figure C22
The design of a three-input NAND gate.

C6f Design a two-input Exclusive OR gate. This is an important circuit for the work that follows on binary adders.

A solution using just NOR or NAND gates would take several gates. By this stage students have extended their list of gates, and they should be encouraged to use them.

It may help students tackling this (and certain other problems) to examine the truth table that they are trying to achieve alongside the truth tables of other gates. (See figure C23.)

inputs		outputs					
A	B	Exclusive OR	Parity	OR	NOR	AND	NAND
0	0	0	1	0	1	0	1
0	1	1	0	1	0	0	1
1	0	1	0	1	0	0	1
1	1	0	1	1	0	1	0

Figure C23

Parity is the inverse of Exclusive OR; the output is high when the inputs are the same. The Parity gate is (NOR) OR (AND). This can be deduced from the table (figure C23). A Parity gate could be made from the circuit shown in figure C24.

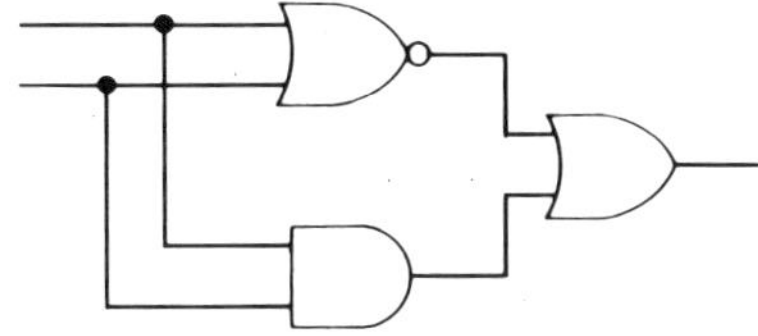

Figure C24
A Parity gate.

Exclusive OR, being the inverse of Parity, can then be made either by putting an inverter on the output of the Parity gate or by replacing the final OR gate with a NOR gate as in figure C25.

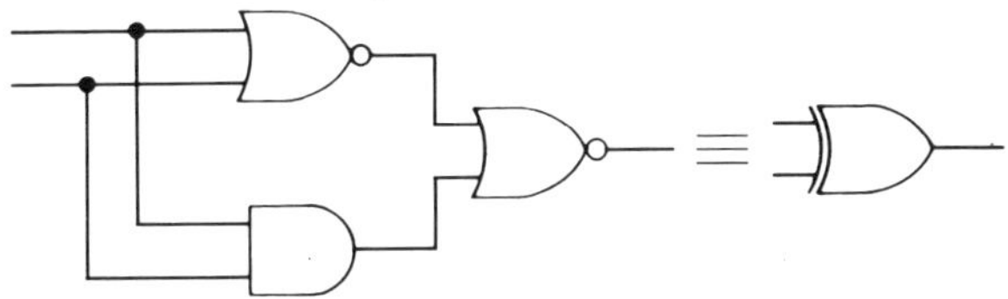

Figure C25
An Exclusive OR gate.

Questions

Questions 8 to 12 could be assigned now, if they have not already been tackled.
Questions 31 to 33 are review and revision questions on combinations of gates.
Questions 13 to 17 pose problems in a practical context.

USING A MICROCOMPUTER TO CARRY OUT LOGIC FUNCTIONS

So far we have chosen appropriate pieces of 'hardware' to carry out the logic functions we require. It is worth showing students that these functions may be carried out by a single piece of hardware – a microcomputer, controlled by appropriate 'software' – the computer program.

DEMONSTRATION

C7 Using a microcomputer to carry out logic functions

microcomputer with BASIC

In BASIC, the logic functions NOT, OR, and AND are standard. Students should understand the following programs which show how the functions are used.

```
10 INPUT A
20 PRINT NOT A
30 GOTO 10
```

```
10 INPUT A
20 INPUT B
30 PRINT A OR B
40 GOTO 10
```

```
10 INPUT A
20 INPUT B
30 PRINT A AND B
40 GOTO 10
```

On some microcomputers (for example, Apple and Sinclair), the inputs A and B should be either 0 or 1. On other machines (for example, Research Machines, B.B.C.) the inputs A and B should be either 0 or −1.

Students are then asked to devise programs that simulate NOR gates and NAND gates. The solutions, which invert the OR and AND functions, are given below.

```
10 INPUT A
20 INPUT B
30 LET Z = A OR B
40 PRINT NOT Z
50 GOTO 10
```

```
10 INPUT A
20 INPUT B
30 LET Z = A AND B
40 PRINT NOT Z
50 GOTO 10
```

Later there will be opportunity for students to carry out more complicated logic functions with the microcomputer.

The functions NOR and NAND are not standard in BASIC because confusion would arise if N were used as part of a variable name, for example, AN OR B can be confused with A NOR B.

Question

Question 18 is concerned with BASIC programs for NOR, Exclusive OR, three-input OR, and three-input AND, that is, software to perform the functions of the circuits of question 10.

THE TRANSFER CHARACTERISTICS OF A LOGIC GATE

Some uses of logic gates require a knowledge of how the output voltage of a gate varies with the input voltage when the latter is intermediate between low and high – the two extremes we have so far considered. In the following experiment, these – the 'transfer characteristics' of the gate – are measured.

EXPERIMENT

C8a Measuring the transfer characteristics of an inverter

ITEM NO.	ITEM
1514	digital electronics kit with power supply
1507	2 voltmeters appropriate to the supply
1510	potentiometer, 1 kΩ
1000	leads

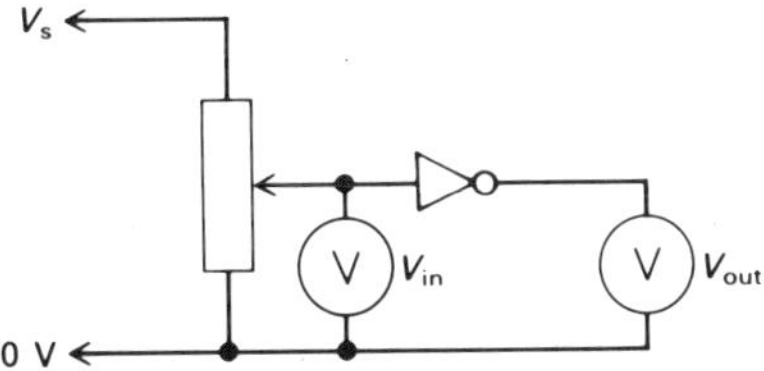

Figure C26
Measuring the transfer characteristics of a gate. (Note that the power supply connections to the gate are not shown in the circuit diagram.)

Students are reminded in the laboratory notes that they can make an inverter by connecting the two inputs of a NOR or NAND gate together.

For the three main types of electronics gate, students should obtain results similar to those in figure C27.

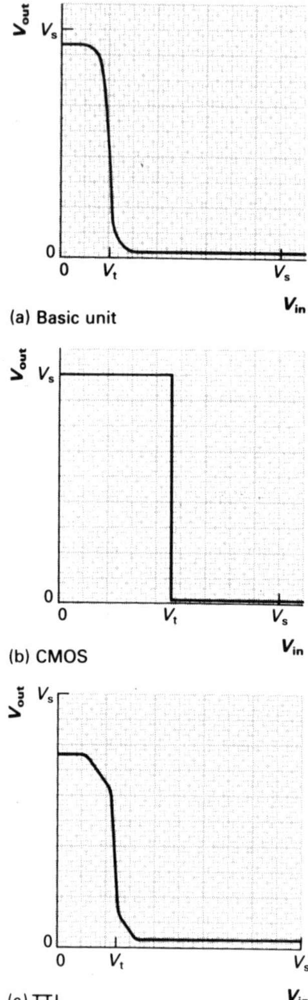

Figure C27
Transfer characteristics. For all gates, there is some input voltage level (V_t) around which the output changes more or less rapidly.

Note that there are two other ways of using the NOR and NAND gates as inverters. All four are given in the diagrams in figure C28. Only small variations in performance will be evident between these circuits. The most noticeably different will be the inverter made by connecting the two inputs of the NOR gate together if the kit uses TTL integrated circuits – figure C28(b). This arrangement draws most current from (or more correctly 'sources' most current to) whatever circuit is driving the gate and may give reduced performance in some later circuits. For that reason it may be better to discuss the matter at this stage.

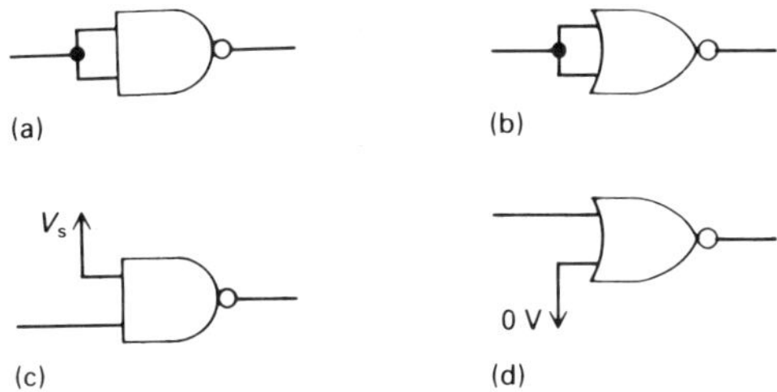

Figure C28
Different ways of making an inverter.

OPTIONAL DEMONSTRATION

C8b Plotting the characteristics of an inverter on an oscilloscope

ITEM NO.	ITEM
1514	digital electronics kit with power supply
1109	signal generator
1551	oscilloscope with X-input
1151	diode, *e.g.*, 1N4001
1000	leads

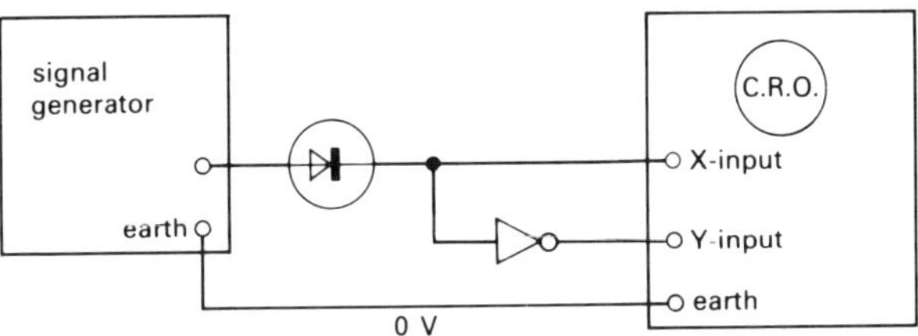

Figure C29
Using an oscilloscope to display characteristics.

Set the signal generator to produce a sine or triangular wave output. This provides a varying input voltage and the diode ensures that this is of the correct polarity. The X-deflection of the spot on the C.R.O. is proportional to the input voltage; the Y-deflection is proportional to the output voltage. Hence a graph of output against input is produced. A frequency of something over 50 Hz produces a steady trace, but it is instructive to start the signal generator at a very low frequency (around 0.1 Hz if it goes that low) and then increase the frequency to show clearly what is happening.

Question 19 is concerned with transfer characteristics.

Students now know enough about the electronics kit to be able to use it in a different manner. A light-operated switch is a circuit that is easy to study.

EXPERIMENT
C9 Making a light-operated switch

ITEM NO.	ITEM
1514	digital electronics kit with power supply
1147	light-dependent resistor (L.D.R.) *e.g.*, ORP 12
1017	resistance substitution box
1507	voltmeter appropriate to the supply
1000	leads

Students are first asked to investigate the behaviour of the circuit given in figure C30. They should find that the output voltage drops as the illumination on the L.D.R. decreases.

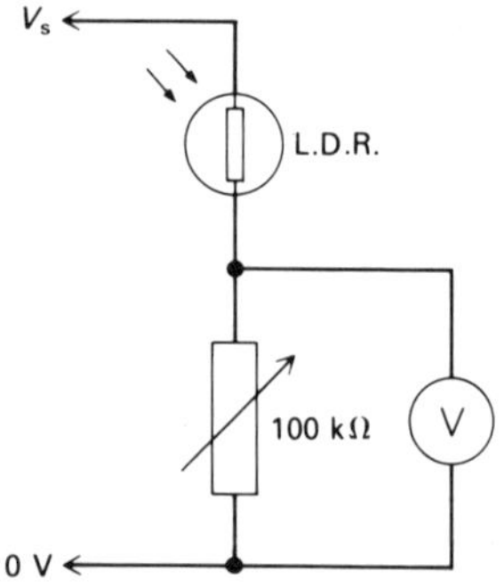

Figure C30
Potential divider with light-dependent resistor.

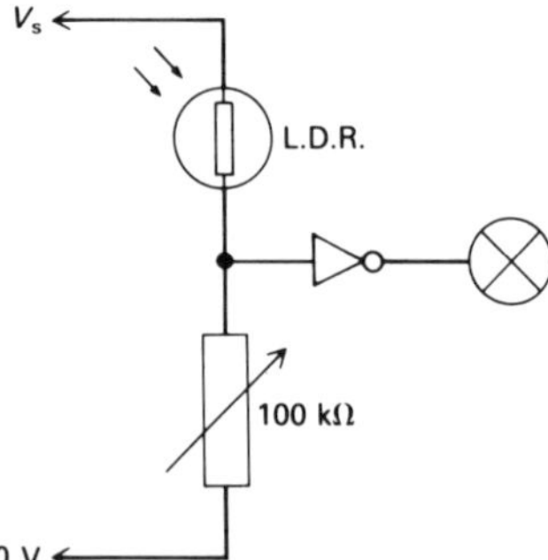

Figure C31
A circuit which turns a lamp on in the dark.

When this potential divider is connected to an inverter, as shown in figure C31, an appreciable current may be drawn if the inverter is a basic unit or TTL gate. This may cause the observed behaviour to be slightly different from that predicted, but appropriate adjustment of the resistance substitution box should make the circuit perform satisfactorily, the indicator illuminating when the L.D.R. is in darkness.

Students are then asked to devise a circuit in which the indicator goes off when the L.D.R. is illuminated. Two possible solutions are given in figure C32. The first circuit accomplishes the inversion by interchanging the positions of L.D.R. and variable resistor. In the second, a non-inverting buffer is driven by the potential divider.

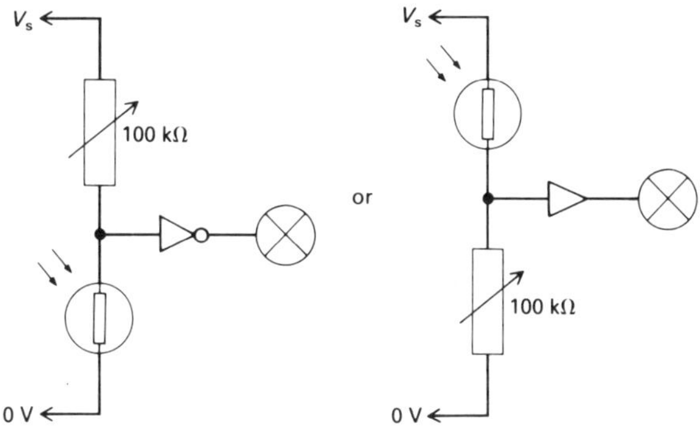

Figure C32
Two circuits which turn a lamp off in the dark.

In all these circuits a gate is needed between the potential divider and the indicator to buffer the potential divider from changes in the load. Indeed we can go further than this by asking students how we might control a 'real' load such as a motor or a large light bulb. In these cases, further stages would be needed between the potential divider and the controlled load – typically a relay might be used.

Questions

Questions 20 to 22, and also 31 and 34 use the ideas in this work.

ARITHMETIC WITH LOGIC GATES

Experiments C10 and C11 are about binary addition. Although it should be possible for a whole class to perform the first experiment (making a 'half adder'), the number of gates required to perform the second (making a 'full adder') means that teachers may wish to demonstrate it, or ask a group of students to demonstrate it to the rest of the class.

EXPERIMENT
C10 Making a half adder

ITEM NO.	ITEM
1514	digital electronics kit with power supply
1000	leads

A system that adds the two digits A and B together to give a SUM and a CARRY digit in binary arithmetic is called a 'half adder'.

A + B = SUM and CARRY

0 + 0 = 0	and 0
0 + 1 = 1	and 0
1 + 0 = 1	and 0
1 + 1 = 0	and 1

Students should recognize that the SUM column is the same as the output of an Exclusive OR gate, while the CARRY column is that of the AND gate. A suitable combination of these two gates should enable us to add two binary digits together.

The system requires two inputs (the two digits) and two outputs. The solution is given in figure C33.

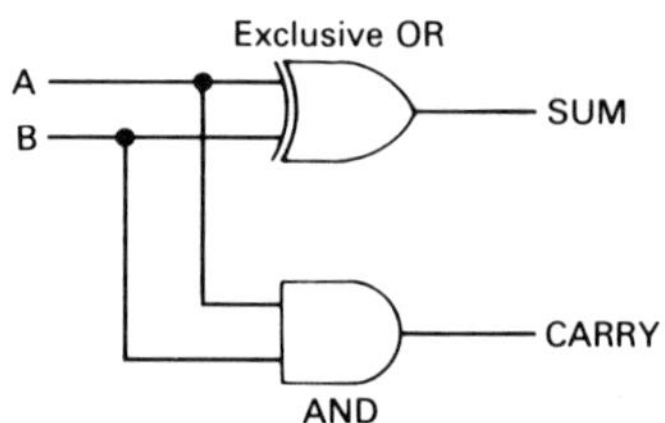

Figure C33
A half adder circuit.

This is fine if all we wish to add together is two binary digits. Unfortunately, that would only allow us to achieve a grand total of decimal 'three', which won't get us very far in the real world. Computers must be able to handle much larger binary numbers.

Experiment C11 asks students to design a 'full adder': one that will add two digits and a CARRY from previous additions.

EXPERIMENT
C11 Making a full adder

ITEM NO.	ITEM
1514	digital electronics kit with power supply
1000	leads

The truth table and the circuit we need are given in figure C34.

Inspecting the truth table shows that a second half adder is needed to add the $CARRY_{in}$ to the sum of the two input digits. The final $CARRY_{out}$ is the output of an OR gate whose inputs are the 'carrys' from the two half adders. This circuit is called a 'full adder'. The *Students' guide* (page 168) gives further information on binary arithmetic.

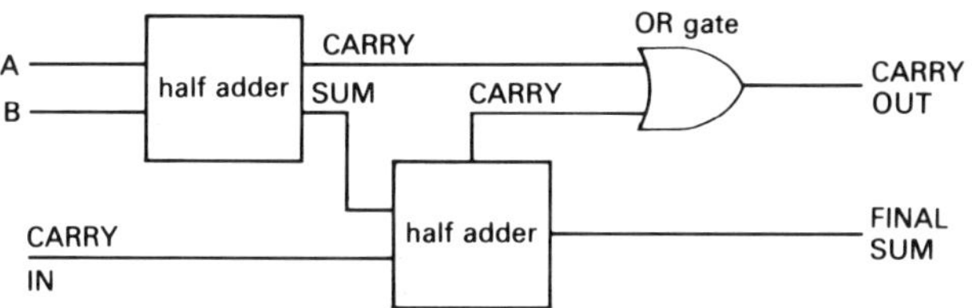

inputs			outputs	
A	B	$CARRY_{in}$	SUM	$CARRY_{out}$
0	0	0	0	0
0	0	1	1	0
0	1	0	1	0
0	1	1	0	1
1	0	0	1	0
1	0	1	0	1
1	1	0	0	1
1	1	1	1	1

Figure C34

Questions 23 and *24* are on binary addition

Reading

The Reader *Microcomputer circuits and processes* goes into some detail about the construction and working of a microcomputer.

Some of the tasks of binary addition may be carried out on a microcomputer, perhaps as an alternative for some students to experiment C11.

OPTIONAL EXPERIMENT/DEMONSTRATION

C12 Using a microcomputer to perform binary addition

microcomputer with BASIC

Figure C35 shows a half adder using only the operations OR, AND, and NOT.

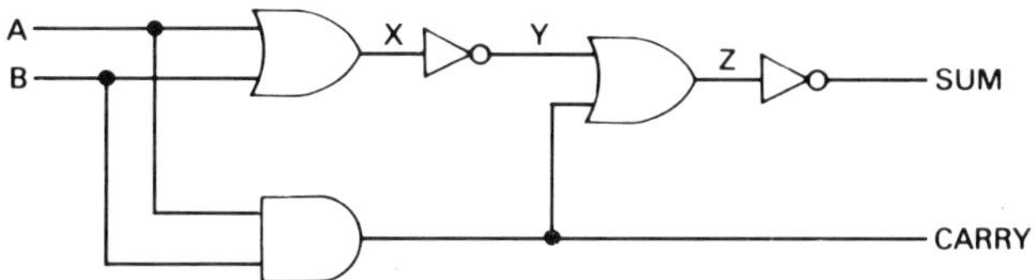

Figure C35
A half adder using only the operations OR, AND, and NOT.

In the process of evaluating the SUM and CARRY outputs, intermediate results X, Y, and Z are evaluated. We may convert this into the following BASIC program. (*Note*: For some microcomputers, the variable names may need to be abbreviated.) Because there are two NOTs in succession in this program, it works on most machines using the digits 0 and 1 as inputs.

```
10  INPUT A
20  INPUT B
30  LET CARRY = A AND B
40  LET X = A OR B
50  LET Y = NOT X
60  LET Z = CARRY OR Y
70  LET SUM = NOT Z
80  PRINT
90  PRINT 'SUM = '; SUM
100 PRINT 'CARRY = '; CARRY
110 PRINT
120 GOTO 10
```

It is worth pointing out to students the difference between the electronics kit which carries out the operations immediately and simultaneously, and the microcomputer which carries out the functions one after the other with a small delay between each successive operation. Most microcomputers are '8-bit' machines – they can carry out operations on up to 8 bits simultaneously.

SECTION C2

SEQUENTIAL LOGIC

MAKING PULSES

Pulses are useful in electronic circuits for timing purposes. In experiment C9, a gate was controlled by a potential divider made from two resistors. A circuit where one of these resistors is replaced by a capacitor allows a momentary change to be conveyed to the input. Experiment C13 first reminds students of earlier work with capacitors and resistors. It is probably best performed as a demonstration. Ask students to predict what will happen when the circuit shown in figure C36(a) is connected to an inverting gate. Students test their predictions when a lamp is connected to this circuit via an inverting gate – figure C36(b). *Note*: The pulse occurs when the switch goes from 0 to V_s, not vice versa.

EXPERIMENT

C13 Making pulses

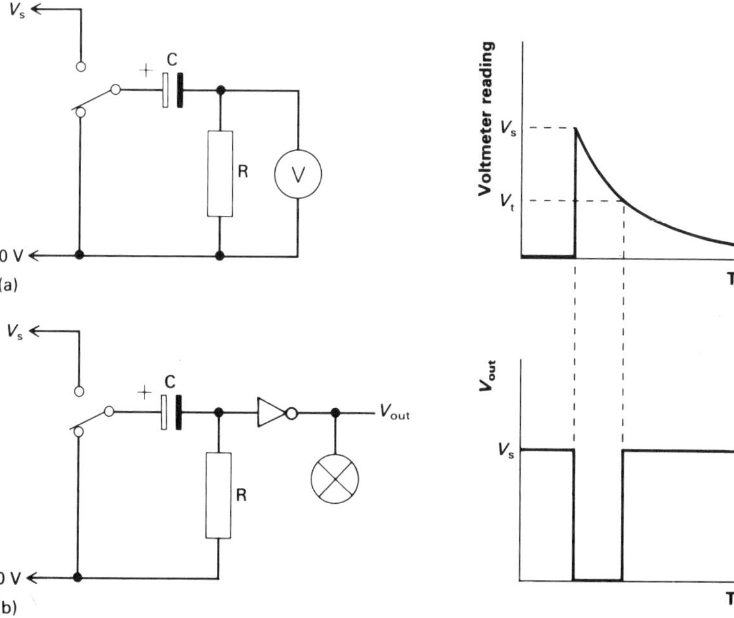

Figure C36
Making pulses.

ITEM NO.	ITEM
1514	digital electronics kit with power supply
1151	capacitor } (1000 μF and 560 Ω are suitable for TTL;
1151	resistor } 47 μF and 10 kΩ for other gates)
1000	leads

The choice of the values of C and R has to be made with some care. For most gates, 47 μF and 10 kΩ perform adequately. For TTL gates where the controlling input currents needed are much higher, 1000 μF and 560 Ω should be used. RC determines the time for which the input to the gate is held above V_t and hence controls the length of the pulse.

Note that the pulse produced by this circuit is negative going. An indicating lamp on the output goes off for a while and then back on when the input lead is moved from 0 V to V_s. Students should know enough about RC circuits to be able to control the length of the pulse by increasing or decreasing RC. Both can be tried within a limited range around the values suggested above. Outside this range, depending on the apparatus used, the effects will be unpredictable because of the loading that the input circuit of the gate presents to the RC circuit. It is best to use high quality electrolytic capacitors – new ones from a reputable supplier – so that the leakage currents are small and do not affect the performance of the circuit.

Students are also asked to make a lamp go on for a short period. To do this, the inverting gate should be replaced by a non-inverting gate (see experiment C6c) as in figure C37.

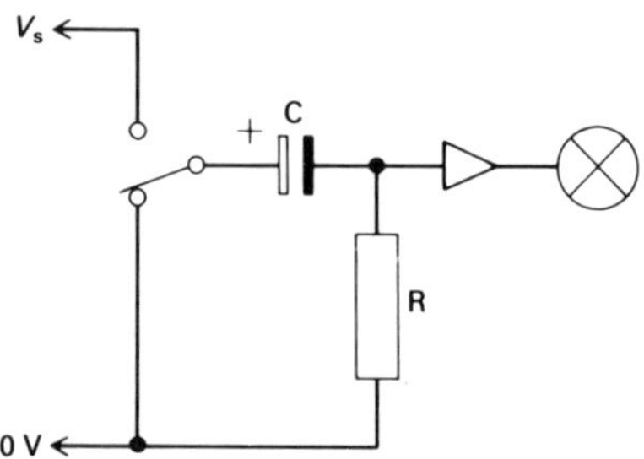

Figure C37
Making a lamp go on for a short period.

In both of these circuits, the gate between the junction of C and R and the indicator both reduces the load that an indicator would present to the RC circuit, and also converts the slowly changing output from the RC circuit to an abrupt pulse.

DEMONSTRATION C14a AND EXPERIMENT C14b
The astable circuit

ITEM NO.	ITEM
1514	digital electronics kit with power supply
1151	capacitors } see note to experiment C13
1151	resistors } see note to experiment C13
1000	leads

C14a The output from one pulse producer may be used to trigger another pulse producer as in figure C38.

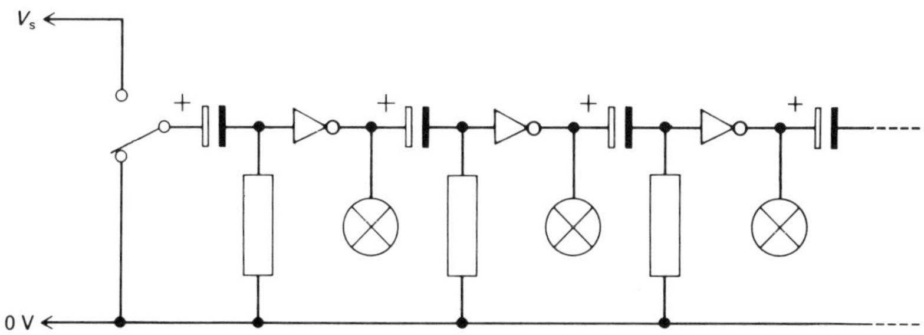

Figure C38
Passing a pulse from one circuit to another.

If the input of the first goes from low to high its output goes from high to low, stays there for a bit and then rises again. This output is the input to the second unit. The fall has no effect but the rise from low to high, arriving at the end of the first unit's pulse, has an effect. The second unit repeats the behaviour of the first. When the second unit's pulse ends, its rising output triggers the third and the pulse is passed along. If the output of any stage is connected back to the input of the first, the pulse will continue round.

C14b The simplest circuit in which this happens contains two stages (figure C39).

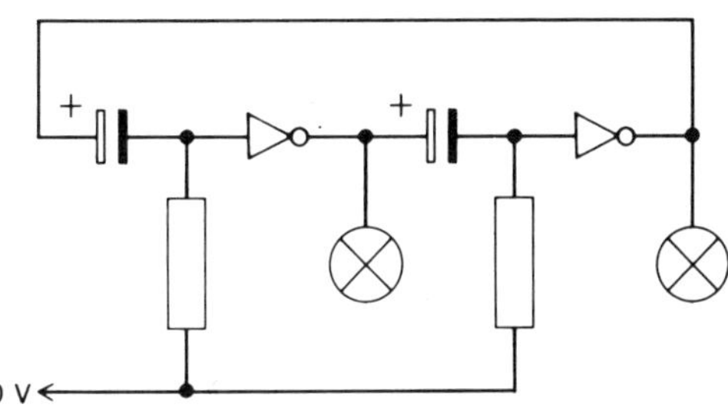

Figure C39
An astable circuit.

This circuit is called astable because it is not stable in any state. Like a pair of quarrelsome men, each stage is busy giving the other a punch on the nose in payment for the last punch received – a process which is notoriously difficult to stop.

When the circuit is first connected, it may not start oscillating of its own accord. It may be started by momentarily shorting across one of the capacitors with a flying lead (the first punch).

Questions

Questions 25 and 26 are simple questions about making pulses.

A device which produces a steady train of pulses can be used to drive a clock, for example, in a digital watch or the internal clock of a microcomputer. The electronics kit contains an astable unit whose frequency of oscillation may be varied over a wide range. The internal circuit may be very different from that of the circuit shown in figure C39, but it will rely on a gate or combination of gates being controlled by an *RC* circuit in the same way as the pulse-producing circuit that the students have investigated. (See Appendix I for ways of making an astable circuit from the basic unit kit.) Students should see the astable output on an oscilloscope and hear it on a loudspeaker. The frequency of the circuit may be made high enough to produce an audible tone.

EXPERIMENT
C15 Investigating the astable module

ITEM NO.	ITEM
1514	digital electronics kit with power supply
1059	high-impedance earpiece
1511	oscilloscope
1000	leads

The output from the astable module may be varied over a wide range and used to drive various devices. (See figure C40.)

An astable game

It may help to play a simple game.

Two students face each other, standing up in front of their seats. Each is told to sit down, count up to five and then stand up again whenever the other student stands up. The astable action is started by pressing one student into his seat.

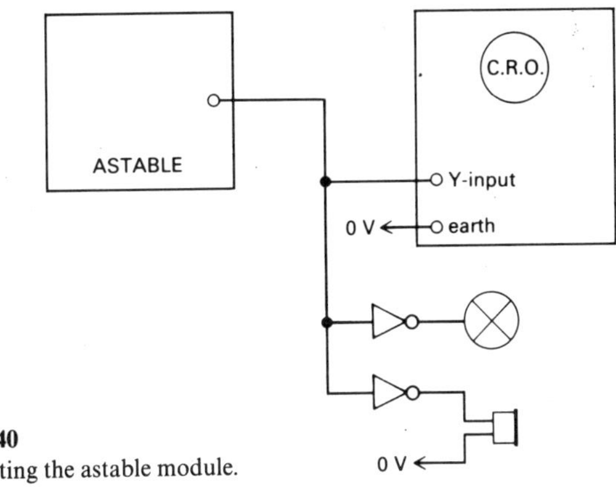

Figure C40
Investigating the astable module.

EXPERIMENT
C16 The bistable circuit

ITEM NO.	ITEM
1514	digital electronics kit with power supply
1000	leads

Students can first investigate the behaviour of two inverters connected as shown in figure C41, without the dotted connection. If the input to A is high, its output and therefore the input to B are low. The output of B is high.

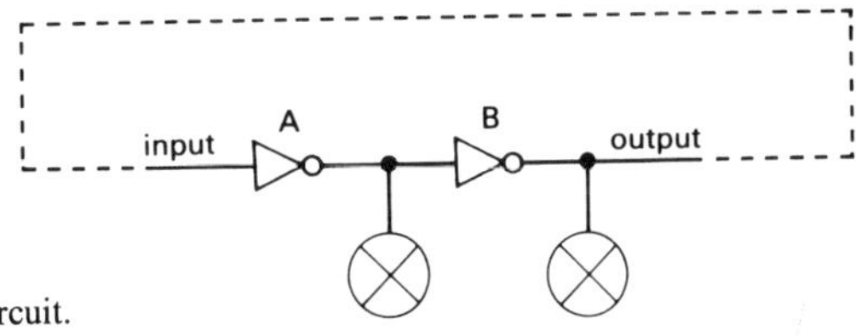

Figure C41
The bistable circuit.

Now suppose that the output of B is fed back to the input of A (the broken line). The new input is high, like the original input to A. So the new input makes sure that the system stays where it started; indeed the original input can now be taken away.

How can the system be made to do anything else? If the input to B is made high for a moment by connecting it to V_s, its output goes low and is fed to A, whose output goes high. This high output now goes to B, and the system has provided just what is needed to keep it in its new state. The high input to B which switched the system over can now be taken away. This new behaviour has been produced by feedback. It is positive feedback; that is, a change at one input is fed back to that input in the same sense as itself, so that the system 'confirms' any change made to it. The circuit has two stable states, so it is called a bistable circuit.

A bistable game

Make two students face each other, each being told to stand if the other is sitting, and vice versa. Seat both and say 'go'. The faster moving one will end up on his feet, with the other sitting down. Delays may introduce oscillation as in the astable module. Pushing one into his seat brings the other to his feet.

The bistable circuit is useful because it remembers what happened to it last; one state can represent the binary digit 0 and the other the binary digit 1. In computers, large numbers of bistable circuits are used in the memory. Two bistable modules are provided ready to use in the electronics kit.

Questions

Questions 27 and 28 are about bistable circuits. Questions 29 to 31 concern feedback.

DEMONSTRATION/EXPERIMENT
C17 The bistable module

ITEM NO.	ITEM
1514	digital electronics kit with power supply
1511	double beam oscilloscope
1059	2 high-impedance earpieces
1000	leads

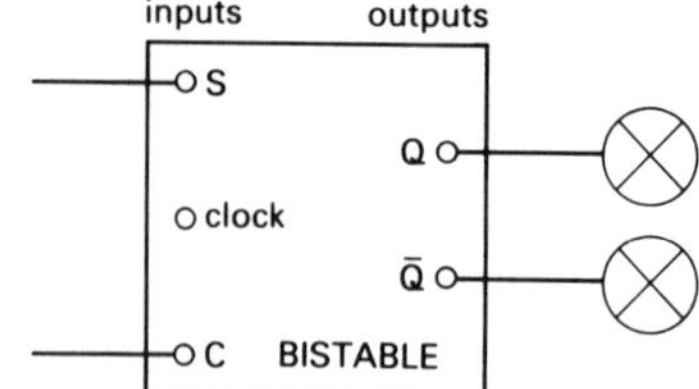

Figure C42
The bistable module.

C17a The bistable module has two outputs, Q and $\overline{Q}$. $\overline{Q}$ is the complement of Q: when Q is 0, $\overline{Q}$ is 1; when Q is 1, $\overline{Q}$ is 0. The two inputs are labelled S (set) and C (clear). When the input to S is logic 1 (high), Q is 'set' (that is, made logic 1); when C is made logic 1, Q is 'cleared' (that is, made low, logic 0).

Show that the bistable module can be switched from one state to the other by making inputs S and C alternately high.

C17b There is a third input to the bistable module called clock (or perhaps trigger or toggle). This input is an extra facility. It is a nuisance sometimes to have to use two inputs to switch the bistable over. The circuit guides successive pulses to the input that was not used last time the module switched over. It is easy to show that this happens though not so easy to say how. *Note*: Some bistable modules (for example, those following the specification in the *Apparatus guide*) change state when the clock input rises from 0 V to V_s. Other bistable modules (for example, that from the basic unit kit) change state when the clock input falls from V_s to 0 V. The difference is not important in most applications.

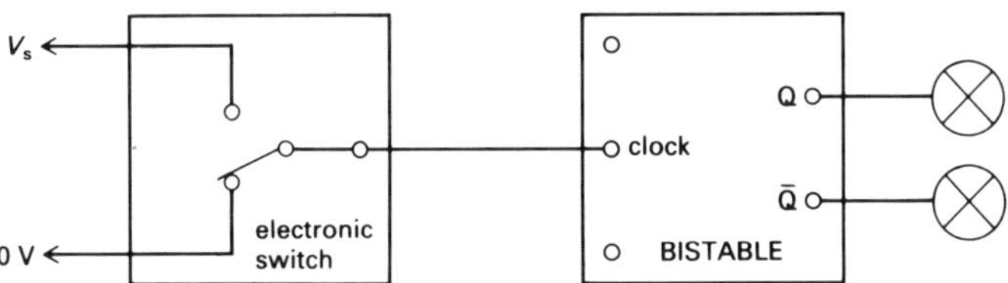

Figure C43
Investigating the clock input.

The electronic switch (figure C43) provides an electrically 'clean' switch signal to the bistable module – that is, an input with a single rising or falling edge rather than several that might result from mechanical and therefore electrical bounce at the contacts of a simple switch used on its own.

C17c Students should investigate the effect of feeding a series of pulses from the astable module into the bistable module. It is good practice to incorporate a gate (an inverter will do) between the astable and the bistable modules, to ensure that pulses fed to the latter are 'square'.

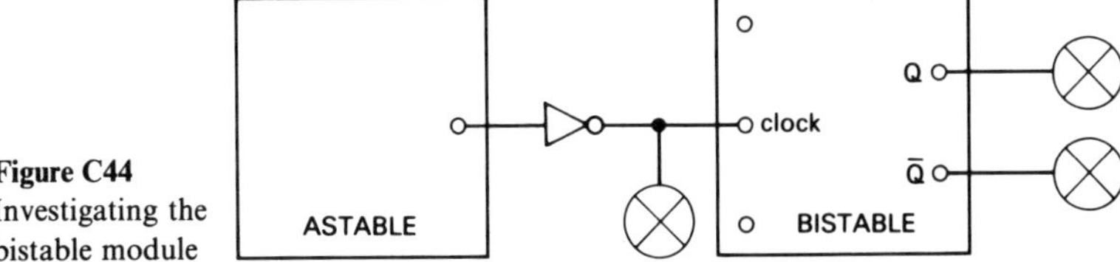

Figure C44
Investigating the bistable module

It is instructive to show the action at higher speed on a double beam oscilloscope, feeding the bistable from an astable module at a few hundred hertz. The earthed side of the oscilloscope goes to the 0 V line as in figure C45. The input to the clock can be shown on one beam and one of the two outputs on the other beam. Finally, the two outputs can be displayed together. Figure C46 shows the waveforms for a bistable module that changes state with a rising input.

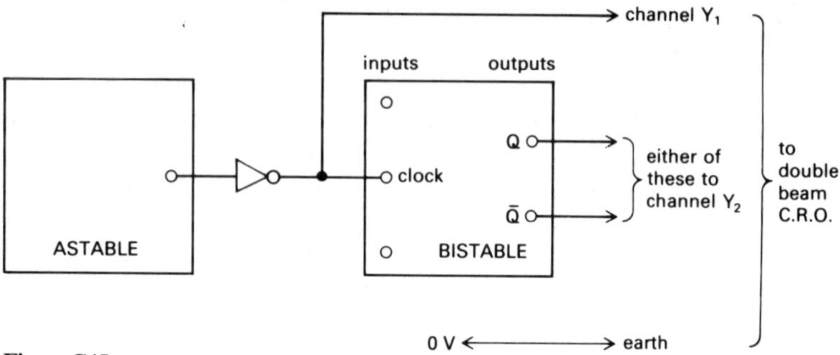

Figure C45
Displaying the behaviour of a bistable module on an oscilloscope.

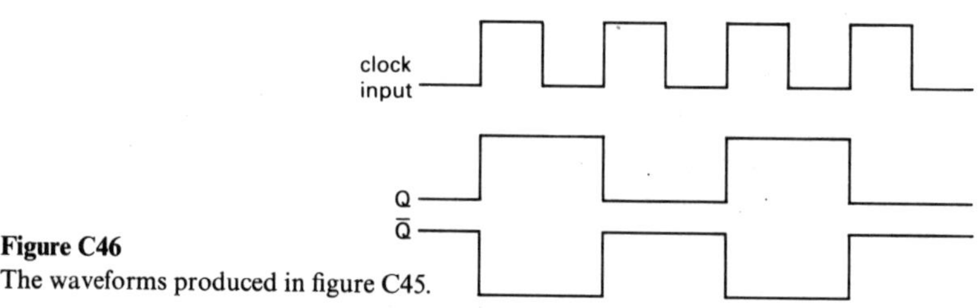

Figure C46
The waveforms produced in figure C45.

There is a sense in which the system does arithmetic. Half as many pulses come out of one output as go into the clock input, so it can be used to divide by two. Later, several of these modules will be used to make counters similar to those used in computers.

C17d If high-impedance earpieces are connected at the input and output of the bistable modules in figure C47, then the factor of two difference in frequency between the input and the output may be noted by the octave difference in pitch between the sounds produced in the earpieces.

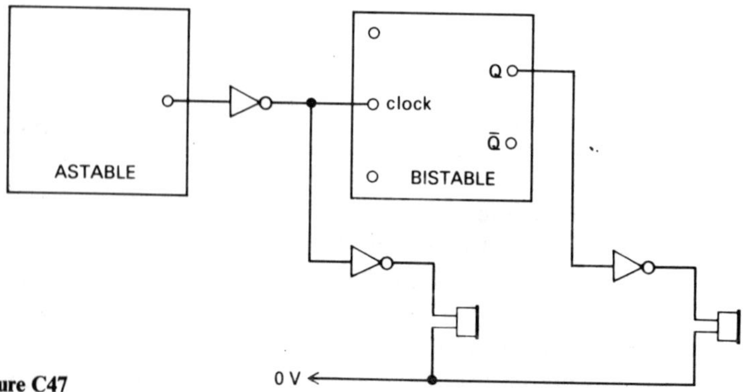

Figure C47
Demonstrating the effect of the bistable module on an audio frequency.

SECTION C3
DESIGNING DIGITAL SYSTEMS

At this point, in an important sense, the real work begins. The task of engineers is to solve real problems, inventively, using materials that are available. Any useful electronic device contains several parts, arranged so as to achieve a particular purpose. Students are given a selection of practical challenges which can be tackled using what has been learned in the Unit. They should be encouraged to *think* about the solution before trying it out practically. Students will need hints and oblique help, but not recipes. If they have recipes given to them, they will work faster and do more, no doubt, but will have little opportunity to use their own initiative and inventiveness. An important lesson from this Unit is that engineering is a matter of being inventive, and it would be a pity to lose that lesson.

The first group of jobs, experiments C18a to C18h, are intentionally rather simple ones. We suggest that each student should manage two or three of these with perhaps another more complicated one, perhaps spending between 2 and 4 double periods altogether on this work.

At least one solution is suggested for each problem, but many more are possible and some more elegant. An attempt has been made to produce 'obvious' solutions rather than ones which minimize the number of gates used. There are standard techniques for minimizing gates (see Thompson *Inside the micro* pages 27 to 31).

Some of the tasks (C18g to C18s) considerably extend the work of the Unit into multiplexing, decoding, and encoding. The students' notes for these (*Students' guide* pages 193 to 196) give more information about the application of such systems and commercially available integrated circuits.

Further reading

For more information about multiplexing, decoding, and encoding, see, for example:

HOROWITZ and HILL *The art of electronics.* Chapter 8.

JONES *A practical introduction to electronic circuits.* Chapter 13.

McWHORTER *Understanding digital electronics.* Chapter 8.

TOCCI *Digital systems: principles and applications.* Chapter 9.

EXPERIMENT

C18 Designing digital systems

Some or all of the following apparatus may be required:

ITEM NO.	ITEM
1514	digital electronics kit with power supply
1151	various capacitors (between 10 μF and 10 000 μF)
1017	resistance substitution boxes
1147	light-dependent resistor
1059	high-impedance earpieces
1151	thermistor
1021	aerosol freezer
1000	leads

C18a Make a lamp go on for half a second.

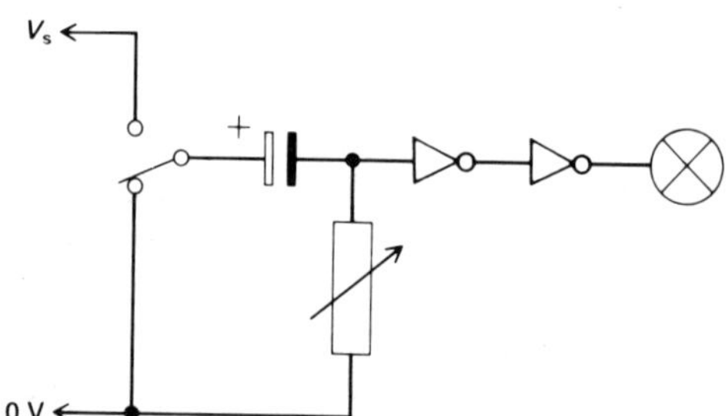

Figure C48

The second gate inverts the output from a pulse producer (figure C48). Control the time constant of the circuit to obtain the desired pulse length.

C18b Make a lamp flash from an input falling from V_s to 0 V.

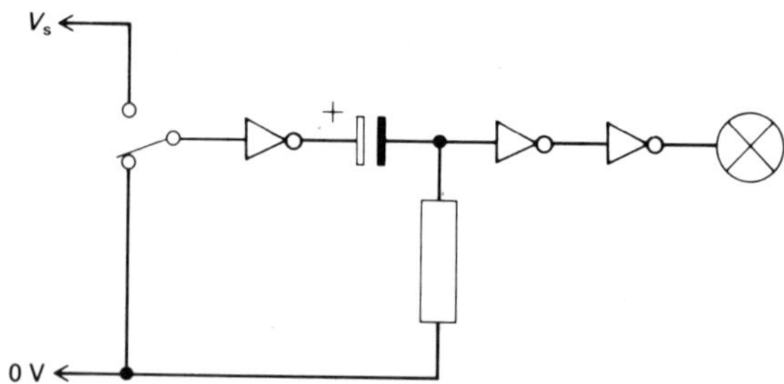

Figure C49

The first inverter turns an input falling from V_s to 0 V, into one which rises from 0 V to V_s, which is what the pulse producer needs (figure C49).

C18c Make a Morse code sender.

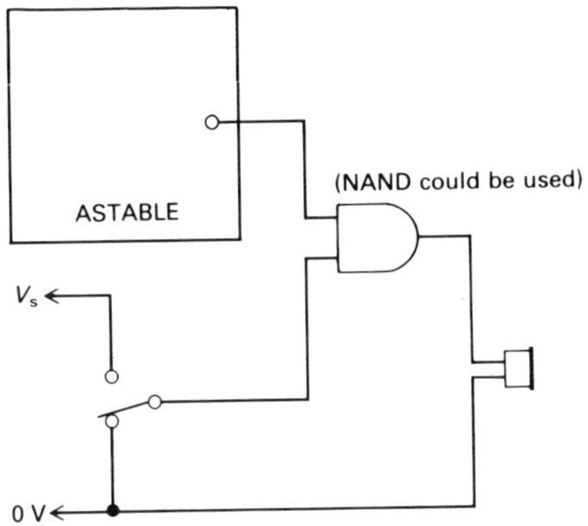

Figure C50
A Morse code sender.

The output of the astable module appears at the output of the AND gate when the switch is connected to V_s.

The action of a NAND gate in this circuit is best understood as an AND gate followed by an inverter. As before, the output of the astable module appears at the output of the AND gate when the switch is high; the inverter inverts this output.

C18d Make a bleeper that emits an audible pulse of sound when a button is pushed.

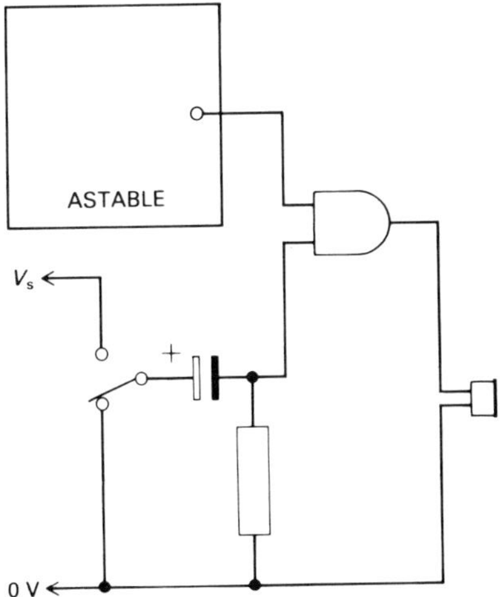

Figure C51
A bleeper.

C18e Make a lamp give six flashes in a row.

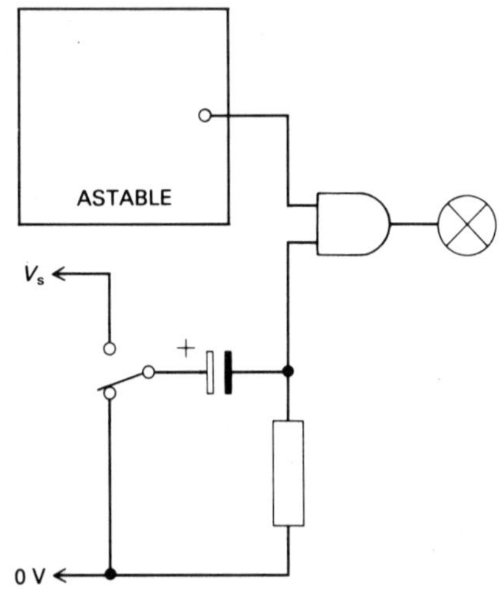

Figure C52
Producing six flashes in a row.

In the system shown in figure C52, a pulse is adjusted until it is long enough to let six shorter pulses through the AND gate.

C18f Make a device that emits six audible pips in a row.

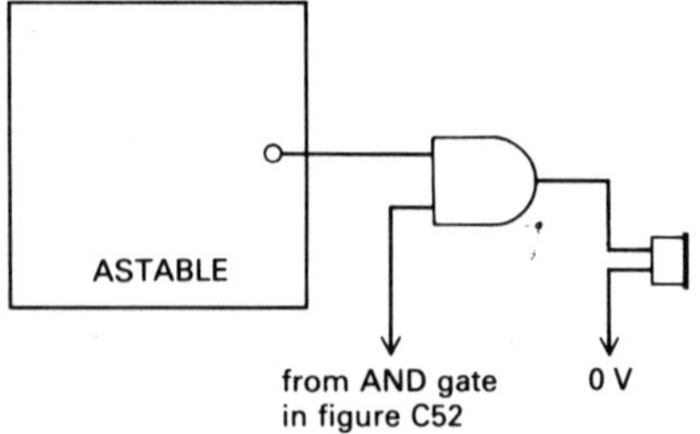

Figure C53
Producing six audible pips in a row.

C18g Make a device that emits a warning tone when a thermistor's temperature becomes high (figure C54). Then try a warning of low temperature.

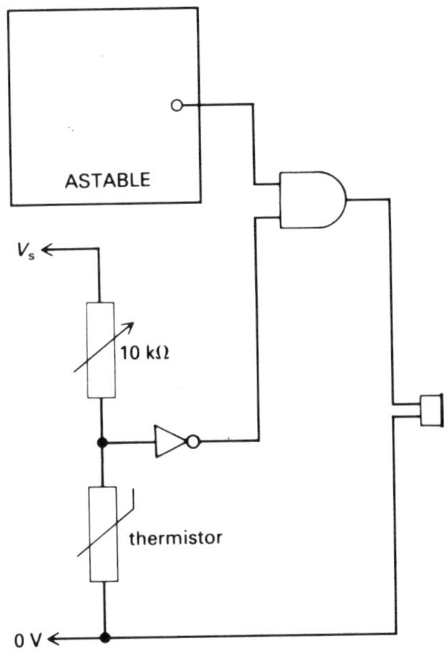

Figure C54
A high temperature warning.

C18h Use a light-dependent resistor to produce a warning tone when the light level is low. Then try high.

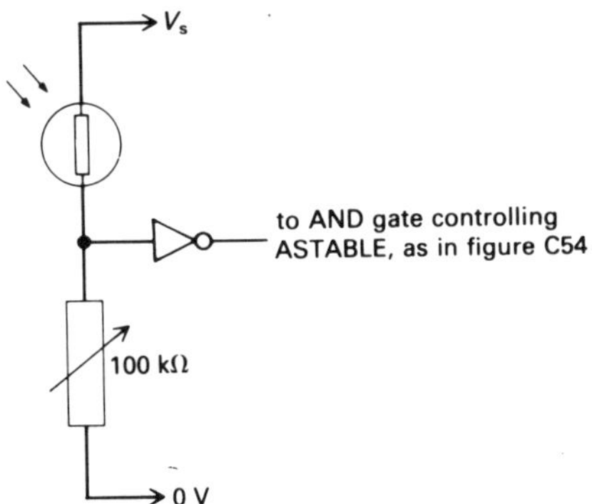

Figure C55
A low light warning.

C18i Make a system that can be used with a counter to time the interval between two successive pushes of a button.

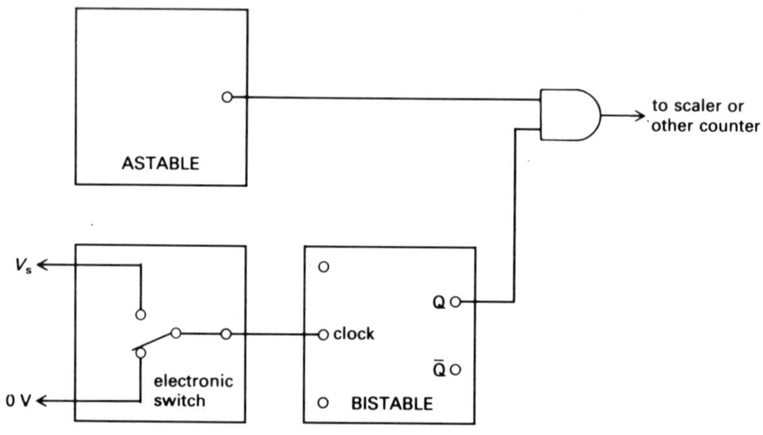

Figure C56
An interval timer.

C18j Make a human response timer to measure the time a person takes to respond to a lamp coming on (or a tone sounding), by pressing a button.

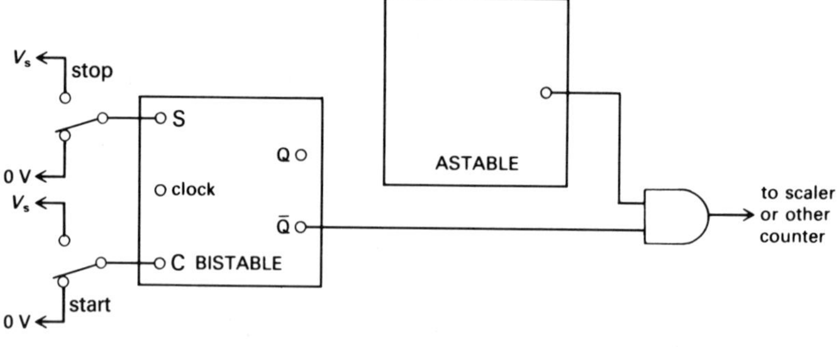

Figure C57
A human response timer.

The circuit is very similar to that of figure C56. Many similar timing systems can be devised: timing the fall of a ball or the time of flight of a bullet, for example.

C18k Make a monostable circuit. Halfway between a bistable and an astable circuit is a circuit that has one stable state but may be briefly switched to another state for a time dependent on C and R (figure C58).

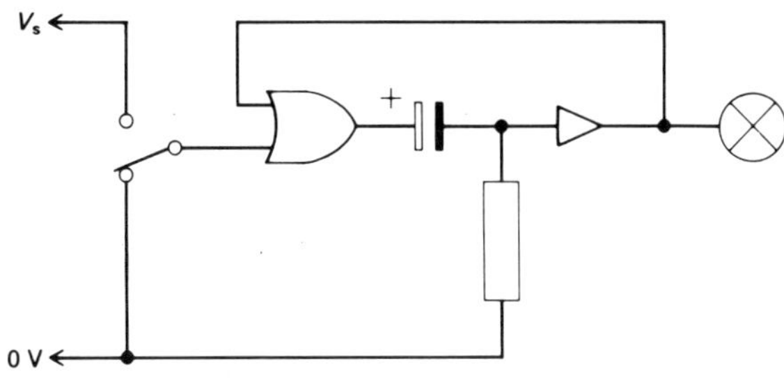

Figure C58
A monostable circuit.

The output pulse that the circuit in figure C58 produces is fixed – depending only on C and R, and not on the time for which the button is pressed because the output, fed back to the input of the OR gate keeps the output of that gate high for a while even if the switch is released. This circuit could provide a replacement for the electromechanical device sometimes used to control lighting for communal areas. When a button is pressed, the light goes on for a predetermined time and then extinguishes.

C18I Make a binary counter. Use three bistable modules to make a counter to count from 0 to 7 pulses. *Extension*: Modify the first circuit so that if all the lamps are on to start with, incoming pulses will turn them off in the binary sequence from 7 down to 0.

The circuit in figure C59 should be used with kits following the specification in the *Apparatus guide*. If the basic unit kit is being used then the clock impulse of the second and third astable modules should be connected to the $\overline{Q}$ outputs instead of the Q outputs.

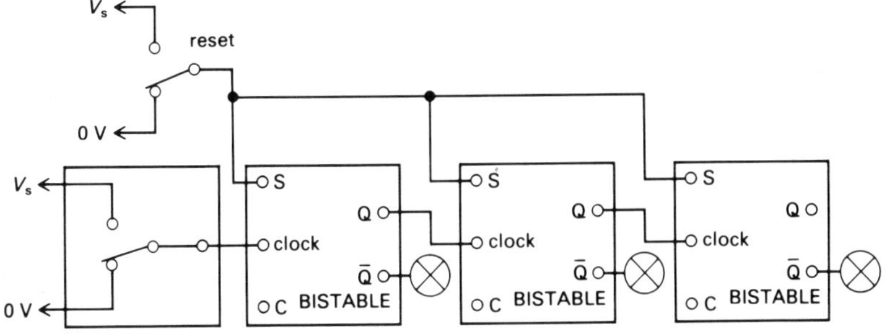

Figure C59
A binary counter.

The second task has the same circuit as the first, but with the indicators connected to the Q outputs of the bistable modules.

C18m Make a digital frequency meter that displays the frequency of a supply in digital form (binary form using the electronics kit, in scale of ten using a counter/scaler).

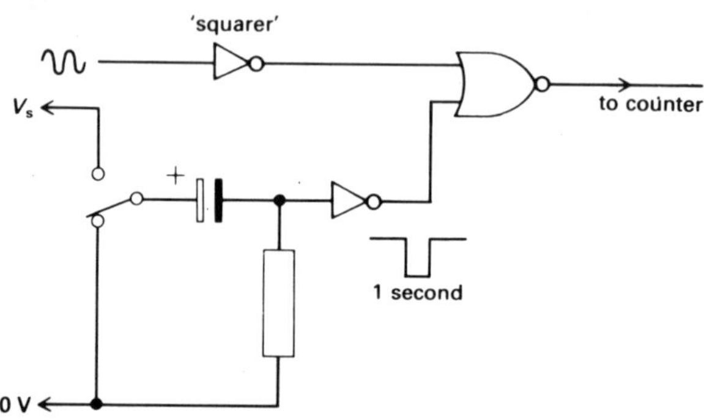

Figure C60
A frequency meter.

Adjust the pulse producer so that it produces a pulse that is one second long. The counter will count the number of pulses getting through the NOR gate in one second so it will indicate the incoming frequency directly. The output may be counted by the electronics kit if it is fed into a chain of bistable modules such as those in figure C59.

C18n Make a safety interlock system for a furnace which is to have a start button and a stop button, but the heating is to be turned off if the temperature is too high or if the safety door is open, and must not come on if the start button is pressed under these circumstances.

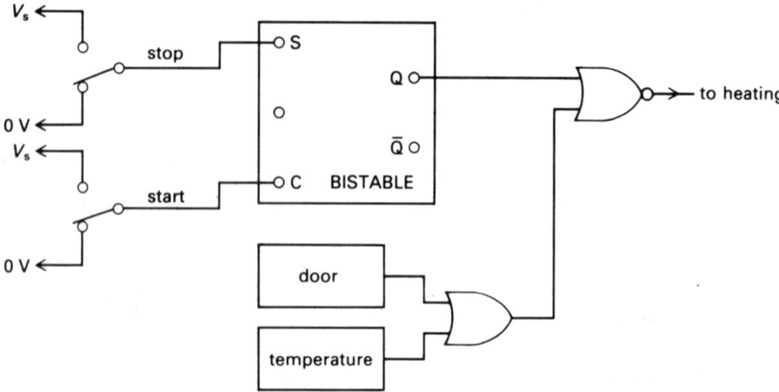

Figure C61
A safety interlock system.

A lamp indicator may be used to show when the system is on.

C18o Design a circuit that compares two one-bit numbers and gives a high output when they are both the same. This circuit is called a 'comparator', and is usually found within the Arithmetic and Logic Unit (ALU) of a microprocessor. See also experiment C6f.

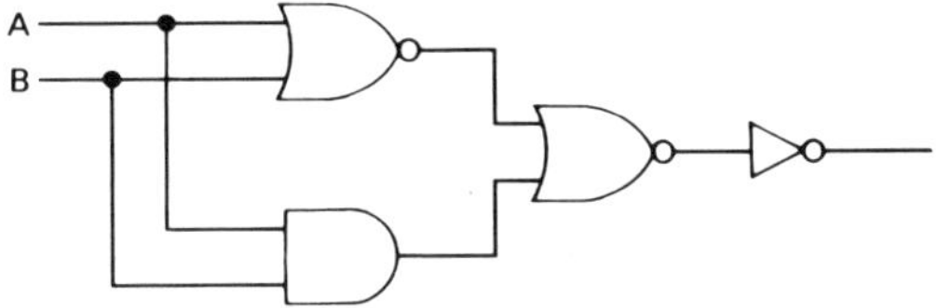

Figure C62
A comparator.

The circuit in figure C62 is the same as the circuit in figure C63.

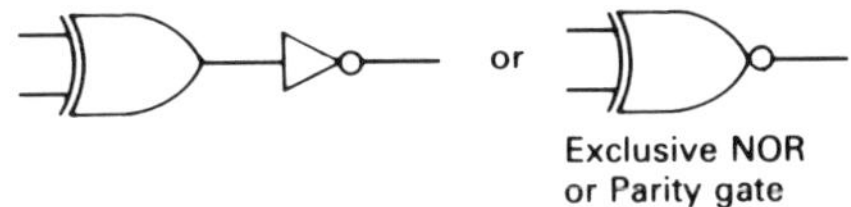

Figure C63
An exclusive NOR or Parity gate.

C18p Extend your solution to experiment C18o to compare two two-bit numbers.

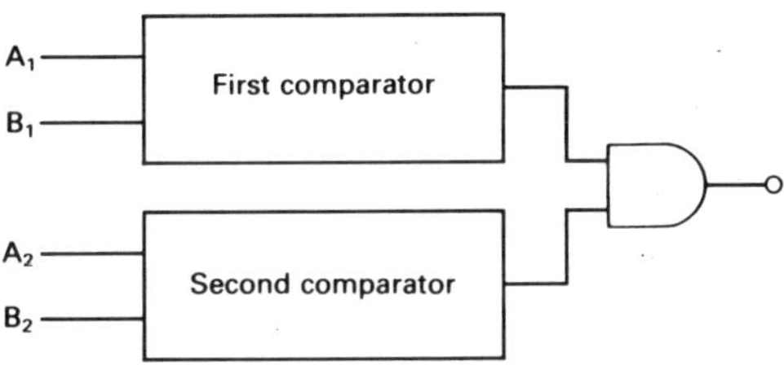

Figure C64
A two-bit comparator.

C18q Design a circuit that will pass one of two digital inputs to an output depending on the state of a single control input (see figure C65).

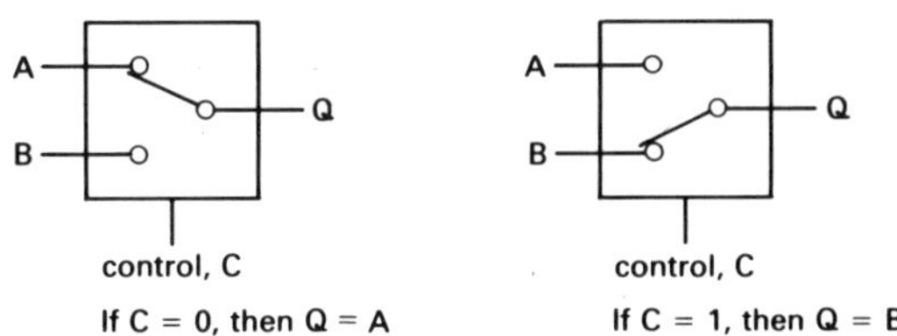

Figure C65

The students' laboratory notes hint at the use of an AND gate as a switch (as in experiments C18d and e) and give examples of two commercial multiplexers and what they can do.

One solution to the problem is given in figure C66.

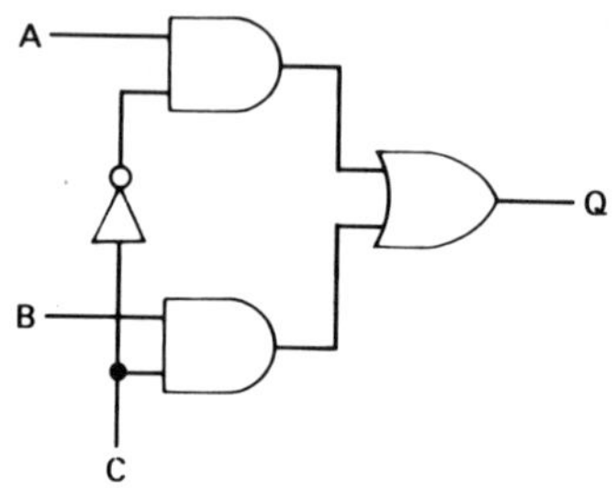

Figure C66
A multiplexer.

C18r Design a circuit with two inputs and four outputs that obeys the truth table in figure C67. Notice that only one of the outputs is high at any time.

inputs		outputs			
A	B	Q1	Q2	Q3	Q4
0	0	1	0	0	0
1	0	0	1	0	0
0	1	0	0	1	0
1	1	0	0	0	1

Figure C67

This circuit is called a decoder. A few examples of its use are given in the students' laboratory notes. One solution is given in figure C68.

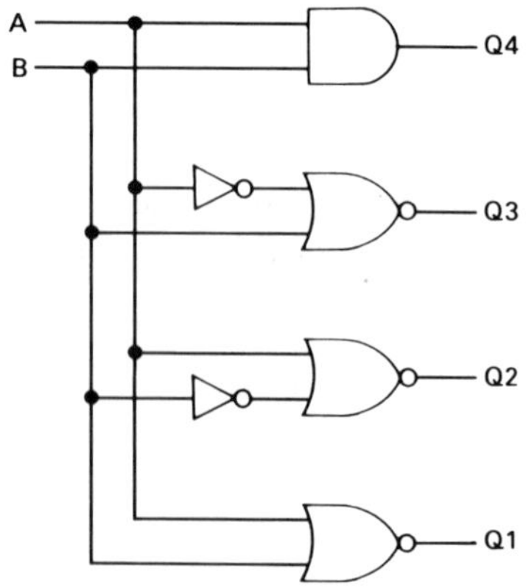

Figure C68
A two to four line decoder.

C18s The opposite of a decoder is an 'encoder'. The problem below illustrates the idea. Design a circuit which produces the correct binary equivalent of a decimal number when one of the keys 1, 2, or 3 is pressed. A circuit and truth table are shown in figure C69.

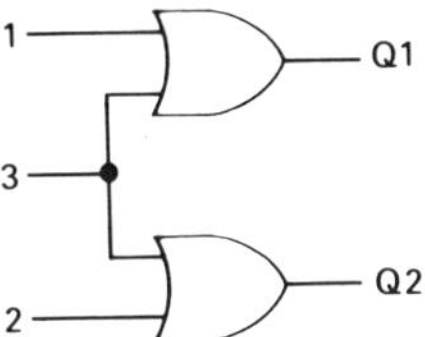

inputs			outputs	
1	2	3	Q1	Q2
0	0	0	0	0
1	0	0	1	0
0	1	0	0	1
0	0	1	1	1

Figure C69
An encoder.

C18t Make a coincidence detector. In an experiment in nuclear physics, pairs of counts from the decay of particles may occur together in time if they have a common origin. Devise a system to indicate if two pulses fall within a fixed time of each other.

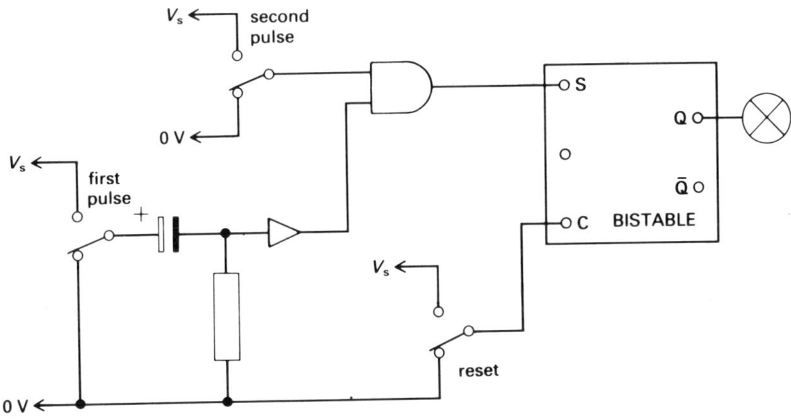

Figure C70
A coincidence detector.

C18u Make a pulse delay system. Start by extending a square pulse by a fixed time. Then add to the system to remove the front part of the extended pulse and so produce a pulse like the original one but delayed by a fixed time.

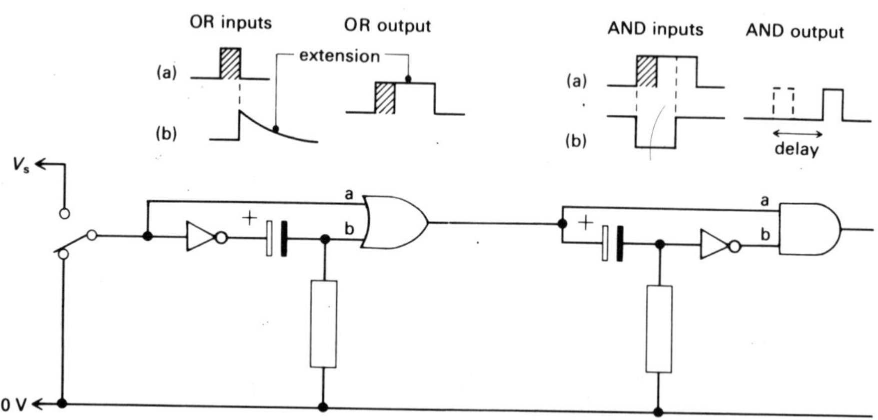

Figure C71
A pulse delay system.

C18v Make a traffic lights system with red, amber, and green flashing in the correct sequence.

Figure C72 shows a simple solution in which 'stop', 'go', and 'caution' time intervals are equal.

Note: 1 The inverter should be omitted with the basic unit kit. Figure C73 shows a more complex system which lights two sets of lamps and makes the 'stop' and 'go' periods seven times longer than the 'caution' period.

2 The 3-input NOR gate should be replaced by a 3-input AND gate with the basic unit kit. In both systems, the green lamp is on when the red and amber are both off. This is done by lighting it from a NOR gate connected to these two lamps.

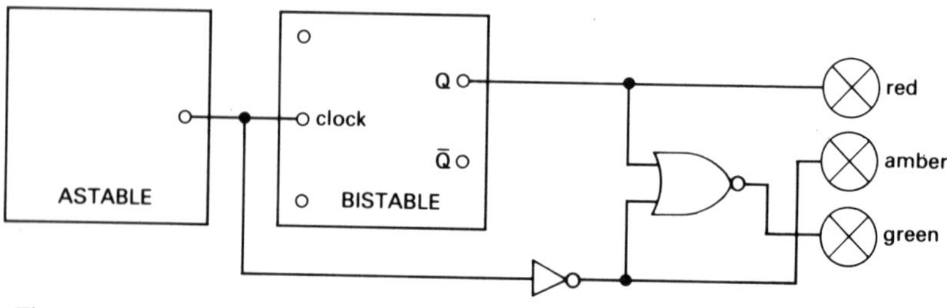

Figure C72
A simple traffic lights system.

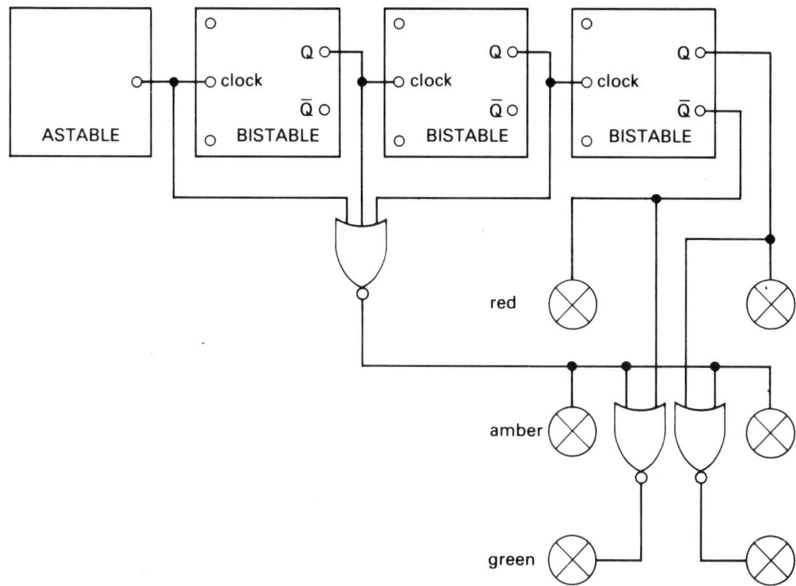

Figure C73
An alternative traffic lights system.

C18w Make a ring counter. A way of counting pulses is to make them light lamps in rotation. If there are ten lamps, numbered 0 to 9, one has a decade counter. Light three lamps in rotation. Now try four (which can be done with only two bistable modules, since two bistables will count up to four in binary arithmetic).

Note: Figure C74 works with kits to the new specification and the basic unit kit. Figure C75 works with the new kit. For use with the basic unit kit the clock input to the second bistable module should be connected to the Q output of the first.

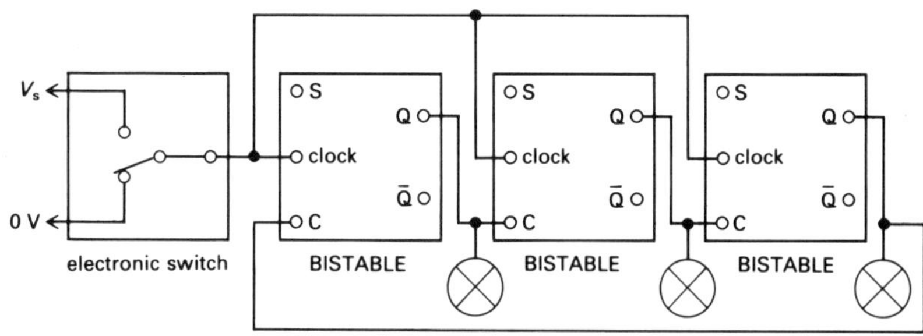

Figure C74
A counter (scale of 3).

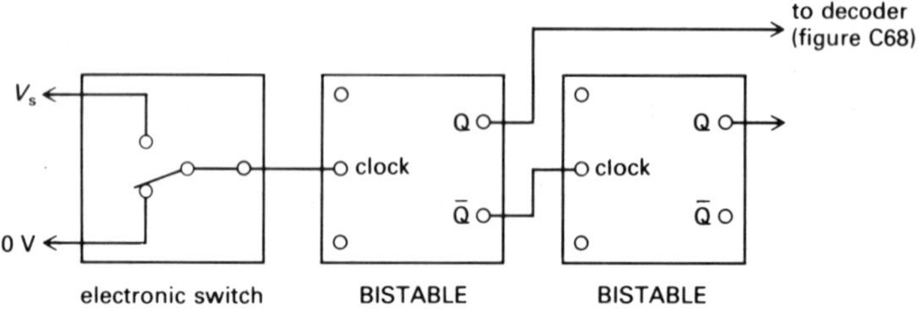

Figure C75
A counter (scale of 4).

C18x Playing tunes. The digital electronics kit can be used to play tunes. One can either make a simple organ with a keyboard (switches or push buttons) controlling a series of notes, or one can arrange that a system itself selects notes of the proper pitch, duration, and sequence to play a tune. Try something like the first bars of 'Three blind mice'.

Figure C76 shows a circuit which plays the beginning of 'Three blind mice'. A slow astable 'clock' drives a counter (cf. figure C75). The outputs of the decoder successively switch, via NAND gates, the signals from astable modules generating the notes E, D, C, successively. When Q4 is high no sound is produced, providing the rest.

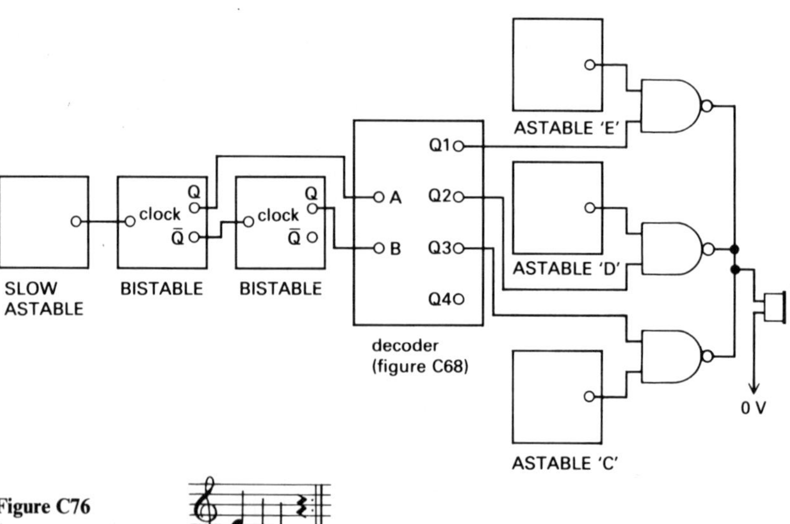

Figure C76
A system to play.

Unit D
OSCILLATIONS AND WAVES

Charles Milward
Royal Grammar School, Worcester

April Bueno de Mesquita
St Paul's Girls' School, London

Susan Ross
Godolphin and Latymer School, London

Suggested time allocation: four weeks

PLAN OF THE UNIT

Section D1
Introduction to oscillators

exponential decrease: Section B3 ► **Introduction to oscillators**

Frequency, displacement, amplitude

Clocks and time

Oscillations and circular motion

Phase differences

Angles in radians

Section D2
Mechanical waves and superposition

GCSE ► **Wavelength, frequency, displacement, amplitude, speed**

Longitudinal, transverse waves ► polarization: Unit J, 'Electromagnetic waves'

Superposition ► superposition of light, X-rays: Unit J, 'Electromagnetic waves'

Path and phase difference

Section D3
Mechanical oscillations

Factors affecting the period of an oscillator

Analysis of an oscillation ► oscillations in *LC* circuits: Unit H, 'Magnetic fields and a.c.'

interatomic force constant: Section A2; electric field: Unit E ► **Application: ions in a crystal**

energy in a spring: Section A1 ► **Energy and damping** ► spreading of energy: Unit K, 'Energy and entropy'; intensity and amplitude: Section J1

interatomic force constant, interatomic spacing: Section A2; Young modulus: Section A1 ► **Speed of a compression wave**

Section D4
Forced vibrations and resonance

Resonance

Quality factor, Q

Standing waves ► standing electron waves in atoms: Unit L, 'Waves, particles, and atoms'

INTRODUCTION

This Unit is about oscillations and waves. In the form and order suggested, it assumes that students will have prior acquaintance with wave concepts from the Nuffield or equivalent GCSE course. Even so, some revision of the principal ideas is suggested. If students have not covered the ideas before, the revision work could easily be extended.

There is much stress in this Unit on using graphs and interpreting their information. Section D1 starts by setting students a challenging practical task: obtain the displacement–time graph for a given oscillator. On the basis of these graphs, the key words displacement, amplitude, and frequency are applied to oscillations.

The analysis of oscillations will be accomplished in Section D3 by reference to the auxiliary circle, shown to be valid empirically. Section D1 concludes by showing a qualitative demonstration of the link between harmonic oscillation and circular motion; and from this early look at the auxiliary circle the important concept of phase difference is introduced.

Section D2, which could be taken at several different points within the Unit, starts by revising, as far as is considered necessary, ideas about waves from the students' earlier physics course; and the link between oscillations and waves is emphasized, with mechanical waves along lines of oscillating particles. Careful observation of pulses along a spring in experiment D4 will be required to bring out detailed points about wave motion; experiment D8 develops ideas about superposition, and allows the possibility of some open-ended experimenting, and also some quite accurate measurement, particularly with the wedge fringes for light. There are some interesting readings available on, for example, tsunamis; the possibility exists for a series of reading and reporting back tasks.

Section D3 analyses simple harmonic motion in some detail. It begins with individual experimental work in which students can investigate what factors affect the period of an oscillator, and in what quantitative way. Then the detailed motion of an oscillator is shown empirically to relate to circular motion, and hence be sinusoidal; and the sinusoidal motion is shown to fit the equation of motion of the oscillator. Two important and complementary mathematical threads emerge: the value of mathematical model-building, its power and limitations; and the method of numerical analysis, with its widespread applicability and its particular adaptability for computer use. It will be of great value to some students to see this latter method in use here: as an introduction to its use both in Unit L, 'Waves, particles, and atoms' to solve a one-dimensional Schrödinger equation, and in later post A-level applications.

Section D4 deals with an important practical consequence of oscillation: resonance. This is investigated with a simple laboratory system, but the wide variety of real-life examples is frequently stressed. The links between damping and energy exchange are mentioned, and brought together by considering the quality factor of an oscillator, Q. Finally more complex situations involving standing waves are considered, treating them both as wave patterns fitting between boundaries, and as special resonant systems oscillating at more than one frequency.

THE PLACE OF THE UNIT IN THE COURSE

The place of Unit D within the course is not tightly constrained. It will certainly help if Unit A has been covered, as several ideas and results from that Unit are put to good use: spring and interatomic force constants, Young modulus, interatomic spacing, energy stored in a stretched spring. There is no detail required from any other Unit, although brief reference is made to exponential decay (Unit B) and electric fields (Unit E). If taken as the fourth Unit it may offer a useful change of topic from the electricity and electronics of Units B and C, 'Currents, circuits, and charge', 'Digital electronic systems', and Unit E, 'Field and potential' which is also much, though not exclusively, concerned with electricity. Ideas from Unit D will be used again in several Units, particularly H, J, and L. Unit H, 'Magnetic fields and a.c.' includes oscillating LC circuits, whose behaviour can be analysed by analogy with the tethered-trolley oscillator. Unit J, 'Electromagnetic waves' develops the wave work, using in particular the ideas from Unit D about superposition. Unit L, 'Waves, particles, and atoms' requires the standing wave ideas developed in the last part of Unit D. Unit D must therefore precede Units H, J, and L.

The order within the Unit can be varied according to individual teaching taste. Sections D1, D3, and D4 form a logical sequence, covering free oscillations qualitatively, then in some quantitative detail, before dealing with forced oscillations. There is latitude, however, in the point at which Section D2, 'Mechanical waves and superposition', is taken: it can be used to begin the Unit; it can be taught after an initial look at oscillations using Section D1, as the Unit is written; or it can be used after the detailed analysis of an oscillator in Section D3. It should, however, come before the later part of Section D3, which deals with the speed of compression pulses.

LIST OF SUGGESTED EXPERIMENTS AND DEMONSTRATIONS

SECTION D1

INTRODUCTION TO OSCILLATIONS

This Section starts with a brief introduction to oscillators of various kinds and the properties they have in common, after which students can discuss the use of oscillators to measure time, and some of the problems and questions this poses. This is followed by a comparison between oscillations and circular motion as a means of understanding phase, phase difference, and angular velocity.

The first set of experiments gives students the opportunity of seeing various oscillators, and of devising a method of producing the displacement against time graph for one of them.

At some stage, perhaps before experimenting starts, check students' understanding of the meanings of the following terms: displacement, amplitude, period, frequency, time trace.

GROUP OF EXPERIMENTS

D1 How do oscillators move?

A choice can be made from the following examples, some of which appear in the Revised Nuffield Physics *Teachers' guide Year 5*, pages 70–74. (The experiment numbers in brackets refer to that book.) You may want to use other examples, but your choice should include at least one non-isochronous oscillator such as **h**, **i**, or **j** below.

a Pendulum (experiment 37)

b Torsion pendulum

c Lath with load

d Inertia balance (experiment 36e)

e Ball rolling on curved tracks (experiment 36f)

f Mass on spring (experiment 35b)

g Undamped light beam galvanometer

h Bar magnet suspended over fixed magnet

i Large-amplitude pendulum

j Air track vehicle running between elastic barriers

k U-tube containing liquid (experiment 36c)

The students should be allowed to see all the oscillators working, and be given a demonstration of how the displacement changing with time can be recorded in one example, perhaps using the pendulum with an ink brush, or the lath and load as shown below.

Working in twos or threes the students can now be given an oscillator and asked to devise a method for showing how its displacement varies with time.

They may want to use some of the following items:

ITEM NO.	ITEM
108/1	ticker-timer
108/3	ticker-tape
507	stopclock
130/2	photodiode assembly with light source
1503	timer (resolution 10 ms)
134/2	xenon stroboscope
134/1	motor-driven stroboscope
133	camera
1033	cell holder with 4 cells
1510	potentiometer, 1 kΩ
1511	oscilloscope
1155	conducting paper

D1a Pendulum

ITEM NO.	ITEM
10F	broomstick pendulum

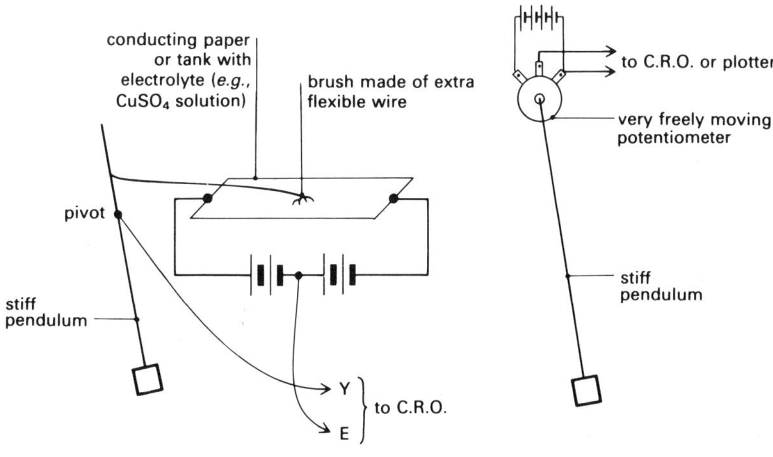

Figure D1 **Figure D2**

The time trace can be obtained by drawing it direct onto moving paper with a brush or sand; using ticker-tape; or by one of the electrical methods illustrated in figures D1 and D2.

D1b Torsion pendulum

ITEM NO.	ITEM
1153	string
503	2 retort stand bases
504, 505	3 retort stand rods and bosses
44/1	G-clamp, large

The rod hangs on a bifilar suspension (two pieces of string) and performs torsional oscillations. This may be better as a demonstration due to the difficulty of producing the time trace, or given to an inventive group of students as a challenge. It can be done by repeated timing, or by multiflash photography.

D1c Lath with load (large-scale version suitable for demonstration)

ITEM NO.	ITEM
	either
1153	long lath, 2.5 m by 75 mm by 10 mm
	or
501	metre rule
55	friction kit
150	fractional horsepower motor
9F	lineshaft unit
59	l.t. variable voltage supply
44/1	2 G-clamps, large
32	2 masses, 1 kg
1000	leads

Also required: a fine pointed felt-tip or ball-point pen; clean smooth paper about 0.5 m by 0.2 m; layers of soft paper such as kitchen paper towelling.

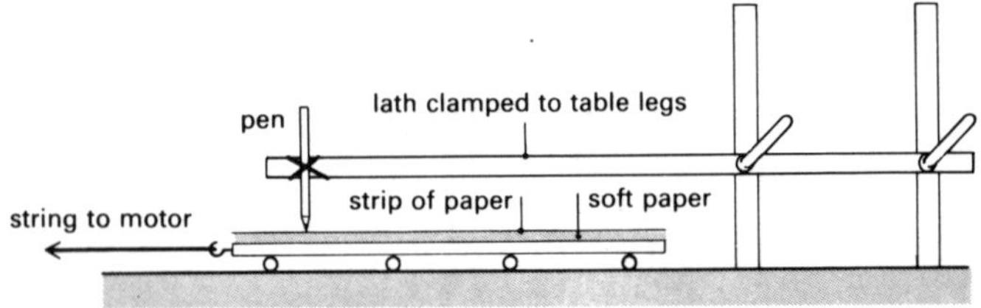

Figure D3
Lath with load.

Clamp the lath firmly at one end to a pair of table legs, about 0.1 m above the floor (see figure D3). Masses can be attached to its free end with rubber bands or string. To obtain a time trace, make a pen on the end of the lath write on a moving paper strip. The paper can rest over several layers of soft paper on a wooden board running on steel rollers. The soft paper allows the pen to write smoothly despite bumpy movement of the moving board, and is needed if the decay of the oscillations is to be anything like exponential.

One way of towing the board and paper is to pull it along with a string wrapped round a shaft turned by the motor.

Students should be asked if they can recognize the shape of the envelope of the oscillation curve from earlier work (demonstration B20, Decay of charge, in Unit B).

Note that exponential decay arises when the damping force is proportional to velocity, and will not necessarily appear in other cases.

D1d Inertia balance (wig-wag)

ITEM NO.	ITEM
146	inertia balance
44/2	2 G-clamps, small

The time trace may be obtained using a ticker-timer or multiflash photography.

D1e Ball rolling on curved tracks

ITEM NO.	ITEM
1153	3 lengths of curtain rail
1153	large ball-bearing, 1 or 2 cm diameter

The three suggested shapes are shown in figure D4.

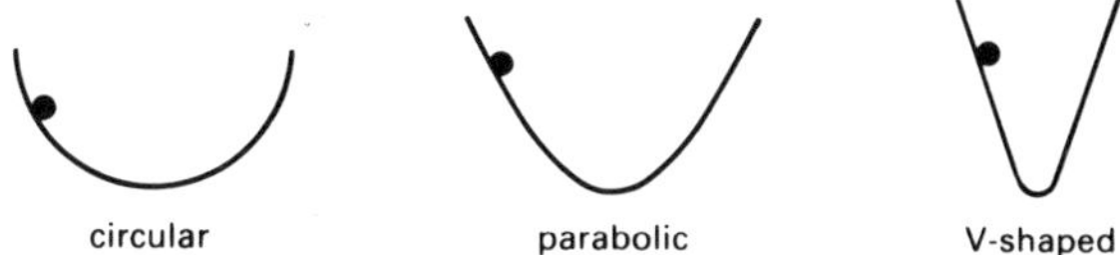

Figure D4
Ball rolling on curved tracks.

The time traces could be obtained using a light beam timing method repeatedly; or by multiflash photography. Students should be invited to *listen* carefully as each oscillation dies away; then they may notice that the circular track gives near isochronous motion; the parabolic one does.

D1f Mass oscillating vertically on spring

ITEM NO.	ITEM
2A	expendable spring
503–6	retort stand base, rod, boss, and clamp
44/1	G-clamp, large
31/2	hanger with masses totalling 400 g

The time trace can be obtained using multiflash photography, or ticker-tape.

D1g Undamped light beam galvanometer

	light beam galvanometer (if available)
1033	cell holder with one cell
52L	switch
1017	resistance subsitution box
1000	leads

Set up a series circuit so that with the switch closed the reading on the galvanometer, at its most sensitive scale before the 'direct' position, is almost full scale. A resistor of over 500 kΩ will be needed. Then, with the galvanometer on its 'direct' setting, after the switch is opened, the light spot will oscillate about its central zero position.

A time trace for one oscillation can be obtained by photography, using a multiple-slit stroboscope. Students can also record how the amplitude dies away, and the isochronous property of the oscillations.

D1h Bar magnet suspended over another magnet

ITEM NO.	ITEM
50/1	cylindrical magnet
50/2	horseshoe magnet
503–6	retort stand base, rod, boss, and clamp
1153	nylon fishing line

Hang the bar magnet on nylon line or cotton so that it is horizontal and lies just over the poles of the horseshoe magnet resting on its back. A small piece of mirror and a suitable lamp should be to hand so that the oscillations can be observed by optical means.

D1i Large-amplitude pendulum

ITEM NO.	ITEM
154/1	turntable (or large gyroscope)
	either
505	boss
	or
44/2	G-clamp, small
503–5	retort stand base, rod, and boss

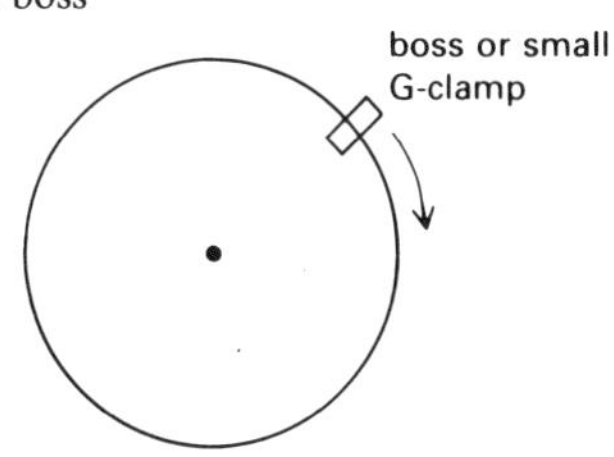

Figure D5
Large-amplitude pendulum.

Fasten the boss or small G-clamp onto the rim of the turntable or gyroscope, clamped with its axis horizontal (figure D5). The boss or clamp executes large-amplitude oscillations about the lowest position.

The time trace can be obtained by multiflash photography, or repeated timing. Students should quickly realize that this motion is not isochronous.

D1j Air track vehicle running freely between elastic barriers

ITEM NO.	ITEM
1019	air track and accessories
1020	air blower

Improvise an arrangement for holding two tightly stretched rubber bands across the track about 0.5 m apart, so that the vehicle rebounds between them with little loss of energy. A magnetic repulsion system can also be arranged, but with greater difficulty. Students should obtain a graph covering a number of cycles. In a sense this is trivial, but it makes the point that not all oscillations are isochronous; in this case, although energy is lost the amplitude remains constant.

D1k U-tube containing liquid

ITEM NO.	ITEM
1155	large U-tube filled with water or potassium manganate(VII) solution

Obtaining this time trace is not easy, since the time period is short and the damping is high. A good student may be able to develop a method using a light beam and scaler, with repeated timings; another possibility is an electrical contact method, using an electrolyte solution to make a connection between electrodes placed at measured heights in the U-tube.

Discussion

The results of each group's attempts to produce a time trace of their oscillator should bring out the fact that many keep steady time (isochronous) though some do not. Those that keep steady time tend to have a characteristic wavy trace, with damping only affecting the amplitude and not the period. On some traces the exponential decay of amplitude due to damping might be visible.

In some cases the factors affecting the period will have become clear – for example, increasing the mass of the oscillator often, though not always, increases its period. Students can be asked about these points as they present their graphs.

The idea of natural frequency or constant period of some oscillators can be used as an introduction to a discussion about time itself. The depth to which this is taken will depend on the group and its interest. One starting point would be to consider which of the oscillators could be used as clocks and how accurate they would be. The importance of being able to measure time accurately can be discussed (see below); this might lead to the more philosophical type of question such as: 'What is time?', 'Does time run evenly?', 'Could we know if it did?'.

Use of clocks

In the middle of the eighteenth century, John Harrison solved the problem of making a clock that would keep good enough time to navigate a ship by, despite the rolling of the ship in the sea and despite changes in temperature. This, along with many other interesting clocks, is on display at Greenwich (National Maritime and Observatory Museums).

It will be necessary to explain that a clock is used in navigation to find the longitude, that is, one's position East or West of a given place. Crudely, if sunrise occurs six hours late or early, one has travelled a quarter of the way round the Earth. An error of one minute in time makes a navigational error of nearly thirty kilometres at the Equator. Between 18 November 1761 and 21 January 1762 one of Harrison's clocks was taken on a voyage from England to Jamaica. On arrival, it was tested and found to be in error by only five seconds. How could such a test be made, there being no other clock that could be taken along for comparison? (By finding the longitude of Jamaica by other, astronomical, means and noting the time of, say, noon in Jamaica.)

Nowadays extremely accurate timekeepers are available for everyone at very low cost – using an oscillating quartz crystal, suitable electronics, and a digital display. Many students in the class probably have such a digital watch. A discussion about the value of the 'hundreths of a second' digit may be heated but useful.

Home experiment

Home experiment DH1, Making a chronometer, challenges students to make an accurate and robust chronometer (*Students' guide*, page 262).

Reading

The *Students' guide* gives some ideas for discussion. The following references are useful:

FEATHER *Introduction to physics*, Volume 1.
FEYNMAN, LEIGHTON, and SANDS *Lectures on physics*, Volume 1.
GOULD *John Harrison and his timekeepers.*
HOWSE *Greenwich time and the discovery of the longitude.*

Questions

Questions 1 and 2 are about the motion of oscillators. Questions 3 to 5 are about the measurement of time; question 5 is about the definition of the second.

Film loop

The film loop 'The measurement of *G*' shows a superb example of damped torsional oscillations.

DEMONSTRATION
D2 Oscillators and circular motion

This demonstration suggests a link between circular motion and oscillations, and can be used to introduce the important ideas of phase difference, angular velocity, and radian measure.

ITEM NO.	ITEM
	either
150, 154/1	fractional horsepower motor, with gearbox, turntable, and band
59	l.t. variable voltage supply
	or
	record-player turntable
1155	2 pendulum bobs
1153	length of string (1 metre)
121	2 metal strips (as jaws)
503–6	retort stand base, rod, boss, and clamp
21	compact light source
102	screen
27	transformer
1000	leads

The length of the pendulum should be adjusted to have the same period as the rotating turntable (approximately 75 cm if the turntable is rotating at 33 r.p.m.). See Revised Nuffield Physics *Teachers' guide Year 5*, Experiment 35a for a convenient way of setting up the pendulum.

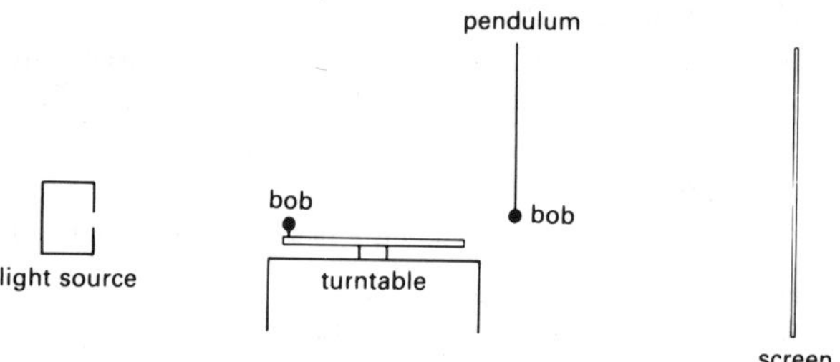

Figure D6
Oscillator and circular motion.

This demonstration brings out the idea of phase angle in terms of the fraction of a complete oscillation that one bob is behind the other.

If the pendulum bob is not exactly in phase with the rotating bob, the phase difference between them can be seen as an angle, as the pendulum passes its equilibrium position. See figure D7.

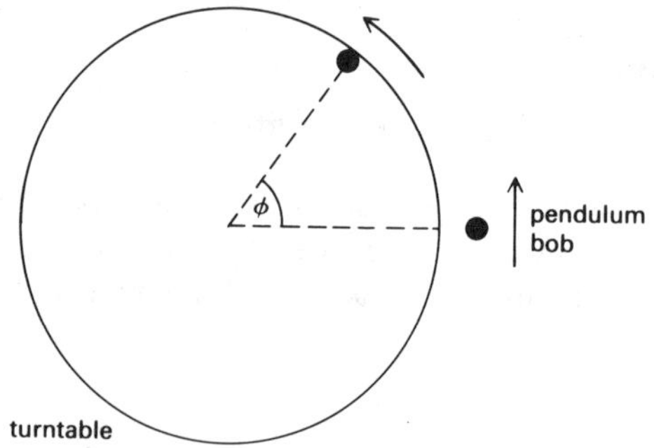

Figure D7
The turntable bob leads the pendulum bob by ϕ.

If the pendulum is too long or too short, the changing phase difference between the two shadows is visible.

Radian measure will need to be brought in here, and can be discussed with angular velocity and its units. Any non-mathematicians in the group will probably not have met the radian, and it is important that they understand it now, before they need to use it in more complicated situations later. The definition of the radian is given in the *Students' guide*, but they will need practice at using it.

It may be worth projecting the shadow of two bobs fixed to the rotating turntable (without the pendulum) to emphasize how a difference in *angle* between the positions of the two bobs corresponds to a difference of *phase* between the two oscillating shadows.

Computer

The program 'SHM' has the same aim as demonstration D2 – to show the relationship between oscillatory and circular motion, and to illustrate phase difference. One or two oscillators (with variable phase difference), the associated circular motions, and time traces can be shown.

Questions

Question 6 is about angles in radians. Questions 7 and 8 show that for small angles $\sin\theta \approx \tan\theta \approx \theta$ in radians.

SECTION D2

MECHANICAL WAVES AND SUPERPOSITION

Section D1 dealt with oscillators; this Section is concerned with a product of oscillators, that is, waves. We suggest starting with mechanical waves that can be seen or felt, like waves on water; waves on taut wires or on strings; sound waves; shocks or vibrations in buildings or in the Earth; and waves of compression and expansion passed along from atom to atom. Later in the Section we look at superposition as a property of all waves, mechanical or electromagnetic.

Mechanical waves have practical importance for ship builders and harbour makers; designers of musical instruments, of electricity cables, and of suspension bridges; acoustic engineers; architects; installers of vibrating machinery; aircraft engine designers; geophysicists; and many others. There is no time to look at all these waves, but luckily there is no great need to: similar arguments about why the wave travels and how fast it goes, and how and why something, whatever it is, moves up and down or to and fro as the wave passes apply to most wave systems.

Many of the students will be familiar with waves and might feel that the opening experiments of this Section are simply revision. It is worth pointing out that familiarity with the properties of simple mechanical waves will enable them to understand and explain more complicated and practically useful wave systems. An example of this will arise later in the course in Unit J, 'Electromagnetic waves'.

DEMONSTRATION

D3 Basic ideas about waves

ITEM NO.	ITEM
	Some method of producing easily visible waves is required such as:
	either
90	ripple tank kit
	or
1013	long spring
	or
101	large Slinky spring
	or
	any other wave machine available

Use the apparatus to show how a single transverse pulse and a continuous transverse progressive wave are produced; then in discussion ensure that all students can answer the following: 'What is a wave?' 'What is it that moves?' 'What is meant by wavelength, frequency, displacement, and amplitude, and in what units are these quantities measured?' 'Why is the speed of a wave given by $c = f \times \lambda$?'

This is intended as a quick demonstration and introduction to simple mechanical waves, and should not take up too much time; nor should it start to answer further questions that the students should answer for themselves in the next group of experiments.

Home experiment

Home experiment DH2, Make your own wave machine (*Students' guide* page 262), describes how to make a wave machine out of simple components at home.

GROUP OF EXPERIMENTS

D4 Properties of mechanical waves

In these experiments students look in more detail at what happens as a single pulse travels. Working in twos or threes, students are asked to observe as much as they can about particular systems, mainly transverse pulses on a long spring. For variety one or two groups can try the other devices (D4b and c).

D4a Transverse waves on a long narrow spring

ITEM NO.	ITEM
1013	long spring
507	stopwatch or stopclock
501	metre rule

D4b Transverse waves on a Slinky spring

ITEM NO.	ITEM
101	large Slinky spring
507	stopwatch or stopclock
501	metre rule

D4c Transverse waves on a trolleys-and-springs model

ITEM NO.	ITEM
106/1	12 dynamics trolleys
2A	44 expendable steel springs
32	12 masses, 1 kg (or 12 more trolleys)
507	stopwatch or stopclock

The model is made of a row of trolleys linked by springs, as shown in figure D8. The trolleys are spaced out so that the springs are in tension. If the model is set up on a table, it is easy for part of it to run off the side, and the rest inexorably follows! So it is best set up on a smooth floor, or on a table with barriers along its edges.

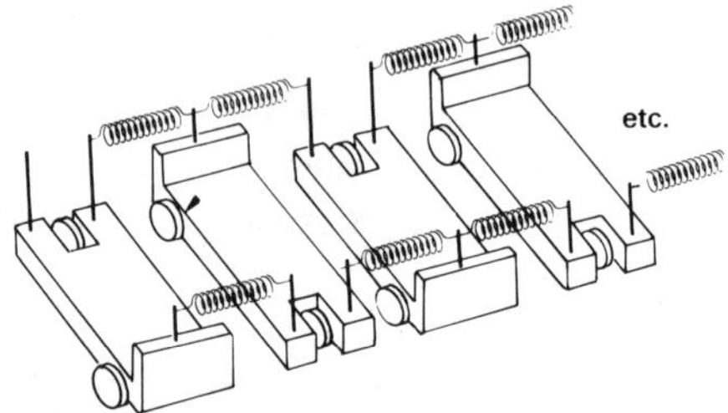

Figure D8
Transverse wave model made of trolleys and springs.

Each pair of students using experiments D4a and b needs a long narrow space to work in, such as a corridor. It is usually best to work on the floor rather than on a bench.

It is suggested that at least half the class use the long springs, as the shape of the pulse is easy to see, and the springs do not become entangled as easily as the Slinky springs.

Even in this simple experiment some questions about instruments and uncertainties can be asked:
'What is the resolution of the stopwatch?'
'But what about personal error?' 'Might we tend to get consistently high or consistently low readings?' 'And how repeatable are our readings in practice?'

Students who make the observations suggested in the laboratory notes should reach the following conclusions:

> The speed of a pulse does not depend on its amplitude, but only on the tension of the spring.
> The pulse speed does not depend on how quickly the spring is jerked to start the pulse (speed independent of frequency), nor on friction. Friction does, however, cause the amplitude of the pulse to decrease as it travels, although the general shape of the pulse does not alter.
> The pulse height depends on the size of the sideways jerk, that is, on the amplitude of the oscillation that causes it.
> The pulse length depends on the wave speed and on the rapidity of the motion starting the pulse, that is, on the frequency of the oscillation.
> Two pulses superpose one on top of another, and pass through one another without effect on either.

Pulses are reflected at a fixed end, but the reflected pulse is upside down, that is, there has been a phase change of π on reflection.

Continuous wave trains can be sent along the spring and reflected. As the reflected wave superposes with the outgoing one, they may combine to form an oscillation pattern that does not travel – a standing wave.

Pulses reflected at a free end do not experience a phase change as they are reflected.

Pulses can be refracted and reflected at a boundary. For example, if a piece of thick string is tied to the end of the spring, as the pulse travels from the spring to the string some of the pulse is reflected and some is transmitted, travelling at a higher speed along the lighter string (*cf.* light travelling from a dense to a less dense substance).

In D4c the mass of each trolley can be doubled by adding loads, and the tension doubled by adding springs in parallel with the original ones. Students should be encouraged to double mass and tension independently, then together. Important observations to bring out are:

Doubling the mass of each trolley reduces the wave speed.

Doubling the tension increases the wave speed.

Both modifications change the speed by the same factor ($\sqrt{2}$), and both made together will restore the speed to its original value.

The trolley model used in D4c can serve as a convenient demonstration around which to focus the discussion of waves on long springs, so that all the students see both systems.

A system like this model, with the mass of the medium concentrated in discrete lumps with forces between each, does not behave in all respects like a smoothly spread out medium would do. The system is dispersive: the speed depends upon the wavelength when the wavelength is not much larger than the spacing between parts of the lumped medium. It exhibits 'cut off': waves of high frequency are not propagated at all. (Try moving an end trolley very rapidly to and fro. The next-door trolley oscillates a little, the next oscillates less, and there is something like an exponential decrease of amplitude along the system. No wave energy propagates down the system.) These problems are discussed in CRAWFORD *The Berkeley Physics Course*, Volume 3, *Waves*.

Graphs: displacement–distance and displacement–time

In Section D1 displacement–time graphs were drawn for oscillators; in this Section students have been drawing displacement–distance graphs for their sketches of wave motions. This is a good moment to make sure that they appreciate the difference between the two types of graph. They should also be able to deduce the displacement–time graph for a point

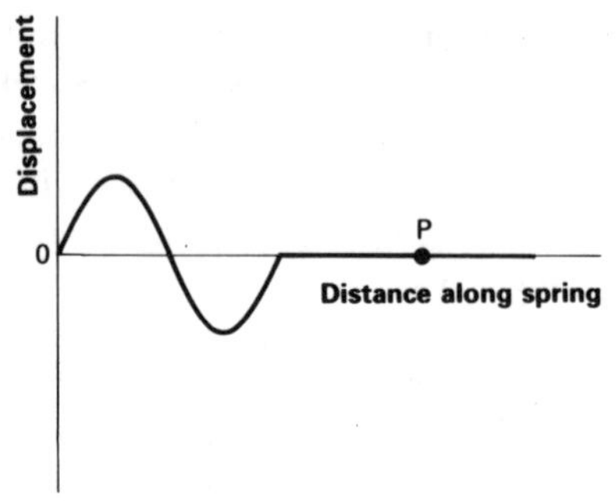

Figure D9
Displacement–distance graph for a pulse on a spring, at time $t = 0$. The pulse is travelling to the right.

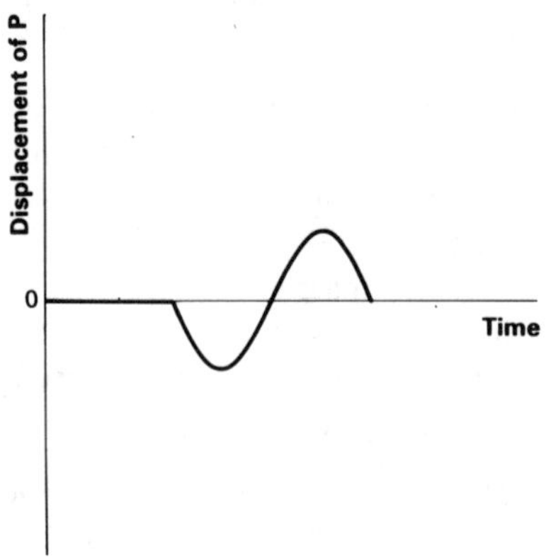

Figure D10
Displacement–time graph for the point P on the spring as the pulse travels along.

on a spring as a pulse passes, from the displacement–distance graph of the pulse. See figures D9 and D10 for an example.

The point can be made that if the displacement–time graph of the oscillator creating the wave and the wave speed are both known (effectively this means that f and c are known) then the wave profile (the displacement–distance graph) and hence λ can be deduced.

DEMONSTRATION

D5 Longitudinal waves on a Slinky spring

Students who used the Slinky spring in experiment D4b may have tried longitudinal waves on their springs, and most of the class will have met them in an earlier course. However, many students do have difficulty in understanding how a longitudinal wave travels, what its wavelength is, and how it can be represented graphically. It is worth spending a few minutes on this now.

ITEM NO.	ITEM
101	large Slinky spring (the Slinky will need to be on a smooth surface)

Show that compression and expansion pulses travel at the same speed. The class should watch the motion of one part of the spring as the pulse goes past. Pull one end of the spring sharply and keep on pulling, moving the end at a steady speed: the stretched region of the spring can be seen to spread along to the far end, which feels no extra pull until the wave front reaches it. In the stretched part all the coils are moving slowly one way, while the wave front is travelling faster the other way.

This demonstration can lead on to a discussion of how sound travels through a solid or a gas. Text books often represent sound waves in a gas (especially standing waves in open and closed pipes) by sinusoidal graphs showing the variation of gas pressure along the pipe. Students can easily misinterpret these as transverse wave profiles; point out that they are in fact graphs which show how the pressure (or perhaps the longitudinal displacement of the gas particles) varies along the length of the pipe at any one instant of time (see figure D11).

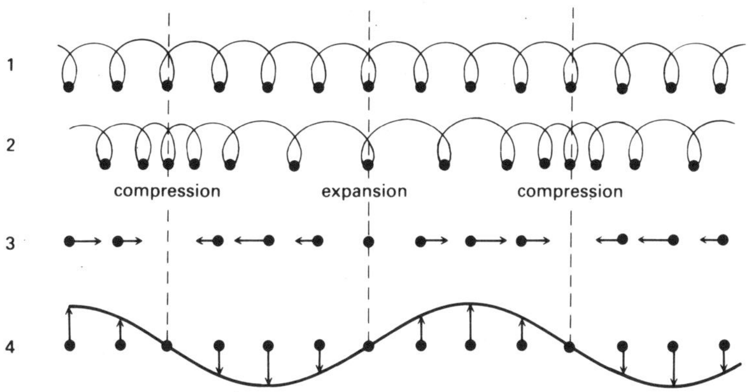

1 Initial position of spring or particles
2 Spring or particles displaced as wave passes
3 Arrows representing the amount and direction of displacement
4 Graph of displacement against distance

Figure D11
Representation of a longitudinal wave.

Resources

A wave machine can illustrate these points well. So too does the computer program 'Longitudinal waves' from Five Ways Software, published by Heinemann Computers in Education.

Questions

Questions 9 to 13 are about basic wave properties and behaviour.

Reading

Some useful background readings on mechanical waves are listed below.

CRAC *Waves and sound.*

TRICKER *Bores, breakers, waves, and wakes.*

CHAUNDY *Waves.*

BASCOM 'Ocean Waves'. *Scientific American.*

BERNSTEIN 'Tsunamis'. *Scientific American.*

GRIFFIN 'More about bat radar'. *Scientific American.*

BULLEN 'The interior of the Earth'. *Scientific American.*

GRIFFIN *Echoes of bats and men.*

OLIVER 'Long earthquake waves'. *Scientific American.*

SCSST Physics plus, *Sonar.*

CARNELL 'Seismic exploration' in *Physics principles at work; a resource book for teachers.'*

OPEN UNIVERSITY 'Earthquake waves and the Earth's interior'.

SUPERPOSITION OF WAVES

Classes following the order suggested here will already have observed pulses on long springs. Most students will also be thoroughly familiar with interference on a ripple tank from earlier courses. It is not necessary for students to handle this equipment again but a quick demonstration might help refresh memories and provide a focus for discussion.

DEMONSTRATION/EXPERIMENT

D6 What happens when waves meet?

D6a Waves on a spring

ITEM NO.	ITEM
1013	long spring

Any student who did not see it in experiment D4 must now see superposition of a wave train on a long pulse, and the pattern of nodes and antinodes which results when a long wave train is reflected at the end of the spring.

D6b Ripples on water

ITEM NO.	ITEM
90	ripple tank kit

Any students who have not already seen superposition of ripples on water must see it now. Others may appreciate a revision of this.

The waves spreading from the two sources superpose to produce a stationary pattern of nodal and antinodal lines, the positions of which may be predicted by consideration of path differences and the associated phase differences. No energy is propagated along the nodal lines. They are lines of near zero disturbance, or minima. Extra energy is propagated along the antinodal lines, or maxima, where the amplitude of vibration is extra large.

Maxima occur at points wherever the path difference $= n\lambda$.

Minima occur wherever the path difference $= (n + \frac{1}{2})\lambda$.

DEMONSTRATION

D7 Path differences and phase difference

ITEM NO.	ITEM
1109	signal generator
183	loudspeaker
157	2 microphones
1511	double beam oscilloscope (sensitivity of 1 cm for 10 mV required)
501	metre rule
1000	leads

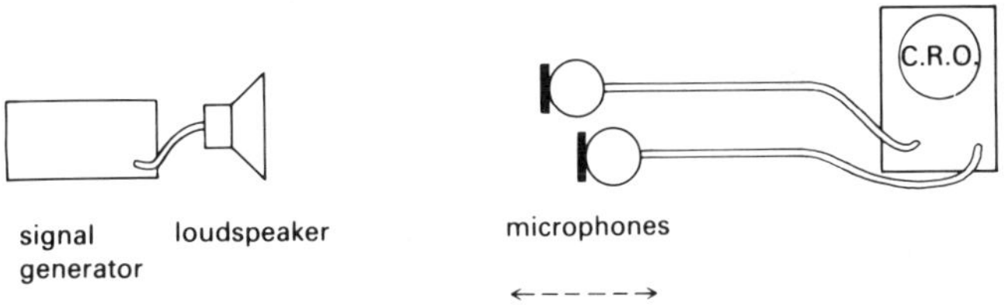

Figure D12
Path differences and phase difference.

Set the frequency of the signal generator at about 1 kHz. Move one microphone (not that connected to the channel triggering the time-base) towards or away from the loudspeaker. Path differences between waves reaching the two microphones may be measured with a ruler. The phase difference between the sound waves reaching the two microphones may be observed on the oscilloscope. Differing amplitudes will arise, not only because one microphone is nearer the source, but also because the sensitivities of the microphones differ. Measure the wavelength of the sound waves by finding how far the microphone is moved between adjacent positions where the signals are in phase.

If the oscilloscope has the appropriate facility for adding signals, you could use it to display the resultant obtained by adding the two signals for various phase differences.

But note that this arrangement is confusingly similar to the superposition experiments which follow later. Here the *electrical* signals arriving at the oscilloscope are being added. There is no superposition of sound waves.

Discussion and conclusions from these demonstrations

When two or more waves simultaneously pass through one place, the displacement at that place is the sum, taking account of phase differences, of the displacements of the individual waves.

This applies so long as there is no discontinuity in the properties of the system – thus an exception arises if, for example, two ordinary waves in shallow water combine to achieve enough height to break.

If the superposing waves have equal constant frequencies, and emerge from stationary sources, then a pattern of maxima and minima is formed which is stationary in space. This is so even if there are phase differences between the sources.

If two sources have different but constant frequencies f_1 and f_2, then beats will occur. At any given point where the waves superpose, the amplitude of the oscillations will rise and fall at a frequency $|f_1 - f_2|$. Beats will generally be noticeable only if the values of f_1 and f_2 are close.

A similar superposition principle applies to the addition of electrical signals in circuits – as, for example, in the combination of carrier and modulation frequencies to form an amplitude-modulated radio signal. The suggestion above to combine the two microphone signals within the oscilloscope is another example. However, this is *not* superposition of waves in the usual sense, and should be mentioned only with caution.

Appendix IV describes a way of phase-shifting a sine wave using an op-amp circuit. It may be useful as an illustration of the meaning of phase differences: the original signal and the shifted one can be displayed on a dual-trace oscilloscope, and the phase-difference altered and discussed. As before, with a suitable oscilloscope the two signals can be combined and their sum displayed.

Superposition effects as a characteristic of wave motion

Given some radiation, how can one tell whether or not it has wave properties? Superposition experiments offer an answer.

In each of the experiments suggested below, superposition effects are observed and the wavelength can be found by measuring a path difference. If the frequency is known, the speed of the waves may also be calculated.

Some of the suggested experiments involve reflection. Reflection often, but not always, results in a phase change of π. If there is a node at the reflecting surface, then a phase change of π is occurring. In calculating the wavelength, either this phase change must be taken into account by adding an extra $\frac{1}{2}\lambda$ to the path difference, or the method employed must be adapted so that it will not affect the results: by, for example, making the path difference *change* by one wavelength, so that the received signal changes from a minimum through a maximum to another minimum.

GROUP OF EXPERIMENTS

D8 Superposition of waves and determination of wavelength

The experiments suggested (for a class of 16 working in pairs) are:

a 1 GHz radiowaves 4 sets

b Microwaves 1 set

c v.h.f. radio or u.h.f. television (optional)

d Light (air wedge) 2 sets (or more)

e Sound 1 set or 2 sets

We advise a time allowance of two long practical sessions so that each pair of students may handle the 1 GHz apparatus and try one other experiment. More time may be needed to allow students to devise a method, try it, and, if necessary, think again. By the end, all students should have seen a demonstration of 2-slit interference using microwaves, and should observe the fringes from an air wedge in D8d; but in this case measurement and the associated theory should be reserved for the most keen and able.

In all the other experiments wavelength should be determined, and if the frequency is known the speed should be calculated from $c = f\lambda$.

The students' laboratory notes encourage them to estimate the uncertainty in their values for wavelength and speed. Some profitable discussions may arise from these estimates.

D8a 1 GHz radio waves

ITEM NO.	ITEM
1050	15 cm dipoles and oscillator, 1 GHz
	either
1507	microammeter
	or
181 and 183	general purpose amplifier and loudspeaker
	or
1101	sensitive galvanometer (*e.g.*, internal light beam)
1153	2 metal screens
501	metre rule
1000	leads

Home Office regulations prohibit the use of this apparatus if it is a nuisance, particularly if it causes interference on television channels. It is desirable to use a television set to discover the range at which there is interference, which will depend on the local frequencies and the nature of the surrounding buildings.

Students may wonder how radiation emitted by the transmitter causes electrical oscillations in the receiver. What is the nature of this radiation? In Unit J, 'Electromagnetic waves', they will learn that it involves fluctuations of electric field travelling from one dipole to the other.

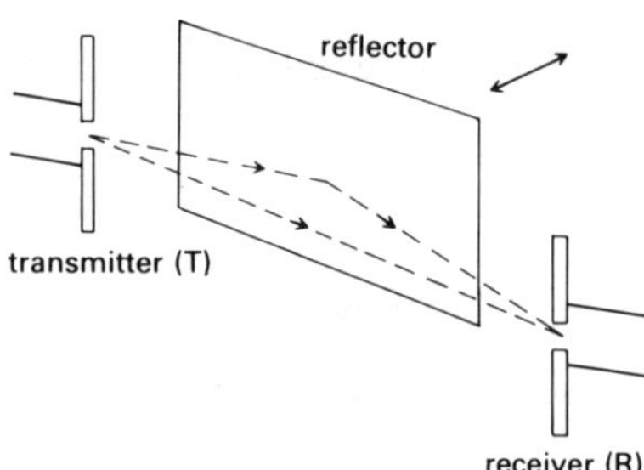

Figure D13

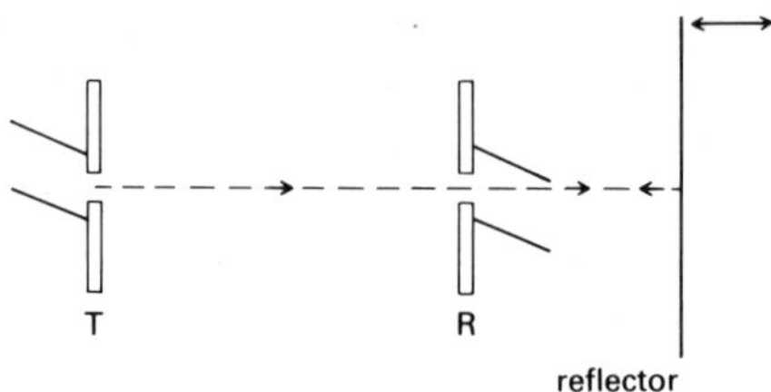

Figure D14

The wavelength (about 30 cm) is long enough for the aerials to look similar to familiar television aerials, but short enough for experiments within limited laboratory space.

Students should establish what types of material transmit or reflect this radiation (radiation will appear behind a metal screen which is too small); and then observe superposition (interference) effects, and measure the wavelength (figures D13, D14).

Faster groups may be encouraged to try further experiments (as suggested in figures D15 to D18), and all may profitably be allowed some time to play with the equipment.

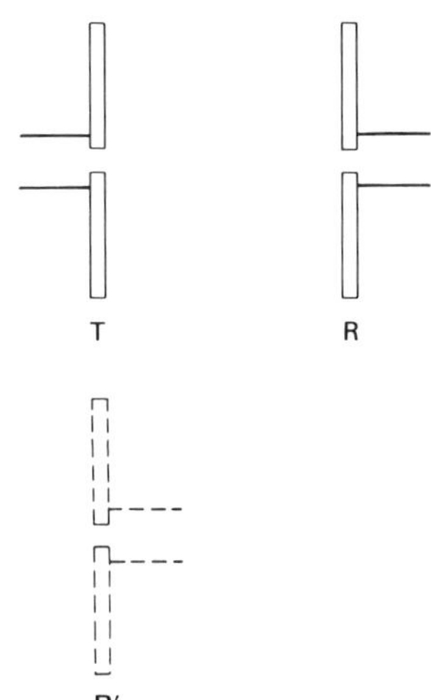

Figure D15
Directional properties of the dipoles: there should be no signal at R′.

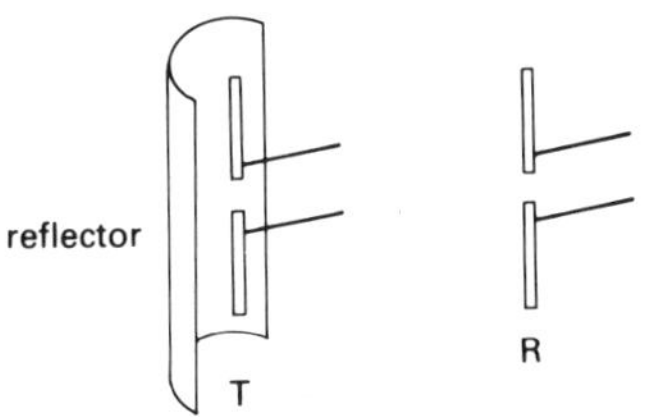

Figure D16
Beaming the radiation using a curved reflector behind the transmitter.

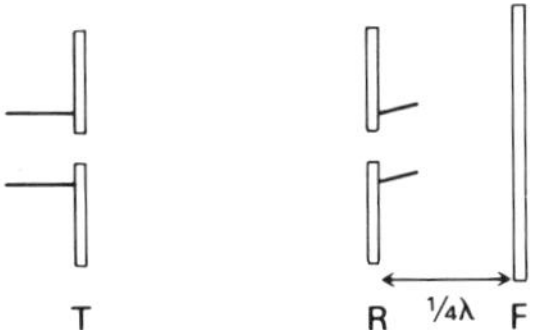

Figure D17
Effect of a 'reflector' (F) as in a television H aerial – a rod a little over $\frac{1}{2}\lambda$ long placed $\frac{1}{4}\lambda$ behind the receiver.

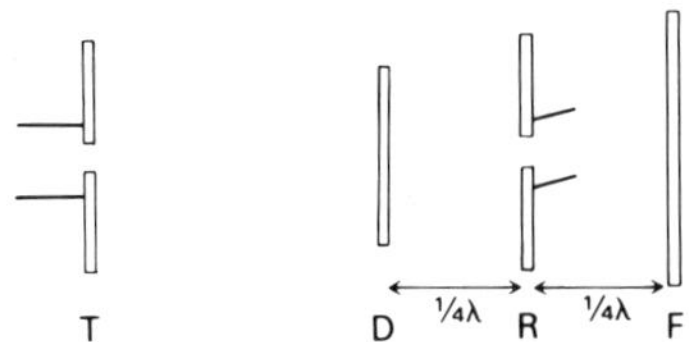

Figure D18
Effect of a 'director' (D) — a rod a little less than $\frac{1}{2}\lambda$ long placed $\frac{1}{4}\lambda$ in front of the receiver.

A director (figure D18) makes the system more directional. If T is moved significantly away from the line DRF, the received signal falls off sharply. Some of these effects are taken up again in Unit J, 'Electromagnetic waves'.

D8b Microwaves

ITEM NO.	ITEM
184/1	microwave transmitter
184/2	microwave receiver
1153	2 metal reflectors, about 0.3 m square
1153	narrow metal plate, 60 mm wide
181	general purpose amplifier
183	loudspeaker (if not with above)
1507	microammeter (if not incorporated in receiver)
1045	diode probe for microwave experiments
501	metre rule
1000	leads

A receiver made of a diode in a waveguide behind a horn is more sensitive than a simple diode on its own, but it is more complicated and its directional properties make it unsuitable for some experiments.

Various experiments are suggested in figure D19.

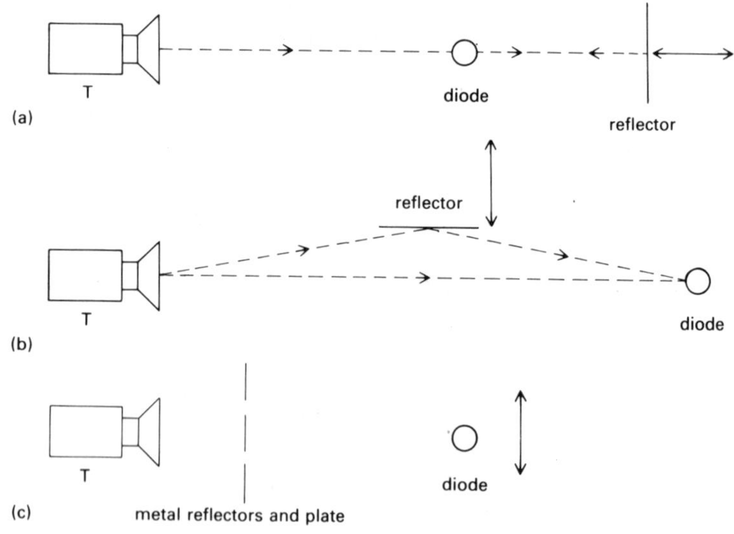

Figure D19
Superposition of microwaves.

It is worthwhile leading students to set up a two-slit experiment in which wavelength is found by direct measurement of the distances from each slit to the diode. Note that modern equipment emits 2.8 cm, rather than 3 cm radiation.

D8c v.h.f. radio waves or u.h.f. television waves

portable v.h.f. radio set (preferably a cheap one with a rod aerial)
u.h.f. television set
1077 television aerial
1501 coaxial cable
1153 metal reflector
501 metre rule

The basic experiment is simple: to arrange a flat reflector facing the transmitter and locate nodes and antinodes at varying distances in front of the reflector. It is important to emphasize that 'interference' in the context of superposition is completely different from the leaking of signals from one channel to another or the noise caused by domestic equipment which sometimes spoils reception.

v.h.f. radio

The v.h.f. band with frequency about 90 MHz is suited to work out of doors. The direction to the transmitting station must be found. If the transmission is horizontally polarized (dipole aerials horizontal) the transmitter lies along the direction of the dipole rods when they are rotated until they are in a position which gives a minimum signal.

If the radio has a rod aerial, this can be used, or a dipole can be improvised from wires taped to a wooden bar. Each wire should be about a quarter wavelength long and be connected to the set by coaxial cable. Problems arise with good radios which keep the output volume constant over a wide range of input strengths, but this can be overcome by tuning slightly off the transmitter and using batteries which are almost spent. A cheap radio is more satisfactory for this experiment.

Arrange a large reflecting screen to face the transmitter. The radio is moved in front of the screen and positions of minimum signal are located. The wavelength can be determined and hence (using the published value of the frequency for the transmitter) the speed may be calculated.

u.h.f. television

The direction of the nearest transmitter can usually be found by looking at aerials on nearby houses or by experimenting with a portable aerial. An improvised dipole aerial, if it receives any signal at all, will be better for this experiment than a complicated 'Yagi' aerial with a reflector and many directors. The dipole will be better at detecting radiation reflected to its back, and the weakness of the signal which it will receive means that the feedback circuits in the receiver which are designed to compensate for signal variations will be less effective. The best results will be obtained when the quality of the reception is very poor.

The wavelength is less than 1 m. Reflecting sheets of modest size may be used and the experiment may be conducted indoors. The effects of the walls of the building may complicate matters and ambitious students might run an aerial out of doors on a length of coaxial cable.

v.h.f. radio is broadcast on frequencies between about 88 and 95 MHz. *Radio Times* gives an indication of the frequency on which local transmitters broadcast the various B.B.C. programmes. u.h.f. television uses frequencies between 470 and 850 MHz. Appendix III lists the frequencies allocated to each channel. Local dealers may be able to say what channel a particular service (B.B.C. 1, I.T.V., and so on) uses.

More detailed information is available from the B.B.C. Engineering Information Department, London W1A 1AA, and IBA Engineering Information Service, Crawley Court, Winchester, Hampshire SO21 2QA.

D8d Light waves

ITEM NO.	ITEM
1155	2 microscope slides
1153	glass plate (preferably fixed at 45° to the horizontal)
1156	sodium flame pencil (or sodium lamp if available)
	thin paper, *e.g.*, cigarette paper
24	hand lens
	travelling microscope (if available)
1155	micrometer screw gauge
503–6	retort stand base, rod, boss, and clamp

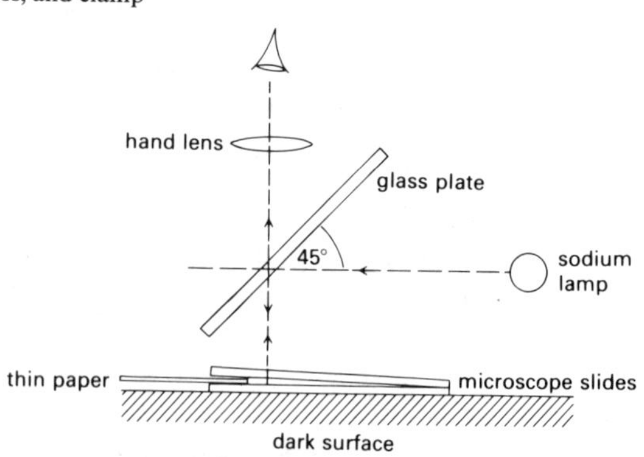

Figure D20
Superposition of light waves by reflection at a wedge.

Safety note: The use of a Bunsen burner and microscope is potentially hazardous. Make sure that the flame is not dangerously close to the experimenter's head, hair, or clothing.

We recommend that one set of equipment should be set up already with the fringes visible through the microscope. Students can use their own equipment to look for fringes using a hand lens, and can compare what they see with the fringes visible through the microscope.

Clean the microscope slides carefully since small specks of dirt will affect the angle of the wedge. Separate the slides at one end with a small scrap of thin paper. Place the wedge thus formed on a dark surface beneath the inclined glass plate in front of a sodium lamp as shown in the diagram. Using the naked eye, view the wedge from above and find a position in which a bright image of the light source is visible. This gives the correct viewing position. Hold the hand lens below the eye in this position and focus on the edge of the scrap of paper. Interference fringes should be clearly visible beyond the edge of the paper, running roughly parallel to the shorter edges of the microscope slides. If a travelling microscope is used it should be focused on the edge of the scrap of paper. Keen students might like to measure the fringe separation, either using a scale in the eyepiece or by racking the microscope sideways across several fringes. The wavelength can then be calculated (see below).

Light incident on the inclined glass plate is partially reflected down towards the microscope slides. At each surface of each slide there is some reflection. The rays which give rise to the interference pattern are those which are reflected at the two surfaces bordering the air wedge enclosed by the slides. The path difference between the two rays at a given position is approximately twice the thickness of the wedge at that position (see figure D21) and increases as the wedge widens, giving rise to the observed fringe pattern.

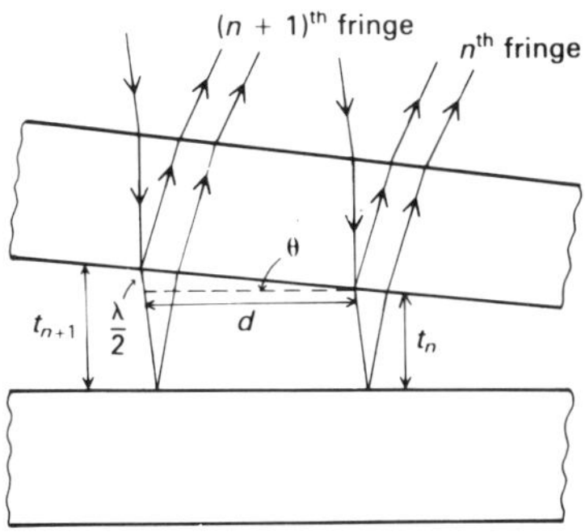

Figure D21

For able students only:
If we move from one minimum to the next, the path differences must have changed by λ so

$$2t_{n+1} = 2t_n + \lambda$$

or

$$t_{n+1} = t_n + \tfrac{1}{2}\lambda$$

If d is the separation of the fringes and θ is the angle of the wedge

$$\tfrac{1}{2}\lambda = d \sin\theta$$

The angle θ may be determined by measuring the thickness of the paper and the length of the slides:

$$\sin\theta = \frac{\text{thickness of paper}}{\text{length of slides}}$$

Thus the wavelength may be determined.

Teachers may like to know that a phase change occurs when light in air is reflected at the surface of glass but not when light in glass is reflected at the air surface. But note that we do not need to know this to work out λ from the fringe spacing.

Reflections from the upper surface of the top slide and the lower surface of the bottom slide do not give rise to interference fringes since the path differences are too great for the light to be coherent. The light from the sodium lamp does not consist of long continuous wave trains; it is emitted in bursts consisting of comparatively short wave trains having no fixed phase relationship. If two reflectors are too far apart, the light reflected from them will be from two different wave trains, with no fixed phase difference, so no consistent interference pattern will develop.

D8e Sound waves

ITEM NO.	ITEM
1109	signal generator
183	2 loudspeakers
157	microphone
1035	pre-amplifier
1511	oscilloscope
1153	metal reflector
501	metre rule
1000	leads

This experiment provides valuable contrast with electromagnetic waves. The frequency for a comparable wavelength is much lower because sound travels about a million times slower in air than electromagnetic waves.

Students can be asked to devise their own arrangements: some possibilities are shown in figure D22. If two speakers are being used, it is worth observing the effect of reversing the connections to one speaker, so reversing its phase relative to the other: consequently the positions of maxima and minima are interchanged.

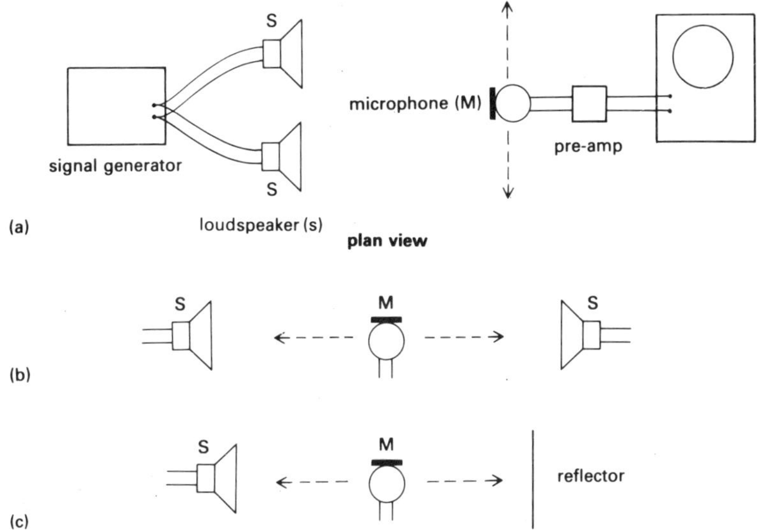

Figure D22
Superposition of sound waves.

Summary: superposition of waves

All students should have handled the 1 GHz equipment, and have seen optical interference fringes and 2-slit interference with microwaves. They see demonstrations or hear reports about the other experiments that have been tried. The points to be made in discussing the whole group of experiments are: wave motions show superposition behaviour, with regions where waves are in phase giving high intensity, and regions where waves are in antiphase giving low intensity. Radiowaves, microwaves, and light are all thought to be waves because they show superposition effects, not because anything can be seen to be oscillating.

Applications of superposition include the use of reflector and director rods in a television aerial; the fading of v.h.f. radio when an aircraft passes overhead and there is superposition of direct and reflected waves; the blooming of lenses to reduce reflections; the Decca system of navigation which uses interference of radio waves to plot the positions of ships or aircraft; the acoustical design of concert halls (it is important to avoid creating 'dead spots' where certain frequencies, or notes, are obliterated by destructive interference; good design will also prevent constructive interference which can locally increase the volume of a particular note).

Further work with superposition of electromagnetic waves

Unit J, 'Electromagnetic waves', deals more thoroughly with superposition effects (diffraction, interference) of light, including Young's slits and the diffraction grating. There is also some discussion of the nature of electromagnetic radiation.

Questions

Questions 14 to 19 are about superposition as experienced or discussed in the context of the laboratory. Questions 20 and 21 are about two applications: the blooming of lenses and the fading of radio signals.

Reading

Most textbooks deal with superposition or interference, for example:
AKRILL, BENNET, and MILLAR *Physics.* Chapter 23.
AKRILL and MILLAR *Mechanics, vibrations and waves.*
BOLTON *Patterns in physics.* Chapter 6.
DUNCAN *Advanced physics: fields, waves, and atoms.* Chapters 6 and 8.
P.S.S.C. *Physics.* Chapters 17 and 18.
CHAUNDY *Waves.*
ROGERS *Physics for the inquiring mind.* Chapter 10.
WENHAM, DORLING, SNELL, and TAYLOR *Physics: concepts and models.* Chapters 18 and 21.
Many interesting examples of interference can be found in WALKER *The flying circus of physics.*

Computer

Among several useful programs giving animated displays representing wave behaviour are:
'Longitudinal waves' (Five Ways Software, Heinemann Computers in Education).
'Transverse waves II' (Five Ways Software, Heinemann Computers in Education).
'Waves' (by R. G. Roscoe, obtainable through MUSE).

SECTION D3
MECHANICAL OSCILLATIONS

In Section D1 students faced the practical challenge of discovering the time trace of an oscillator. This Section explores the detailed motion of one oscillator whose motion is isochronous and sinusoidal. The sinusoidal property is established empirically by associating the motion with that of a point moving round a circle with constant angular velocity (previewed in demonstration D2). The motion is analysed through alternative routes, using either calculus or graphs. The energy aspects of oscillations are considered. The speed of a compression wave along a line of trolleys and springs is predicted by considering each individual trolley as an oscillator; and finally this prediction is applied to the case of sound in a solid.

INTRODUCTION: DISPLACEMENT, VELOCITY, AND ACCELERATION OF OSCILLATORS

The nature of the time traces (displacement–time graphs) obtained from experiments D1 may already have been discussed: in particular, the fact that some of the oscillations are isochronous, while others are not. Since later work concentrates on sinusoidal oscillations, it is worth stressing here that the time traces are *not* all similar: not all are isochronous, and not all are 'nice curvy lines' (for example, that of the oscillating air track vehicle is neither). To emphasize this further, invite the class to sketch the time trace for one or more of the following:

a tennis-ball in the course of a rally,

a bus, train, or aircraft shuttling between two termini,

a ball performing successive bounces after being dropped on to a surface.

Remind students that the gradient of the displacement against time graph gives the velocity at the time chosen: and invite them to sketch graphs of velocity against time underneath their displacement against time graphs. Similarly, acceleration against time graphs can be deduced from velocity graphs, and plotted. The velocity and acceleration graphs could also be drawn from the air track vehicle time trace. The exercise is an opportunity to ensure that students appreciate the significance of the signs, positive or negative, of the gradients.

The nature of oscillators

Introductory discussion should bring out these general points, most of which apply to most oscillators: oscillations are the successive movements of an object or system to one side then the other of an equilibrium position; there is a force directed towards the equilibrium position acting on the object; the force is related to the object's displacement from the equilibrium position in some way; this force accelerates the object towards the equilibrium position; the object or system has inertia, which means that it continues through the equilibrium position, rather than immediately coming to rest there; the object possesses kinetic energy as it passes through the equilibrium position; the object possesses potential energy when at the extreme ends of its motion; the object possesses both K. E. and P. E. at points between the extremes and the equilibrium position; there are resistive forces against which the object must do work; as a result the total energy of the object decreases.

Discussion of these points should, of course, be accompanied by frequent reference to examples. As well as the oscillators in experiment D1, there is a list of other favourite examples in the *Students' guide*, Section D1. Particularly for the non-sinusoidal ones, some care will be needed in justifying all the general points listed above.

Oscillations and waves

Section D2 pointed out that a mechanical wave has its origin in an oscillating object. The particles in the medium themselves oscillate, in a similar way to the originating object, but a certain time later.

Brief mention might be made now of electrons oscillating in a transmitting radio aerial, giving rise to fluctuating fields which travel and cause electrons to oscillate in a receiving aerial.

EXPERIMENT

D9 Factors affecting the period of an oscillator

D9a Mass on spring

This experiment gives practice in isolating and controlling one variable at a time, in taking measurements accurately, and in drawing numerical conclusions by comparing sets of measurements. It may be presented as a puzzle to challenge the faster students; others may need more help in finding an answer. One approach might be: 'You may assume that the frequency is either directly or inversely proportional to m, or to $\sqrt{m}$, or to m^2, and that the same applies to variation with k. Do the simplest experiments you can, to decide which functions apply'. Careful experimenting is reasonably certain to produce a 'right' answer.

ITEM NO.	ITEM
2A	4 expendable steel springs
31/2	hanger with 8 slotted masses, 100 g
503–6	retort stand base, rod, boss, and clamp
507	stopwatch or stopclock
1501	stiff wire
1155	pliers

It is worth asking students to work individually on this experiment, if enough equipment is available.

A note on uncertainties in measurement

It may be desirable to review (or introduce) ideas about using a stopwatch, and about timing repetitive events. A mechanical stopwatch may advance in intervals of 0.2 s, and a digital one will probably use intervals of 0.01 s. A human reaction time is typically 0.2 s. Students may argue that this error is self-cancelling, since it occurs at the beginning and the end of the timing; but a further point to make is that a person's reaction time itself is not consistent: it may well vary by perhaps 0.1 s. So quoting a time from a stopwatch more accurately than ± 0.1 s is unjustified. Provided the oscillations are isochronous, accuracy can of course be improved by timing a number, say 10: the percentage uncertainty in the value obtained for the period is then reduced by a factor of 10, so one may be justified to quote the time of one oscillation to ± 0.01 s.

On the matter of strict validity, it might be pointed out that adding springs entails adding mass. However, a moment's thought shows that the difference this could be expected to make could not lead to a wrong answer to the problem as it was posed.

Warn the students when varying m, not to use more than 800 g (or whatever mass is known to damage the springs).

The spring constant, k, of the system need not be measured. Students need only know that they can alter it by adding further springs to the initial one, either in series or parallel; then if the k for one spring is called 1 unit: k for two in series is $\frac{1}{2}$ unit, k for three in series is $\frac{1}{3}$ unit, k for two in parallel is 2 units, etc.

Students should aim to find out how T, the time of one oscillation, depends on

i the amplitude, A, of the motion,

ii m,

iii k.

The first task is relatively easy; but it is an essential preliminary to the other two, since if the motion is isochronous (it should be!) the light damping should not affect T significantly.

Candidates who are skilful and numerate may be able to find the

relationships $T \propto \sqrt{m}$ and $T \propto 1/\sqrt{k}$ on their own. Others, however, may need some prompting from questions like
'How do you have to change m so that T doubles?'
'How do you have to change k so that T doubles?'

Students may be slow to grasp that it is *factor* changes which are being considered (two times, three times, half, etc.), not *numerical* differences (1 s longer, 100 g more, etc.).

Another difficulty students often encounter is recognizing that one time may be, say, twice another *within the uncertainty of the experiment*: for example, when $T_1 = 0.98\,\text{s} \pm 0.02\,\text{s}$ and $T_2 = 0.51\,\text{s} \pm 0.02\,\text{s}$, then $T_1 = 2 \times T_2$ within the experimental uncertainty.

Home experiment

In Home experiment DH3, A mechanical oscillator, students are invited to make a mass oscillate at exactly 5 Hz. It could be used as an extension of D9a. Additional materials, including perhaps a balance and Plasticine, may then be required.

ALTERNATIVE EXPERIMENT
D9b Simple pendulum

ITEM NO.	ITEM
501	metre rule
503–6	retort stand base, rod, boss, and clamp
1153	string, 2 m length
121	2 metal strips (as jaws)
31/2	hanger with 8 slotted masses, 100 g
507	stopwatch or stopclock

See Revised Nuffield Physics *Teachers' guide Year 5*, Experiment 35a, for a convenient way to set up the pendulum.

For teachers who want a change from masses and springs for a while (experiment D10 is a detailed study of the motion of a trolley tethered between springs), a simple pendulum offers a good alternative. Ask students to find out how T depends on amplitude, mass of the bob, and length of the string. Note, however, that the qualitative discussions below and in the *Students' guide* all concern the mass-on-spring system.

Strictly, of course, the pendulum's motion is only isochronous for small angles of swing. But students who notice any discrepancies at large angles will be doing well. (T for swings of 90° amplitude is about 20 per cent longer than for small swings.)

Discussion of *T* being independent of *A*

Students will probably have found that T is independent of A (for given values of k and m), by starting with displacements of say 10 cm and 5 cm, and finding the same times for 10 oscillations in each case. It is worth asking 'Since damping visibly altered the amplitude during each set of 10 oscillations, how can you be sure that each individual oscillation took the same time?' It is also worth asking how they can be sure that every combination of k and m gives an isochronous oscillation, and not just the one (or more) which they tried. This latter point is a philosophical one, of course: we can never test all possibilities experimentally; but theories generalize the experimental evidence available, and stand until they are disproved or replaced by better ones.

A qualitative argument shows why an oscillator having the property

restoring force $\propto$ − displacement

should have T independent of A. The argument is summarized in the *Students' guide.*

An object moving with the property

acceleration $\propto$ −displacement,

that is, with a force acting on it as above, is defined as having *simple harmonic motion.* Few real oscillators behave so exactly; on the other hand, many systems behave approximately like this, particularly for small amplitudes.

Arguments for the dependence of *T* on *m* and *k*

(These arguments are developed in question 24.)

i m varies, k is constant.

Suppose m has been changed in such a way that T has doubled, the amplitude remaining fixed;

the same distance has been covered in twice the time

$\Rightarrow$ average speed halved;

$\Rightarrow$ maximum speed halved.

This speed has taken twice as long to attain

$\Rightarrow$ average acceleration quartered.

But force at each displacement the same (same k),

$\Rightarrow m\ (=F/a)$ must have increased four times.

But T has doubled,

$\Rightarrow T \propto \sqrt{m}$ is reasonable.

ii k varies, m is constant.

The argument is identical to the above, except for the last four lines, which would read:

But mass the same,

$\Rightarrow F(=ma)$ must have quartered;

Since displacement is the same, $k\ (=F/s)$ must have quartered,

$\Rightarrow T \propto \dfrac{1}{\sqrt{k}}$ is reasonable.

It is not easy to know whether this kind of semi-quantitative argument really suits beginners. It may be that they please the expert more than they impress the beginner, for the expert can see how they cut through the formalism of algebra to the essence of the ideas behind it, while the beginner has yet to find out what ideas are of the essence.

Nevertheless, they seem worth pursuing. Arguments like this are used by physicists and engineers during the first stages of considering the feasibility of a new proposal, or in estimating orders of magnitude. For instance, the fact that scaling up a load-bearing structure (whether an animal's leg or a bridge component) will raise the dead weight by the cube of the linear dimensions, but the strength by only the square of these dimensions, lies at the heart of the limitations imposed by existing materials on the size that things may be.

Another example comes from E. R. Laithwaite, who argues in *Propulsion without wheels* that bigger electromagnetic machines are usually better machines, because an increase in size reduces both the resistance and the reluctance within the machine. (Both increase in proportion to circuit length, but decrease in proportion to circuit cross-section, so the net effect of an increase in scale is a decrease in both.)

Questions

Question 23 recaps the 'T independent of A' argument; question 24 goes through the argument for the dependence of T on k and m. Questions 25 to 27 are general questions about oscillators.

EXPERIMENT

D10 Oscillation of a tethered trolley

This experiment relates harmonic motion to the mathematical cosine function, using the idea of a spot moving round a circle. The angular velocity of the spot, ω, is given a tangible meaning. Later, the mathematics is developed to consider the velocity and acceleration of the trolley.

ITEM NO.	ITEM
106/1	dynamics trolley
503–4	2 retort stand bases and rods
44/1	2 G-clamps, large
107	runway for trolley
2A	6 expendable steel springs
81	newton spring balance, 10 N
108	ticker-tape vibrator, carbon paper disc, and gummed ticker-tape
27	transformer
501	metre rule
31/2	slotted masses, 100 g
31/1	slotted masses, 10 g
1153	Plasticine
1000	leads

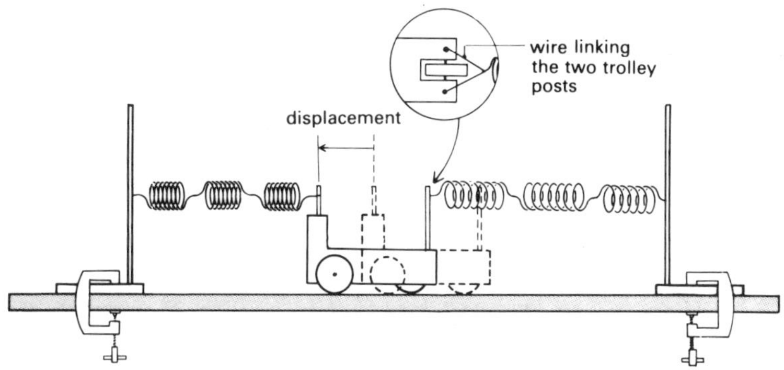

Figure D23
Oscillation of a tethered trolley.

The trolley runway should be compensated for friction in the direction the trolley will travel. It will help students to visualize the analysis if the trolley runs from right to left in front of them. The force constant, k, is measured by pulling the trolley aside by a measured distance using a spring balance; $20\,\mathrm{N\,m^{-1}}$ is typical. The measurement must of course be made with all six springs in position, as in figure D23.

The mass required to give the ratio k/m a simple value, preferably $10\,\mathrm{s^{-2}}$, is then calculated and the necessary mass added to the trolley. If k/m has the same simple numerical value for everyone in the class, the numerical analysis that follows goes more smoothly.

Each student should obtain a tape of half an oscillation. A suitable amplitude is 10 cm but this is not critical. k and m should be recorded, then T should be measured by direct timing.

Analysis

It is a useful exercise for students to plot a time trace for the half oscillation from the tape. (If this has already been done by one or more groups during experiment D1, a review of the graph may be sufficient.) The graph is shown in figure D24.

Students should check that the value for T from the graph agrees with that which they measured directly.

The next part of the analysis will probably proceed best with step-by-step instructions from the teacher. The stages are illustrated in figure D25 and could be as follows.

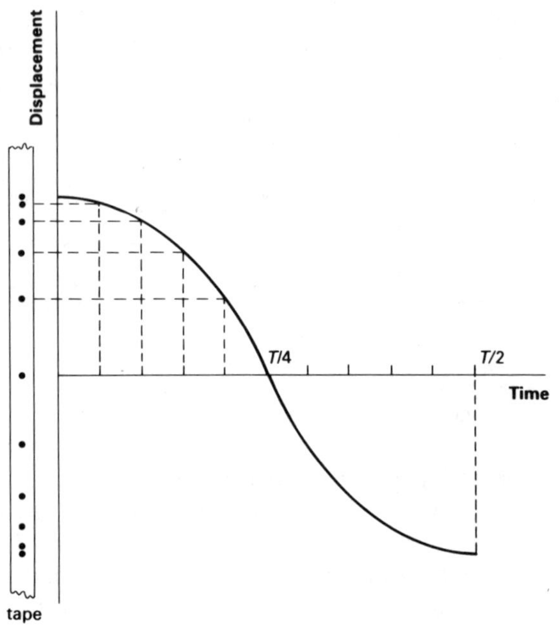

Figure D24
Displacement–time graph for a tethered trolley.

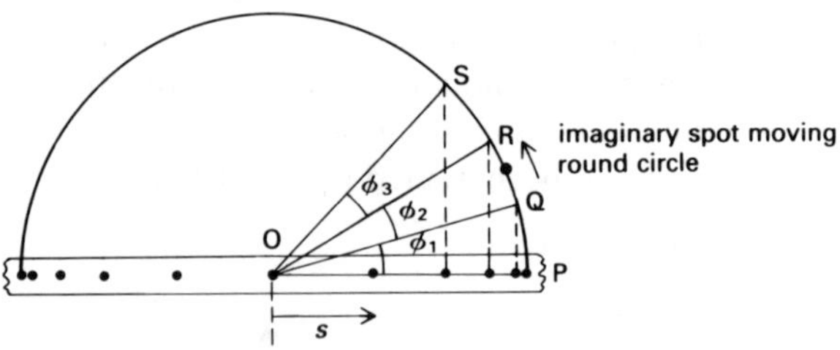

Figure D25
Mapping simple harmonic motion on to a circle.

Stick the tape onto the edge of a sheet of paper.
Highlight every third dot (or whatever is suitable to give about 10 highlighted dots along the tape).
Draw a semicircle, centre at the midpoint of the half-oscillation (O in figure D25), radius equal to the amplitude, so that the line of dots on the tape forms a diameter.
Draw a line from each dot perpendicular to this diameter, to intersect the circle (at Q, R, S, etc. in figure D25).
Draw lines joining these intersection points to O.

Measure the angles between these various radii (ϕ_1, ϕ_2, ϕ_3, etc. in figure D25).

It may come as a significant '*aha*' moment when students realize that the angles are all equal, within a degree or so.

Now invite students to imagine a spot moving steadily round the circle, P, Q, R, S, etc. covering the appropriate number of degrees in the appropriate time interval. When the trolley is released, the spot leaves P. The shadow, or projection of the spot will always lie exactly on the trolley: a harmonic oscillation may be visualized as the projection of a spot moving round a circle at constant speed. This should be familiar from demonstration D2 and/or the computer program 'SHM' (see page 217).

Students should calculate ω for the spot from their measurements, in degrees per second and then radians per second. They can check that $2\pi/\omega$ is equal to their measured value of T.

Question 28 goes through an example of mapping a harmonic motion on to circular motion, using the same steps as suggested above.

THE MATHEMATICAL DESCRIPTION OF HARMONIC OSCILLATION

The following work illustrates the possibility and power of mathematical model building in physics.

We choose the harmonic oscillator as an example because it has such a wide variety of later uses, some within the course and others in later stages of education or in practical tasks. In addition, the model of the harmonic oscillator develops further the ideas about rates of change and their representation by derivatives.

A possible route of four stages is summarized below; it is expanded in later pages. Teachers may wish to modify the suggestions considerably, to take account of their students' interests and mathematical competence, and also their own experience.

i Derive $s = A\cos\omega t$ empirically.

ii Show $\left.\begin{array}{l} v = -A\omega\sin\omega t \\ a = -A\omega^2\cos\omega t = -\omega^2 s \end{array}\right\}$ by calculus and/or graphs and find maximum values of v and a; measure maximum v $(=\omega A)$ from tape.

iii Dynamics of situation: $a = -\dfrac{k}{m}s$

Equations are similar in form, with $\omega^2 = \dfrac{k}{m}$

Calculate $\omega^2, \frac{k}{m}$ from measurements.

Equations: $\omega = \sqrt{\frac{k}{m}}, T = 2\pi\sqrt{\frac{m}{k}}$

iv Solve $a = -\frac{k}{m}s$ by a numerical method to obtain s against t graph (time trace).

i How does displacement vary with time?
Empirically, we have discovered that the projection of the spot generates harmonic motion along the diameter. The oscillator possesses the same harmonic motion (figure D26).

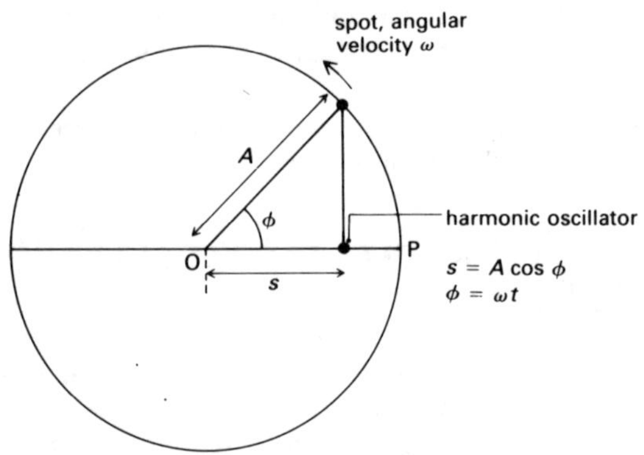

Figure D26

From the meaning of ω, $\phi = \omega t$, t being measured from the time when spot and oscillator leave P. Thus the displacement of the oscillator is described completely by $s = A \cos \phi = A \cos \omega t$.

Note that the time trace obtained from the tape (figure D24) resembles half a cycle of cosine graph.

Stress that A is constant, and that $\cos \omega t$ varies with time, its extreme values being $+1$ and -1; thus s varies between $+A$ and $-A$. Stress also that positive and negative values of s mean displacements right or left of the equilibrium position.

Conclusion: $s = A \cos \omega t$

ii Velocity and acceleration equations
Using calculus is the most powerful way to proceed for students who understand its meaning. It is desirable, however, to keep returning to the physical significance of the results, by reference to graphs and measurements.

Calculus

$$s = A\cos\omega t$$

$$\Rightarrow v = \frac{\mathrm{d}s}{\mathrm{d}t} = -\omega A\sin\omega t$$

$$\text{and } a = \frac{\mathrm{d}v}{\mathrm{d}t} = -\omega^2 A\cos\omega t = -\omega^2 s$$

Points to stress are:
these expressions tie up with the graphs in figures D27 to D29;
positive values of v and a mean velocity and acceleration to the right;
negative values mean to the left;
the maximum value of v is ωA (when the displacement is zero).

This can be checked from the tethered trolley tape by measuring the separation of the dots round O (see figure D25) and comparing the value of v obtained from them with the value of ωA.

Alternative arguments using graphs

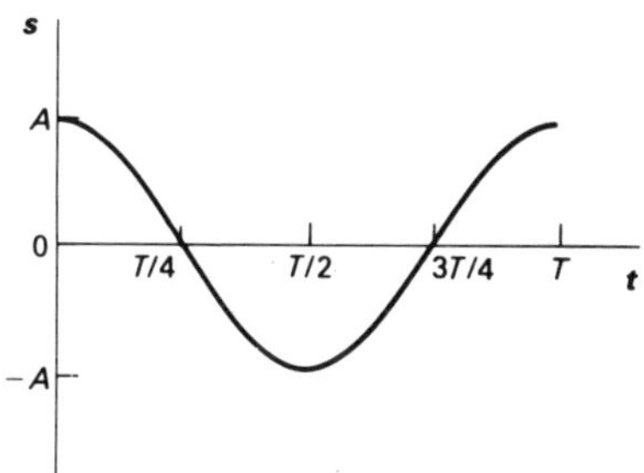

Figure D27
Time trace of the oscillator from figure D24, extended to one complete cycle.

The *velocity* of the oscillator at any moment is the gradient of the time trace. The maximum value of v must equal the speed of the spot in its movement round the circle. (When the oscillator is moving fastest, at O in figure D26, the spot is moving parallel to it.)

The spot takes time T to travel a distance $2\pi A$ around the circle,

$$\Rightarrow v_{\max} = \frac{2\pi A}{T} = \omega A.$$

This fixes the limiting values of v.

So the velocity against time graph must be like the one shown in figure D28.

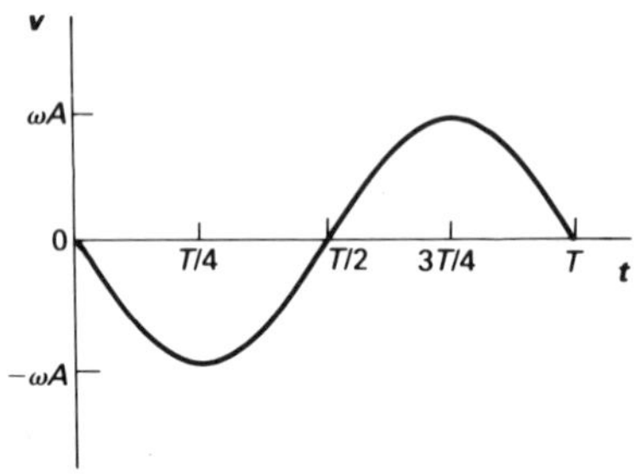

Figure D28
Velocity–time graph for the oscillator.

The *acceleration* of the oscillator at any moment is the gradient of the velocity graph.

The argument that leads to actual values of a depends on an understanding of the acceleration involved in uniform circular motion.

The maximum value of a must equal the centripetal acceleration of the spot, since when the oscillator is accelerating fastest, at P (see figure D26), the centripetal acceleration (centripetal acceleration is introduced in questions 36 and 37 of Unit E, 'Field and potential') of the spot is directed exactly along the diameter as well:

$$a_{max} = \frac{{v_{max}}^2}{A} = \frac{\omega^2 A^2}{A} = \omega^2 A.$$

With students who have not studied circular motion yet, it may be best to use a more qualitative argument, that a_{max} is likely to be proportional to $\frac{v_{max}}{T}$, that is, to ωv_{max} or $\omega^2 A$.

So the acceleration against time graph must be like the one shown in figure D29.

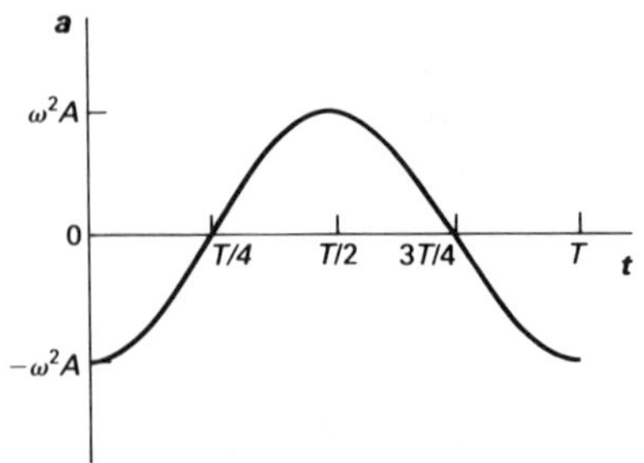

Figure D29
Acceleration–time graph for the oscillator.

Conclusions:

$v = -A\omega \sin \omega t \qquad v_{max} = \omega A$

$a = -A\omega^2 \cos \omega t \qquad a_{max} = \omega^2 A$

iii The dynamics of the oscillator
If the oscillator is displaced from the equilibrium position a distance s to the right, the unbalanced force on it is $-ks$ (to the left).

$\Rightarrow a = \frac{F}{m} = -\frac{k}{m}s$, from consideration of the force. equation [1]

Now $s = A \cos \omega t$

and $a = -\omega^2 A \cos \omega t = -\omega^2 s$ from stage *ii* equation [2]

Equations [1] and [2] are the same, provided

$$\omega^2 = \frac{k}{m}$$

Students should now calculate ω^2 from the tethered trolley motion, and compare the value with k/m to test the validity of this relationship.

Conclusion: $\omega = \sqrt{k/m}$
$T = 2\pi\sqrt{m/k}$

iv Numerical solution of $a = -\frac{k}{m}s$ equation [3]

It was shown in stage *i* above that the relationship $s = A \cos \omega t$ represents the motion of a harmonic oscillator. Stages *ii* and *iii* show that $s = A \cos \omega t$ is a solution of the mechanical equation of motion, $a = -\frac{k}{m}s$, if $\omega^2 = \frac{k}{m}$

The following piece of work shows how the displacement–time relationship can be found from the mechanical equation of motion by an iterative numerical method. The value of this approach is that it can be applied to a situation which is much more complicated, and for which the displacement–time relationship cannot be expressed as a simple equation. Since the method involves a succession of identical calculations, it can easily be adapted for computer use. The harmonic oscillator is a simple problem, with a simple solution: the numerical analysis is worth carrying out here as an illustration of its power, even though the harmonic oscillator as a simple problem has already been solved.

A typical problem is to construct a graph of s against t, knowing the initial values of s and v, and having a recipe for finding a at any s. One generally calculates what happens in successive short time intervals (Δt). During each short time interval the velocity is assumed to be constant. The starting conditions (initial values of s and v) must be known.

If the acceleration is a, then the change in velocity during one time interval is

$$\Delta v = a\Delta t$$

and a can be obtained from s. If the average velocity during the preceding time interval is v_{old}, then the average velocity during the new time interval will be

$$v_{\text{new}} = v_{\text{old}} + \Delta v \qquad \text{equation [4]}$$

The distance travelled during the new time interval is

$$\Delta s = v_{\text{new}}\Delta t$$

From Δs the new displacement can be found

$$s_{\text{new}} = s_{\text{old}} + \Delta s \qquad \text{equation [5]}$$

and we can now calculate the acceleration at the new position.

The whole procedure is now repeated, using the newly calculated value of v as v_{old} in equation [4], and the new value of s as s_{old} in equation [5].

These equations are close to being a computer program, but it is important for students to work through at least a few steps of the iterative process by hand.

There is one small trick that is important if an accurate result is to be obtained. The values of v we need are values for the *average* velocity during the time intervals between the instants of time at which s and a are calculated. If we start from $t=0$ with a time interval Δt, then we must work out v at $t=\Delta t/2$, $3\Delta t/2$, $5\Delta t/2$, ... and use these values of v as the required average velocities needed to work out s at $t=\Delta t$, $2\Delta t$, $3\Delta t$....

For a detailed and clear account of the numerical method see FEYNMAN, LEIGHTON, and SANDS. *The Feynman lectures on physics*, Volume 1, Chapter 9.

Questions

Question 29 leads students through the iterative method numerically. It is probably worth starting students on this in class, so that any difficulties or uncertainties can be cleared up before they go too far.

Question 30 suggests that students write a short computer program to solve the harmonic oscillator problem.

Computer

Teachers or students may not want to write their own computer programs. The 'Dynamic modelling system' is ideally suited to deal with this problem, and easily allows extensions such as damping. And it helps to emphasize that many of the ingredients needed to solve this problem (Newton's Second Law and the equations of kinematics) are common to all dynamics problems. What distinguishes one problem from another are the force laws involved: in this case Hooke's Law; in a satellite problem the inverse square law. To solve the problem of a damped oscillator means adding an equation for the damping law.

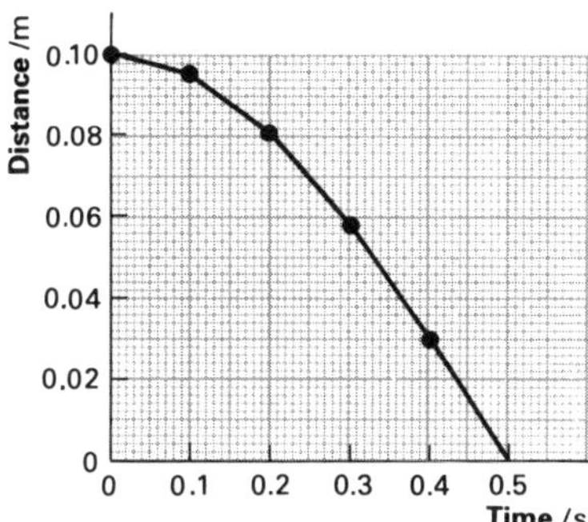

Figure D30

Figure D30 shows a completed solution to question 29, using the numbers suggested on page 243. Points to note are:

i The graph resembles the cosine graph found by experiment (figure D24).

ii The period predicted by this solution can be compared with that found experimentally. Agreement should be gratifyingly close.

iii The method is a series of approximations, since it is assumed that v remains constant for the entire interval Δt, then changes abruptly to a new value, with an acceleration characteristic of only one value of displacement. Considerable discussion may ensue at this point, with the teacher being called upon to justify making approximations, and hence not getting the 'proper', 'right', or 'exact' answer. Further ideas are given below, but two simple points are: the method has (one hopes!) yielded a good answer for T; and by making Δt as small as you like, and having correspondingly more steps, the method can be made accurate to within any limits one pleases.

Question 31 asks students to test whether a motion shown as a multiflash photograph could be simple harmonic.

Application of numerical methods of calculation

Some students may feel that numerical methods are a cheat, not deserving serious attention. They would be wrong. Even in pure science, where exact solutions are highly valued, approximate methods are widely used. Complex X-ray analysis (of proteins, for example) uses computers to compute maps of the molecules, whose shape is fitted by no simple equation. Numerical methods are used for predicting the structure of all atoms or molecules other than the very simplest, as the equations (Schrödinger's equation) cannot be solved exactly by analytic methods for more than two particles. No analytic equations have been found for astronomical problems involving several comparably sized bodies, and numerical methods rather than exact equations are used to guide space probes (and very exactly too). On the other hand, approximate analytic solutions are used to guide numerical methods. If they were not, purely numerical methods would often defeat the largest computer imaginable.

In applied science, the role of approximate numerical methods is even wider. The airflow over an aircraft, the distribution of heat or sound in a building, the stresses in a proposed dam or bridge, the magnetic field around a new design of motor armature, or the effect of changing the shape of the hull of a ship are all examples where such methods, using computers, offer the only possible approach in practical problems. Of course, the exact methods of differential and integral calculus are still valued, though more for their elegance, generality, and compactness than for their precision, for numerical methods can be made as exact as one pleases if one takes enough trouble.

Because computers are so much used for this kind of problem, there is merit in handling some problems in physics teaching in a computational manner rather than an analytic one. Students who meet such methods later on may find them less strange than do others who meet only analytic methods.

Reading

FEYNMAN, LEIGHTON, and SANDS *The Feynman lectures on physics* Volume 1. Chapters 21 and 22 are good value for teachers.

ROGERS *Physics for the inquiring mind.* Chapter 10.

The usefulness of the results $\omega = \sqrt{k/m}$ and $T = 2\pi\sqrt{m/k}$

It would be disastrous if students supposed that all that has been done is to explain the oscillation time of a trolley between springs, when there are many problems that can be handled using the result obtained. There is room here for only one example: it concerns the absorption of light by oscillating ions in sodium chloride.

Order of magnitude calculation of the frequency of atomic oscillations

The average mass, m, of sodium and chloride ions is close to 5×10^{-26} kg. The spring constant, k, of the ionic bond has the order of magnitude $100\ \mathrm{N\,m^{-1}}$. Thus k/m is about $20 \times 10^{26}\ \mathrm{N\,m^{-1}\,kg^{-1}}$. Since $\omega^2 = (2\pi f)^2 = k/m$, an estimate for f^2 is about $5 \times 10^{25}\ \mathrm{s^{-2}}$, giving an order of magnitude of 10^{13} Hz for the frequency, f, with which the ions might oscillate. A closer estimate is not worth making, because the frequency will depend on the particular directions in which the ions oscillate, and on which way neighbouring ions are moving.

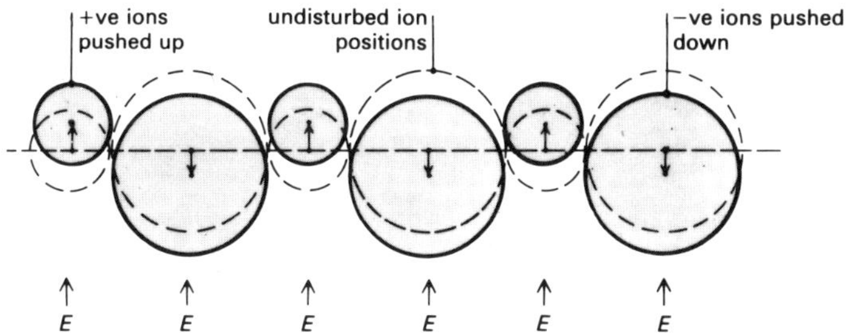

Figure D31
A row of ions in the electric field of long wavelength radiation.

Suppose now that light is an oscillating electrical disturbance. If light shines on sodium chloride, the ions might be driven into oscillation as suggested in figure D31. If the light has just the right frequency, energy would be absorbed from it by the oscillating ions. (This is of course an example of resonance, and looks forward to work in Section D4.) The required frequency is of the order 10^{13} Hz, corresponding to a wavelength of 3×10^{-5} m, which is in the infra-red region of the spectrum. Experiments show that sodium chloride does indeed absorb infra-red radiation at one particular wavelength, close to 6×10^{-5} m. The order of magnitude of the calculation is right.

The wavelength is long compared with the inter-atomic spacing, so the field can be considered as acting on a complete row simultaneously.

ENERGY OF AN OSCILLATOR

The oscillating ions in the preceding example remove energy from the infra-red radiation shining on the crystal.

How much energy is stored by an oscillating mass? What happens to the energy during one oscillation? Such questions will bring out the changes from potential energy to kinetic and back again.

The potential energy stored in a spring, $\frac{1}{2}ks^2$, was discussed in Unit A. Figure D32(a) shows how the potential energy varies with displacement. If the total energy is constant (no damping) the kinetic energy is the difference between total and potential energy. Figure D32(b) shows how the kinetic energy varies. If A is the amplitude (the maximum displacement), the total energy is equal to $\frac{1}{2}kA^2$. It is useful to list values of potential, kinetic, and total energy at several places in the oscillation cycle, as in figure D33.

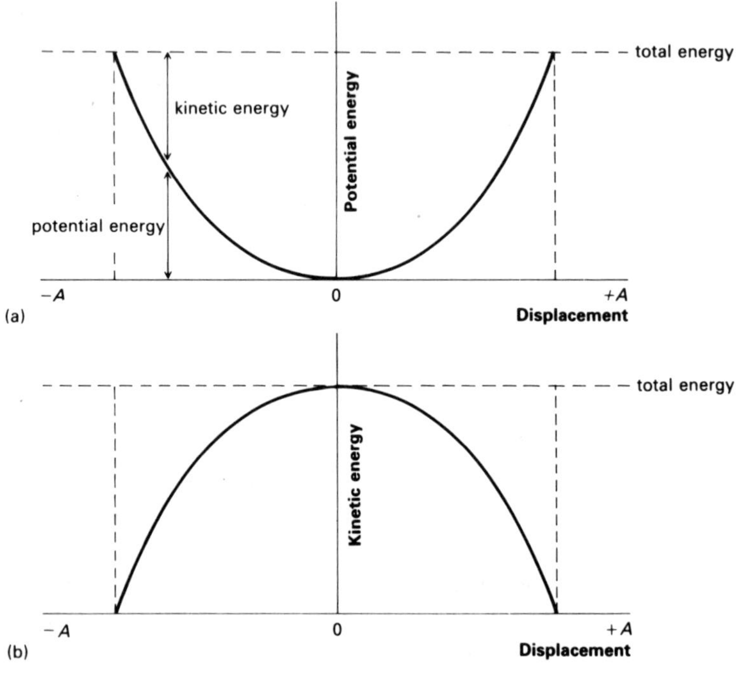

Figure D32

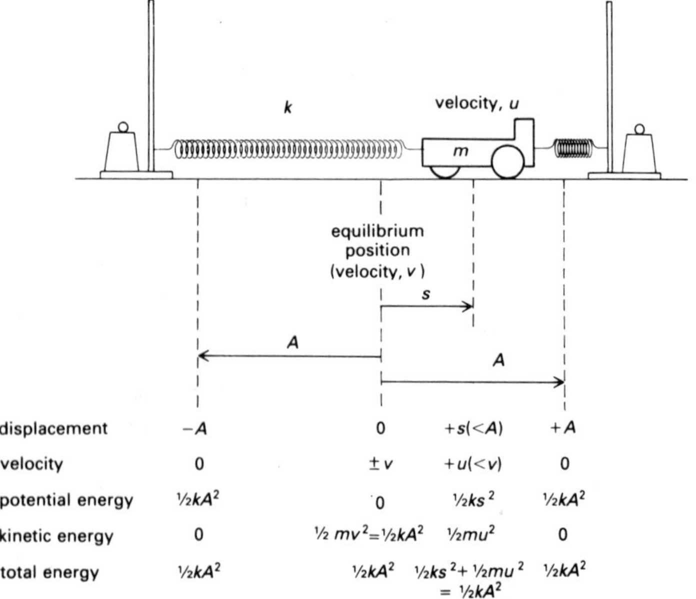

displacement	$-A$	0	$+s(<A)$	$+A$
velocity	0	$\pm v$	$+u(<v)$	0
potential energy	$\frac{1}{2}kA^2$	0	$\frac{1}{2}ks^2$	$\frac{1}{2}kA^2$
kinetic energy	0	$\frac{1}{2}mv^2 = \frac{1}{2}kA^2$	$\frac{1}{2}mu^2$	0
total energy	$\frac{1}{2}kA^2$	$\frac{1}{2}kA^2$	$\frac{1}{2}ks^2 + \frac{1}{2}mu^2 = \frac{1}{2}kA^2$	$\frac{1}{2}kA^2$

Figure D33
Energies at various stages of an oscillation.

In general, the K. E. and P. E. at time t (assuming the oscillator loses no energy) are:

$$\text{K.E.} = \tfrac{1}{2}mu^2 = \tfrac{1}{2}m\omega^2 A^2 \sin^2 \omega t$$

$$\text{P.E.} = \tfrac{1}{2}ks^2 = \tfrac{1}{2}kA^2 \cos^2 \omega t$$

Mathematically-inclined students may like to see the effect of adding K.E. and P.E.:

$$\text{Total energy} = \text{K.E.} + \text{P.E.} = \tfrac{1}{2}m\omega^2 A^2 \sin^2 \omega t + \tfrac{1}{2}kA^2 \cos^2 \omega t$$

But $k = m\omega^2$

$$\Rightarrow \text{total } E = \tfrac{1}{2}kA^2 \sin^2 \omega t + \tfrac{1}{2}kA^2 \cos^2 \omega t$$

but $\sin^2 \omega t + \cos^2 \omega t = 1$ always,

$$\Rightarrow \text{total } E = \tfrac{1}{2}kA^2, \text{ always.}$$

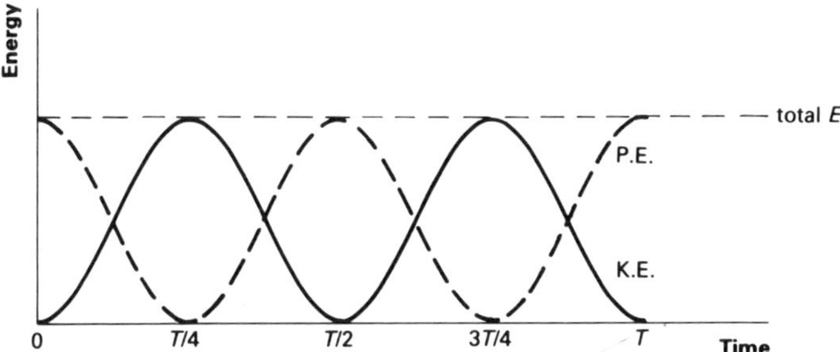

Figure D34
K.E., P.E., and total E of a harmonic oscillator.

This topic has a brief mention here, partly for completeness, and partly for later use. Work on a.c. circuits in Unit H, 'Magnetic fields and a.c.' will use the idea that the power delivered by an alternating current is proportional to the square of the maximum current, and work on optics (Unit J, 'Electromagnetic waves') will involve the idea that the intensity of light on a screen is proportional to the square of the wave amplitude. Finally, in Unit L, 'Waves, particles, and atoms', an account of wave–particle duality will require the idea that the probability of arrival of photons or electrons at a screen depends on the square of a wave amplitude.

For these purposes, the main emphasis here should be on energy proportional to (amplitude)^2, with a mention of the variation of potential energy as $\tfrac{1}{2}ks^2$. The kinetic energy as the difference between total energy and potential energy will also be needed for Unit L.

DAMPING AND *Q*

It may be useful at this stage, before getting on to resonance, to introduce the quantity Q. There are several equivalent ways of defining Q, most of which relate to resonance; but it is also useful to realize that it gives a measure of how slowly free oscillations die away.

Refer students to one of the time traces from experiment D1 which clearly shows amplitude diminishing with time – the lath oscillations may be suitable. Or alternatively set up a simple oscillator, say a mass-on-spring, and watch the amplitude decrease. The oscillations demonstrated may or may not have an exponentially-falling amplitude, depending on the nature of the damping force. This is not a point to labour.

Ask questions like:
'Why do the oscillations die away?'
'Where does the "lost" energy go?'
'Where does the opposing force arise?'
'What factors determine how quickly the oscillations die away?'
'What factors determine how many oscillations occur before they die away?'

Introduce Q by saying it is *approximately* equal to the number of oscillations which occur before they die away – and is thus a number, without units, which is large if the damping is small, and vice versa. (In theory of course the amplitude of an exponentially damped oscillation will never become zero. Any student who raises this objection shows good understanding of the exponential function.) Typical values might be:

a car suspension	$Q \approx 1$
trolley and springs	$Q \approx 10$
pendulum	$Q \approx 1000$

In Unit K, 'Energy and entropy', the tendency of energy to spread throughout a system will be looked at in some detail, and seen to be a consequence of chance alone.

Questions

Questions 32 to 39 give opportunities to apply the ideas of harmonic oscillation to a variety of situations.

THE SPEEDS OF COMPRESSION WAVES

We return to mechanical waves at this point, to apply our knowledge of oscillations to the motion of one particle in a lumped system transmitting a longitudinal wave.

DEMONSTRATION
D11a Longitudinal waves on a trolleys-and-springs model

ITEM NO.	ITEM
106/1	11 dynamics trolleys
32	11 masses, 1 kg (or 11 extra trolleys)
1080/1	20 compression springs
1080/2	20 spring holders
501	metre rule
81	newton spring balance, 10 N

Eleven trolleys are laid end to end in a row and each trolley is linked to the next by a compression spring, using a spring holder on each end of each trolley (figure D35).

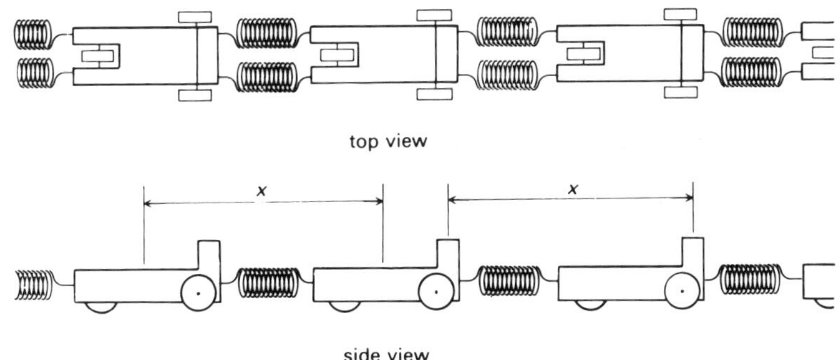

Figure D35
Trolleys linked by compression springs.

In Unit A, 'Materials and mechanics' the Young modulus, E, for steel was related to a simplified atomic model of atoms spaced in a cubic array with spacing x and separated by springy bonds with spring constant k. (The relation was $E=k/x$.) How such an arrangement of atoms can propagate a compression wave can be demonstrated by the trolleys-and-springs model. Each trolley represents an atom and each spring represents an atomic bond. Compression and expansion pulses can be sent along the row of trolleys.

The distance x (see figure D35) is measured with the trolleys in equilibrium. The spring constant, k (the constant for one pair of springs linking adjacent trolleys) can be measured by pulling on one end of the row of trolleys with a spring balance, the other end of the row being fixed. (Divide the total extension obtained by the number of pairs of springs in the row, to obtain the extension per pair.) The mass, m, of one of the trolleys is also needed. It is worth taking these measurements now, as the results will be needed shortly to calculate the theoretical speed for pulses travelling along the model.

Demonstrate that expansion and compression pulses travel at the same speed. This speed is reduced if extra masses are added to each trolley.

Theoretical argument for the speed of the wave

The pulse discussed is one in which the lefthand trolley is moved at a steady speed to the right. One by one, each trolley in turn acquires the same speed, *u*, to the right.

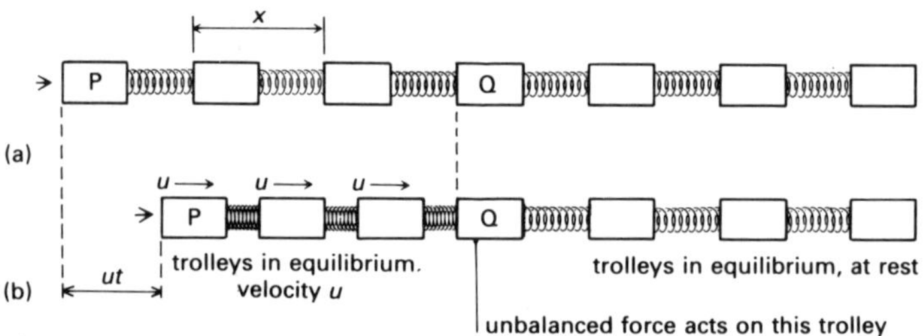

Figure D36
Pulse on a row of trolleys.

The speed with which the disturbance passes along the row depends on how quickly a trolley, initially at rest in equilibrium, accelerates when its neighbour is displaced from its equilibrium position. This in turn, of course, depends on the mass of the trolley, and the force acting on it. Now the force depends on the compression of the spring and its force constant. (One can show that adding mass to the trolleys slows the pulse down; and making the springs stiffer – by adding a second spring in parallel – speeds it up.)

By applying Newton's Laws to this event one can derive an expression for the speed of the pulse in terms of the masses of the trolleys, their spacing, and the stiffness of the springs linking them (see for example Bolton, *Patterns in physics*, page 130). But we do not suggest a detailed derivation in this course, preferring a more qualitative argument (see Tabor, *Gases, liquids and solids*, Chapter 8).

When the pulse travels along the row each trolley in turn begins to move, as the disturbance passes. The speed with which the disturbance moves along the trolley train depends on how quickly this movement is passed on from one trolley to the next.

The time it takes a trolley to respond to the push of its neighbour depends on the period of oscillation of the trolley: the longer the period the greater the time delay between one trolley starting to move and reaching the speed *u* and the next doing so.

It may help to think about other oscillating systems. The time it takes for a pendulum or child's swing to respond to a sudden impulse depends on its period. The response of a moving coil meter to a sudden change in current depends on the meter's response time, which is governed by 'mass' and 'stiffness' factors. For a rapid response 'mass' should be low and 'stiffness' high. A reasonable guess might be that the maximum deflection is reached after a time delay of a quarter of a period of oscillation.

The period of oscillation of a trolley is $T = 2\pi\sqrt{m/2k}$ (springs 'in parallel' – see figure A9, page 20). In fact the 'hand-on' time for the pulse is rather less than $T/4$, it is $T/\sqrt{2}\pi = \sqrt{m/k}$. This is the time it takes the pulse to travel the distance between trolleys, x. So the disturbance travels at an average speed

$$c = \frac{x}{T/2\pi} = x\sqrt{\frac{k}{m}}$$

The argument has been presented in terms of a step impulse. Experiment shows that waves travel with this speed, and that for low frequencies (that is, wave frequency $\ll$ natural frequency of oscillator) the speed does not depend on wave frequency. The relationship $c = x\sqrt{\frac{k}{m}}$ applies to continuous waves as well as the idealized step impulse we have considered.

This qualitative argument is valid as long as the wave frequency is small compared with the natural frequency of oscillation of the particles of the medium. The argument breaks down at higher frequencies for which the wavelength begins to get as small as the spacing between particles. And by oscillating the end trolley rapidly one can show that high frequency waves are not transmitted at all.

A student who has followed this argument well might wonder why the natural oscillation frequency (f_{nat}) should control the speed of propagation of a sinusoidal wave of much lower frequency (f_{wave}). *If*, but only if, such an objection is voiced one might argue as follows. The displacement of an individual particle and therefore the force on it varies sinusoidally at frequency f_{wave}.

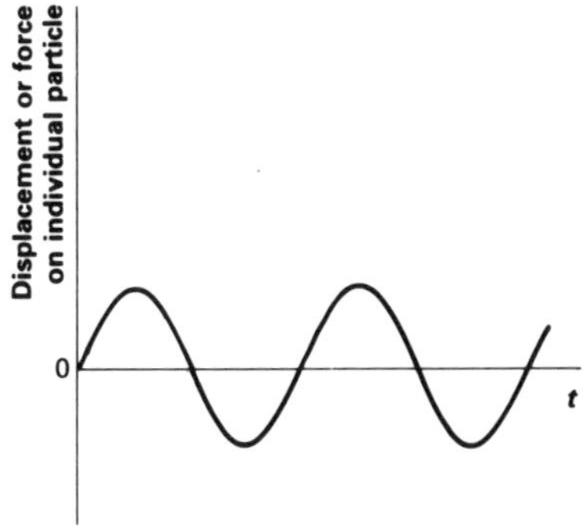

Figure D37
Displacement, or force, on individual particle.

We could consider this continuous variation in force as being made up of many tiny step impulses (see figure D38).

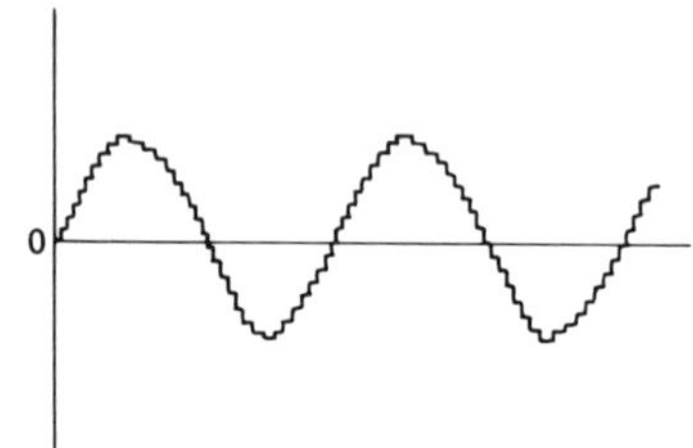

Figure D38
Sinusoidal variation made up of a series of steps.

The reaction of each particle to a step impulse is limited by its natural response time, that is, by f_{nat}, not by the wave frequency.

It may be useful to show that an oscillating system can respond to a continuously varying input, if the frequency is low enough: a moving coil meter will show *slow* a.c. quite faithfully. But at high frequency it hardly responds at all.

The formula $c = x\sqrt{\frac{k}{m}}$ can now be tested experimentally, using the method described below.

DEMONSTRATION

D11b Measuring the speed of the wave

This is a continuation of demonstration D11a, but only requires a row of four trolleys. It should help students understand how the speed of sound in a metal rod is measured in demonstration D12. The two could be done side by side.

ITEM NO.	ITEM
	Apparatus as for demonstration D11a with:
77	aluminium block
1503	timer, resolution 10 ms
52K	2 crocodile clips
1000	leads
1501	insulated copper wire, 0.45 mm diameter

The wall in figure D39 can be an aluminium block held firmly in place by several 1 kg masses piled on and behind it. The block should not move. One lead to the timer 'make to count' terminals goes to the block. The other, which should be very flexible, goes to a contact on the front of the front trolley. Microprocessor based instruments (for example, VELA, GiPSI) could replace the method of timing suggested here.

The trolleys must be set travelling towards the wall as a whole. This can be done by spreading a pair of hands over the whole row of trolleys and pushing all of them together. The wave travels twice along four sections of medium. The distance travelled is thus twice the overall length of the whole system.

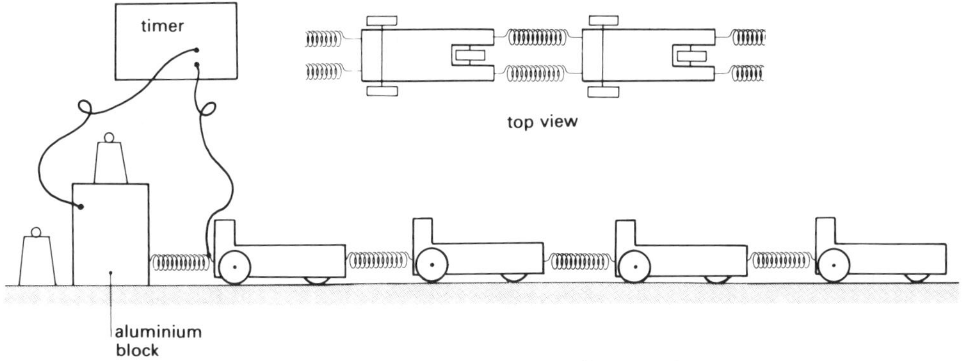

Figure D39
Timing a pulse up and down a row of trolleys.

Send the row of four trolleys, all travelling at the same speed, towards a rigid wall as shown in figure D39. When the front trolley hits the wall it stops. The others come up behind it and stop one after the other, so that a compression wave front travels along the row to the rearmost trolley.

This last trolley stops, and then moves away from the wall, followed one by one by the other trolleys. An 'expansion' wavefront, in which the trolley spacing returns to normal, travels along the row back to the front trolley. This trolley has meanwhile been pressed firmly against the wall by the compression in the spring attached to it. Finally, the 'expansion' wave front reaches the front trolley and it moves away from the wall.

The time the front trolley spends in contact with the wall, which is equal to the time needed for a wave to travel twice along the row of trolleys, is recorded. The wave speed follows from the time and from the distance travelled by the wave.

The value calculated should be in good agreement with the predicted speed and be of the order of $3\,\mathrm{m\,s^{-1}}$. (For example, values of $x = 0.35\,\mathrm{m}$, $k = 50\,\mathrm{N\,m^{-1}}$, and $m = 0.95\,\mathrm{kg}$ give a predicted speed of $2.5\,\mathrm{m\,s^{-1}}$.)

Prediction of the speed of sound in steel

We assume that the atoms in steel are arranged in a cubical array (figure D40), and that when a wave travels each row of atoms moves in unison. The speed of the wave is then the same as the speed of a wave along a single row of atoms.

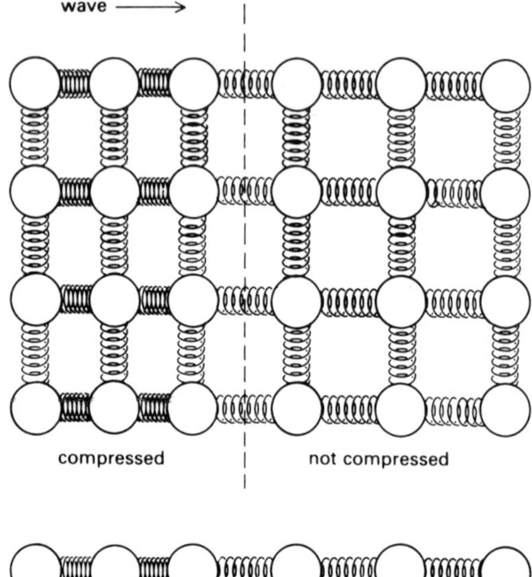

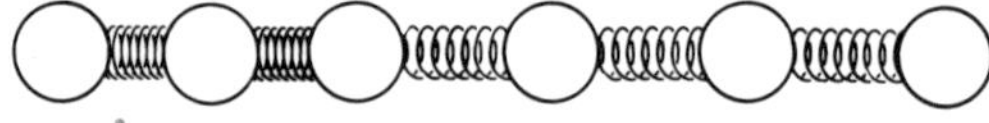

Figure D40
A plane wave in an array of atoms.

From Unit A, 'Materials and mechanics', for steel:

E (the Young modulus)	$20 \times 10^{10}\,\mathrm{N\,m^{-2}}$
x (atom spacing)	$2.5 \times 10^{-10}\,\mathrm{m}$
k (spring constant $= Ex$)	$50\,\mathrm{N\,m^{-1}}$
m (mass of atom)	$9.3 \times 10^{-26}\,\mathrm{kg}$
ρ (density)	$7.8 \times 10^{3}\,\mathrm{kg\,m^{-3}}$

(The mass, m, is obtained from the density, ρ, and the spacing, x, using $m = \rho x^3$, supposing the atoms to be in a cubic array.)

Then $c = x\sqrt{k/m}$ gives a speed close to $6000\,\mathrm{m\,s^{-1}}$. But k and m were obtained from the Young modulus, E, and the density respectively; and x can be found from the density and the Avogadro constant. So the speed can also be expressed entirely in terms of the large-scale quantities E and ρ, as follows:

$$c = x\sqrt{\frac{k}{m}} = x\sqrt{\frac{Ex}{\rho x^3}} = \sqrt{\frac{E}{\rho}}$$

Note the natural oscillation frequency of an atom in the steel:

$$f = \frac{1}{2\pi}\sqrt{\frac{k}{m}} = \frac{1}{2\pi}\sqrt{\frac{50}{9.3 \times 10^{-26}}} \approx 3.7 \times 10^{12}\ \mathrm{Hz}$$

This is much higher than the frequency of any sound, and so the 'lumpiness' of the medium does not invalidate the derivation.

The same value for the speed of sound in a rod can be derived without making any assumptions about the atoms and how they are arranged. See, for example, Duncan, *Advanced physics: fields, waves, and atoms*, Appendix 5.

DEMONSTRATION

D12 Speed of sound in a metal rod

The speed of sound in steel (about $5100\,\mathrm{m\,s^{-1}}$) can be found in tables; it can also be measured by a method exactly analogous to that used in demonstration D11b to time the wave along a row of trolleys and springs.

ITEM NO.	ITEM
1511	oscilloscope
1109	signal generator
504	2 retort stand rods, 1 m long
503–5	retort stand base, rod, and boss
1153	2 rubber bands, about 10 cm long
	either
52K	crocodile clip
	or
1153	adhesive tape
1153	hammer, club or claw head, at least 0.5 kg
1000	leads

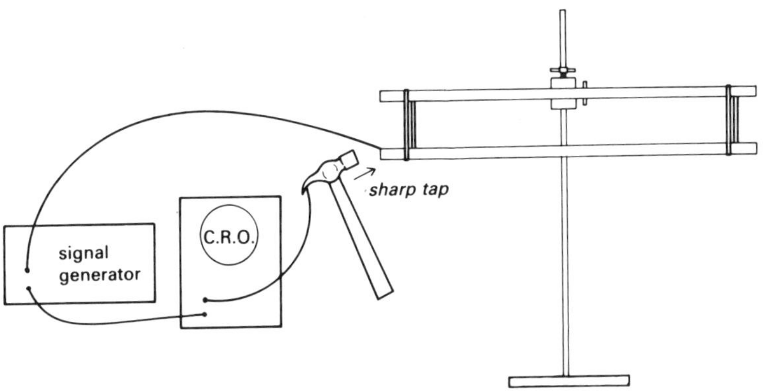

Figure D41
Speed of sound in a metal rod.

One 1 m long rod is hung on rubber bands below another supported on a retort stand. The suspended rod is connected to the output of the oscillator; it is convenient to have a 4 mm hole drilled in this rod. The earthed oscillator output and oscilloscope input terminals should be joined.

Connect the hammer head by a short lead to the oscilloscope input. This lead should not touch anything else. Using the oscilloscope settings below, check that the oscilloscope trace shows little or no pick-up from the mains or from the oscillator, when the hammer is held in the hand. Insulate the hammer handle if necessary. A rubber handled, steel shafted hammer is ideal; it also helps to choose a hammer with a head that is not rod-shaped.

Instrument settings:

Signal generator	frequency	25 kHz
	output	6 V amplitude sine wave, high impedance
Oscilloscope	input switch	a.c.
	brightness	maximum
	trigger	automatic, positive going triggering
	input sensitivity	$5\ \mathrm{V\ cm^{-1}}$
	time base and X-gain	$100\ \mu\mathrm{s\ cm^{-1}}$, calibrated positions
	X-shift	trace to start on the screen, at left

Turn the stability control just so far anti-clockwise that the 25 kHz trace appears when the hammer is held in contact with the suspended rod, but vanishes when this contact is broken. This adjustment is critical.

Tap the end of the suspended rod smartly with the hammer, hitting it end on. A train of oscillations should appear, lasting as long as the contact is made. The train should be about 40 mm long on the screen. Repeat as often as is needed to note (or mark with a wax pencil on the screen) the start and finish of the train of oscillations allowed to pass from signal generator to oscilloscope through the contact between hammer and rod. This contact remains closed while a pulse travels up the rod and down again, that is, for *twice* the time it takes the pulse to travel the length of the rod.

The hammer should be a heavy one and held loosely so as to damp the wave that travels across the hammer head and back again.

The time-scale can be calibrated by holding the contact closed and counting the number of 25 kHz oscillations between the two marks on the screen. This method eliminates uncertainty due to the rather variable speed of the time-base near the start of the trace.

Typically, there may be ten oscillations, giving a time of 0.4 ms for the pulse to travel 2 m, a speed of $5000\ \mathrm{m\ s^{-1}}$. Frequencies above 25 kHz can be used, but the trace is then fainter. At lower frequencies, the end of the train is hard to pick out.

There are other possible ways of measuring the very short time of contact between hammer and rod. One could use a binary counter made up from the digital electronics kit (experiment C181) to time the number of pulses from the signal generator – that is, to replace the oscilloscope in the method described above. At 25 kHz four stages will count up to 0.6 ms. Another possibility is to replace the signal generator and oscilloscope with a light beam galvanometer, cells, and resistance substitution box. The galvanometer must have a calibration for use as a ballistic instrument: then the charge, Q, which passes during the short contact can be measured; the current, I, can be measured at leisure; and the time, t, is obtained from $t = Q/I$. Undoubtedly the simplest alternative is to measure the time directly using one of the timers or multi-purpose microprocessor based devices now available; a resolution of at least 0.1 ms, but preferably 0.01 ms, is needed.

For details, see these references:

FORD, and SOPER 'Velocity of sound in metallic rods'. *School Science Review.*
MACE 'Speed of sound in a steel rod'. *School Science Review.*
ERICSON 'Velocity of sound in a steel bar'. *School Science Review.*

One perceptual problem for students is that in demonstration D11b the row of trolleys was moved towards the wall. In this case the hammer (equivalent to the wall) moves. The equivalence of these situations should be easy to justify. A more difficult problem is: 'Are we timing the wave which travels along the rod, or one which travels along the hammer?' The simple answer is that with a suitably shaped and massive hammer, the initial wave amplitude is small (because the hammer, like the aluminium wall, is massive), and the wave energy dissipates in different directions (because of the irregular hammer shape).

Questions

Questions 40 to 42 are about the speeds of compression waves, first along a line of trolleys-and-springs, then in solids.

SECTION D4

FORCED VIBRATIONS AND RESONANCE

One example of a system which can vibrate when a periodic force is applied to it was mentioned in Section D3: the response of ions in sodium chloride to an oscillating electric field. Other examples abound and students should be able to suggest several. If the forced vibrations are large the results can be catastrophic; in other contexts a large response can be usefully exploited.

A brief demonstration with a mass on a spring will illustrate the variation of response with forcing frequency.

The destructive potential of resonance may be dramatically demonstrated using a hanger with slotted masses.

DEMONSTRATION

D13 Forced vibration of a mass on a spring

ITEM NO.	ITEM
2A	2 expendable springs
31/2	hanger and slotted masses, 10 g
1153	light string
1155	Perspex tube or glass tube (wide enough just to fit outside slotted masses)
	either
1060	vibrator
1109	signal generator
	or
150	fractional horsepower motor having wheel with pin offset 1 cm (less for a less violent response)
38	single pulley
59	l.t. variable voltage supply
1000	leads

Alter the frequency of the forcing vibrations by adjusting the speed of the motor which drives the wheel, shown in figure D42(a). Alternatively, attach the spring to the vibrator, which is driven by the signal generator, as in figure D42(b).

At most frequencies the mass on the spring oscillates gently, but at the resonant frequency the vibration is violent. The masses may leap off the hanger, illustrating the destructive potential of resonance.

The Perspex tube prevents the 'pendulum' mode of oscillation; removing the tube shows this second mode, and illustrates that systems in general can have several modes.

Discussion about this demonstration should give students ideas for their own investigations of resonance, which follow. It may be useful to introduce damping at the end of the demonstration, using the additional apparatus in experiment D14b.

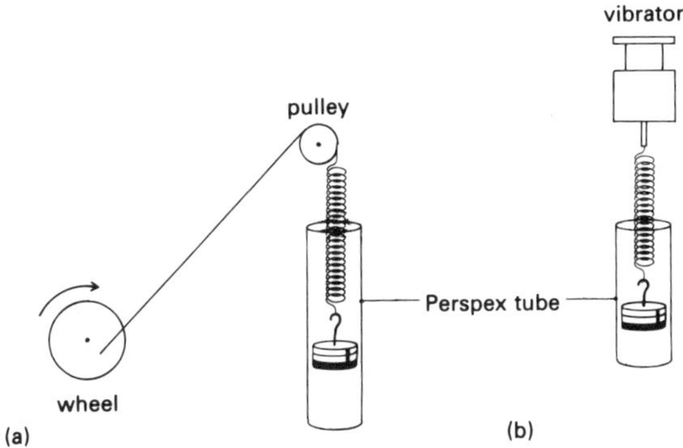

Figure D42
Forced vibrations.

EXPERIMENTS

D14 Investigations of resonance

There are many practical arrangements suitable for investigating resonance. We suggest two, but others are possible. It should be possible for students to measure the amplitude of forced vibrations over a range of frequencies for both lightly damped and heavily damped vibrations (in systems which are too lightly damped transient oscillations are persistent and can cause confusion). In some cases phase relationships may also be observed.

Ask students to devise experiments to answer the question: 'What happens when a system which can vibrate has a periodic force applied to it?' This is an exercise in thinking what to do, and doing it. It matters less what is or is not observed so long as some reasonable experiment is devised and tried. Good students, though, should be encouraged to plot graphs of amplitude against driver-frequency, if they are progressing well.

Adequate time is essential. A single long practical session may not be enough.

The subsequent demonstration, D15, Barton's pendulums, can cover all the resonant phenomena students must see.

D14a Resonance of a pendulum

ITEM NO.	ITEM
1124	resonance kit
501	metre rule
503–6	retort stand base, rod, boss, and clamp
507	stopclock
1153	card

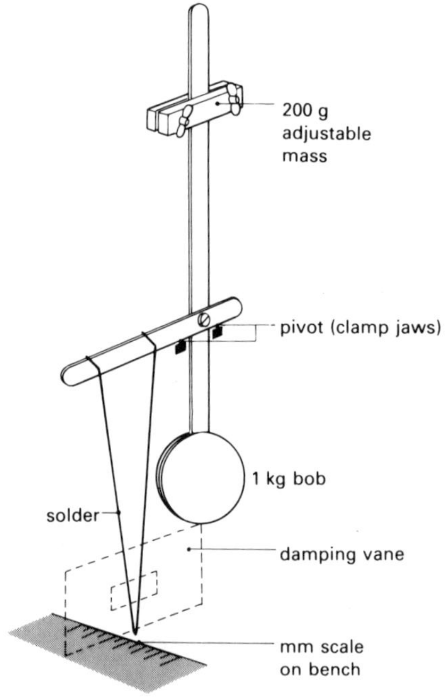

Figure D43
Resonance of a pendulum.

The apparatus should be set up already. The two strips are lock-nutted together about 20 cm from the top of the vertical strip. The solder is bent as shown, so that it is supported by the horizontal strip, but free to swing. (Cored solder is easily bent into the required shape.) The horizontal strip rests on the open jaws of a clamp, which acts as a pivot.

The pendulum of solder is the driven oscillator. The driving force is provided by the horizontal strip; its frequency can be varied by moving the adjustable mass up or down. A damping vane made of paper may be attached to the solder with adhesive tape. (See *School Science Review* **61**, 1980, pages 537–538.)

D14b Resonance of a mass on a spring

ITEM NO.	ITEM
2A	2 expendable springs
31/2	mass, 50 g
1153	thread
38	single pulley with suitable support
513	large beaker or measuring cylinder
501	metre rule
	either
1109	signal generator
1060	vibrator
	or
150	12 V fractional horsepower motor and wheel with pin offset about 1 cm
59	l.t. variable voltage supply
1153	drinking straws
1155	Perspex tube (or wide glass tube)
1000	leads

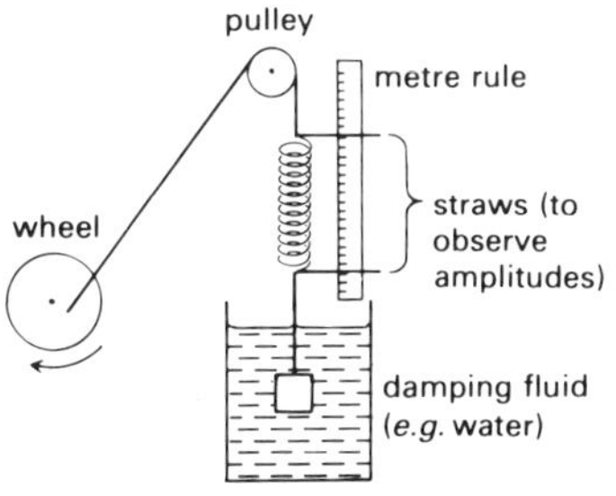

Figure D44
Resonance of a mass on a spring.

Using the straws, the amplitudes of the driven oscillator and the driving force can be observed; also the phase relationships may become apparent.

This experiment is not as easily controlled as D14a, but it is more obviously related to the many mass-on-spring-type resonance problems which confront engineers. It may be difficult to achieve a sufficiently fine adjustment of the motor speed: a vibrator is easier to use, as in figure D42b.

When the mass is not damped the system oscillates as a pendulum in addition to its vertical oscillations; this can be eliminated as in demonstration D13, by using the Perspex tube.

DEMONSTRATION

D15 Barton's pendulums

This demonstration gives the chance to find out what students have observed in experiment D14 without them having to make formal reports. Many students will have difficulty in transferring observations from the first situation to this new one, so the practice in doing so is likely to be worth while. Being able to use ideas in new problems is a good test of understanding.

ITEM NO.	ITEM
	slide projector or other light source
	Materials for constructing the pendulums and support:
1153	wooden rod, horizontal, 1.5 m long
1153	screw eyes or nails
1153	thin string or thread
1155	pendulum bob, mass about 0.04 kg
1153	nylon fishing line
1153	plastic curtain rings
1153	paper cones

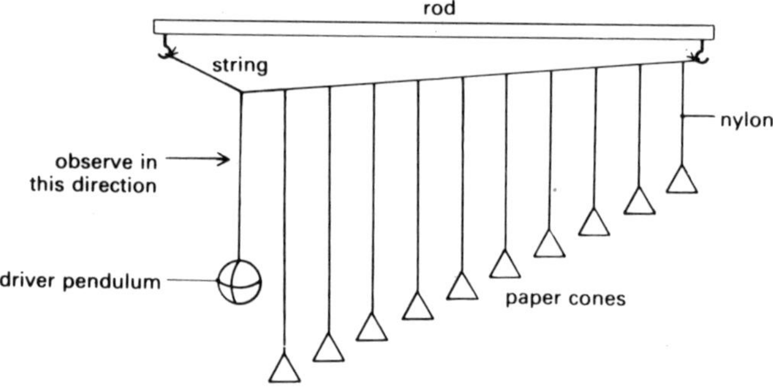

Figure D45
Barton's pendulums.

The construction is shown in figure D45. The wooden support rod should be firmly clamped so as to leave an unobstructed view along the line of pendulums. The lengths of the pendulums can be from about $\frac{1}{4}$ m to $\frac{3}{4}$ m, with the driver pendulum $\frac{1}{2}$ m long.

The nylon thread of the cone pendulums may be attached to the string by a half-hitch or slip-knot; this makes it easy to adjust the lengths. The pendulums should be as close together as possible. Teachers have also used thread, securing the cones with a blob of Plasticine at the end of the thread.

The demonstration is most effective in a darkened room with the cones brightly illuminated by the slide projector.

Students look along the line of pendulums and observe what happens when, with the paper cones at rest, the driver pendulum is released from a widely displaced position.

When the class and teacher look together at Barton's pendulums, resonance, phase relationships, transients, and damping effects may all be seen again in this fresh situation. Ask 'What are the similarities between this and the two examples of resonance studied so far?' 'Which pendulum is in resonance with the driver?' 'Are the pendulums damped?' (The effective damping may be reduced by slipping the plastic curtain rings over the cones. This is easily done if the rings are first cut.)

Important points which should emerge are listed in the *Students' guide* (page 226):

> The amplitude of the forced oscillations depends on the forcing frequency of the driver and reaches a maximum when
> forcing frequency = natural frequency of driven oscillator.
> The amplitude also depends on the degree of damping.
> Once any transient oscillations of varying amplitude have died away, a driven oscillator oscillates at the forcing frequency. At resonance the driver is one quarter of a cycle ($\pi/2$) ahead of the driven oscillator.
> If $f_{nat} < f_{driver}$, driver and driven are nearly in antiphase.
> If $f_{nat} > f_{driver}$, driver and driven are nearly in phase.

Photographs of Barton's pendulums in the *Students' guide* (page 227) help bring out some of these points.

Resonance curves

A useful description of resonance is a graph of the amplitude of forced vibration against forcing frequency: a resonance curve. See figure D46 and the *Students' guide* page 228, where the effects of damping on resonance are also summarized.

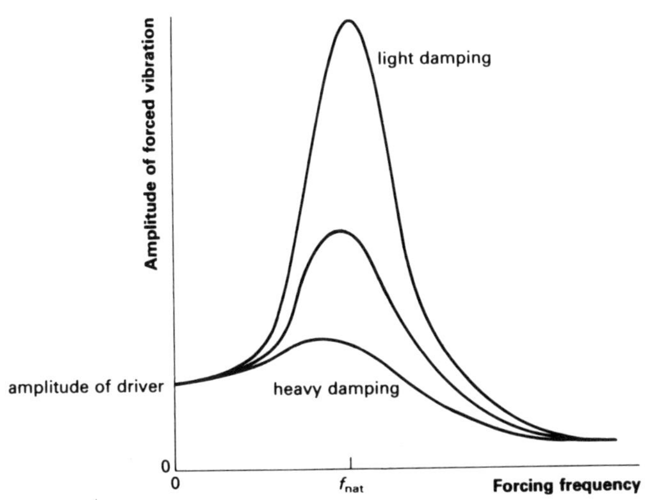

Figure D46
Resonance curves.

Energy in forced oscillations

During each cycle of its oscillation the driver delivers some energy to the driven oscillator. Some of this energy may be returned to the driver later in the cycle (a spring which has been forced to stretch can pull back, helping the driver as it returns in its oscillation); but at resonance the phase difference of $\pi/2$ prevents this. Some energy is dissipated in overcoming damping. The rest is stored in the forced oscillator, increasing its amplitude.

The amount of energy lost to damping in each cycle increases with the amplitude of the oscillation. The final amplitude of oscillation is that for which the energy delivered by the driver in each cycle is equal to the energy used to overcome damping. With heavy damping this happens at a small amplitude, and even the resonant oscillations are not violent. With light damping large amplitudes are achieved. A lot of energy from the driver is stored in these oscillations, particularly at resonance.

Questions

Questions 43 to 53 are about resonance in a variety of different systems.

The quality factor, *Q*

The quality factor, Q, gives a measure of the degree of damping and the sharpness of resonance. It can be defined in several ways including:

$$Q = 2\pi \times \frac{\text{energy stored in oscillator}}{\text{energy lost per cycle}}$$

For the purpose of this course, however, a simple non-rigorous idea of Q will suffice: Q is roughly the number of cycles of free oscillation which will occur before the oscillator in question has lost all its energy.

A *high value of Q* corresponds to a *lightly damped* oscillation with a sharp resonance peak.

An oscillator for which the instantaneous resistive force is proportional to its speed loses its energy exponentially. In this case, Q is strictly $2\pi \times$ (the number of cycles during which the energy left falls to $1/e$ of its original value). That means that the time-constant of the exponential is $\dfrac{Q}{2\pi f_{nat}}$.

The value of Q may be calculated from the resonance curve (figure D47). Find the frequencies f_1 and f_2, on each side of the resonant frequency, f_{nat}, for which the amplitude equals the amplitude at resonance divided by $\sqrt{2}$. Then $Q = \dfrac{f_{nat}}{f_2 - f_1}$.

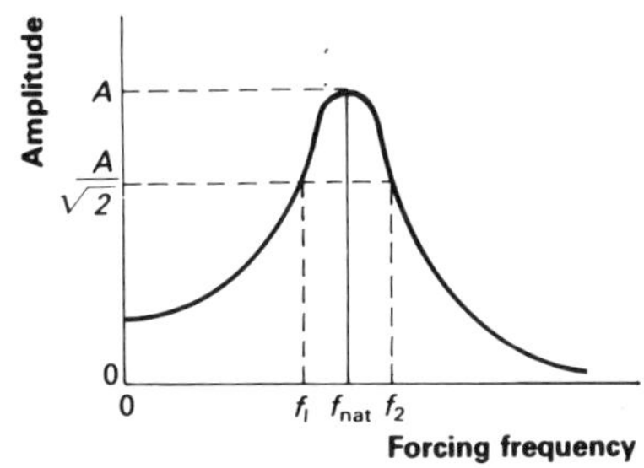

Figure D47

Students will not be expected to memorize this equation and we do not deal with its derivation here, but some students may like to know how Q can be calculated from the resonance curve. Values of Q for various systems are given in the *Students' guide*.

Questions 54 and *55* are about Q.

Dynamic modelling system

Interested students could use the dynamic modelling system for further exploration of forced vibrations with various amounts (and forms) of damping, perhaps as a theoretical ingredient in an Investigation.

STANDING WAVES AND RESONANCE

A concluding discussion of standing waves, which may be looked at from the point of view of oscillations, resonance, and superposition, offers a chance to draw together the threads in this Unit. A series of demonstrations is suggested, preceded by a demonstration and discussion of standing waves on springs and strings.

Many oscillations of practical importance are modes of standing waves (rather than pendulums, or masses on springs). References to applications should be made here: for example, musical instruments and microwave resonators.

When a wave train a dozen or so wavelengths long is sent down a long spring, a standing wave appears briefly where the reflected waves travel back through the last part of the outward-going wave train.

It may help to recall, or show again, standing waves along the line joining a pair of dippers in a ripple tank.

The standing wave is so called because it does not look as if it is travelling in either direction along the spring. Yet it is the result of two similar waves travelling in opposite directions. Detailed discussion in the *Students' guide* (page 229) shows how a standing wave develops as two such waves come to overlap each other.

DEMONSTRATION

D16 Standing waves on a rubber cord

ITEM NO.	ITEM
1109	signal generator
1060	vibrator
134/2	xenon flasher
1153	rubber cord (0.5 m long, 3 mm cross-section)
503–6	2 retort stand bases, rods, bosses, and clamps
121	4 metal strips (as jaws)
44/1	2 G-clamps, large
1000	leads

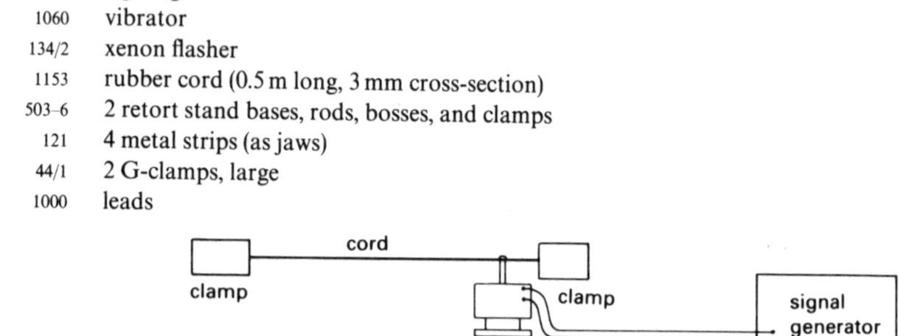

Figure D48
Standing waves on a rubber cord.

The ends of the rubber cord are held by the metal strips in the retort stand clamps. Clamp the retort stands to the bench so that the rubber cord is stretched to about 1 m length. Link the vibrator to the cord, a few cm from one end, by a short length of wire (0.71 mm diameter) twisted round the cord and fastened to the vibrator.

Set the signal generator on 2 V sine wave output (low impedance), and slowly increase the frequency from 10 Hz to 100 Hz. There should be 4 or 5 resonant frequencies in this range. It helps to have white bands painted on the cord at regular intervals along it, and to observe the motion under stroboscopic illumination.

As the signal generator's frequency is raised, the cord resonates first in one, then in two, then in three sections, and so on. What is the relationship between the frequencies at which these resonances occur?

The wave produced by the vibrator must take a certain time to travel to one end of the cord and then back to the vibrator again. If this time coincides with the period of the vibrator, we can expect resonance, because the vibrator is just sending off a second wave when the first is about to go on its next trip. But if the vibrator sends off exactly two or exactly three (or exactly any number) of waves in this time, each wave will be reinforced when it passes again travelling in the same direction. So we may expect a fundamental resonant frequency f and other resonant frequencies $2f$, $3f \dots nf$ (called harmonics).

Readings on the scale of the signal generator should support this. And for, say, the fifth harmonic $5f$, the cord vibrates in five sections (figure D49), separated by four motionless points (nodes).

Standing waves are an example of superposition, and occur where two similar trains of waves pass through one another going in opposite directions.

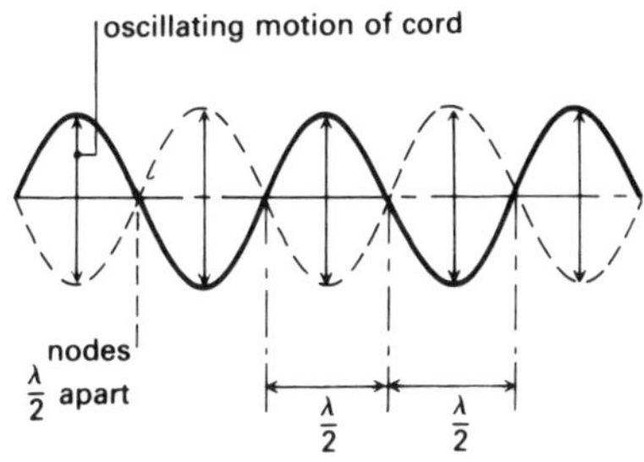

Figure D49

Resonance, in the sense of a slow building up of a big localized store of energy, does not occur if the waves can travel on indefinitely. But if the waves are confined to a limited space, and the two trains are simply reflections of the same original waves, the energy is confined and all the normal effects of resonance are seen. There can be standing wave resonance not only in springs but in many other situations, sometimes in two or three dimensions, with boundaries which involve or do not involve phase changes on reflection.

Many waves mutually cancelling

The discussion above suggests why the resonating cord responds, but not so clearly why, off resonance, there is practically zero amplitude, despite the motion of the driving force. The point need not be pursued, unless students raise it.

If the distance along the cord and back is *not* an exact whole number of wavelengths, a wave which has completed a few return trips will be out of phase with the vibrator. If the reflected wave has not lost much energy through damping, destructive superposition occurs. Each new wave is cancelled by an earlier one persisting in the cord.

DEMONSTRATIONS

D17 More complicated standing waves

A selection from the following suggestions can be used to illustrate that standing waves are to be found in many situations. The waves are in general more complex than those on a cord, but certain features remain the same. These, which should be pointed out as the demonstrations are shown, are:

i There are definite modes of oscillation, at each of which the response is large (resonance).

ii The patterns depend on the frequency, there being more nodes or nodal lines for high frequencies (short wavelengths).

iii The standing waves have to 'fit' into the system, whether it has one or more dimensions. They are the result of superposition of reflected waves from the boundaries and waves travelling towards the boundaries. As a general rule, wave-carrying systems with edges exhibit standing waves. Physicists now think of electrons held inside atoms as being like waves held inside the atom. (See Unit L, 'Waves, particles, and atoms'.)

D17a The Kundt dust tube

ITEM NO.	ITEM
	Any of the usual arrangements may be used, but the following is convenient:
1109	signal generator
1151	small loudspeaker, about 60 mm diameter
1155	measuring cylinder, 100 cm^3, or longer closed tube
	cork dust (see below)
	paper

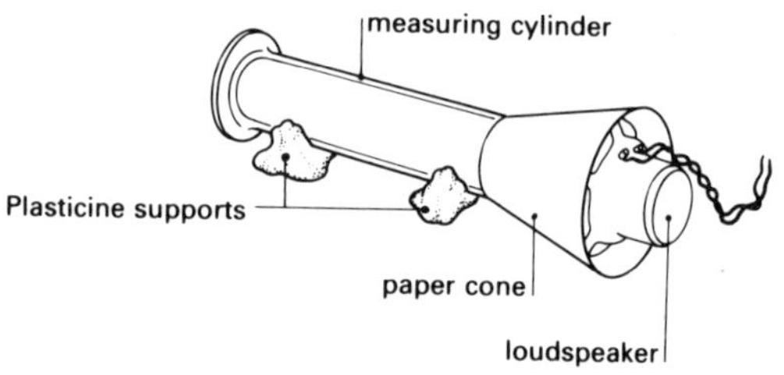

Figure D50

Dry the measuring cylinder and put some cork dust (made by filing a cork) into it. Arrange the cylinder horizontally, and tap it so that a *thin* layer of cork dust forms along the bottom. Set the signal generator to give an output of about 1 W, and tune it through the 1 kHz to 10 kHz range. The cork dust will show the positions of nodes and antinodes. Resonant frequencies can also be detected by ear: they sound distinctly louder.

The resonant frequencies are in arithmetic sequence. (An interested student could calculate whether the velocity of sound in a tube differs appreciably from that in open air.)

D17b Longitudinal standing waves in rods

ITEM NO.	ITEM
504	a rod, about 10 mm diameter, and about 1.5 m long, of glass, steel, or brass, *e.g.*, retort stand rod
44/1	G-clamp and wooden blocks or other clamping devices
1153	cloth
1153	rosin (for metal)
1156	alcohol (for glass)

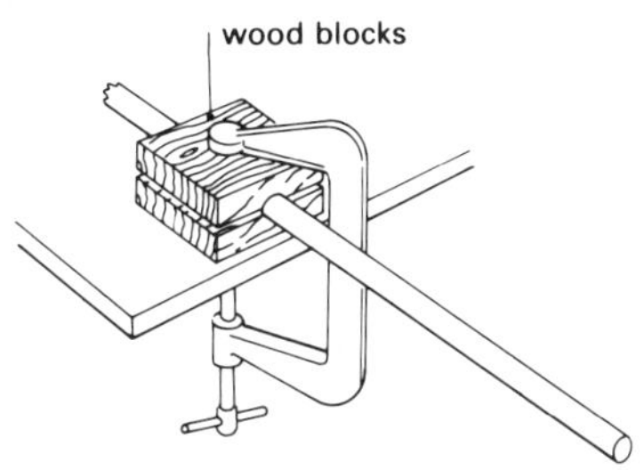

Figure D51

Clamp the rod at the middle and excite it, stroking with a wet or rosined cloth. (The velocity of sound in the rod may be calculated if a student can solve the problem of measuring the frequency.)

D17c Vibrations of circular wire rings

ITEM NO.	ITEM
1109	signal generator
1060	vibrator
134/2	xenon flasher
1501	copper wire, 0.90 mm diameter, or thinner steel wire
1000	leads

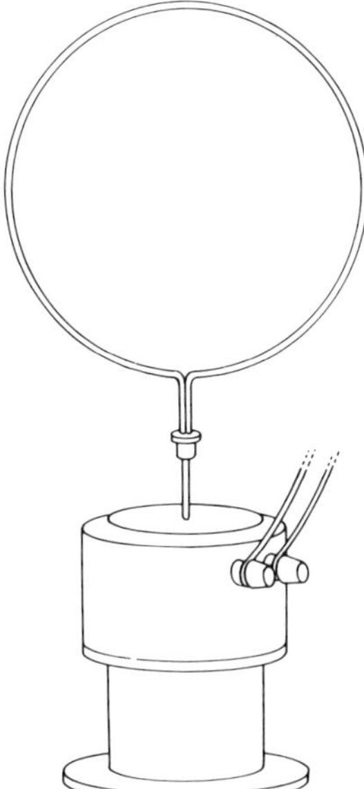

Figure D52

Form a length of wire (about 1 m) into a circle and attach it to a vibrator so that it stands vertically. One method is to form the two ends of the wire into loops and anchor them between washers on the vibrator shaft.

Rings of various diameters may be vibrated along a diameter. The frequencies of normal modes can be found. A stroboscope is useful.

D17d Longitudinal standing waves

ITEM NO.	ITEM
1109	signal generator
1060	vibrator
134/2	xenon flasher
1013	long spring
501	metre rule
1000	leads

Put the vibrator on its side, attach one end of the spring to the vibrating element using string or a wire loop. A length of about 0.3 m of spring should be stretched to about 0.50 m (these distances are not critical). The hand holding the spring may be rested on a metre rule, the other end of which acts as a stop to prevent the vibrator sliding along the bench. The signal generator low impedance output is used, at full output. The frequency is increased from about 20 Hz to several hundred hertz. The standing waves should be viewed stroboscopically, as well as by eye, to give students a clear understanding of their nature.

The spring shows standing waves at frequencies in arithmetic progression.

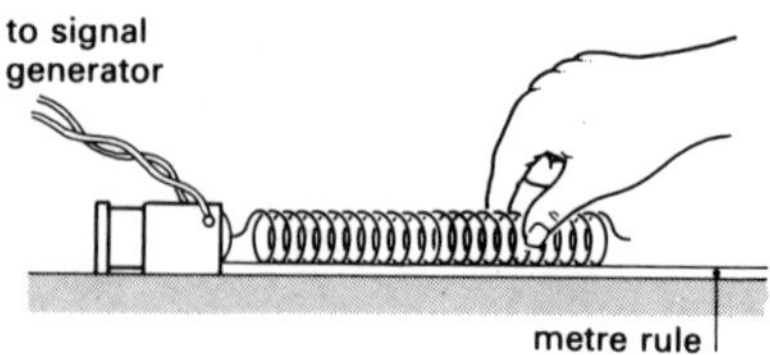

Figure D53

D17e Vibrations in a rubber sheet

ITEM NO.	ITEM
1109	signal generator
1044	large loudspeaker
134/2	xenon flasher
1153	sheet of rubber
503	2 retort stand bases and rods
1000	leads
1076	big wire ring, *e.g.*, embroidery ring, about 25 cm in diameter

Stretch the rubber sheet over the ring. Then place it over the loudspeaker driven by the signal generator. Lines drawn with a ball point pen on the rubber sheet help to show the changes in vibration pattern as the oscillator frequency is altered. It is worth viewing these oscillations under strobe illumination. Details of a demonstration of these standing waves will be found in the *School Science Review*, **50**, No. 173, June 1969, page 930.

It is useful to repeat this demonstration in Unit L, 'Waves, particles, and atoms'.

D17f Chladni figures

ITEM NO.	ITEM
1153	square or round metal plate
1109	signal generator
1060	vibrator
1156	sand
1000	leads

Thin metal plates, square or round, are driven at the centre by a vibrator, and sand is used to observe the vibration patterns of Chladni figures. The resonant oscillations of car door or body panels are of this general kind.

D17g Vibrations of a loudspeaker cone

ITEM NO.	ITEM
1109	signal generator
1044	large loudspeaker
134/2	xenon flasher
1000	leads

The loudspeaker is connected to the signal generator so that the cone vibrates in a vertical direction. A few grains of semolina placed in the cone should show the resonances up clearly, as the generator frequency is changed. The cone should also be viewed under stroboscopic illumination as the frequency alters.

D17h Standing waves in a round bowl

ITEM NO.	ITEM
1109	signal generator
1060	vibrator (with some form of dipper attached)
1155	Petri dish from electric fields apparatus
1153	plastic washing-up bowl
	wooden block
1000	leads

Ring patterns can be produced by using a vibrator with a dipper attached to it. It should be used at a low frequency to generate waves in a Petri dish containing a little water. The dish may be supported with a screen about 0.1 m below it and with a 12 V lamp above it, as a miniature ripple tank. Water in a Petri dish may be excited into several kinds of standing wave motion. A vibrator can be used to produce ring patterns of waves.

A wooden block can be used to make the water in a larger plastic bowl oscillate with several kinds of sideways 'slopping' motions, or waves may be sent round the perimeter of the bowl.

D17i Musical instruments

ITEM NO.	ITEM
1511	oscilloscope
157	microphone
1035	pre-amplifier (if necessary)
	assorted musical instruments (provided by students)

Exhibit the waveforms of the musical notes produced by various instruments on the oscilloscope. It may help to tape record the instruments beforehand, though the recorder may modify the characteristics of the sounds.

A flute is almost a simple pipe, while other wind instruments are more complex. The effective length is changed by opening holes (stops). Stringed instruments have standing waves on the strings, but the wooden structure and air inside also resonate, giving the instrument its particular quality by reinforcing many but definite frequencies.

Home experiments

More standing wave systems can be investigated by students at home, for example Home experiments DH4, Standing waves in a rectangular tank, and DH5, Standing waves under a running tap.

DH6, Step wave under a running tap, is not, it must be stressed a standing wave in the usual sense of the term. However, it is an interesting physical phenomenon which, like DH5, might form the basis of an Investigation.

The usefulness of the standing wave idea

Engineers continually deal with standing waves of one sort or another. Anything that can vibrate and has edges may have a standing wave on it. So the shaft driving a ship's propellers, or turning a turbine, can go into a standing wave oscillation, flexing as it turns.

The wings of an aircraft also flex like a springy ruler. Two-dimensional standing waves are likely wherever flat panels can vibrate, so they matter to the motor and to the building engineer. Three-dimensional standing waves are a problem for acoustic engineers. A good example is a loudspeaker cabinet enclosing a volume of vibrating air.

Physicists, too, continually deal with standing waves. At the end of this course, in Unit L, 'Waves, particles, and atoms', standing waves are used to explain why atoms have definite energy levels, by thinking of electrons trapped in an atom as like a standing wave. But standing waves appear in many other parts of physics. For example, radio waves can form standing waves inside metal cavities, and they have been used both to make very accurate measurements of the velocity of the waves, and in the design of powerful high frequency generators of microwaves.

Film loops, video, and television

'Tacoma narrows bridge collapse' Ealing Scientific.
'Vibrations of a drum' Ealing Scientific.
'Soap film oscillations' Ealing Scientific.
'Wind-induced oscillations' Penguin.
B.B.C. television Physical Science: 'Oscillations' (This programme illustrates very nicely many of the points made in this unit.)
Open University television 'Vibrations of music' is relevant.

Reading

Several textbooks have good sections on resonance, for example:
AKRILL, BENNET, and MILLAR *Physics.* Chapter 22.
AKRILL, and MILLAR *Mechanics, vibrations and waves.*
BOLTON *Patterns in physics.* Chapter 4.
WENHAM, DORLING, SNELL, and TAYLOR *Physics: concepts and models.* Chapter 20.
DUNCAN *Advanced physics: materials and mechanics.* Chapter 8.

Further examples of resonance:
BISHOP *Vibration.* Excellent on applied problems concerned with resonance.
CRAC *Waves and sound.*
CRAC *Vibrations.*
FEYNMAN, LEIGHTON, and SANDS *The Feynman lectures on physics* Volume 1. Chapter 23.
Physics Reader *Physics in engineering and technology.* 'Buildings, bridges, and wind.'
WALKER *The flying circus of physics.* This has many examples.
SHIVE and WEBER. *Similarities in physics.*

Questions

Questions 56 to 59 are about standing waves.

Unit E
FIELD AND POTENTIAL

Trevor Sandford
Henbury School, Bristol

Suggested time allocation: four weeks

PLAN OF THE UNIT

Section E1
The uniform electric field

forces between +ve and −ve charges	►	**Forces on charges**
		Uniform field between parallel plates
$g = F/m$ Millikan experiment energy = Fd p.d. = energy/charge (Unit B)	►	$E = F/Q$ **(** $= V/d$ **for a uniform field)**
		Patterns of fields
		Flame probe
meaning of p.d.; choice of zero (Unit B)	►	**Potentials in a field; equipotentials**
		$E = -\Delta V/\Delta x$
		Parallel plates: Q, V, A, $1/d$**; effect of medium**

Prerequisite		Topic		Link
use of coulombmeter to measure charge; $C = Q/V$ (Unit B)	▶	**Reed switch and coulombmeter experiments**		
		$\sigma = \varepsilon_0 E$, $C = \varepsilon_0 \frac{A}{d}$, ε_r	▶	Unit J, 'Electromagnetic waves'
		Applications		
		Section E2 Gravitational field and potential		
energy = force × distance	▶	**Gravitational potential difference in a uniform field**		
gravitation	▶	$g = -\frac{F}{m}$, $F = -G\frac{m_1 m_2}{r^2}$, $g = -\frac{GM}{r^2}$		
Newtonian dynamics	▶	**Data used to justify** $g \propto -\frac{1}{r^2}$		
energy = area under force–distance graph (Unit A)	▶	$\frac{1}{r^2}$ **field** $\Rightarrow \frac{1}{r}$ **potential** $\left(-\frac{GM}{r}\right)$		

continued

Prerequisite		Content		Leads to
vectors (Unit A)	▶	**Vector field, force; field = − potential gradient**		
		$\frac{1}{r}$ **potential hill, well**	▶	nuclear atom: Unit F, 'Radio-activity and the nuclear atom' electrons in atoms: Unit L, 'Waves, particles, and atoms'
dynamics	▶	**Centripetal acceleration; orbits**		

Section E3
The electrical inverse-square law

Prerequisite		Content		Leads to
flame probe: Section E1	▶	$\frac{1}{r}$ **potential** $\Leftrightarrow \frac{1}{r^2}$ **field**		
		$V = \frac{kQ}{r}$; $E = \frac{kQ}{r^2}$		
		$k = 9 \times 10^9\ \mathrm{N\,m^2\,C^{-2}}$		
		$k = 1/4\pi\varepsilon_0$		
		Sizes of electrical and gravitational forces		
		Field of dipole, array of charges	▶	materials: Unit A, 'Materials and mechanics'

INTRODUCTION

Students will probably have met the notion of a field (magnetic, gravitational, or electric) before. Here it is more clearly defined and the important concept of potential is introduced. Using the analogy between gravitational and electric fields the relationships between force, field, energy, and potential can be drawn out. The emphasis is on a sound *physical* understanding, and applications cover a wide range, from atoms to the Solar System.

Concepts emerge, as elsewhere in the course, from an experimental base wherever possible. A minimum of mathematical terminology and techniques is employed; the emphasis is always on grappling with the difficult physical concepts.

Starting with forces on charges in a uniform field, these Sections develop the idea of potential difference towards that of potential and its gradient. The inverse-square field is introduced with gravity, where it may be more familiar. The link between $\frac{1}{r^2}$ fields and $\frac{1}{r}$ potentials is forged here and then used in the analogous electrical case for charged spheres.

Practical applications concerning gravity include the orbits of satellites, which can be tackled after circular motion is understood. Electrical applications range from industrial processes and problems to the microscopic world of atomic structure.

THE PLACE OF THE UNIT IN THE COURSE

Essential topics which must have been studied *before* this Unit include the following:

Basic Newtonian dynamics (GCSE and Unit A);

Potential difference (GCSE and Unit B);

Charge and capacitors (GCSE and Unit B).

The work on circular motion ties up a little with simple harmonic motion from Unit D, 'Oscillations and waves' which provides a useful break between the electrical work of Units B and C and Section E1, though it would be possible to complete almost all of Unit E before Unit D.

Other topics which use ideas developed in this Unit include:

Forces between ions (Section A2);

The nuclear atom (Unit F, 'Radioactivity and the nuclear atom');

Electrons in the atom (Unit L, 'Waves, particles, and atoms').

The Unit also has its own 'end-points'. A theoretical one is the appreciation of the analogy between electrical and gravitational fields and the links between the key concepts of force, energy, field, and potential. Another is an appreciation that such abstract ideas have practical applications, ranging from the orbits of spacecraft round the Earth to an understanding of electrons within atoms, from photocopying and printing to painting and prospecting.

LIST OF SUGGESTED EXPERIMENTS AND DEMONSTRATIONS

SECTION E1
THE UNIFORM ELECTRIC FIELD

Forces on charges

'Field and potential' – what is a 'field'? What fields have we already met and studied – gravitational, magnetic, electrical ...? What does a field do? What sorts of things does it affect? (For example, gravity affects masses, magnetism affects magnets or currents ....) These are the sorts of question which can open up discussion leading towards the idea that a field exists where we can observe a *force* on something. Depending on what that something is, then, we say a different type of field exists. If students take kindly to this sort of debate, it can form a useful introduction. Discussion can continue around the following demonstration which may have been met in an earlier course, but can be introduced as a puzzle to be explained.

DEMONSTRATION
E1 Forces on a charged ball between charged plates

ITEM NO.	ITEM
14	e.h.t. power supply
65	2 metal plates with insulating handles
57L	table tennis ball coated with colloidal graphite
1153	nylon sewing thread
1101	sensitive galvanometer (*e.g.*, internal light beam)
51G	polythene strip
503–506	2 retort stand bases, rods, bosses and clamps
1000	leads

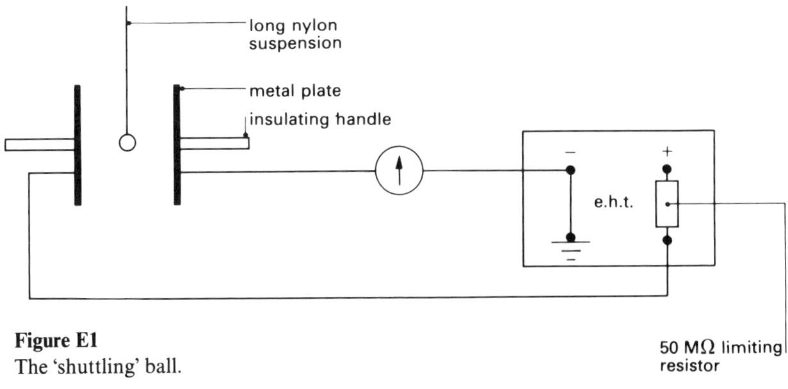

Figure E1
The 'shuttling' ball.

The galvanometer may at first be left out of the circuit and the p.d. increased to about 4 kV, with the plates about 0.1 m apart. The ball, on making contact with one or other of the plates, will 'shuttle' back and forth continuously.

Students may suggest that:
– forces between like and unlike charges are causing the effect;
– the direction of the force depends on the charge on the ball;
– the ball is 'ferrying' charge between the plates.

This is a good point to revise the idea that current is simply 'charge in motion' and that potential difference (here provided by the power supply) is what causes the charge to move.

Induced charges

The effect may begin spontaneously with the ball initially not touching either plate. The idea of induced charge (that is, charge separation) on an object with no net charge may be mentioned, and examples (for instance a charged comb picking up paper or attracting a stream of water) may be shown in passing.

The main thrust, however, should be towards establishing the idea that the ball moves because there is a *force* on it, and that this force arises because the ball carries an *electric charge*. So we say that an electric field exists where electric charges experience a force.

What factors affect this force?

Students may suggest, or have noticed, that the p.d., V, affects the force on the ball and they can also see the effect of varying the separation, d, between the plates.

Quantitative results are not expected at this stage as it is probably only the frequency of oscillation, or perhaps the current, which is readily observed rather than the force on the ball. There is no clear evidence yet that decreasing d, say, does increase this force. Nor is the effect of the ball's own charge observed as this too is changed if V is altered. These factors are dealt with in the next demonstration.

Question

Question 1 discusses the 'shuttling ball'.

Timing

The first two demonstrations, E1 and E2, serve to focus attention on vital concepts rather than being ends in themselves. Therefore, whilst adequate time must be allowed for discussion, they need take no longer than one or two periods.

DETECTING AN ELECTRIC FIELD

'How can we detect where there is a force on a charged object, and how can we see how it varies from place to place?' The answers to these questions may be found using some sort of field detector, basically a charged object on which the electric force has some observable effect.

DEMONSTRATION

E2 Using a charged foil strip as a field detector

ITEM NO.	ITEM
1025	1 pair capacitor plates
30	2 slotted bases
51M	2 square polythene tiles
51G	polythene strip
	foil (see note below)
1153	adhesive tape
1153	razor blade or scissors
14	e.h.t. power supply
1000	leads
	means of projection (optional)

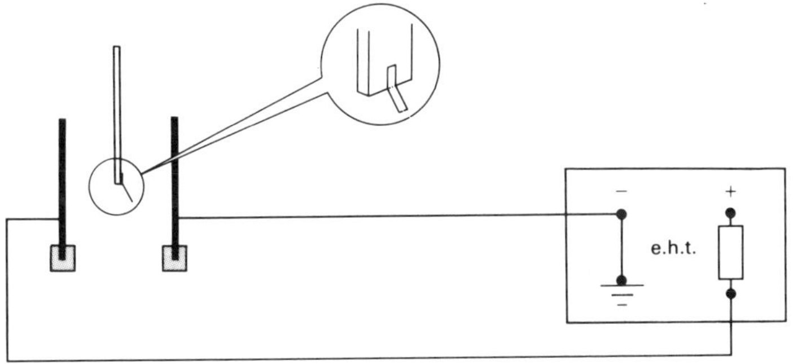

Figure E2
Charged foil field detector.

Making a suitable foil detector

The best foil to use is aluminized plastic film (25 gauge), for example 'Melanex'. (Normal cooking foil is too thick.) It needs to be made rather narrow (approximately 1 mm × 10 mm of foil exposed) and stuck carefully onto the polythene strip with tape.

Very good results have been obtained with the very thin aluminium leaf supplied for electroscopes. A rather larger strip (approximately 8 mm × 40 mm) may be cut with scissors by first sandwiching the foil between two sheets of paper. The end of the polythene strip is moistened using a licked finger and the damp surface offered towards the end (about 15 mm) of the foil strip whereupon it sticks quite firmly and permanently. The larger movement is more clearly seen but the foil is more fragile and should be stored carefully. (See figure E3.)

Figure E3
Storage of foil detector.

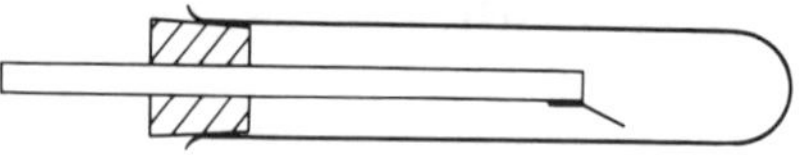

Set up the plates a few centimetres apart. The foil is charged by contact with the positive plate and the angle at which it hangs gives an indication of the force on it. The p.d. between the plates can be adjusted to give a convenient angle, say 45°. For large groups projection on to a screen may be a help, but students must be close enough to see the foil being moved about in the space between the plates.

The foil may first be used to show
– that there *is* a field in the region between the plates, and that it even extends somewhat outside this region,
– that increasing the p.d. whilst holding the charged foil steady increases the force on it,
– that decreasing the separation of the plates has the same effect,
– that the angle at which the foil hangs, and therefore the force on it and so the field, is almost constant over much of the region between the plates (shown by moving the detector in 3 directions mutually at right angles, parallel and perpendicular to the plates),
– that the force on a positively charged foil is directed away from the positive plate, towards the negative.

The force on the foil may not remain constant very near the plates. A shorter foil avoids this problem, but then it must be narrow (as described) so as not to be too stiff.

DEFINITION OF STRENGTH OF FIELD

Students should be familiar with gravitational field strength as the force per kilogram of an object or 'force per unit mass'. (REVISED NUFFIELD PHYSICS *Pupils' Text Year* 4, Chapter 2.) Similarly, we can now define the strength of an electric field as the 'force per unit charge' or $E = \frac{F}{Q}$, where E represents the electric field strength, F the force, and Q the charge on which it acts. Note here that field, like force, is a *vector* quantity, possessing both magnitude and direction. Its units are clearly $\mathrm{N\,C^{-1}}$, and students should get used to this unit as the fundamental unit for field since it reinforces the definition.

The foil experiment shows that between parallel plates the electric field strength is uniform and affected by the p.d. between and separation of the plates.

A more strict definition

A strict definition regards field as a limit:

$$E = \lim_{Q \to 0} \frac{F}{Q}$$

If the field is due to a charged conductor, the 'test' charge can cause movement of charge in the conductor, thereby altering the field itself. This effect becomes less as Q falls to zero. Students might be given some hint of this problem, though teachers must be careful not to introduce further confusion: a large test charge does give a true measure of the field it is testing, but this field might be quite different if the test charge were removed. The situation might be compared with that of using a 'bad' (that is, low resistance) voltmeter to measure p.d. in a circuit. If the original field were due to charges which could *not* move, then the test charge, Q, could have any magnitude without affecting the field due to those charges.

Problems in using F/Q to measure fields

There are two clearly *practical* objections to using $\frac{F}{Q}$ as a measure of the field: F and Q are both very small and therefore difficult to measure accurately. A larger, more measurable force on the foil might be gained by giving it a much greater charge, say 1 coulomb. But this would be very much bigger than the charge on either plate and would radically distort the field.

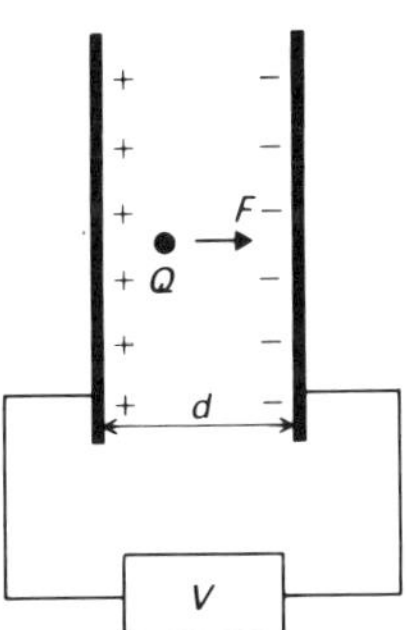

Figure E4
Relationship between F, Q, V, and d.

The way out of this problem relies on looking again at the relationships between F, Q, V, and d (figure E4). The ratio $\frac{V}{d}$ is equivalent to $\frac{F}{Q}$ for a uniform field. In this limited but very common context then we have a very simple method of measuring the field strength – with a voltmeter and a ruler – which avoids all the practical problems of measuring F and Q. Stress that $E = \frac{V}{d}$ is a special case for uniform fields only, but $E = \frac{F}{Q}$ is always true. (Strictly speaking, perhaps we should write $E = -\frac{V}{d}$. At this stage students must know that the field is in the direction of the force on a *positive* charge.)

Questions

Question 2 discusses the foil used above.
Question 3 shows the equivalence of $\mathrm{N\,C^{-1}}$ and $\mathrm{V\,m^{-1}}$ as field strength units.
Question 4 gives practice in calculating field strengths.

PATTERNS OF ELECTRIC FIELD FOR DIFFERENT GEOMETRIES

In the same way that iron filings align themselves to 'show up' a magnetic field, small poorly conducting particles may be used to reveal the shapes of electric fields.

DEMONSTRATION

E3 Patterns of electric field for different geometries

ITEM NO.	ITEM
	either
14	e.h.t. power supply
	or
60/1	Van de Graaff generator
149	electric field apparatus
1501	bare copper wire, 2 mm diameter, for electrodes
1153	castor oil
1153	semolina (chopped hair will do)
1156	1,1,1-trichloroethane
1000	leads

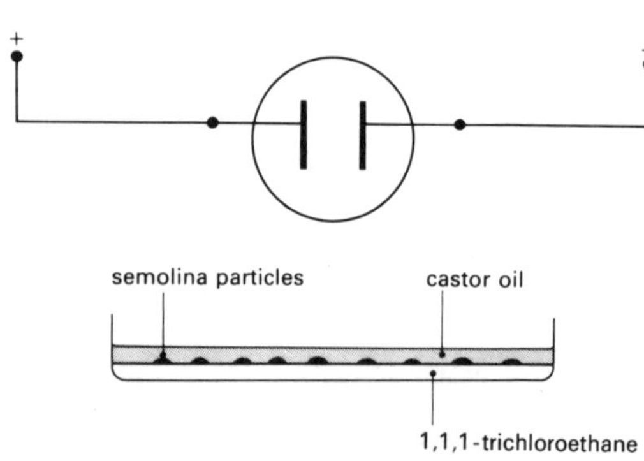

Figure E5
Using semolina to show electric fields.

Semolina powder floats on the surface of 1,1,1-trichloroethane. (Tetrachloromethane (carbon tetrachloride) has a toxic vapour and must not be used for this or any other experiment in the open laboratory. 1,1,1-trichloroethane is considered safer.) Put a small amount (about $\frac{1}{2}$–$\frac{3}{4}$ teaspoonful) of oil in the Petri dish, along with a sprinkling of semolina and enough trichloroethane to about half cover the electrodes, which may be of thick (2 mm) copper wire or simply welding rods. Stir the mixture and allow it to settle for about 3 minutes, when the semolina becomes quite mobile in the interface between the trichloroethane and the castor oil (which prevents the other liquid from evaporating). Semolina in oil alone gives satisfactory results, though since it is then less mobile and tends to sink, higher p.d.s may be needed and a Van de Graaff generator is required. Also, instead of semolina, the hair from a clean dry paintbrush cut to less than 1 mm in length, or grass seed, make acceptable substitutes.

A p.d. of a few kV from the e.h.t. supply is sufficient to cause the semolina particles to orientate themselves in accordance with the electric field between the electrodes. A little agitation or stirring may be necessary. When the pattern is clear the p.d. may be switched off. Different shaped electrodes may be used to show both uniform, radial, and other configurations of field in two dimensions. Overhead projection makes the effects most clearly visible, though heating may cause the liquid to evaporate, and after 5–10 minutes or more the oil may begin to conduct, giving spurious results. Otherwise the demonstration gives good results and rewards the little trouble it requires.

Field lines and induced charges

The semolina particles form into lines which follow the direction of the electric field. Since 'field lines' are just one way of representing a field we do not stress them. They can lead to misunderstanding (see Appendix V); here they are just a useful way of showing the direction and shape of a field. Nor do we deal here with induced charges and polarization, or attempt to explain why the experiment works. Interested students could follow this up for themselves.

Questions

Questions 5 and 6 involve estimating the shapes and strengths of electric fields.

FIELD AND POTENTIAL

Teaching sequence The next discussion based around demonstrations continues the argument towards the relationship between field strength and potential gradient, after which follows the first of the Section's class experiments (E5) on plotting equipotentials. If practical work is required at an earlier stage, the work based on exploring the charge on parallel plates (Experiments E6 and E7) could be taken first, then returning here to complete the Section.

The strategy here is to move from $E = \frac{V}{d}$ for the uniform field towards the more general $E = -\frac{\Delta V}{\Delta x}$ for any region of any field where changes in V and x can be measured. The flame probe is introduced as a tool to measure V.

Measuring electric fields 'in space'

The discussion might proceed as follows:

'So far we have seen that $\frac{V}{d}$ may be used as an alternative to $\frac{F}{Q}$ as a measure of the strength of the uniform field between charged parallel plates. We know that there is a certain potential difference between the plates; however, we know very little about the space between them. Could we somehow measure a p.d. between isolated points in space and thereby deduce the field (if it is uniform) between those points?' This is what we set out to do using the flame probe, which will not only introduce the concept of potential and its variation with distance but also give more meaning to the abstract talk about field as an effect 'in space'.

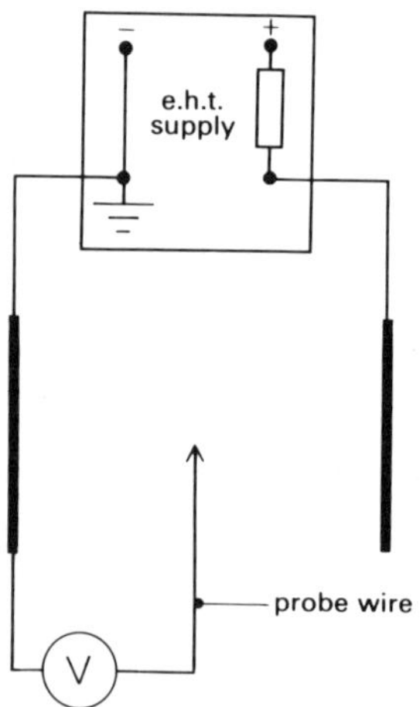

Figure E6
A p.d. probe.

The circuit of figure E6 will not work. If the wire probe is near, say, the positive plate, electrons will be drawn up the wire to its tip. This collection of negative charge will disturb the field. Recalling the strict definition of field, perhaps it can be seen that this problem is minimized if the charge on the tip is small, and indeed solved if it can be completely neutralized. This may be achieved by surrounding it with a collection of ions, which will carry away any excess charge acquired by the tip; this is most readily arranged in practice using a small gas flame. Thus the 'flame probe' may be introduced as a device to measure potentials within the field.

DEMONSTRATION

E4 Measuring potentials in a uniform field using a flame probe

ITEM NO.	ITEM
51A	gold leaf electroscope
51J	hook for electroscope
14	e.h.t. power supply
94A	lamp, holder, and stand
27	transformer
52K	crocodile clip
1025	pair of capacitor plates
51M	2 square polythene tiles
30	2 slotted bases
51G	polythene strip
501	metre rule (or Perspex rod)
1501	PVC covered copper wire
1155	hypodermic syringe, 1 cm^3
1155	25 gauge hypodermic needle
1155	PVC tubing, 2 m, 6.5 mm bore
1153	adhesive tape
1153	razor blade
552	Hoffman clip
503–6	retort stand base, rod, boss, and clamp

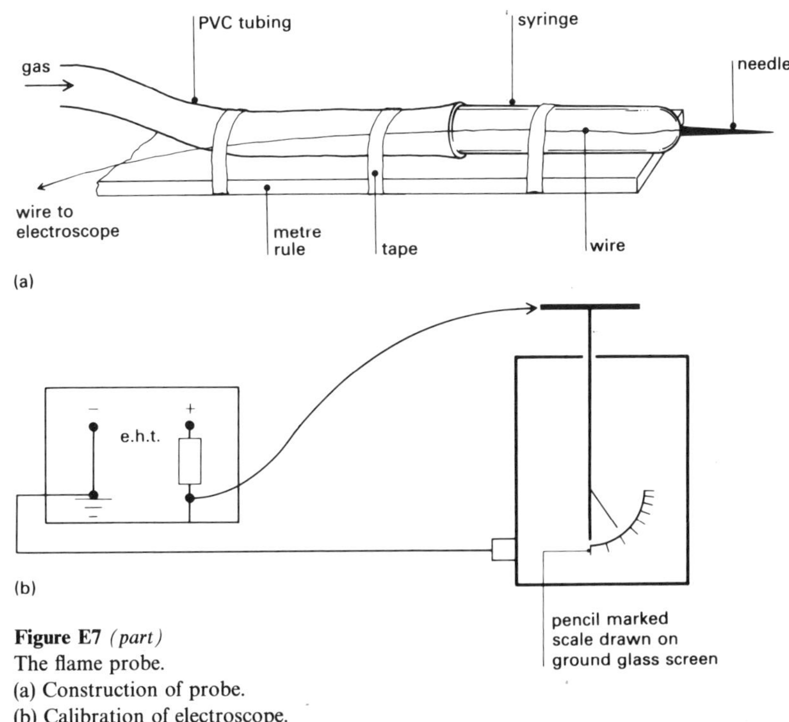

Figure E7 *(part)*
The flame probe.
(a) Construction of probe.
(b) Calibration of electroscope.

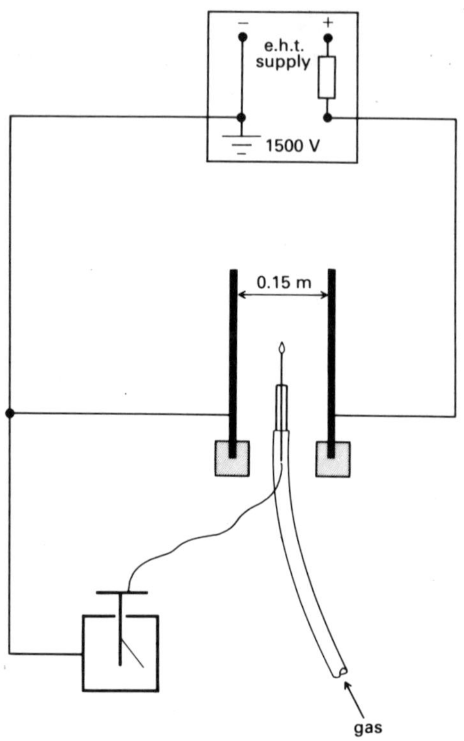

(c)

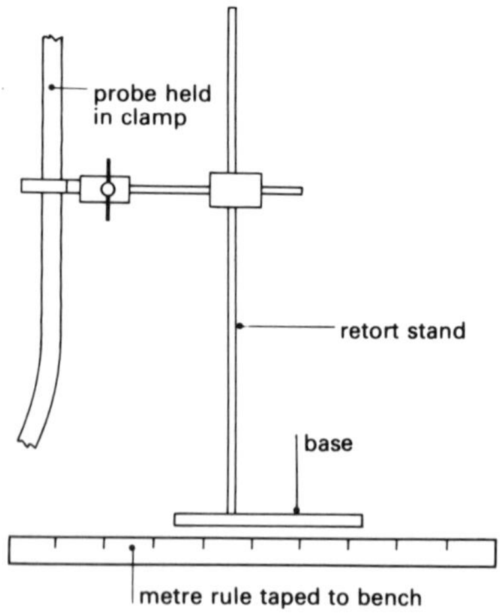

(d)

Figure E7 *(continued)*
The flame probe.
(c) Measuring potentials.
(d) Measuring the movement of the probe.

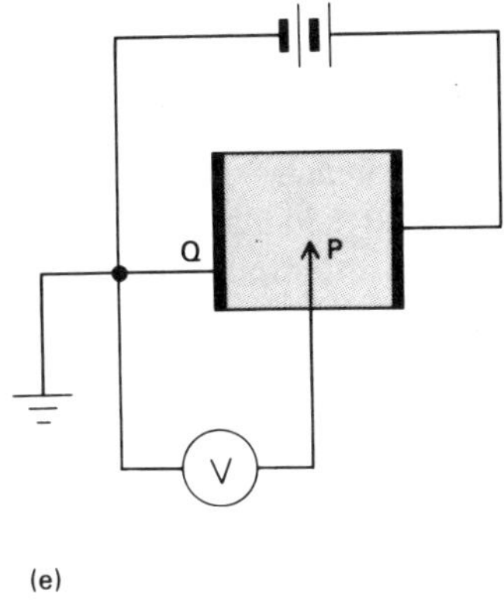

Figure E7 *(continued)*
The flame probe.
(e) 'Equivalent electrical circuit'.

Questions

Questions revising the potential divider and the idea of potential difference may help in leading towards the concept of potential at this point. Question 7 discusses the flame probe experiment.

Construction of the probe

The probe may be made from the body of a plastic disposable hypodermic syringe over which PVC tubing makes a gas-tight fit. An insulated wire is soldered to the base of the needle and the whole is taped to a long Perspex rod or metre rule which is held in a clamp. (See figure E7(a).) (Use additional plastic as insulation between the jaws if a wooden rule is used.) A Hoffman clip on the gas supply tubing enables the pressure to be adjusted until a small flame can be maintained at the tip (too high a pressure will 'blow away' the flame, which should be as small as is practicably possible, but stable). The wire is attached to the cap or hook of a gold leaf electroscope using a crocodile clip.

Alternative construction A much simpler construction using a cigarette lighter has been suggested. Although the lighter itself and the flame are both larger than the needle and its flame, this apparently cruder flame probe may well suffice for our purposes. Tape the lighter to a metre rule or Perspex rod. Some means of holding the lighter 'on' without touching it is needed – more tape. Solder an insulated wire to the metal part of the lighter (near the flame) to connect it to the electroscope.

Calibration of the electroscope This must first be done using the circuit of figure E7(b), applying a p.d. which varies from 0–1500 V across the instrument, and using a pencil to mark the ground glass screen with a scale which indicates the position of the shadow of the leaf when projected from behind. (A small bulb permanently attached to the clear glass 'back' of the electroscope gives consistency in the position of this shadow.) A good leaf will move smoothly at the higher voltages, but the scale is not at all linear and there are quite considerable uncertainties, particularly with the lower readings.

Use of the probe

The complete circuit is as in figure E7(c). A p.d. of 1500 V and plate separation of 0.15 m are suitable. Slide the retort stand along a metre rule taped to the bench from a zero mark determined by touching the probe on to the earthed plate with the flame extinguished – figure E7(d). Take readings of potential at points in the space, in particular along a line joining the plates and at right angles to them.

Potential

The circuit of figure E7(e) may be recalled from Unit B, 'Currents, circuits, and charge', and seen to be equivalent to the present arrangement. We are effectively going to measure the potential difference between point P and point Q which is *earthed*, and therefore at our chosen *zero* of potential. So, as in Units B and C, we can talk about the 'potential at P' rather than the more long-winded 'potential difference between P and 0 V': the potential at P is, say, 500 volts if there is a p.d. of 500 V between P and Q. (Potential can therefore be negative or positive, though it is clearly a *scalar*, not a vector, having, like energy or mass, no sense of direction. In this experiment the potentials will all be positive.)

Start without the flame by first touching the probe on to each plate, and showing that 0 V or 1500 V is indicated. With the probe about half way between the plates only a very small deflection is obtained until the flame is lit. Then move the lighted probe around a surface parallel to both plates: the potential should be unchanged. (The 'biofeedback' obtained by watching the electroscope whilst doing this is probably the best means of achieving the correct movement!) Explore the potential outside the immediate vicinity of the plates. Last and most important, measure the variation of potential with distance along a line perpendicular to the plates. Draw a graph of potential against distance. (It is convenient, though not necessary, to arrange for a potential gradient of 100 V per centimetre.) See figure E8.

Two important concepts emerge from the experiment: *potential gradient* and its relation to field strength, and *equipotentials*.

End effects Measurements of potential with the flame probe very near (about 1 cm from) the plates will yield values almost equal to those on the plates themselves. The ionized air around the flame effectively increases the size of the probe tip making electrical contact with the plate. So there is considerable inaccuracy here, as well as in the indication of the electroscope; nevertheless the graph is almost linear over most of the space.

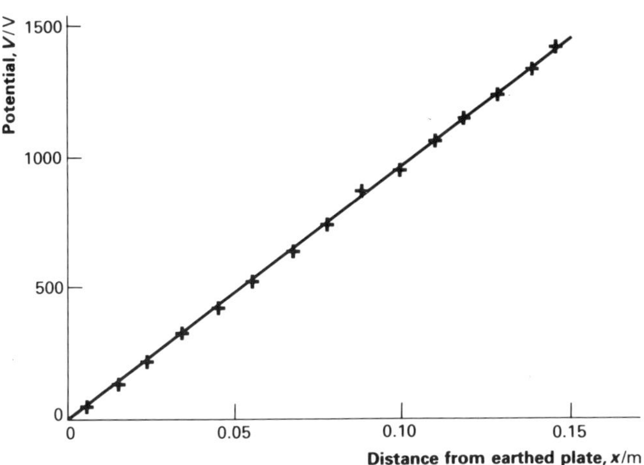

Figure E8
Potential against distance in a uniform field.

Field and potential gradient

Perhaps during the demonstration and certainly on plotting the graph, students will see the linear variation of potential with distance. They can measure the gradient of their graph $\frac{\Delta V}{\Delta x}$ and see that it is nearly equal in magnitude to the ratio of the overall p.d. to the total separation, *i.e.* $\frac{V}{d}$, which should be recognized as equal to the field strength, E, between the plates. Now it can be seen that potential difference/separation gives the electric field strength between the plates, and that p.d. and separation can refer to *any* two points on a perpendicular line joining the plates – not only points on the plates themselves.

This suggests the equation $E = \frac{\Delta V}{\Delta x}$ or, in words, 'field strength equals potential gradient'. Further discussion, based perhaps around question 8, indicates that $E = -\frac{\Delta V}{\Delta x}$ or 'field strength is the negative of potential gradient' is a truer statement and one important enough to commit to memory.

Questions

Question 8 is mentioned above; question 9 is about potentials and electric fields in an electronic tube.

$$E = -\frac{\mathrm{d}V}{\mathrm{d}x}$$

Students who are familiar with the mathematical notation $\frac{\mathrm{d}V}{\mathrm{d}x}$ as 'rate of change of V with distance' may like to use it from here onwards, though we prefer the Δ notation, which is used less in mathematics. Strictly, we interpret Δ as meaning 'a gain in', rather than just 'a change in', that is, ΔV will be a *negative* quantity if V decreases. Still, we must always *think physically* about the change and its meaning, and beware of students using mathematical formulae with no physical understanding.

EQUIPOTENTIALS

Following the flame probe investigation, students should be aware that there are surfaces (in three dimensions) in which the potential does not vary. These are called equipotentials, which for the uniform field are equally spaced parallel planes in the region between the plates. Equipotentials every 250 V, say, may be drawn in cross-section to represent the field as in figure E9, or indeed at any (arbitrary) chosen interval. Clearly the potential gradient over any small region can be worked out from these; their spacing is some visual measure of the field strength.

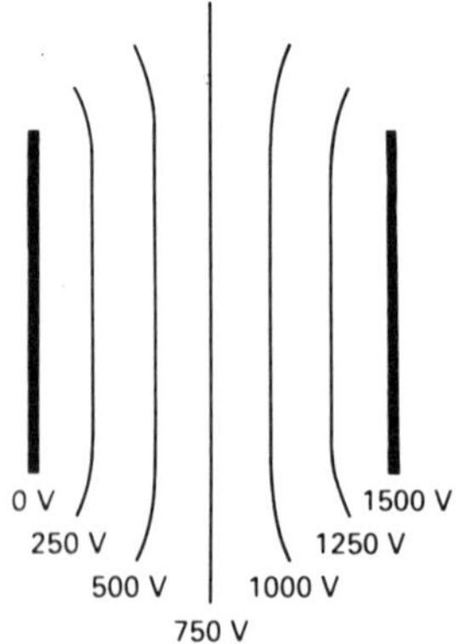

Figure E9
Equipotentials in a uniform field.

The analogy with contour lines on a map may be fruitful here and useful later on. Closely-spaced contours indicate a steep slope and therefore large downhill forces. Also the direction of the force (and therefore the field, in the electrical case) can be seen to be at right angles to the equipotentials (that is, 'down the slope').

Equipotentials in non-uniform fields

It may have been noticed, for positions near a plate, that the equipotential surface becomes curved in the region outside the plates. This may be demonstrated with the flame probe but is best explored in two dimensions by the students themselves plotting their own equipotentials for different configurations of field.

EXPERIMENT

E5 Plotting equipotentials in two dimensions for various shapes of field

ITEM NO.	ITEM
	either
1033	cell holder with 4 cells
	or
59	l.t. variable voltage supply
1064	low voltage smoothing unit
	either
1511	oscilloscope
	or
1507	voltmeter
	pencil or ball-point pen
1153	copper or aluminium sheet, 0.1–0.5 mm thick, 10 mm × 100 mm strips
1155	conducting paper ('teledeltos') cut into approximately A5 or A6 sheets
1155	conducting putty (see notes)
1153	staples and stapler or bulldog clips
1153	carbon paper and white paper, or 'no-carbon-required' paper
1153	drawing board or hardboard
	electrically conductive paint, *e.g.* 'Silverdag' (if available)
1000	leads

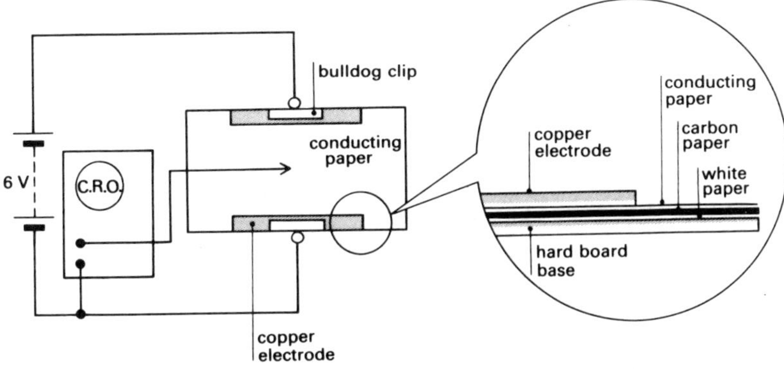

Figure E10
Plotting equipotentials on conducting paper.

A potential difference of 6 V (any p.d. up to about 12 V will do) is set up between two electrodes in good electrical contact with the conducting paper. Potentials are measured using the voltmeter or, preferably, an oscilloscope. A conducting probe can be touched on to the conducting paper surface and the potential at that point read off the oscilloscope. By moving the probe to different positions, points at equal potential are marked on paper beneath and subsequently joined up to reveal equipotentials at chosen intervals.

Good electrical contact with the conducting paper is essential, especially at the edge of the electrode. Colloidal silver paint (*e.g.* 'Silverdag', from RS Components) may be used to paint electrodes directly on to the paper but since this is expensive a means of re-using the arrangement is recommended. A drawing board base is used and a wire may be stapled to the painted electrode at one point only; the contact is 'touched up' using Silverdag, Aquadag, or conducting putty. This is particularly useful if irregular shapes are to be used (figure E11(a)). Using carbon paper or an equivalent, marks are made on white paper beneath rather than on the conducting paper itself, so that many permanent records can be taken from the same set of painted electrodes.

Copper or aluminium electrodes are cheaper. These can be the strips supplied for bending in materials kits or cut from copper sheet used to make electrodes for electrolysis. All electrodes must be cleaned with emery paper before use. If a drawing board base is used the electrodes must be carefully stapled every 2–3 cm to maintain good contact, but they may be placed anywhere on the conducting paper (figure E11(b)). With a hardboard base, bulldog clips holding down the electrodes ensure very good contact but generally restrict the electrodes to the edge of the paper (figure E11(c)). This, however, is the cheapest, simplest method.

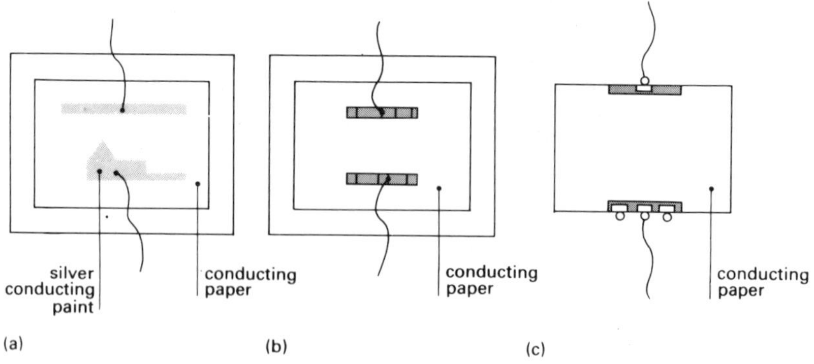

Figure E11
Types of electrodes.
(a) Painted electrodes using silver conducting paint.
(b) Stapled electrodes on a drawing board.
(c) Electrodes held by clips on a hardboard base.

Poor contact at any electrode will produce irregularities in the potential gradient nearby. A thin layer of conducting putty has a resistance very much lower than the paper for most arrangements and ensures better contact if placed under an electrode. A strong drawing pin makes a good 'point' electrode if the rim of its head grips the paper well enough and contact can be made underneath the board. The shaft of a pin is so narrow that it produces a very large potential gradient nearby making equipotential plotting very difficult but providing a very useful teaching point. A clean coin, or a bulldog clip works well.

Probes may be made from graphite pencils, used ball-point pens (either all-metal or with a wire soldered to the metal tip), test probes from a meter, or insulated stiff wire (figure E12). Metallized paper used by some computer printers may replace teledeltos paper.

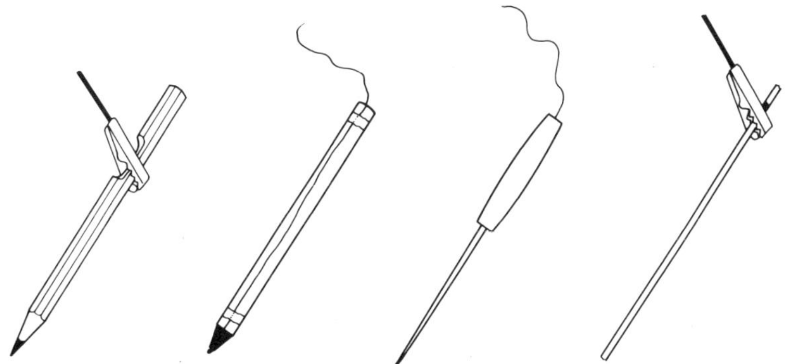

Figure E12
Types of probe.

Students should plot equipotentials at, say, 1 V intervals, for parallel electrodes and at least one other configuration: some suggestions are given in figure E13. From their equipotential patterns they should construct some field lines at right angles and see how the potential gradient varies with distance along these lines, perhaps plotting a graph of this. It is instructive to see how nearly uniform a field between quite small parallel electrodes really is. (This ties in with the computer program 'EFIELD', which might be used now or later.)

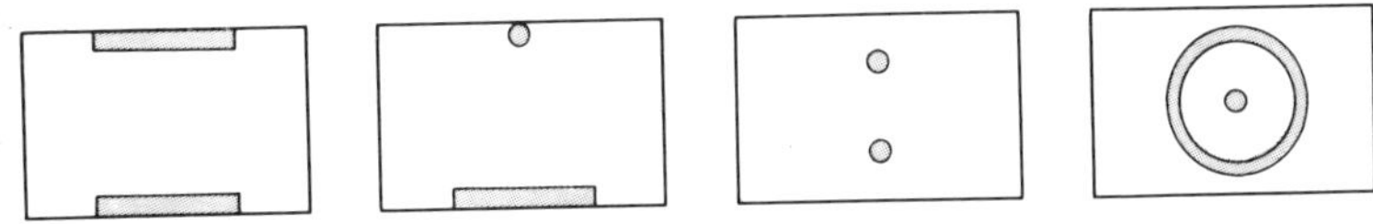

Figure E13
Possible electrode configurations.

Students may notice that if they sketch in several field lines these are closer together where the potential gradient (*i.e.*, the field strength) is larger. This happens near sharply curved edges, corners, and points. Two free probes connected to the oscilloscope can be used to measure the potential difference between *any* two points on the conducting paper and hence deduce the potential gradient.

ALTERNATIVE EXPERIMENT

E5a Equipotentials in an electrolytic tank

ITEM NO.	ITEM
1109	signal generator
1511	oscilloscope
1153	rectangular transparent Perspex tank or 'lunchbox'
1153	copper sheet for electrodes
52K	crocodile clips
1501	bare copper wire, 0.2 mm diameter
1156	copper sulphate solution, concentration not critical
1155	Petri dish
1000	leads

This alternative uses copper electrodes, which may be cut from sheet flexible enough to be bent if desired, dipped into copper sulphate solution in a shallow transparent tank. A 'square wave' alternating p.d. at 1–10 kHz is applied from a signal generator and an oscilloscope used as before with the time-base on so that two horizontal lines appear on the screen. The probe wire must be copper to prevent electrolytic action and its position may be noted on graph paper below the tank and recorded on identical paper alongside – this is easy if students work in pairs (figure E14(a)). There are no problems with contact resistance and different shaped electrodes are easy to improvise. A Petri dish or larger circular tank can be used for radial fields (figure E14(b)).

An overhead projector can be used, as in Experiment E3, for a qualitative demonstration, provided students can see the oscilloscope screen as well.

Two probes may be used to measure potential difference, as in Experiment E5.

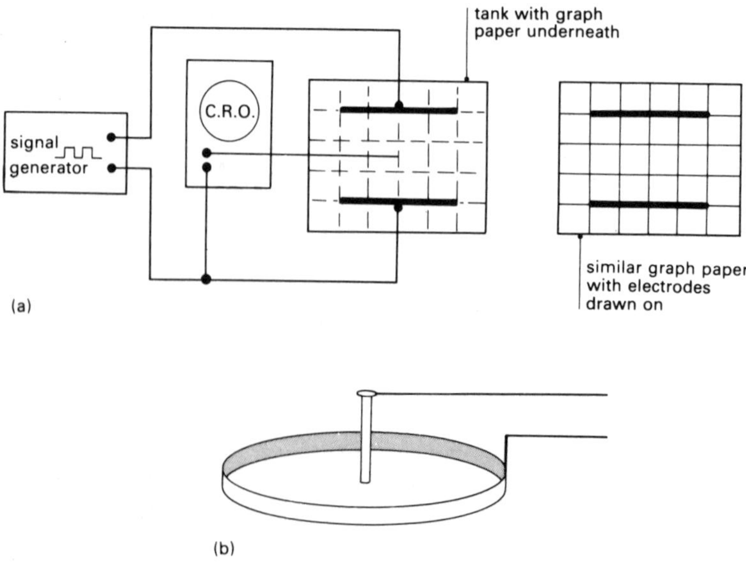

Figure E14
Plotting equipotentials in an electrolytic tank.

Direction of electric field vector

Probes at A and B (figure E15) indicate a certain potential difference. The same p.d. exists between A and C, but the distance AC is greater than AB. Hence the potential gradient and therefore the field strength is less in this direction. So the electric field has components in all directions (except perhaps AD and into the paper) but we say its direction is from A to B. Like the 'fall line' of a slope, this is the direction in which the force of the field acts.

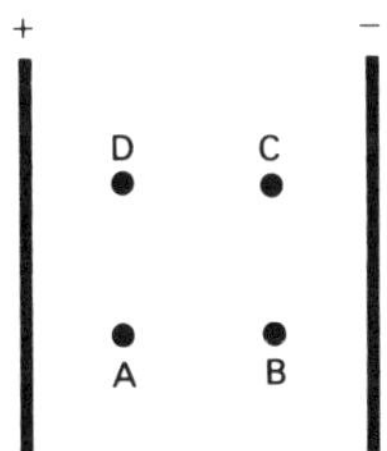

Figure E15
Points in an electric field.

The concept of the field being conservative could be brought in here by a very slight adaptation of the argument in Section E2 (pages 325–6). Also, the gravitational analogy can be exploited further using a rubber sheet to simulate potential gradients, probably for uniform fields at this stage.

Questions

Questions 10 and 11 are about equipotentials and fields.

Timing, sequence

Section E1 should take under 2 weeks. It should not be rushed, as the ideas are difficult and fundamental, but time must be allowed for the work on gravity and circular motion. The ideas of potential and its gradient will be taken up again in Sections E2 and E3; the following passage, on factors which affect field strength, is needed when the inverse-square electric field is met in Section E3. For this introduction, no more than one practical session plus one period and a homework can be spared, unless time is saved elsewhere.

FACTORS AFFECTING THE FIELD STRENGTH BETWEEN PARALLEL PLATES

The electric field between the plates is clearly associated with the charge stored on them. What factors affect how much charge can be stored, and how does this charge affect the field? These questions can lead to reasonable suggestions on the basis of students' present knowledge followed by empirical tests.

V? Parallel plates look, and perhaps act, like a capacitor; it is hardly necessary to test $Q \propto V$, though it can easily be done if Q can be measured. A reed switch or coulombmeter may be suggested.

A? Larger plates can be thought of as smaller plates in parallel and we know, from Unit B, that charge stored adds up in this situation. So $Q \propto A$ is not difficult to justify, and could be easily checked.

d? A smaller separation certainly increases the field strength, for fixed V. Perhaps this arises from more charge being crammed on to the same plate; maybe the closeness of an oppositely charged plate assists this process. $Q \propto \frac{1}{d}$ is plausible, but then so is $Q \propto \frac{1}{d^2}$ for instance, so this one really does need to be investigated.

The medium? Few real capacitors have air between the plates. Many have plastic or paper. This may affect the charge stored, all other factors being constant.

Any other suggestions the students make can be argued through. Although tests on all four factors above are easily made, it is suggested that only the latter pair need be done if time is short.

EXPERIMENT
E6 Investigating factors affecting the charge on parallel plates using a coulombmeter

ITEM NO.	ITEM
1512	coulombmeter, 100 nC
65	2 metal plates with insulating handles
14	e.h.t. power supply
51G	polythene strip for use as insulating handle
503–5	2 retort stand bases, rods, and bosses
501	metre rule
1000	leads

Coulombmeter See demonstration B16, Spooning charge (page 128 of this *Guide*) for a discussion of this instrument – use of high impedance voltmeter, or electrometer/d.c. amplifier.

One of the pair of plates is connected to earth. The other is charged by a flying lead from the e.h.t. positive, the negative being earthed. The flying lead *must* then be removed.

The charge stored is measured by bringing up the coulombmeter so that the probe rod touches this plate. The ratio of capacitances should be such that nearly all ($\approx 99\,\%$) of the charge is given to the coulombmeter. (Students can check this later when they know $C = \varepsilon_0 \dfrac{A}{d}$ and the value of ε_0; at this stage, discharging the coulombmeter and touching the plate a second time confirms that very little charge is left on it.)

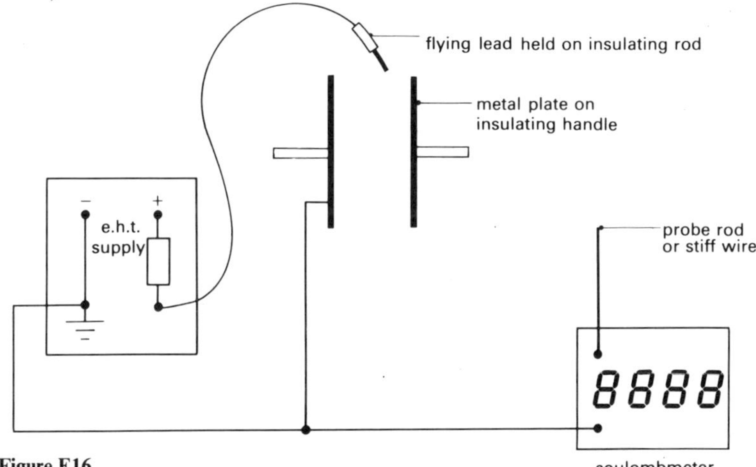

Figure E16
Using a coulombmeter to measure Q.

The effect of a medium (polythene or paper) can be tested by placing a sheet between the plates and repeating the experiment.

The effect of separation is tested using $V \approx 500$ V and $d \approx 5$–25 mm. Large separations may also be tried and certainly Q should be measured for an 'isolated' plate ($d \approx 0.5$ m or more).

Stray capacitance and fluctuations in the coulombmeter reading may complicate matters. Hands well away and fresh batteries (if appropriate) may help, but students should consider the range of uncertainty in their readings, representing these as error bars on a graph. For instance, how accurately repeatable is any given reading; does the coulombmeter faithfully return to zero when shorted; does it absorb all the charge on the plate?

Students should plot a graph of Q against $\dfrac{1}{d}$ and think carefully about the meaning of the very definite intercept. (Charge stored when d is large.)

This apparatus may also be used to test $Q \propto V$ for p.d.s up to 500 V and $d \approx 10$ mm.

$Q \propto A$ is less easy to check unless plates of various sizes are available.

Experiment E6, as described above, could be done as a demonstration.

Alternatively, it can be done as a class experiment in combination with experiment E7 which uses the reed switch as a charge measuring device, and may be more accurate than this one. All the suggested relationships can be tested in experiment E7.

Question 12 tests understanding of charge transfer in experiment E6.

EXPERIMENT
E7 Investigating factors affecting the charge on parallel plates using a reed switch

ITEM NO.	ITEM
1010	reed switch
1109	signal generator
1025	1 pair capacitor plates with polythene sheet and 16 polythene spacers, 10 × 10 × 1.5 mm
	either
59	l.t. variable voltage supply
1064	low voltage smoothing unit
	or
1033	three cell holders with four cells
1507	voltmeter, 100 V and 10 V
1101	sensitive galvanometer (*e.g.* internal light beam)
1017	resistance substitution box
1511	oscilloscope
501	metre rule
32	mass, 1 kg
1000	leads

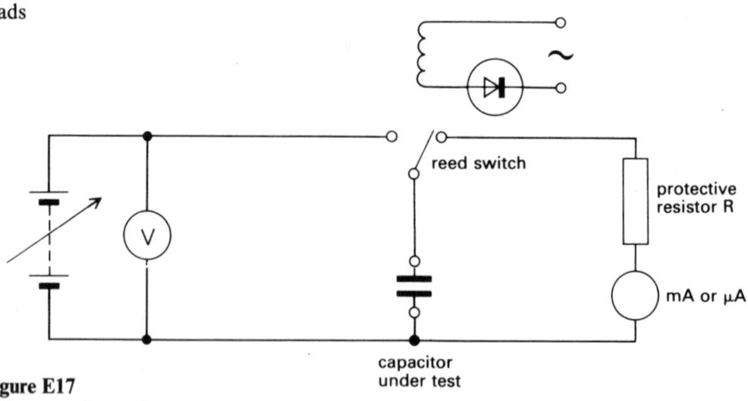

Figure E17
Using a reed switch to measure Q.

The charge stored, Q, is calculated from the signal generator frequency f, and the apparently steady discharge current I using $Q = \frac{I}{f}$. f should be about 400 Hz and R about 100 kΩ. A light beam galvanometer should be used if possible, though a digital meter can be used if it has better than 1 μA resolution. The capacitor plates must not touch whilst the reed switch is working, nor should the switch's operating p.d. be exceeded. The output p.d. of the signal generator should also be the minimum consistent with audibly clean vibration of the reed. The values of R and f suggested allow complete discharge: this can be checked using an oscilloscope across R and raising its value. Stray capacitance, from hands, leads, etc., should be avoided whilst taking readings.

$Q \propto V$ can be tested for p.d.s of up to 25 V.

$Q \propto \frac{1}{d}$ can be tested using increasing numbers of spacers to separate the plates.

$Q \propto A$ can be tested by varying the effective area of overlap (figure E18).

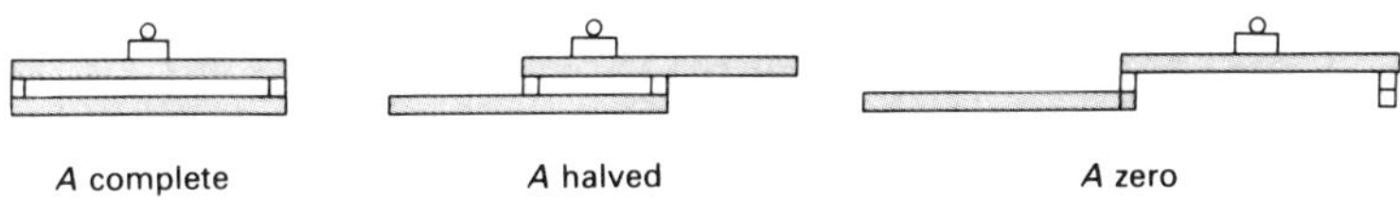

Figure E18
Changing the area of overlap.

The effect of the medium can be tested using polythene and other materials of equal thickness in place of an air gap.

Question 13 is about the reed switch experiment.

Summary of results from Experiments E6 and E7

Graphs such as those of figure E19 can be drawn. A definite intercept indicates that some charge is stored even when a plate is isolated ($d = \infty$), or not overlapping with another plate ($A = 0$). The capacitance of an isolated object is not discussed here, nor can we properly account for the effect of the bench, connecting wires, etc. Q clearly varies linearly with both A and $\frac{1}{d}$ and in the absence of the above effects would be proportional to both, as well as to V.

We may combine these into one relationship:

$$Q \propto \frac{AV}{d} \quad \text{or} \quad Q = \varepsilon_0 \frac{AV}{d}$$

where ε_0 ('epsilon nought') is the appropriate constant for air (strictly, for a vacuum). The effect of the medium is considered in a moment.

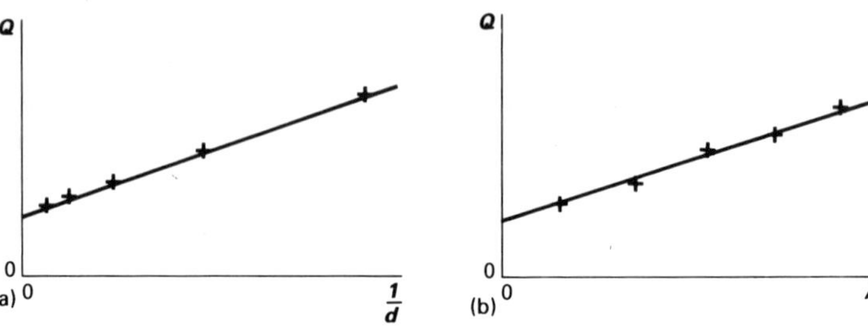

Figure E19
Graphs
(a) Q against $1/d$.
(b) Q against A.

Field and charge density Rearranging the above equation gives

$$\frac{Q}{A}=\varepsilon_0\frac{V}{d}$$

$\frac{V}{d}$ is identified as the field strength E, and $\frac{Q}{A}$ as the surface density of charge, σ ('sigma').

Hence

$$\sigma=\varepsilon_0 E$$

The value of ε_0 can be obtained from the gradient of the best straight line obtained (for example, the slope of the Q against $\frac{1}{d}$ graph will be $\varepsilon_0 AV$). The accepted value is $\varepsilon_0 \approx 8.85 \times 10^{-12}\,C^2\,N^{-1}\,m^{-2}$ (units should also be derived).

Teachers may wonder what effect a decision to assign a value to c, the speed of light *in vacuo*, making it rather than the metre one of the base units of the system of measurement would have on the validity of this 'experiment'. Since we also assign the value $4\pi \times 10^{-7}\,H\,m^{-1}$ to μ_0, the permeability of free space, we have implied a value $\varepsilon_0 = 1/c^2\mu_0$. In a sense then, experiments 'to measure ε_0' become experiments to check our metre rules. (See, for example, *Phys. Educ.*, **19**, 215, 1984.)

Meaning of ε_0 The accepted term 'permittivity of free space' suggests that ε_0 is some property of a vacuum. Here we understand it as defining a fundamental relationship between the charge density on a plate and the field around it, and it is a universal constant in the sense that this relationship is independent of other conditions (for example, material of plates, actual values of σ, E, etc.).

Effect of medium The medium increases the charge stored for a given field strength $\left(\frac{V}{d}\right)$ by some factor. Since it is a simple ratio it has no units and is referred to as the relative permittivity, ε_r. Our equation above then becomes

$$\sigma = \varepsilon_r \varepsilon_0 E$$

For air ε_r is almost unity, for plastics about 2–3, for water about 80.

Students might be interested to find out why ε_r varies, and perhaps read about 'polarization' and induced charges. Measurement of ε_r for water proves rather tricky. A neat way of doing it for paper uses two sheets of foil sandwiched between pages of a book (figure E20).

Figure E20
Measuring ε_r for paper.

Capacitance Another rearrangement of the relationship between Q, A, V, and d is

$$\frac{Q}{V} = \varepsilon_0 \frac{A}{d}$$

$\frac{Q}{V}$ stands out as the capacitance C of the arrangement. Including the effect of the medium we then have

$$C = \varepsilon_r \varepsilon_0 \frac{A}{d}$$

This is a useful expression which enables the capacitance of the plates in either experiment to be estimated (neglecting the extra effects due to charge stored where either A or $1/d$ is zero). In this context also the units of ε_0 appear in the more palatable, but equivalent form of $\mathrm{F\,m^{-1}}$.

$$\mathrm{F\,m^{-1}} = \frac{\text{farad}}{\text{metre}} = \frac{\text{coulomb}}{\text{volt} \times \text{metre}} = \frac{\text{coulomb} \times \text{coulomb}}{\text{joule} \times \text{metre}}$$

$$= \frac{\text{coulomb}^2}{\text{newton} \times \text{metre} \times \text{metre}} = \mathrm{C^2\,N^{-1}\,m^{-2}}.$$

Home experiment

Home experiment EH1 challenges students to construct, at home, from readily obtainable materials (for example, cooking foil and paper), a capacitor of about the size of a matchbox. A prize might be given for the highest capacitance to volume ratio. The capacitance can be measured using the reed switch or a reactive bridge circuit; students should use the relationships developed in this Section to estimate the capacitance beforehand.

Questions

Question 14 is about a parallel plate arrangement.
Question 15 is about capacitor construction.
Question 16 tests understanding of the parallel plate relationships.
Questions 17 and 18 are review questions.

Applications

There may not be much time to dwell on these in class, but in a largely theoretical Unit it is important to mention the breadth of applications of electric fields. A few examples are included in the *Students' guide* with some questions to start discussion and thinking. Teachers and students should add their own ideas to those presented. Topics for reading might include the following:

1 The Xerographic process.
2 The 'ink jet' printer.
3 Electrostatic generators.
4 Electrostatic dust precipitator.
5 The capacitor microphone.
6 The electrostatic loudspeaker.
7 Electrostatic paint and crop spraying.
8 Thunderstorms.
9 A capacitor 'tiltmeter'.
10 Polarization, dielectrics, and design of capacitors.
11 Field plotting and design of cathode ray tube electrodes.
12 Problems caused by 'static' in everyday life.
13 Generation and use of electricity by fish.
14 Electrostatic prospecting.
15 Ionizers and their effects.
16 Electrets and their uses.

SECTION E2
GRAVITATIONAL FIELD AND POTENTIAL

INTRODUCTION

The approach here is to develop the idea of gravitational potential difference and its relationships to energy and field in the context of the uniform field at the Earth's surface. Then the gravitational inverse-square law is introduced and is shown to describe the way the field strength of the Earth varies with distance. Gravitational potential difference is explored in this context and leads to the definition of gravitational potential and a consideration of energy.

A separate part introduces, or revises, circular motion so that satellites and planetary orbits may be considered. This piece could be studied before the rest of the Section (indeed, historically this was the order Newton followed) but the treatment should allow time for the difficult concepts of gravity in a $\frac{1}{r^2}$ context to be understood.

The Section emphasizes physical thinking rather than a potentially more concise mathematical approach, in the belief that such thinking is essential to the work of a physicist at any level. The computer is used as a versatile calculating and graphical tool.

Students have arguably more experience of gravity than of electricity and may even have met the inverse-square law in their pre-A-level studies. Beginning with the uniform field helps strengthen the links between electricity and gravity which are heavily used in Section E3, for the inverse-square field.

A UNIFORM GRAVITATIONAL FIELD

The definition of gravitational field strength (g) as the force on unit mass ($N\ kg^{-1}$) and a way to measure it should emerge readily, and a short journey around the laboratory with a kilogram on a force meter shows the field to be uniform there.

Gravitational potential energy

The field strength around the kilogram is still the same if it is raised, say, by 1 m. But its gravitational potential energy has increased. Objects of different masses (students standing on a bench for instance) clearly gain different amounts of energy when raised by 1 m, though one kilogram of each gains just the same energy (about 10 J).

Gravitational potential difference

In this or some other, preferably visual, way the usefulness of the idea of potential energy per kilogram can be brought out. Marks at 1 m intervals on the wall could be labelled in steps of $10\,\mathrm{J\,kg^{-1}}$ and some quick calculations made of energy changes for moving different objects. Broadening the discussion, marks at 10 m or 100 m vertical intervals around the countryside can be seen to give contours, as on a map. Since these link points where a mass has the same potential energy they can be called gravitational *equipotentials*.

Some electrical examples might be considered to reinforce the analogy with gravity and the idea of a zero for gravitational potential energy could be discussed (the laboratory floor? sea level?). This is a useful place to introduce *negative* potential energy, particularly if a basement or lower floor is available.

Summary

Gravitational potential difference between two points is the change in gravitational potential energy when 1 kilogram moves between the two points.

Questions

Questions 19 and 20 introduce changes in gravitational potential energy and the concept of gravitational potential difference.
Question 21 uses the idea.

THE INVERSE-SQUARE LAW FOR GRAVITATIONAL FIELDS

Visual aid

It is useful to represent the Earth by some spherical object (for example, a football) during the following discussions, and to try to illustrate the examples visually.

Newton's inverse-square law for the gravitational force, F, between two masses, m_1 and m_2, a distance r apart will be a new idea for some, a reminder for others.

$$F = -G\frac{m_1 m_2}{r^2}$$

G is a universal constant (value 6.67×10^{-11} N m^2 kg^{-2}), and the negative sign indicates a vector *inwards, i.e.*, an attractive force. If $m_1 = M$, a large mass (for example, the Earth) and $m_2 = m$, a smaller 'test' mass nearby, then since

$$g = \frac{F}{m}$$

this leads to

$$g = -\frac{GM}{r^2}$$

as the expression for the field strength of a mass M.

Teachers may be surprised to see negative signs in the expressions for F and g. This is consistent with the treatment in Unit A, where attractive forces between molecules were considered negative and repulsive forces positive. Of course, it is more important for students to understand whether forces are attractive or repulsive, whether potential energy decreases or increases for movement in a given direction, than to learn rules about sign conventions. See also the note about signs on page 319.

Background to Newton's work

This is not part of the course, but students might be interested in the unfolding of the inverse-square law and the contributions of Galileo, Kepler, and others: it is a marvellous example of the process of building and refining a model. The puzzle of how 'action at a distance' could occur remained intriguing to Newton, but his mathematical scheme, which so accurately described the motions of heavenly bodies, went so far to fulfil Kepler's vision of 'a celestial physics based on causes' that we see here perhaps the beginnings of modern science.

Reading

BRONOWSKI *The ascent of Man.* Chapter 7.
COHEN 'Newton's discovery of gravity'. *Scientific American.*
KOESTLER *The sleepwalkers.*
ROGERS *Physics for the inquiring mind.*

Measurement of *G*

This is painstaking and difficult, because of the small size of the constant. Reference to Cavendish's experiment of 1798 may be made and students may watch a modern version on video, taking results from the screen to find their own value of G. This, however, is not an essential part of the Unit.

Television and video

'The determination of the Newtonian constant of gravitation' in the Granada television series *Experiment: Physics.*

Question 22 is based on Cavendish's experiment.

Testing the inverse-square law using spaceflight data

Newton's explanation of the motion of the Moon was a triumph for the inverse-square law. Nowadays, scientists are so confident that it holds over a vast range of distances to a high degree of accuracy that it is used to plan the trajectories of spacecraft. Students can use data from one such spaceflight to verify that the inverse-square law does apply.

Questions

From the data in question 23 students deduce the acceleration of the spacecraft and equate this to the gravitational field strength. ($a = \frac{\text{force}}{\text{mass}}$, from Newton's Second Law; and $g = \frac{\text{force}}{\text{mass}}$, from its definition.) This equivalence may need to be stressed beforehand.

In question 24, by taking a series of pairs of points, g can be found at different values of r. A graph of g against $\frac{1}{r^2}$ gives a straight line (figure E21). Note that both acceleration and field strength have negative values.

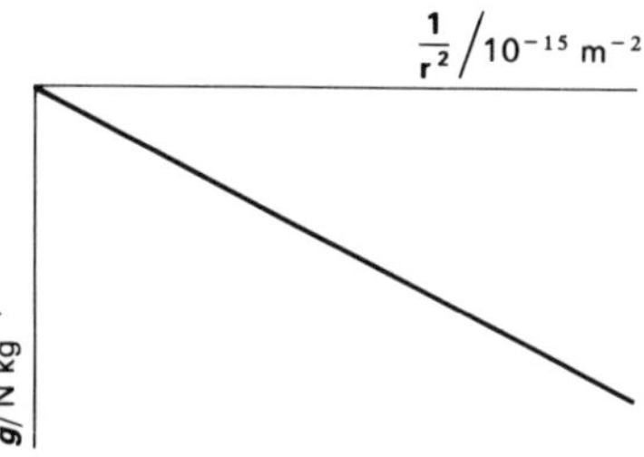

Figure E21
$g \propto -1/r^2$.

GRAVITATIONAL POTENTIAL DIFFERENCE, ΔV_g, AND GRAVITATIONAL POTENTIAL, V_g

Timing, structure, and sequence of this work

This work is central to the Section and will take a few lessons based either on the computer program 'GFIELD', and/or on the series of structured learning questions (25–29) in the *Students' guide*. Students without access to a computer must not be disadvantaged and all the results they need are reproduced in the questions. Those with a computer should use it to perform the tedious tasks such as repetitive calculations; the students themselves, or with the teacher, have to decide *what* calculations to do and *which* graphs to plot, so they need to be thinking about the physics all the time.

First the means of calculating ΔV_g from a g against r graph (by adding areas of strips under it) is introduced. Then the computer may be used to explore how accurate the method may be (question 25). The energy needed to launch spacecraft may be calculated (question 26). Values of ΔV_g for escape from various positions (question 27) lead to the concept of gravitational potential V_g (question 28) and a precise expression for it (question 29). Then data from an actual spaceflight are used to confirm this expression (question 30).

Calculating energy changes from force–distance graphs

Students should remember that change in energy = force × distance, and that this can be applied to a uniform field, where the force is constant. If the force does vary then the energy can be deduced from the area under a force–distance graph (as for a spring in Unit A, 'Materials and mechanics'; see figures E22 and E23).

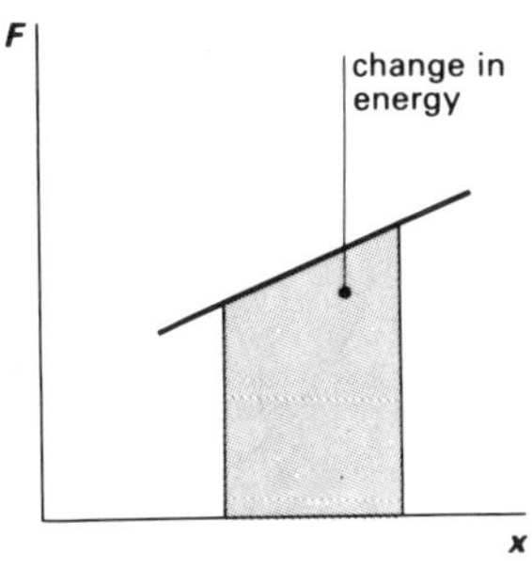

Figure E22

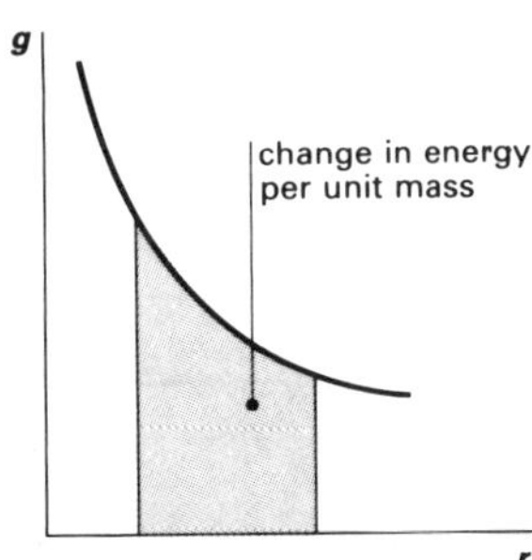

Figure E23

Since field strength, g, is force per unit mass, then the area under a g against r graph will be the change in energy per unit mass. This was previously introduced as the gravitational potential difference to which we now give the symbol ΔV_g (anticipating the later use of V_g for gravitational potential).

Integration

Mathematicians may suggest that integrating an expression for field strength with respect to r between specified limits will yield a value for ΔV_g. This can be done now or later (Appendix VI) but care must be taken to sort out the signs and limits to get the correct answer. We proceed with a longer but more physical argument which we believe will be of greater benefit to understanding. As it involves areas under graphs and (later) gradients, students who are not familiar with computing these might be given practice beforehand.

Sign of *g*

The computer program, and the argument here, use a g against r graph *above* the r axis, *i.e.*, depicting g as a *positive* quantity. We feel that this simplifies the consideration of adding up areas *under* the graph. In any case, at first, we are interested only in gravitational potential *difference*, ΔV_g and not in the sign of potential. When this is introduced later, students may be reminded that g is negative (indicating an attractive field) as indeed it must be to give negative potential.

Calculating ΔV_g from a g against r graph

Figure E24(a) shows that, for a uniform field, $\Delta V_g = g\Delta r$.

Where g varies with r, as in figure E24(b), we can consider small steps, Δr, over which g remains roughly constant. The sum of the areas of all the strips $g\Delta r$ as shown approximates to the gravitational potential difference between two limits and is almost equal to the area under the graph between those limits. Adding up such areas is a tedious job if the approximation is to be good, but it can be done easily using a computer.

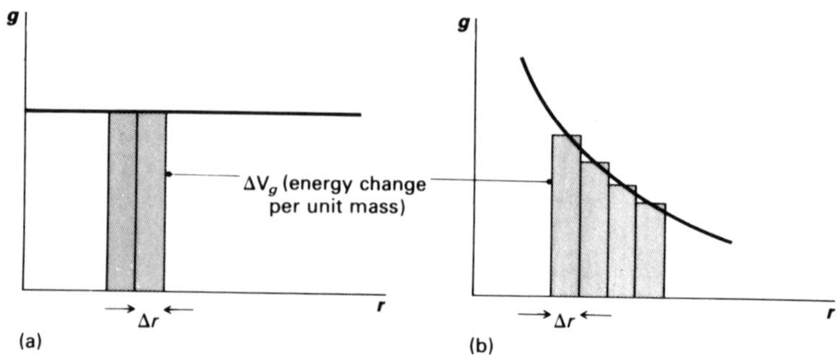

Figure E24
ΔV_g from field–distance graphs.

Using the computer program 'GFIELD'

Students are now ready to use the computer program, and to work through the questions based on it. (More information about the program itself is given in the *Notes* accompanying the package 'Software for Nuffield Advanced Physics', Longman's Micro Software.)

Questions

Question 25 shows how calculations of gravitational potential difference between the Earth's surface and a distance of 50×10^6 m are used to try out different step sizes, or values of Δr (figure E25).

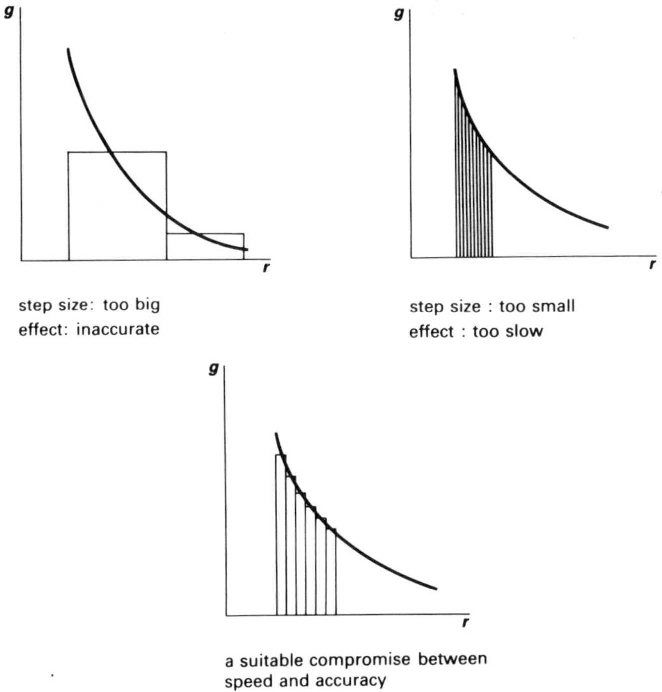

Figure E25
Effect of step size on calculation of ΔV_g.

Question 26 shows how the launch of satellites from a fixed value of r_1 (R, the radius of the Earth) to variable distance r_2 can be explored (figure E26(a)).

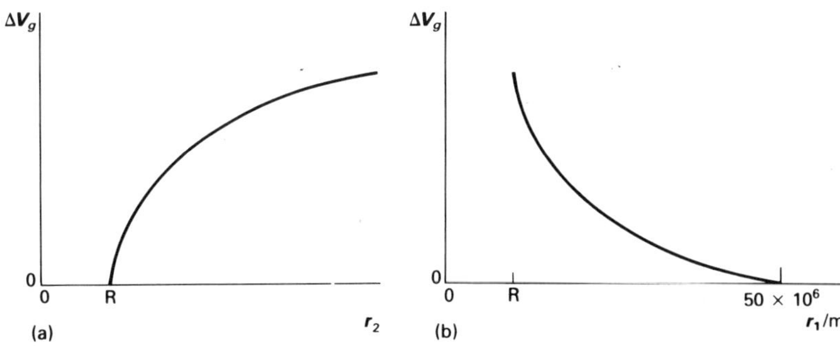

Figure E26
Potential energy change per kilogram for journeys:
(a) from the Earth's surface (R) to variable distances r_2;
(b) from variable r_1 to 50×10^6 m from the Earth's centre.

In question 27, an interplanetary space convention is to be held on a satellite 50×10^6 m from the Earth. (Strictly speaking, this satellite should be 'massless', though this complication is not mentioned in the question.)

The energy per kilogram (ΔV_g) needed for delegates to travel from various distances r_1 to the convention at $r_2 = 50 \times 10^6$ m can be calculated and displayed graphically (figure E26(b)). Students may guess at the way in which ΔV_g varies with r_1 and plot suitable graphs to test their guesses. It turns out that ΔV_g is linearly related to $\frac{1}{r_1}$ (figure E27(a)) and is, in fact, proportional to $\left(\frac{1}{r_1} - \frac{1}{r_2}\right)$ (figure E27(b)). The gradient of either graph can be measured, for later reference.

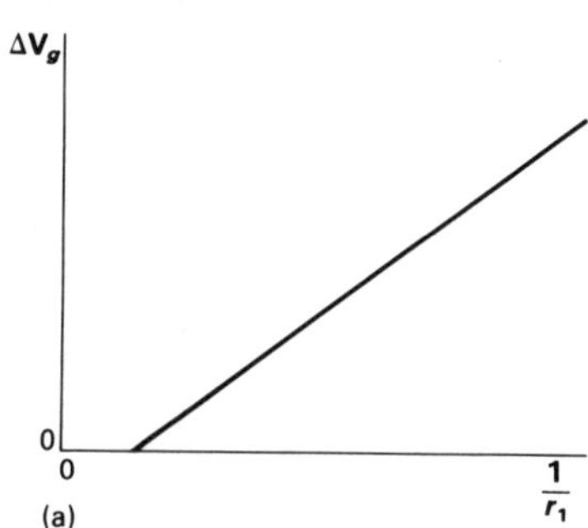

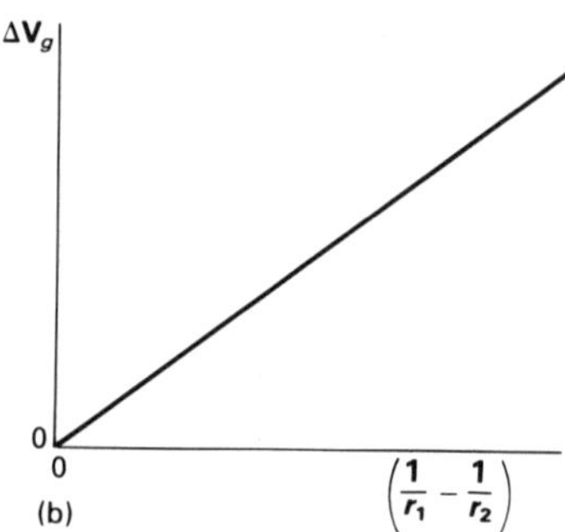

Figure E27
Variation of ΔV_g with
(a) $\frac{1}{r_1}$
(b) $\left(\frac{1}{r_1} - \frac{1}{r_2}\right)$

Question 28 considers journeys beyond 50×10^6 m to a place 'as far away as one could wish' (*i.e.* $r_2 = \infty$). The two graphs of figure E27 become one and the same, and ΔV_g now represents the minimum energy per kilogram required to escape completely from the Earth's pull. This then leads to the concept of gravitational potential.

A zero of potential

Question 28 raises the question of choosing a place where the gravitational potential energy of all masses is agreed to be zero.

Then we can simply refer to the gravitational *potential* at a point. This would be the gravitational potential difference between that point and our chosen zero, that is the energy required to bring 1 kilogram from our zero to the point in question. (Remind students of earlier discussions about electrical potential and electrical potential difference.)

We might regard the Earth's surface (sea level) as a suitable zero. However, in dealing with situations involving other planets as well it is most helpful to regard some neutral, far away point as our zero. The only place which is sufficiently neutral, however, is nothing like a 'point' at all but a 'place' which we call 'infinity'. In this context, this means 'as far away as you would like to go', or 'as far from the influence of any massive object as you can be'. For our example, 50×10^6 m is nearly 90 % of the way to infinity in energy terms. The energy required to push 1 kilogram 'the rest of the way' is small. (The areas of the extra strips under the g against r graph beyond 50×10^6 m add up to a small, finite limit.)

Summary

Gravitational potential difference (ΔV_g) between two points is the energy per unit mass needed to move between the two points.

Gravitational potential (V_g) at a point is the energy per unit mass needed to move from infinity to that point.

Gravitational potential energy (E_p) of an object at a point is the energy required to move the object from infinity to that point.

Both the latter are inversely proportional to the distance from the centre of the Earth (or whatever planet is considered).

Consequences of the convention for zero potential

Up to now in dealing with ΔV_g we have been concerned only with *changes* in gravitational potential, worrying little about the signs of those changes. We can, and indeed should, always refer to the physical situation: moving away from the Earth 'requires' energy, so constitutes a *gain* in gravitational potential energy, and vice versa.

But if potential energy is gained in moving away (because work has to be done against an inwards attracting force) and yet an object as far away as possible has zero potential energy, then when it is nearer the Earth its potential energy must be *negative*. Indeed this is so for any system of attractive forces: potential energy, and therefore potential, must be negative. The rocket is *lacking* the energy it would need to escape from the Earth until fuel is burned. Repulsive forces, on the other hand, would give positive potential energy – an object, once released, could 'shoot off to infinity'.

The energy required to escape the Earth's influence may be calculated, but first a precise expression for the gravitational potential at a point must be established.

Questions

In question 29, the expression for ΔV_g obtained from the graphs of question 27 is compared with that found from mathematical integration (Appendix VI) and seen to be identical. It follows that the gravitational potential at distance r from mass M is given by:

$$V_g = -\frac{GM}{r}$$

In question 30 data from an actual spaceflight confirm the above expression for V_g.

FIELD AND POTENTIAL GRADIENT

Since graphs of gravitational potential against distance can now be drawn it is useful to remind students of the relationship between field strength and potential obtained for the uniform electric field in Section E1.

$$E = -\frac{\Delta V}{\Delta x}$$

The corresponding relationship for gravity is

$$g = -\frac{\Delta V_g}{\Delta r}$$

This has been met before in the uniform field situation ($\Delta V_g = -g\Delta r$) but it can also be applied to the inverse-square field if we consider small enough steps Δr. In the limit, as Δr becomes infinitesimal, we can write:

$$g = -\frac{\mathrm{d}V_g}{\mathrm{d}r}$$

using $\frac{\mathrm{d}V_g}{\mathrm{d}r}$ to represent the rate of change of potential with distance or 'potential gradient'. The field strength can therefore be found easily from a V_g against r graph by measuring the gradient at any point. Similarly, the precise relationship for electricity $\left(E = -\frac{\mathrm{d}V}{\mathrm{d}r}\right)$ applies *however* the field varies.

Students should note here again that $g = -\frac{\mathrm{d}V_g}{\mathrm{d}r}$ holds for all values of r, and that this simplifies to $g = -\frac{\Delta V_g}{\Delta r}$ if the field is uniform. However, it is always a mistake to find g from a V_g against r graph by dividing the value of V_g at a point by the value of r (although this happens to give the same result for $\frac{1}{r}$ graphs).

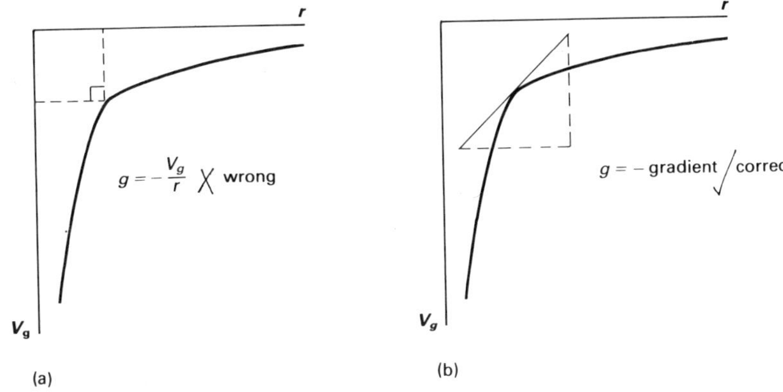

Figure E28
(a) Wrong, (b) right way to find field from a potential–distance graph.

Questions

Question 31 is mainly about escaping the Earth's pull.
Question 32 involves the *total* potential of the Earth–Moon system.

Formulae and relationships

A set of gravitational formulae and the links between them is given in the *Students' guide*, figure E21. This is worth looking at now so that students become familiar with the formulae and understand the links between them.

Equipotentials around the Earth

It is also worth looking at the drawing of equipotentials around the Earth (*Students' guide*, figure E18). This brings out visually the $\frac{1}{r}$ variation in potential contrasting with the uniform potential gradient of a uniform field (for example, the contours of question 19).

A conservative field

We have assumed that the kinetic energy 'lost' by a space probe on the way up may be fully regained on its return, regardless of the route taken. Perhaps this needs to be justified.

All journeys between two points can be broken up into small steps like AB (figure E29) *along* equipotentials which require no energy, and steps at right angles like BC, which give energy $F\Delta r$. This is equal to the energy gained for the single step AC since the force component in this direction is $F\cos\theta$, but the distance AC is $\dfrac{\Delta r}{\cos\theta}$. So to calculate the energy transformed per unit mass on a journey between two points it is necessary only to know the potentials at both points: the route taken is irrelevant.

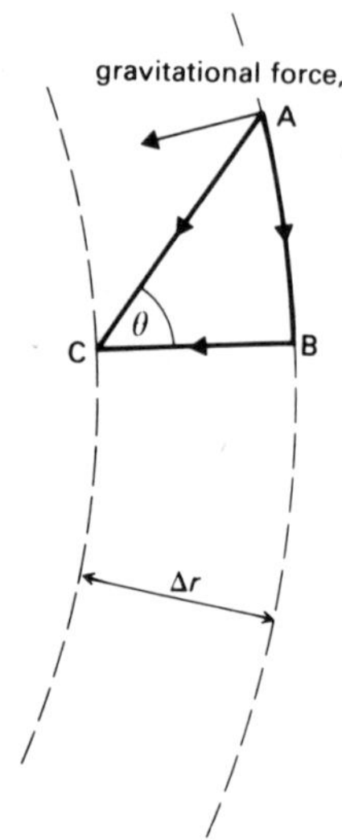

Figure E29
A conservative field.

The argument of course assumes that no other forces (such as friction) act on the mass. Also the force does not have to be inverse-square for it to work: any radial field would be conservative.

A model potential well (or hill)

Drawings of equipotentials are useful, two-dimensional visual models, but a tangible model is perhaps more useful still. A graph of potential against distance has a $\dfrac{1}{r}$ profile. Extending this to another dimension gives the $\dfrac{1}{r}$ potential well or upturned 'Chinese hat' (figure E30).

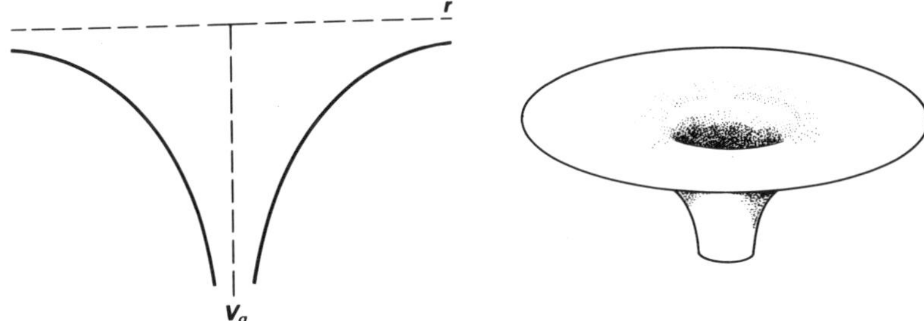

Figure E30

A $\frac{1}{r}$ potential well: the 'Chinese hat'.

Its profile is part of a curve whose depth varies as $\frac{1}{r}$ where r is the distance from the centre. The potential energy of a marble rolling in the well is proportional to its depth, and therefore to $-\frac{1}{r}$. Therefore the marble models the behaviour of an object moving in a gravitational field (at least in two dimensions). Paths of comets near the Sun are easily simulated; to demonstrate elliptical planetary orbits requires more skill – spin and friction are clearly complications. Students should realize that the *slope* of the well at any point indicates the field strength there.

The energy of the system is clearly negative, the marble being 'trapped' in the well, lacking the energy required to escape. This anticipates the use of the well to model the hydrogen atom in Unit L, 'Waves, particles, and atoms'. Positive potentials are easily simulated by inverting the well, turning it into a hill. This will be used to model the paths of α-particles near a nucleus in Unit F, 'Radioactivity and the nuclear atom'.

Home experiment EH2 gives directions for making a $1/r$ shaped hill or well from paper. A more flexible model can be made using a thin rubber sheet fixed to a circular frame and pulled down at a point. Positive potentials (useful later for electricity) can be obtained by pushing up and the combined potential of two bodies (for example, the Earth and Moon) can be easily simulated.

Field outside a solid sphere

The inverse-square law relates to point masses, but we have seen it work close to the Earth, which is large and approximately spherical. This is far from obvious for a good deal of the Earth is closer than the centre (though pulling at a wide angle) and a good deal is further away (though pulling more directly) (figure E31).

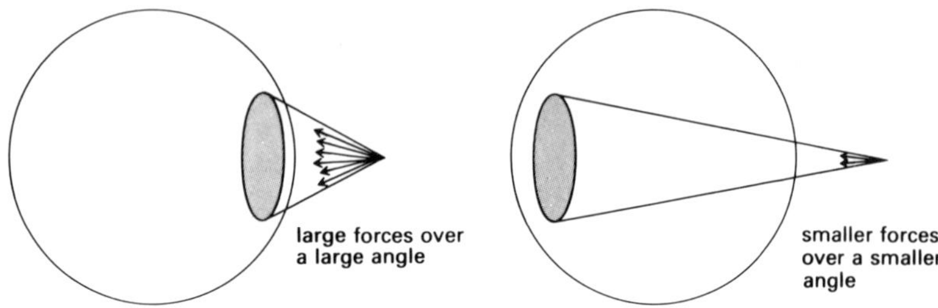

Figure E31
Effect of a large sphere.

When all the contributions to the field are added it turns out, that, if there is spherical symmetry, the overall effect is the same as if all the mass were concentrated at the centre. This result, first derived by Newton himself, is important since it applies to all inverse-square fields. It is assumed throughout this and the next Section.

Details of the integration are not important here, though it appears in some textbooks, and some students might like to see it.

Field inside a sphere

This is an interesting, if theoretical, problem for gravity but one of vital importance when electric fields are considered. Question 33 uses σ for the 'surface density' which can apply equally well to mass or charge. A few other small changes convert the question to the electrical situation. As written, though, it leads on to other interesting 'underground' ideas and applications (questions 34 and 35).

Circular motion

REVISED NUFFIELD PHYSICS *Year 5* contains a fairly full treatment of this topic, so it will be revision for many. A physical understanding of the ideas is vital and a wide range of applications should be discussed. The importance of orbital motion in leading towards Newton's Law of gravitation may have emerged from reading; Newton's triumph was to show that the most general orbit under an inverse-square law is an ellipse. For many purposes circular motion is a good approximation, and the mathematics is certainly simpler.

CENTRIPETAL FORCE

'What force keeps the Earth in orbit around the Sun?' 'What force causes a car to turn a corner?' 'What force causes a ball on a string to move in a circle, not a straight line?' Such questions can introduce the discussion. In all these cases the speed may be constant but the velocity (a vector) is not. It is constant in magnitude but not direction. The change in velocity (and therefore the acceleration) is directed towards the centre and is provided by a centripetal or 'centre-seeking' force (for example, the pull of the string or the Sun's gravitational pull). (See figure E32.)

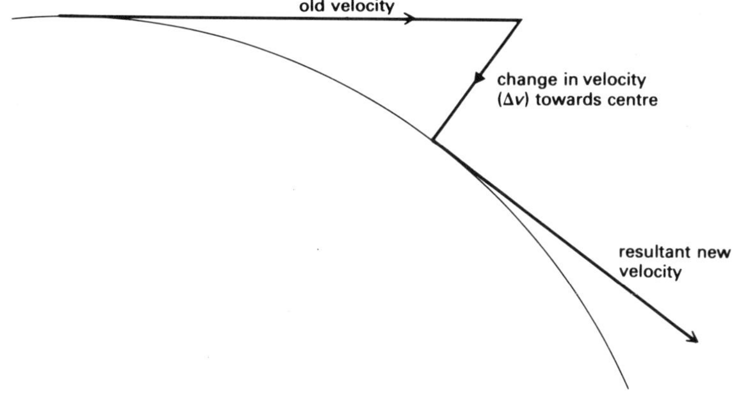

Figure E32
Constant speed, changing velocity.

Centrifugal force

This concept is only valid when the situation is viewed from a rotating frame of reference and whilst used by the layman and some engineers is only likely to confuse students; it is best avoided. If it does arise, students can be asked 'on what object does it act?' for the key point is that any object moving in a circle requires an overall *inwards* force, not an outwards one.

Question 36 applies Newton's Laws to circular motion.

Formulae for centripetal acceleration and force

Question 37 leads to the $\frac{mv^2}{r}$ formula for centripetal force in stages; the last of these (part **d**) is a more familiar derivation. Since for gravity a negative sign was used to denote an attractive, inward force $\left(\text{for example, } F = -G\frac{m_1 m_2}{r^2}\right)$, the same convention is used here $\left(\text{centripetal force} = -\frac{mv^2}{r}\right)$. This convention may not be used elsewhere, but students should always be aware of the direction in which a force acts when using mathematics to solve a physical problem.

Questions

Question 38 is about geostationary satellites.
Question 39 is about energy of satellites.
Question 40 links Kepler's Third Law of planetary motion with the inverse-square law.

Elliptical orbits

Kepler, from the observations of Brahe, deduced that the orbits of planets were elliptical, with the Sun at one focus. Newton later demonstrated that this was the expected orbit if the force was inverse-square. However, circular orbits are a close approximation for most planets. For instance, there is only a 0.02% difference between the major and minor axes of the Earth's orbit.

For Pluto, the planet with the most eccentric orbit, this difference is only 3%, but this is enough to take it at times closer to the Sun than Neptune, its less eccentric nearest neighbour.

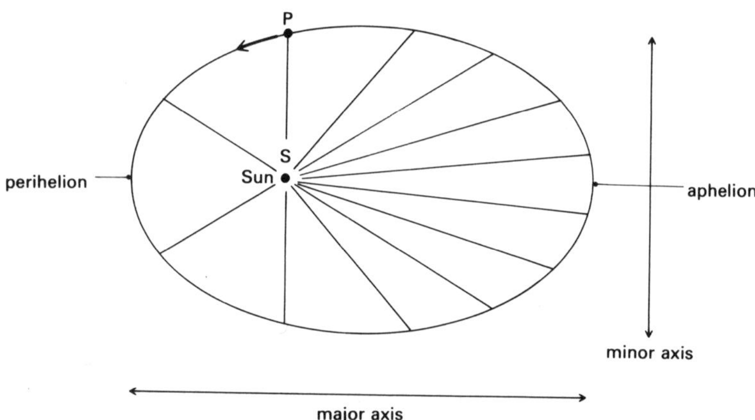

Figure E33
An elliptical orbit.

Since the total energy of a planet (potential and kinetic) is constant, a gain in P.E. caused by moving further away must be offset by a loss in K.E. Hence a planet moves more slowly near aphelion than perihelion (figure E33). This is expressed by Kepler's equal area Law (the line PS traces out equal areas in equal times).

Question 39 will have shown that the total energy is negative, as in any bound system (for example, the electron in the hydrogen atom: Unit L, 'Waves, particles, and atoms').

Computer

Dynamic modelling system

Satellite orbits and space probe trajectories can be modelled using the system. The ingredients needed to solve such a problem are as follows:

Newton's Law of Gravitation
Newton's Second Law } each in two dimensions
Kinematics }

A suitable set of equations is:

```
R = SQR(X^2+Y^2)         displacement
F = -G*M1*M2/R^2         gravitation
FX = F*X/R               } x, y components of force†
FY = F*Y/R               }
AX = FX/M1               } Newton II
AY = FY/M1               }
VX = VX+AX*DT            }
VY = VY+AY*DT            } kinematics, in two dimensions
X = X+VX*DT              }
Y = Y+VY*DT              }
T = T+DT
```

Note the negative sign in the second equation which expresses the fact that the force is *attractive*.

Constants are:

G gravitational constant (6.67×10^{-11} N m^2 kg^{-2})

M1 mass of satellite (arbitrary)

M2 mass of central body (6×10^{24} kg for Earth; 2×10^{30} kg for Sun)

Other variables which must be given initial values are: X, Y, VX, VY, T, DT.

Other useful data:

1 Earth's radius $= 6.37 \times 10^6$ m.

2 A satellite launched tangentially at the Earth's surface with a speed of 8000 m s^{-1} will go into approximately circular orbit with a period of about 90 minutes. (Suitable starting values are X $= 6.4 \times 10^6$ m; Y $= 0$; VX $= 0$; VY $= 8000$ m s^{-1}; T $= 0$, DT $= 300$ s.)

3 Radius of Earth's orbit around Sun $= 150 \times 10^9$ m.

$$\text{Speed of Earth in its orbit} = \frac{2\pi \times 150 \times 10^9 \text{ m}}{365 \times 24 \times 60 \times 60 \text{ s}} \approx 3 \times 10^4 \text{ m s}^{-1}.$$

By changing only one equation (and of course the starting values) this set of equations can be used to model the behaviour of an alpha particle near a nucleus, so it is well worth saving for use later. For more information see the booklet *Dynamic modelling system*.

†x and y components of force:
$F_x = F \cos\theta$; $F_y = F \sin\theta$; $\cos\theta = x/r$; $\sin\theta = y/r$ so $F_x = Fx/r$; $F_y = Fy/r$ (figure E34)

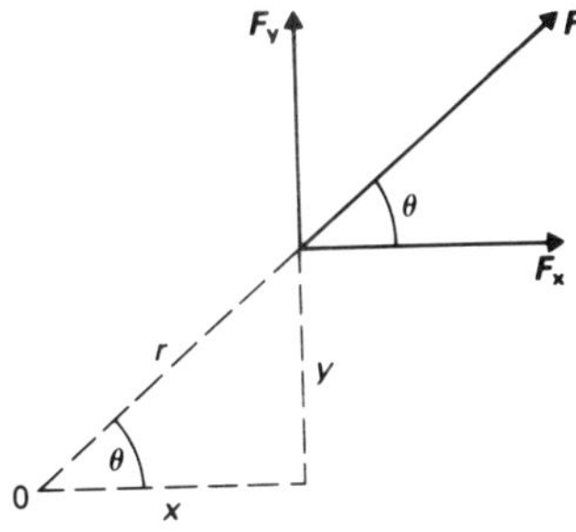

Figure E34

SECTION E3

THE ELECTRICAL INVERSE-SQUARE LAW

Having started with the uniform electric field between parallel plates and then moved on to the gravitational field and potential near a mass, the Unit ends with the electric field and potential around a point charge. The relationship between field strength and potential has already been developed: in particular, students should know that a $\frac{1}{r^2}$ field has a $\frac{1}{r}$ potential. Experimental tests of Coulomb's Law are often tricky; it is easier to explore the potential around a charged sphere using the flame probe. The potential varies as $\frac{1}{r}$; therefore the field strength must be inverse-square.

The importance of forces between charges

If this course were not concerned at all with atomic physics, we could stop at the uniform electric field. However, in this and other Units (F and L) we explore further the size, structure, and energies of atoms. In doing so we treat protons and electrons as point charges; therefore we must know how the electric field and potential around them vary.

POTENTIAL NEAR A CHARGED SPHERE

The shape of the field around a point electrode may be recalled from Experiment E3. Students may see that equipotentials will be spherical around a point charge but to explore the variation of potential with distance it has to be measured. The flame probe used in Experiment E4 can be used again.

In practice a charged *sphere* must be used, not a *point* charge. In the last Section it was established that as long as the field *is* inverse-square, the effect outside the sphere is the same as if all the charge (in this case) were concentrated at the centre. This assumption will be vindicated when the $\frac{1}{r}$ potential is found and the $\frac{1}{r^2}$ field deduced, but is perhaps best not mentioned yet, lest students imagine the argument to be circular or the test of potential superfluous.

DEMONSTRATION

E8a Investigating the variation of potential around a charged sphere using a flame probe

ITEM NO.	ITEM
1053	plastic football (see below for means of support)
10Y	colloidal graphite
14	e.h.t. power supply
51A	gold leaf electroscope
94A	lamp, holder, and stand
27	transformer
52K	crocodile clip
	flame probe (see experiment E4)
501	metre rule
1000	leads

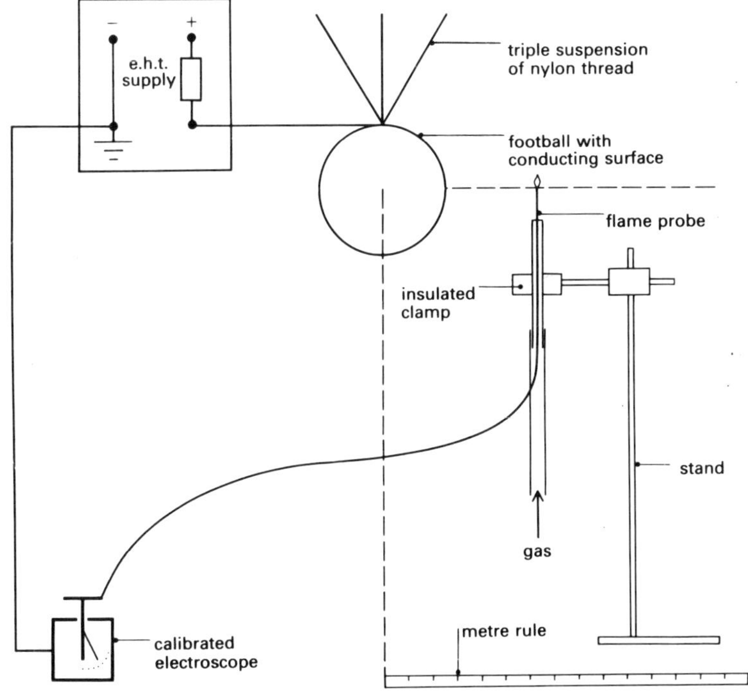

Figure E35
Using a flame probe to explore potential near a charged sphere.

The flame probe is set up and the electroscope calibrated as in experiment E4. Here the electroscope indicates the p.d. between the tip and distant walls, floor, bench, and ceiling which are at earth potential. The ball should be suspended as far as possible from these surfaces using a triple nylon suspension. It is painted with colloidal graphite or covered *smoothly* with aluminium foil. (The radius, about 0.1 m, can be deduced using string or a shadow method.) A metal screw in the valve socket can provide both support and electrical contact. A potential of +1500 V applied to the sphere is suitable.

The potential at the surface of the sphere can be measured by touching with the *unlit* probe. A ruler scale allows the potential to be measured at various distances from the centre, with the flame lit. (Take care not to melt the ball when taking close measurements.) A vertical flame reduces lateral uncertainty in the probe's position.

First the probe can be used to show that equipotentials are concentric spheres. It is easiest to do this by watching the electroscope and moving the probe so that the reading remains steady. As for gravity, there is no force component along the spherical surface so no change of energy (see 'A conservative field' page 325). The potential increases as we get closer to the sphere, as a positive test charge must be pushed against a repulsive force so gaining energy. Compare gravity, where a mass is attracted and loses potential energy, and therefore potential, as it approaches the Earth.

If the electric field obeys an inverse-square law, as for gravity, then the potential V should vary with distance r from the centre as $\frac{1}{r}$. A quick test of V at the surface of the sphere and, say, double or treble this distance should show that $V \times r$ is constant. A graph may be plotted to see if $V \propto \frac{1}{r}$ for a range of r.

Some points of experimental design might be raised, for example:

How reliable is the electroscope reading?

What is the resolution of its scale?

How big is the flame? Will the experiment give an average value for V over a region of space or perhaps a highest or lowest value?

What is the uncertainty in r?

Nevertheless the results should support a $\frac{1}{r}$ variation rather than, say, $\frac{1}{r^2}$.

THE ELECTRICAL INVERSE-SQUARE LAW

A $\frac{1}{r^2}$ gravitational field had a $\frac{1}{r}$ potential. Here the electrical potential varies as $\frac{1}{r}$ so it is reasonable to assume that the electric field strength varies as $\frac{1}{r^2}$. This will be tested later.

If this is so then it was reasonable to use a charged sphere in place of a point charge since for $\frac{1}{r^2}$ fields the effect outside a sphere will be the same as if all its charge were concentrated at the centre (as stated for gravity in Section E2).

The electrical force constant

For gravity, we have the formulae for field strength and potential:

$$g = -\frac{GM}{r^2}, \quad V_g = -\frac{GM}{r}$$

So far, for electricity, we have established the $\frac{1}{r}$ variation in V. Earlier experiments, or a test now, show that V is proportional to the charge Q on the sphere $\left(\text{it has constant capacitance } C = \frac{Q}{V}\right)$. Hence $V \propto \frac{Q}{r}$ or $V = k\frac{Q}{r}$.

There is no negative sign in the expression for V, as there is for gravity. All gravitational potentials are negative because all gravitational fields are attractive. Electrical fields can be attractive or repulsive; so electrical potential can be negative or positive. This is determined in the formula by the sign of Q. Positive charged spheres have positive potentials, negative charges negative potentials.

But to calculate the potential, and therefore the energy of a charge near the sphere we must know the value of the constant k. This is analogous to G and fixes the sizes of forces, fields, etc. for given values of Q and r.

The simplest way to find k is to record V and r at the sphere's surface (though any pair of V and r values would do). The charge Q can be measured using a coulombmeter.

DEMONSTRATION

E8b Measurement of the constant *k* in $V = k\frac{Q}{r}$

Additional apparatus required:

ITEM NO.	ITEM
1512	coulombmeter, 100 nC, with probe rod or stiff wire
51G	polythene strip for use as insulating handle

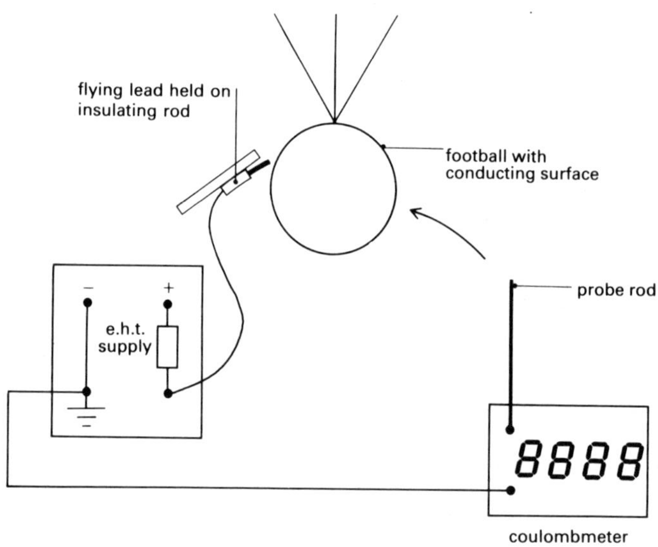

Figure E36
Measuring the value of *k*.

Using a method similar to that of experiment E6, the sphere is charged by touching with a flying lead to a potential of about 1000 V. This gives a charge of nearly 10^{-8} C for a ball of 0.1 m radius. The charge is measured by bringing up the coulombmeter so that the probe rod touches the ball. (1000 V on the ball and about 1 V across the coulombmeter shows that charge transfer is complete to within about 0.1 %.)

An accurate measurement gives k as 8.98×10^9 units. 9×10^9 is good enough for our purposes. The units can be deduced from $k = \frac{Vr}{Q}$.

$$\frac{\text{volt} \times \text{metre}}{\text{coulomb}} = \frac{\text{joule} \times \text{metre}}{\text{coulomb} \times \text{coulomb}} = \frac{\text{newton} \times \text{metre} \times \text{metre}}{\text{coulomb} \times \text{coulomb}}$$
$$= \text{N}\,\text{m}^2\,\text{C}^{-2}$$

An astute student may spot these as the reciprocal of the units of ε_0.

THE ANALOGY BETWEEN ELECTRICITY AND GRAVITY

Now we are ready to complete the expressions for electrical field strength and force:

	Force	Field strength	Potential
Gravity	$-\frac{Gm_1m_2}{r^2}$	$-\frac{GM}{r^2}$	$-\frac{GM}{r}$
Electricity			$+\frac{kQ}{r}$

Comparing the expressions we now know for potential, it is a good guess that the electric field strength will be $\frac{kQ}{r^2}$, and the force $k\frac{Q_1Q_2}{r^2}$, if the analogy holds good. More rigorously we can apply the relationship $E = -\frac{dV}{dr}$, which does not depend on the force law involved, to confirm the expression for E. Then using $F = QE$, we have $k\frac{Q_1Q_2}{r^2}$ as the force between two charges Q_1 and Q_2.

Signs

If teachers have been consistent up to now, students should not be confused by the negative signs in the expressions for gravitational force, field, and potential, and their absence in the corresponding expressions for electricity. For gravity only one kind of field (an attractive one) is possible with inwards forces (represented by negative force and field expressions), and negative potential and potential energy. The electric field is directed away from a positive charge, Q_1 (*i.e.* it is positive), and the potential is also positive. So another positive charge, Q_2, in the vicinity experiences an outwards, repulsive force $\left(k\frac{Q_1Q_2}{r^2}\right)$ and has positive potential energy $\left(k\frac{Q_1Q_2}{r}\right)$. However, a negative charge, $-Q_2$, would be attracted $\left(\text{by a force } -k\frac{Q_1Q_2}{r^2}\right)$ and the overall system would have negative potential energy $\left(-k\frac{Q_1Q_2}{r}\right)$. The expressions are summarized in the *Students' guide* (page 302). They do not have to be memorized, but a physical understanding of their forms and relationships is of course needed.

Understanding of analogy

Students should be assured that the use of a formal analogy does not mean that gravitational and electrical forces are in any other way related. Physicists continue to search for some deeper theory which includes both, without success. Nevertheless the similarity in the mathematics that describe them is extremely useful for once we have solved a problem in one field, we can transfer the result directly to the other, without having to do the mathematics all over again. Such analogies are quite common in physics. For instance between the energy of a stretched spring ($\frac{1}{2}Fx$) and that stored in a capacitor ($\frac{1}{2}VQ$); current carried by charged particles in a wire and traffic carried by occupied vehicles on a road; flow of 'heat', of water, of current, and of magnetic 'flux'.

Constants

The difference in magnitude between the two constants
$G \approx 7 \times 10^{-11}$ N m^2 kg^{-2} and $k \approx 9 \times 10^9$ N m^2 C^{-2}
is worth remarking on now, as it has important consequences which will be referred to later.

Question 41 is about 'flux' and the inverse-square law.

COULOMB'S LAW

The electric force between charges Q_1 and Q_2 at separation r is

$$F = k\frac{Q_1Q_2}{r^2}, \quad k \approx 9 \times 10^9 \,\mathrm{N\,m^2\,C^{-2}}$$

This is regarded as one of the fundamental laws of physics, like Newton's Law for Gravity, but unlike others, such as Hooke's, or Ohm's, which, though simple and extremely useful, fall very far short of universal application. In discussion of what makes a *fundamental* law, factors mentioned may include
- generality, having no known exceptions
- power to explain many other things
- simplicity, lacking special conditions, particular qualifications
- having no deeper explanation of which it is only a special case.

Still the law attempts only to explain *how* charges behave, not *why*.

Testing Coulomb's Law

In 1785 Coulomb created a quantitative basis for electrostatics by establishing this Law (Cavendish's experiment for gravity was done in 1798). Students have approached the force law via the $\frac{1}{r}$ potential, so testing the $\frac{1}{r^2}$ force law here is confirmation rather than discovery. Such tests are subject to considerable inaccuracy. One admires Coulomb for his painstaking efforts and students may respond to a challenge to their practical expertise. The experiments are tricky but reasonable results have been obtained even by groups of 18 on a foggy winter morning.

EXPERIMENT
E9 Experiments to test the inverse-square law for electric forces

ITEM NO.	ITEM
51D	2 metallized polystyrene balls
51E	nylon thread for suspension
51L	2 proof planes
94A	lamp, holder, and stand
27	transformer
	either
14	e.h.t. power supply,
	or
60/1	Van de Graaff generator
	or
51K, I, M	electrophorus plate, 'rubber', and tile
503–6	retort stand base, rod, boss, and clamp
	graph paper
1153	glue (Durafix or Evo-stik 863)
1153	adhesive tape
1504	balance, resolution 10 mg
1515	hair dryer
1000	leads
	additional apparatus for parts a, b, c as specified below

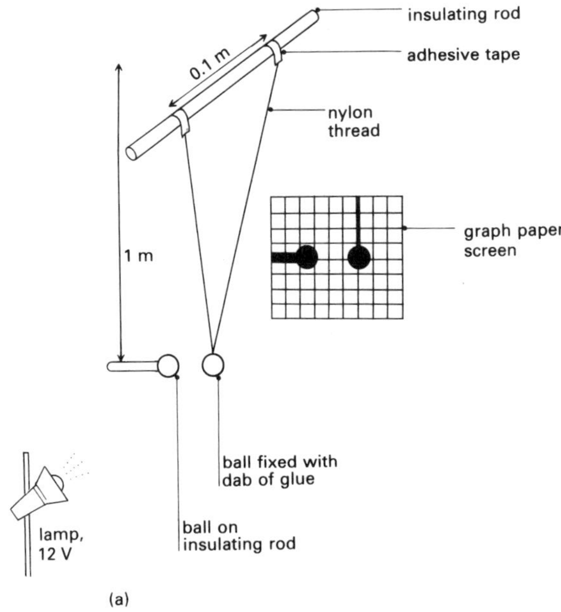

Figure E37
Testing Coulomb's Law.

One of the small balls is glued to an insulating rod (the small proof plane or the barrel of a used ball-point pen will do); the other is glued to a nylon 'trapeze' up to 1 m long, which is stable yet sensitive. All the apparatus should be clean, dry, and free of grease. (Use care in setting up and keep a hair dryer to hand throughout.)

A projection system allows measurements to be made without touching the balls. A distant lamp and close screen eliminates scale factors, or the distances can be magnified if desired.

The spheres are charged using the Van de Graaff dome (see figure E38 for methods of + and − charging) or the e.h.t. supply (with one terminal earthed). The former method gives a larger but less consistent charge.

The experiments assume knowledge (from Units A and D) that sideways displacement of a suspended object (if small) is proportional to the sideways force on it. This can be demonstrated using a rather massive pendulum and a newton meter if necessary.

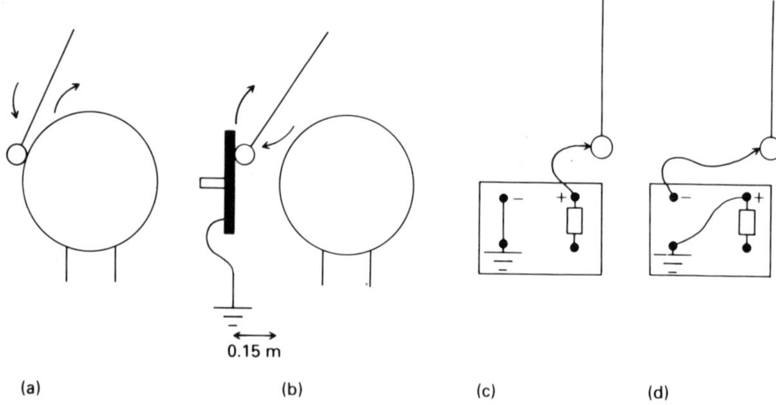

Figure E38
Charging a ball:
(a) positively, using a Van de Graaff generator;
(b) negatively, using a Van de Graaff generator;
(c) positively, using an e.h.t. supply;
(d) negatively, using an e.h.t. supply.

Alternative demonstrations

Using the glass spring suspension designed for simulating Millikan's experiment, forces between oppositely charged spheres are easily measured if the spring is calibrated (using small pieces of card of known weight to stretch it), see figure E39(a). Difficulties over sideways displacements and forces are thus avoided.

One suspended ball and a fixed, oppositely charged ball on an insulated stand on a top pan balance is another alternative, see figure E39(b).

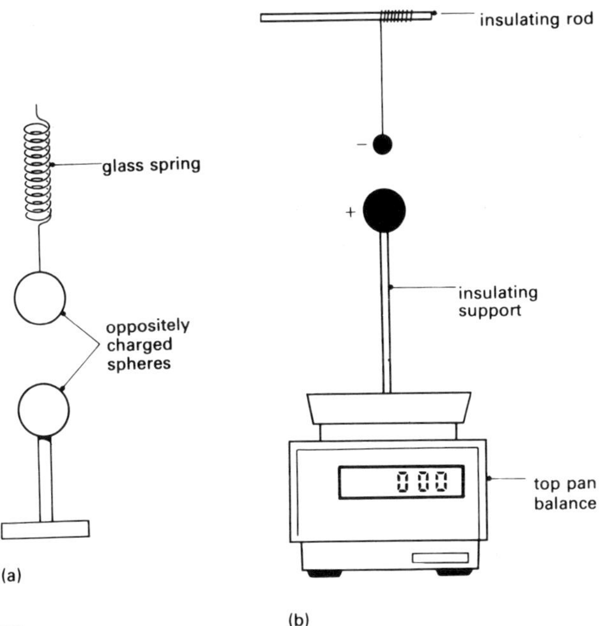

Figure E39
Alternative Coulomb's Law demonstrations.

E9a $F \propto \frac{1}{r^2}$ This is the easiest test to make. The rest position of the shadow of the suspended ball is marked. The other ball is brought near, both being charged, and the positions of both shadows are marked (figure E40). Students can take a series of d and r readings, plotting d against $\frac{1}{r^2}$.

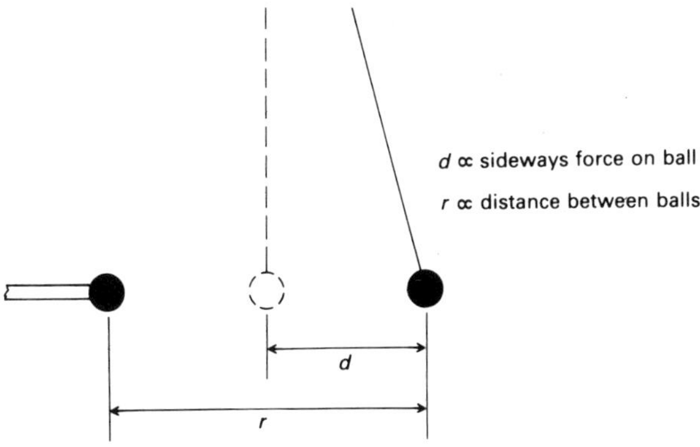

Figure E40
Deflection of a suspended ball.

E9b $F \propto Q_1, Q_2$ Coulomb's original elegant method of halving the charge on a ball by touching it with an identical uncharged ball leads to very small charges. If using the e.h.t. supply gives large enough deflections, its potential may be varied to alter the charge on one ball ($Q \propto V$ from Experiment E8b). Otherwise the charges have to be measured as in **c** below.

E9c $k \approx 9 \times 10^9\,\mathrm{N\,m^2\,C^{-2}}$ It is possible, if not easy, to obtain a result of the right order of magnitude. Charges on the balls can be measured using a coulombmeter with a range of 10 nC. The force is deduced from the weight of the ball and the angle at which it hangs (figure E41).

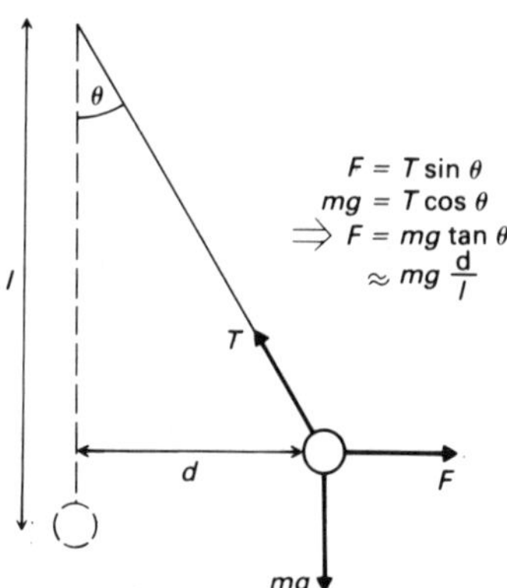

Figure E41
Calculating the force on a displaced ball.

Questions

Question 42 gives results for the above experiment as a series of photographs.
Question 43 discusses why the experiments often fail.
Question 44 is about forces between charged spheres.

THE RADIAL FIELD AND THE UNIFORM FIELD

The field of a point charge is radial and inverse-square. It seems odd, then, that when a collection of charges is arranged on a large flat plate the field above it does *not* fall off as $\frac{1}{r^2}$ but is uniform. This is the result of the way in which contributions from each bit of charge add up to give an overall effect. The precise integration is rather involved, though some students may like to follow it through.

Question 45 shows how, in principle, a uniform field may be obtained by adding contributions from charges on a flat sheet.

Appendix VII gives the full integration and shows that the constant $9 \times 10^9\,\mathrm{N\,m^2\,C^{-2}}$ is, in fact, $1/4\pi\varepsilon_0$.

Computer plots of field strength near charged single and parallel plates

The computer program 'EFIELD' calculates the field strength on the axis of a circular sheet of charge or between two oppositely charged parallel plates. It plots the field strength as a function of distance and can be used to explore how nearly uniform the field is between parallel plates; the effect of plate size and separation.

The user chooses the radius of the sheet of charge, which is then displayed. The field strength at positions on the axis is calculated and plotted or displayed as a table. The transition between an inverse-square force law far from a small sheet of charge, and the uniform field near a large sheet is seen.

Alternatively, two parallel plates may be set up. The field strength is calculated at points along the line joining their centres and displayed as before.

Further details are given in the *Notes* for the package 'Software for Nuffield Advanced Physics', Longman's Micro Software.

The force constant $1/4\pi\varepsilon_0$

The adding up process described above gives an expression for the field strength, E, between parallel plates in terms of the force constant k ($\approx 9 \times 10^9\,\mathrm{N\,m^2\,C^{-2}}$). By comparing this with our previous expression $\left(E = \dfrac{\sigma}{\varepsilon_0}\right)$ we can deduce that $k = \dfrac{1}{4\pi\varepsilon_0}$. The value and units of the constant can be checked; it takes its place among other 'universal' constants in determining the sizes of forces between all charges everywhere (as G does for the forces between masses). The electrical formulae and relationships can now be 'updated' by replacing $9 \times 10^9\,\mathrm{N\,m^2\,C^{-2}}$ by $1/4\pi\varepsilon_0$:

Force	Field	Potential	Potential energy
$F_e = \dfrac{1}{4\pi\varepsilon_0}\dfrac{Q_1Q_2}{r^2}$	$E = \dfrac{1}{4\pi\varepsilon_0}\dfrac{Q}{r^2}$	$V = \dfrac{1}{4\pi\varepsilon_0}\dfrac{Q}{r}$	$E_p = \dfrac{1}{4\pi\varepsilon_0}\dfrac{Q_1Q_2}{r}$

Questions

Question 46 is about a Van de Graaff generator.
Questions 47 and 48 apply the ideas of electrical potential and energy in an inverse-square field to nuclei and atoms.
Question 49 involves drawing electrical equipotentials.
Questions 50 and 51 are about charged spheres.

SIZES OF ELECTRICAL AND GRAVITATIONAL FORCES

Students might be excused for thinking that electrostatic forces are rather weak, especially if they had difficulty over the Coulomb's Law experiments. However, a comparison of the constants in Coulomb's and

Newton's Laws reveals a striking contrast: $\frac{1}{4\pi\varepsilon_0} \approx 9 \times 10^9\,\mathrm{N\,m^2\,C^{-2}}$ whilst $G \approx 7 \times 10^{-11}\,\mathrm{N\,m^2\,kg^{-2}}$. Electrical forces on objects carrying unit charge will be enormously greater than gravitational forces between objects of unit mass at the same separation. A few comparisons might illustrate this.

Question 52 shows that the ratio F_e/F_g for an electron and proton in a hydrogen atom is of the order of 10^{39}. For two electrons it will be even greater, for two protons only about 2000 times less. So within atoms electricity is definitely the dominant force.

The charges in the Coulomb's Law tests were about 10^{-8} C; the masses were about 0.001 kg (1 g), and the distances 0.1 m at most. A comparison of electric and gravitational forces here gives a ratio of about 10^{10}.

$$\frac{F_e}{F_g} = \frac{QQ/4\pi\varepsilon_0 r^2}{G\,mm/r^2} = \frac{1}{4\pi\varepsilon_0 G}\frac{Q^2}{m^2} = \frac{9 \times 10^9}{7 \times 10^{-11}}\frac{(10^{-8})^2\,\mathrm{N}}{(10^{-3})^2\,\mathrm{N}} \approx 10^{10}$$

Note that the separation does not matter. The *ratio* F_e/F_g is the same at any distance.

Further estimates show that 10^{-8} C represents about 6×10^{10} electrons and if we have, say, 6×10^{21} atoms in the ball, this means that only 1 atom in 10^{11} or so in the ball is charged. But even this charge leaked away quickly unless we were very careful.

This is why electrical forces between charged objects are small: the balance of + and − charge can only be disturbed very slightly. Although very large forces exist between individual ions in a grain of salt, say, the almost perfect balance of + and − charges ensures that the whole grain is neutral and exerts very little force on other grains.

Question 53 argues from the field of a dipole towards the effect of a neutral array of charges as in a crystal.

If we take more massive neutral objects the gravitational force begins to dominate so that electrical forces are irrelevant when it comes to satellites and planetary orbits. It should be noted, though, that the range of both electrical and gravitational fields is infinite as both obey an inverse-square law.

Questions

Questions 54 and 55 test understanding of the similarities and differences between electricity and gravity.

Reading

ASIMOV *The collapsing Universe: the story of black holes.* Uses obsolete units, but Chapters 1 and 2 on gravity are interesting.
CALDER *The key to the Universe.* Takes the story from quarks to black holes.
DAVIES *The forces of nature.* Chapters 1 and 4 give an especially readable survey.
FEYNMAN, LEIGHTON, and SANDS *The Feynman lectures on physics* Volume 1. Pages 2 and 3 on the balance of positive and negative charges.

THE KNOWN INTERACTIONS

Of the four kinds of force known, electrical and gravitational are two. Question 52 introduced the 'strong' nuclear force which, with the 'weak' nuclear force, operates over very short ranges within nuclei ($\gtrsim 10^{-15}$ m). The tremendous energy associated with interactions involving these forces can be released in nuclear fission (Unit F, 'Radioactivity and the nuclear atom', and Unit G, 'Energy sources').

Students who ask about magnetic forces can be told that these arise between charges in motion (that is, currents) and that they follow as a result of transforming Coulomb's Law to a moving frame of reference.

Everyday 'contact' forces, pushes and pulls in springs, friction, and forces in fluids can be attributed to the interaction of atoms via the electric fields of the electrons which surround them (Unit A).

Much has been achieved in 'boiling down' forces to these four and one of the chief preoccupations of contemporary physics is to find some deeper unifying theory which links them all. So far progress has been made with linking electricity and the weak force. Many more questions remain unanswered: for example why is there only one kind of mass but two kinds of charge, why does every electron have just the same amount of charge, and so on?

Students may wonder why electrons, pulled by such enormous forces, do not collapse into the nuclei they surround. This is one question we can begin to answer, at least in a simple way, in Unit L, 'Waves, particles, and atoms'.

Reading

'The particles and forces of nature' in the Reader *Particles, imaging, and nuclei* deals with the four known forces, the particles associated with them, and attempts to produce a unified theory.

Unit F
RADIOACTIVITY AND THE NUCLEAR ATOM

Paul Jordan and Peter Harvey
Highfields School, Wolverhampton

Suggested time allocation: four weeks

PLAN OF THE UNIT

Links from		Topic		Links to
		Section F1 The Rutherford model of the atom		
GCSE	▶	**Properties of α, β, and γ radiation**		
		α-particle scattering		
Coulomb's Law (Unit E)	▶	**The nuclear atom**	▶	the nucleus: Section F3
		Section F2 Exponential decay		
		Half-life		
		Chance, randomness	▶	Unit K, 'Energy and entropy'
numerical methods (Units B, D)	▶	**Exponential growth and decay**	▶ ▶	Unit G, 'Energy sources' the Boltzmann factor: Unit K, 'Energy and entropy'

Section F3
The nucleus

Links from		Topic		Links to
chemistry	▶	**Periodic Table**		
collisions (Unit A)	▶	**Neutrons**		
		Proton number, *Z*		
		Nucleon number, *A*		
		Isotopes and their uses		
		Changes in *Z*, *A* on α and β decay		
		Ionization by collision	▶	Unit L, 'Waves, particles, and atoms'
potential in attractive field is negative (Unit E)	▶	**Electron binding energy**		
chemistry	▶	**Ionization energy**		
		Nuclear binding energy		
		Mass and energy		
		Fusion and fission	▶	Unit G, 'Energy sources'

INTRODUCTION

The major aims of the Unit are:

1 to examine the properties of ionizing radiations;

2 to investigate the ionization of atoms;

3 to provide a satisfactory understanding of radioactive decay and of exponential change;

4 to develop an understanding of the nature of the nucleus;

5 to provide knowledge of the energy changes associated with nuclear reactions required for any consideration of the principles of nuclear energy generation.

Such a study of radioactivity provides much of the experimental evidence upon which the modern understanding of the structure of atoms is built and provides essential knowledge for citizens of the uses and the dangers of radioactivity. Such knowledge is likely to become of increasing importance for our civilization.

The first Section examines the properties of the three radiations and considers their energies. It proceeds to develop the Rutherford model of the nuclear atom. In the second Section the concern is with the process of decay and with the measurement of half-life. This is followed by a consideration of the equation for decay and its implications. In the third Section, the story of the nucleus is taken up again with specific reference to some questions which were previously left open; for example, where in the atom are the electrons to be found? what holds the nucleus together?

Before considering the development of the Rutherford model from the scattering experiment of Geiger and Marsden, students will require access to additional background information. This is conveniently acquired if students are asked to find answers to some specific questions whilst they are doing experiments F1 to F4. Individual students may be set specific tasks, such as those suggested on pages 359–360, make their own notes on their reading and be asked to report back to the group as a whole. This might occupy a single homework since the answers are easy to find from standard texts or from the brief history given in the *Students' guide* for this Unit.

Teachers may find it necessary to re-order the sequence of the experiments in this Unit in order to suit the availability of the equipment.

Teachers should know that Revised Nuffield Advanced Chemistry, Topic 4, 'Atomic structure' deals with the nuclear atom, the neutron, isotopes, and ionization energy.

THE PLACE OF THE UNIT IN THE COURSE

A theme running through the course is the attempt to explain large-scale happenings in terms of small-scale behaviour. Within this theme, understanding about the nature of the atom plays a big part. It requires students to fit together ideas from Unit E, 'Field and potential' and Unit B, 'Currents, circuits, and charge' into new patterns to explain new things.

It also provides ideas essential for later work and, in particular, Unit G, 'Energy sources' and Unit L, 'Waves, particles, and atoms'.

LIST OF SUGGESTED EXPERIMENTS AND DEMONSTRATIONS

SECTION F1
THE RUTHERFORD MODEL OF THE ATOM

Using radioactive sources in schools

The attention of teachers is drawn to DES Administrative Memorandum 2/76 *The use of ionizing radiations in educational establishments* and the accompanying *Notes for guidance.* This memorandum gives the regulations governing the type, quantity, and use of radioactive sources in schools. Of particular significance for first-year sixth forms is the regulation forbidding students under the age of 16 to use radioactive sources other than the naturally occurring isotopes of potassium, uranium, and thorium.

An introductory passage to the students' laboratory notes for this Unit draws their attention to sensible precautions to take when handling sealed sources or naturally occurring radioactive substances. It also suggests that any experiment involving uranium or thorium salts should be set up in a spill tray lined with absorbent paper. There are also some notes on the use of Geiger–Müller tubes.

Experiments on alpha, beta, and gamma radiations

The work provides an opportunity for reporting back and the results of each experiment will be required by all. The reading tasks (pages 359–360) may also be undertaken at the same time as these experiments.

Questions

Questions 1 to 7 concern these experiments and it is suggested that these could be assigned as the experiments are completed.

A number of the experiments in this Unit occur in earlier courses but, of course, as demonstrations. For example, the Revised Nuffield Physics course, *Year 5*, includes:

Radiations and counters (Demonstration 83);
Experiments with α-particles: range and stopping (Demonstration 84);
Experiments with cloud chambers (diffusion and expansion types) (Demonstration 86 and Experiment 87);
The deflection of beta particles in a magnetic field (Demonstration 88);
The decay of protactinium (Demonstration 89).

Reading

Students can find useful information about the experiments in:
AKRILL, *et al. Physics.*
CARO, *et al. Modern Physics.*
DUNCAN *Advanced physics; fields, waves and atoms.*
LEWIS and WENHAM *Radioactivity.*
WENHAM, *et al. Physics: concepts and models.*

THE PROPERTIES OF ALPHA, BETA, AND GAMMA RADIATIONS (EXPERIMENTS F1 TO F4)

These four experiments lead to the following conclusions:

i beta particles carry negative charge;

ii alpha particles are strongly ionizing and have energies in the order of MeV;

iii beta particles ionize much less strongly and therefore are more penetrating than alpha particles;

iv the intensity of gamma radiation is subject to an inverse-square law of distance.

EXPERIMENT
F1 The deflection of beta radiation in a magnetic field

ITEM NO.	ITEM
130/1	scaler
130/3	GM tube holder
130/5	thin window GM tube
195/2	beta particle source
196	source holder
92I	mild steel yoke
92B	2 Magnadur magnets
503–6	2 retort stand bases, bosses, rods, and clamps
1M	2 lead blocks
92D	plotting compass

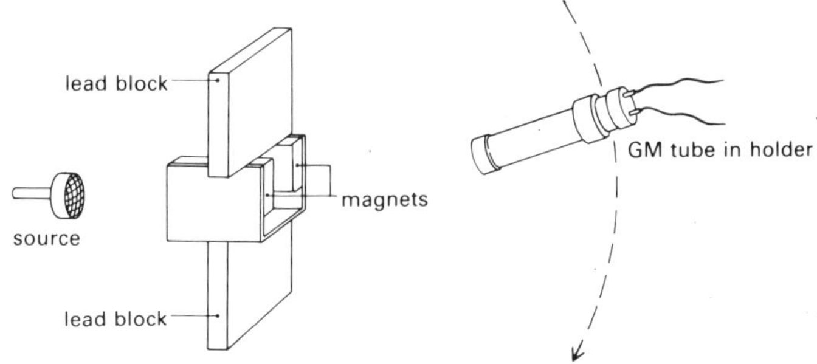

Figure F1
Arrangement of apparatus for the magnetic deflection of beta particles (seen from above).

Figure F1 shows the arrangement. Only the GM tube holder, not the tube, may be gripped by a clamp. The other clamp can be used to hold the source. The lead blocks will serve to prevent radiation reaching the GM tube directly. A ratemeter may be substituted for a scaler.

This experiment is essentially the same as Demonstration 88 in the Revised Nuffield Physics *Pupils' text* and *Teachers' guide, Year 5*. Brief instructions are given in the students' laboratory notes.

This simple experiment may be better suited to slow or inexperienced students.

An understanding of how magnetic deflection gives evidence for the sign of the charge carried by the particles will be helpful in discussing the nature of alpha radiation, which is essential in the development of the Rutherford model of the atom.

Having done this experiment, students should know that:

i beta particles are charged, and

ii the direction of the deflection suggests that the charge is negative.

Questions

Questions 1 and 2 are relevant to this experiment.

EXPERIMENT

F2 Measuring the ionization current to find the number of ion pairs produced by an alpha particle, and an estimation of the energy of the alpha particles emitted

ITEM NO.	ITEM
1516	picoammeter (see note below)
1108	ionization chamber
195/3	pure alpha source
14	e.h.t. power supply
196	source holder
1000	leads

Picoammeter Currents of the order of 10^{-9} A to 10^{-11} A (depending on the source used) are to be measured in this experiment. One way of achieving this sensitivity is to use the electrometer/d.c. amplifier with an input resistance of $10^9\,\Omega$ or $10^{11}\,\Omega$, since the p.d. across such a resistance will be of the order of 1 V. Alternatively, a digital meter could be used. A typical value for the resistance of such a meter is $10^7\,\Omega$. So on its 200 mV setting it has a range of 0 to 2×10^{-8} A and a resolution of 10^{-11} A.

The source is mounted inside the ionization chamber. The picoammeter measures the current between the central electrode and the can (figure F2). The output of the power supply, connected between the chamber and the earthed side of the meter, is increased until, at a potential difference of about 750 V, the current flowing in the meter, which is equal to the ionization current in the chamber, is steady and increasing no further.

If a 4×10^3 Bq (0.1 μCi) source is used then a range of 0–10^{-11} A may be needed. Teachers should try this beforehand and determine the ionization current caused by the source used. When the p.d. across the chamber is increased, there is a current surge before the constant current is achieved.

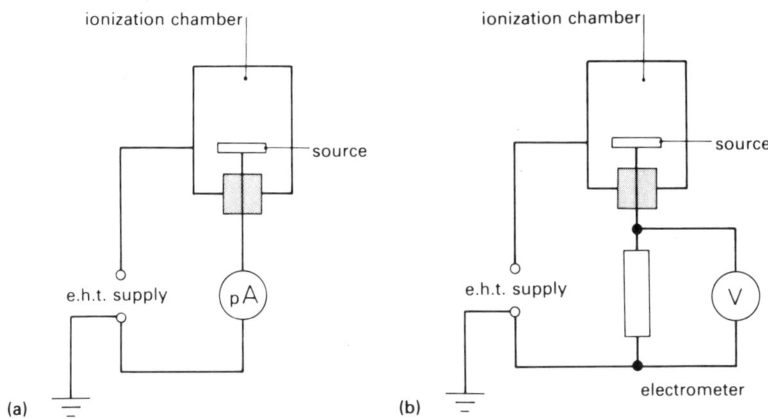

Figure F2
Measuring ionization current:
(a) with a picoammeter;
(b) by measuring p.d. across a very high resistance.

The idea is to measure the ionization current produced by alpha particles from the alpha source. As the charge on an ion is of the order of the charge on an electron, the number of ions produced per second can be estimated from the ionization current. (Note that although each ionizing event produces a pair of charged particles, for example, a single charged positive ion and an electron, the total charge transported across the ionization chamber by the pair is equal to one single charge. Each of the oppositely charged particles makes only part of the journey across the chamber – they travel in opposite directions from the ionizing event to either the chamber wall or to the central electrode. This is equivalent to one particle doing the complete journey.)

Estimation of the number of ions produced by each alpha particle requires knowledge of the number of alpha particles produced by the source per second. This can be estimated from the stated source strength. Students will need to know that, on average, an alpha particle loses about 30 eV of its energy for each ion pair produced. It may be necessary to remind students of the meaning of the electronvolt, which was introduced at the end of the Unit B, 'Currents, circuits, and charge'.

The experiment can give evidence for the energy of a typical alpha particle. This will be needed when estimating how close such a particle might approach a nucleus in a head-on collision. It also brings out the meaning of the unit of activity, the becquerel (Bq) and its relation to the earlier unit, the curie (Ci), as well as illustrating the great ionizing power and the short range of alpha particles. Nevertheless, there are several sources of uncertainty in the experiment. For example, do all the alpha particles produced by the source cause ionization in the air within the chamber? If not, what happens to them? Is the volume of the chamber significant? What might happen if it were smaller? Or bigger?

Having done this experiment, students should know:

i how to use an ionization chamber;
ii how to estimate the number of ion-pairs produced by an alpha particle;
iii that when the ionization current in an ionization chamber is constant all the ion pairs are being collected;
iv that the energy of an alpha particle is in the order of MeV.

Questions 3 to *6* are relevant to this experiment.

EXPERIMENT
F3 The penetrating power of alpha, beta, and gamma rays

ITEM NO.	ITEM
195/1–3	alpha, beta, and gamma sources
130/1	scaler
196	source holder
130/3, 5	GM tube and holder
507	stopwatch or clock
503–6	2 retort stand bases, rods, bosses, and clamps
501	metre rule
1052	set of absorbers (see below)
1155	Vernier callipers or micrometer screw gauge

Item 1052, the collection of absorbers, should include cigarette paper, thin glass and Perspex sheets, aluminium cooking foil, aluminium sheets of various thicknesses sufficient to cover the range 1 to 5 mm, lead sheet or blocks to cover the range of thickness from 5 to 20 mm.

This experiment has a number of parts (listed below) and details are given in the students' laboratory notes.

F3a Range of alpha particles in air

The energy of the alpha particles can be found from the graph of range against energy given in the laboratory notes. These notes also give the air equivalent of the windows of the MX 168 and MX 168/01 GM tubes (30 and 17 mm respectively).

An alternative method uses the ionization chamber fitted with a gauze lid and a central rod that reaches nearly to the top of the chamber. With a p.d. of about 500 V across the chamber (from e.h.t.

supply with 50 MΩ limiting resistor) the ionization current caused when an alpha particle source is within about 5 cm of the gauze can be detected by a picoammeter (*i.e.*, digital meter, or electrometer with $10^{10}\,\Omega$ input resistor – see experiment F2).

F3b Range of beta particles in aluminium

The energy of the particles can be found from the graph provided in the laboratory notes.

F3c 'Half-thickness' of lead for gamma radiation

This gives the energy of the radiation.

F3d The relation between the thickness of the absorber and the radiation transmitted for gamma and beta radiation

Students draw a graph of radiation transmitted against absorber thickness. For gamma radiation this curve is exponential in form. This graph and the data could be used in Section F2 as an example in a discussion of exponential behaviour.

F3e The relation between intensity and distance for gamma radiation

This provides practice in handling data in order to formulate a relationship (here an inverse-square law). Students should read the discussion of zero error in their laboratory notes.

A ratemeter can be substituted for a scaler in parts **b**, **c**, and **d**.

Parts **d** and **e** illustrate some of the factors involved in controlling dosage from radiation.

Before starting these experiments remind students of the need to deal with the background count.

These experiments will provide students with knowledge of:

i the range of alpha particles in air and the energy of each alpha particle;

ii the range of beta particles in aluminium and the maximum energy of the beta particles;

iii the meaning of 'half-thickness' and the energy of the gamma radiation;

iv the exponential relationship between absorber thickness and count rate for gamma radiation;

v the way in which the intensity of gamma radiation falls off with distance from the source (inverse-square law).

Questions

Questions 3 to 5 are relevant to experiment F3. Question 7 is about precautions to be taken when handling radioactive material.

EXPERIMENT
F4 Photographic detection of radiation

ITEM NO.	ITEM
16	radium source
	either
1155	dental X-ray film
	or
	bromide paper with developer and fixer

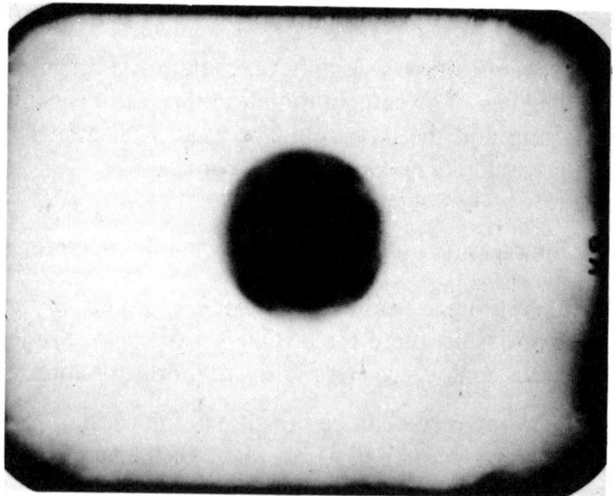

Figure F3
The fogging of dental film by radium; source of strength 2×10^5 Bq (approximately).
Mark Jeffery.

Dental X-ray film is more convenient to use than photographic paper. In one form the film is in a pod which also contains sachets of developer and fixer, so that no darkroom is needed for either exposure or processing. Another (less expensive) form of X-ray film consists of small, individually wrapped plates. This film contains a yellow filter on top of the emulsion which means that it can be processed in subdued light. Both types of film can be exposed by placing the source on top of the film packet for 20 to 30 minutes (the packet may contain a lead foil behind the film – the source must of course be on the opposite side to this). Most of the effect is due to beta radiation which penetrates the wrapping. Fast bromide paper can also be used ('hard' paper is better than 'soft'). If the source is placed face down on the emulsion side of the paper (in the dark, of course), exposure times should again be 20–30 minutes; the effect is almost entirely due to alpha particles. If the bromide paper is wrapped in black paper, a much longer exposure time will be needed – hours or even days.

Reading tasks

As mentioned in the Introduction to this Unit, students will need additional background information before discussing the Geiger and Marsden scattering experiment and Rutherford's interpretation of the results. A single homework might suffice for this, the students being asked to find answers to specific questions whilst doing Experiments F1 to F4; they may be directed to the brief history in the *Students' guide* (page 373) and to the references which follow.

Questions

a *In 1904, J. J. Thomson, the discoverer of the electron, suggested a model for the atom which is frequently called the 'plum-pudding model'. Describe this model and indicate how it incorporated the facts known at the time.*

The students' reports or notes should give a brief description of the model and at least some of the evidence incorporated: for example, atoms are neutral but since they contain negative charges they must also contain equal positive charge; electrons obtained from all sources and by all methods are alike; few atoms disintegrate naturally so there must be a balance of forces within most atoms; radiations from radioactive elements appear to pass through matter without deflection.

b *How was radioactivity discovered and how were the radiations sorted out into the three different types?*

The replies should indicate that absorption in matter suggested three radiations: one which was readily absorbed; one which passed through several millimetres of aluminium; and one which was quite difficult to absorb. The first two of these were deflected in electric and magnetic fields but in opposite directions; the last was undeflected in such fields and was very penetrating in the way that Röntgen's X-rays were penetrating. See Rutherford and Soddy, *The cause and nature of radioactivity* in Phil. Mag. S 6. Vol. 4. No. 2, Sept. 1902; Rutherford, *The magnetic and electric deviation of the easily absorbed rays from radium*, Phil. Mag. S 6. Vol. 5. No. 26, Feb. 1903. Both papers are reprinted in WRIGHT *Classical scientific papers – physics.*

c *What is the evidence for the identification of*
i an alpha particle with a helium ion, and
ii a beta particle with an electron?

i The experiment of Rutherford and Royds provides the evidence for this identification although estimates of the specific charge had already indicated that this was so; it may be worth reminding students of the right angle between the tracks produced when an alpha particle collides with a helium atom (Revised Nuffield Physics *Year 5*, and Section A4 of this course).
ii The similarities between the beta particles and the electrons produced in vacuum tubes and especially the possession of the same specific charge.

d *(An additional question) How does a Geiger–Müller (GM) tube work and what are its limitations?*

One might expect reference to the initial ionizing event and the consequent 'avalanche' of ions causing a pulse; to the threshold and operating voltages, and the 'dead-time'.

References

Reading section of *Students' guide*, page 373, 'Radioactivity and the nuclear atom: a brief history'.
AKRILL, BENNET, and MILLAR *Physics.*
BENNET *Electricity and modern physics.*
BOLTON *Patterns in physics.*
CARO, McDONELL, and SPICER *Modern physics.*
DUNCAN *Advanced physics: Fields, waves, and atoms.*
LEWIS *Electrons and atoms.*
LEWIS and WENHAM *Radioactivity.*
ROGERS *Physics for the inquiring mind.*
WENHAM, DORLING, SNELL, and TAYLOR *Physics: concepts and models.*
WRIGHT *Classical scientific papers – physics.*

THE DEVELOPMENT OF RUTHERFORD'S NUCLEAR MODEL

The discussion might follow the following lines. 'What was the Thomson model of the atom?' 'If one accepts this model, what might be expected to happen to an alpha particle which is fired towards a metal foil?' (Some scattering through small angles would be quite likely but large-angle scattering, which could only result from many chance encounters with atoms, would be improbable.)

Now Geiger had already shown (1910) that the vast majority of the deflections were by only a few degrees when alpha particles fell on such a foil. This could be explained by assuming that the deflection of a single alpha particle was the resultant of a large number of very small deflections caused as the particle passed through successive individual atoms of the foil. In this case theory suggests that the most probable number scattered through a particular angle depends on $\sqrt{\text{thickness}}$; Geiger's results confirmed this.

But scattering through quite large angles, even over 90 °, was also observed. Results collected later by Geiger and Marsden for scattering through particular angles by gold foil are given in the *Students' guide* (page 363) and it can be seen that the number of alpha particles scattered through the large angles, although small, is significant. For these cases, the experiments showed that the number scattered at a particular angle depends directly on the thickness of the foil. Thomson's model could not account for this.

Rutherford, in whose laboratory Hans Geiger was working, realized that this scattering through large angles required an enormous force to be exerted on the incoming alpha particle as it travelled through the foil. He suggested that something like the force of repulsion or attraction between the charge on the particle and a charge equivalent to about 100 electron charges concentrated into a tiny space at the centre of a gold atom could turn the alpha particle back where it came from.

'What would be the shape of the path followed by the alpha particle if the force were **1** one of attraction or **2** one of repulsion?' (For **1** the path is rather like the nearly hyperbolic path followed by many comets as they fall in towards and then recede from the Sun, and for **2** it is a hyperbolic and symmetrical path in which the particle approaches the charge but does not pass round it. Teachers might like to introduce the 'Chinese hat model' at this point. Roll a ball towards the hat when it is in the inverted position and the path models that of the comet; turn the hat the other way up and the path models that of the alpha particle.)

Rutherford reasoned that such a 'nucleus' would be very small compared with the size of the atom (so that close encounters by alpha particles from the incident stream would be rare events), and massive. (Why? Alpha particles themselves have a small mass and are travelling very fast: any obstacle of comparable mass would be brushed aside or knocked on.)

The force responsible might well be the electric force between the two charges. Perhaps thinking, as we have done, about the motion of comets, Rutherford was at first inclined to think that an attractive force was responsible. This, however, didn't seem very likely. (Why? A negatively charged nucleus still leaves the problem of the positive charge which must neutralize the negative charges known to be present.) However, the theoretical analysis is the same in both cases; only the sign has to be changed.

'How does an electrical force depend on the distance between the charges?' (Inverse-square.) 'What is the force law?' $\left(F = k\dfrac{Q_2 Q_2}{r^2}\right)$.

Rutherford, assuming that this *was* the force which determined the scattering behaviour of the alpha particles, then developed the mathematics which could be applied. This led him to an equation which could be tested experimentally. He asked Geiger and Marsden to undertake this critical experiment. His account of his nuclear model was published in 1911 and the experimental results of Geiger and Marsden which followed a year later confirmed the hypothesis.

Both Thomson's and Rutherford's models of the atom predicted scattering in ways which could be tested experimentally. Rutherford's model to account for the unexpected large-angle scattering of some alpha particles may well have been derived from the analogy with the behaviour of a comet in the gravitational field of the Sun. It is of interest to see if we can take the gravitational model already used and develop it to provide some laboratory check on the way in which the inverse-square law is applied in the Geiger and Marsden experiment.

We might try using the repulsion between two magnets instead, as suggested in Revised Nuffield Physics *Year 5*, Demonstration 90X. But this cannot be made quantitative because the force between the magnets does not follow an inverse-square law. An electrostatic model (Demonstration 90Y) using a Van de Graaff generator would be suitable, but very difficult to control in a repeatable way. Instead, we intend to develop the gravitational hill model already mentioned.

'What happens to the alpha particle as it approaches a nucleus?' (It slows down, losing kinetic energy and gaining potential energy.)

A ball, given some kinetic energy and aimed at the hill, behaves in this way and might offer the model we are looking for.

'What shape must the hill have?' To answer this, consider how the potential energy of the alpha particle depends on its distance from the nucleus (the force varies as $1/r^2$, so the potential energy varies as $1/r$).

The hill must be constructed so that the potential energy of a ball rolling on it is proportional to $1/r$. The height must vary as 1/(distance from the centre). See figure F4.

At this stage teachers may wish to ask a student to verify that the hill to be used conforms to this requirement.

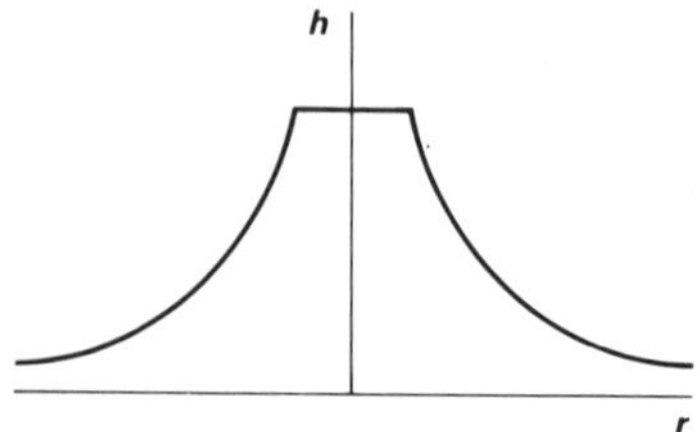

Figure F4
Section through alpha scattering analogue: $h \propto 1/r$.

We are now in a position to use the hill in a qualitative way as an aid to thought.

DEMONSTRATION
F5 Qualitative test of the gravitational hill model

ITEM NO.	ITEM
1028	alpha particle scattering analogue
1153	drawing board

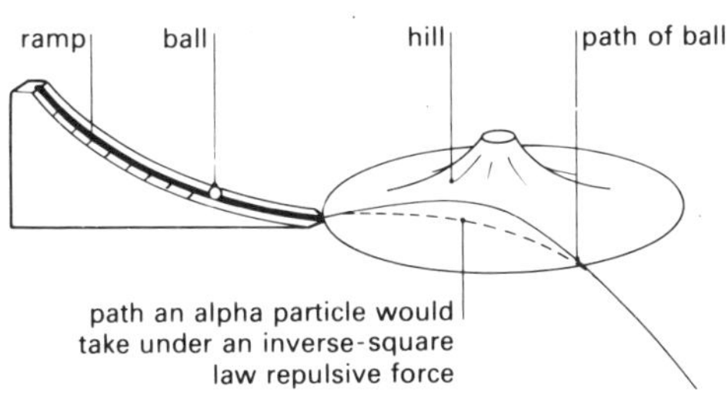

Figure F5
Gravitational analogue of alpha particle scattering.

Use of the hill model
Allow the ball to roll down the ramp and on to the hill. To avoid a bounce at the lip of the hill, it may be necessary to place the bottom of the ramp against the edge of the hill. First, roll the ball directly along a radius of the hill to establish the kinetic–potential–kinetic energy interchange. Then release the ball from the top of the ramp positioned for a deflection of about 30 °. Without moving the ramp, release the ball from a lower starting point to show that the deflection depends on the speed. Finally, repeat with the ramp positioned for a deflection of about 15 ° to show how the deflection depends on the aim.

The analogue enables us to study the path of a single ball rolling on the hill. Geiger and Marsden were unable to study the motion of a single alpha particle. Instead they allowed a random hail of alpha particles to fall on the many atoms in a foil and observed how many were detected at various angles to the incident beam (figure F6).

On the hill, a ball turns through a bigger angle if it runs more slowly. Geiger and Marsden set their detector at a fixed position and slowed their alpha particles down by inserting mica sheets in the incident beam. Table F1 summarizes their observations. It is evident that the slower the speed of the alpha particles, the greater is the number scattered through the set angle.

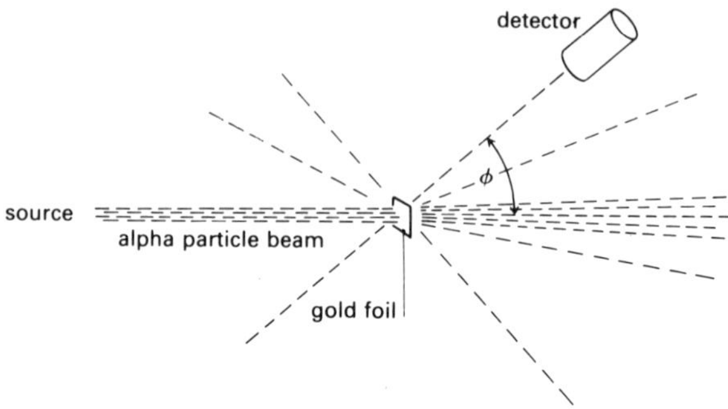

Figure F6
Rough sketch of Geiger's and Marsden's experiment.

Number of mica sheets		Number of alpha particles counted per minute at a particular angle
0		24.7
1		29.0
2		33.4
3		44
4		81
5	speed	101
6	decreasing	255

Table F1
From table VII of GEIGER, H. and MARSDEN, E. 'The laws of deflexion of alpha particles through angles'. Phil. Mag. **25**, 6th series, *1913, pp. 604–623.*

A quantitative test?

It is easy enough with the hill to measure the scattering angle, ϕ, for various values of the aiming error (impact parameter), p. But comparing these results with the results of a scattering experiment with alpha particles is not straightforward. There are two difficulties. With alpha particles one can neither choose nor measure p; and alpha scattering is three-dimensional, whereas with the hill the scattering is restricted to the plane of the table. The steps needed to compare the two sets of results, in order to test whether alpha particle scattering is governed by an inverse-square law, are, briefly:

i the alpha particle results are manipulated so that a graph can be plotted showing how the number scattered at angles greater than ϕ depends on ϕ;

ii measurements of ϕ for various values of p on the hill are used to plot a graph of p^2 against ϕ. If the hill is a good model, these two graphs should have the same shape – but some scaling is necessary before they can be compared directly.

The argument is complex and is presented here chiefly for teachers, though it may be worth pursuing with at least some students, as an example of how a model can be tested. Even without the full argument it is worth demonstrating with the hill that the larger the aiming error, the smaller the scattering angle. The fact that in an alpha particle experiment using foils many atoms thick, only a small fraction of particles are scattered at large angles suggests that very few have small impact parameters: the nucleus is very small.

The relation between the number of particles scattered at more than a certain angle and the aiming error *p*

When the ball is rolled along paths which would pass a distance *p* from the centre of the hill, the smaller *p*, the larger the angle ϕ through which the ball's path is turned.

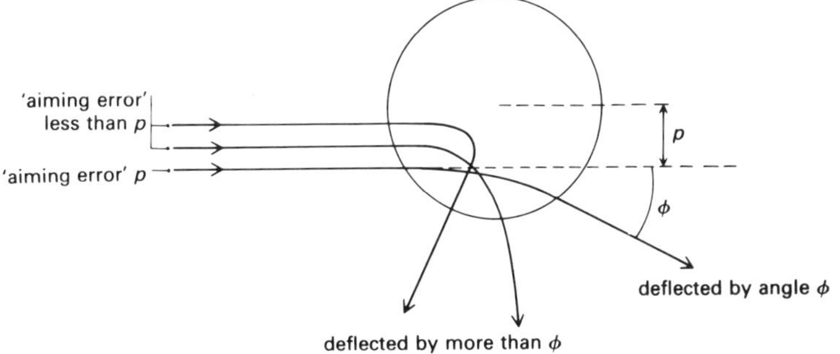

Figure F7

Figure F7 shows the paths of balls on the hill as *p* is varied. If the deflection of the path is ϕ when the 'aiming error' is *p*, an aiming error less than *p* will give a deflection which is greater than ϕ. The path of a ball is essentially two-dimensional, but an alpha particle can go above and below, as well as to the side of a nucleus. Figure F8 illustrates alpha particle paths in three dimensions.

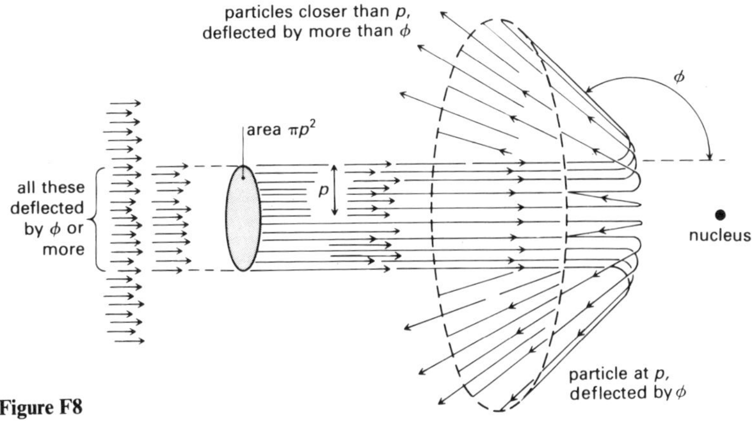

Figure F8

In the three-dimensional situation it is still true that a particle with aiming error less than p will be deflected through an angle greater than ϕ. These particles will be those that fall within an *area* of radius p. (For the hill, those scattered by more than ϕ fall within a *width* $2p$ on either side of the centre.) The area of the circle of radius p is πp^2. If the incoming hail of particles allows equal numbers to fall everywhere, the number passing through area πp^2 will be proportional to p^2.

So the number of particles scattered at more than ϕ is proportional to p^2.

Geiger and Marsden actually measured the number of particles arriving at a particular angle, ϕ. Their results can be treated to find the number scattered by more than ϕ and these are shown in table F2.

Angle ϕ in degrees	**Number of particles detected in fixed time at angle ϕ (Geiger and Marsden)**	**Numbers proportional to number of particles scattered at more than angle ϕ**
180	—	0
165	—	8
150	33.1	32
135	43.0	79
120	51.9	154
105	69.5	266
90	—	448
75	211	767
60	477	1 384
45	1 435	2 811
37.5	3 300	—
30	7 800	7 725
22.5	27 300	—
15	132 000	45 800

Table F2

This table is the end result of fairly complex calculations which involve, among other things, the geometry of the particular experiment. (For details see Appendices A and B of Nuffield Advanced Physics *Teachers' guide Unit 5*, Atomic structure. 1st edn, Longman 1971.)

Values of the 'aiming error' p and the corresponding scattering angle ϕ can be obtained for balls rolling on the hill. If the hill is a good model, these values should be related in the same way as the values of p and ϕ for alpha particles are related – if only these could be obtained. So a graph of p^2 against ϕ for measurements made with the hill should have the same shape as that of a graph of the numbers (given in table F2) of alpha particles scattered at more than ϕ (proportional to p^2) against ϕ.

Summary

We now have the Rutherford model of the atom: an object with a small, massive, positively charged nucleus with electrons somewhere in the space around the nucleus. The whole assembly is electrically neutral.

As so often in physics, we have answered one question, about the structure of the atom, but many more arise immediately. An obvious one concerns the size of the nucleus compared with the atom. This is considered in questions 8 and 9.

Questions

Questions 10 to 13 deal with alpha particle scattering and the inverse-square law.

Film

'The Rutherford model of the atom' is a filmed reconstruction of Geiger and Marsden's experiment using modern apparatus.

Computer

Dynamic modelling system

The system can be used to model the behaviour of an alpha particle near a nucleus. The problem is very similar to the problem of a satellite or planetary orbit (see page 330). This problem needs:

Coulomb's Law Newton's Second Law Kinematics	in two dimensions

A suitable set of equations is:

`R = SQR (X^2+Y^2)`	displacement
`F = K*Q1*Q2*E^2/R^2`	Coulomb's Law
`FX = F*X/R` `FY = F*Y/R`	x and y components of force†
`AX = FX/M` `AY = FY/M`	Newton II

`VX = VX+AX*DT`
`VY = VY+AY*DT`
`X = X+VX*DT`
`Y = Y+VY*DT`
	kinematics in two dimensions

`T = T+DT`	

Note that it is only Coulomb's Law which distinguishes this from a set of equations for a gravitational problem, and so this set could easily be got by editing a set prepared and stored earlier.

Constants are:

E charge on electron (1.6×10^{-19} C)
K Coulomb's Law constant (9×10^{9} N m^2 C^{-2})
M mass of alpha particle (6.6×10^{-27} kg)
Q1 charge on alpha particle (2)
Q2 charge on nucleus (*e.g.*, 79 for gold)

†see footnote on page 331.

Other variables which must be given initial values are

```
X, Y, VX, VY, T, DT
```

Note that a typical alpha particle energy is 5 MeV which corresponds to a speed of $\sqrt{2 \times 5 \times 1.6 \times 10^{-13}\,\text{J}/6.6 \times 10^{-27}\,\text{kg}} = 1.6 \times 10^7\,\text{m s}^{-1}$.

Try VX $= 1.6 \times 10^7$ m s^{-1}; VY $= 0$; X $= -1 \times 10^{-13}$ m; and an initial y-displacement of, say, 0.5×10^{-13} m. A suitable time increment is 10^{-21} s. Explore the effect of varying the aiming error (Y), the alpha particle's energy (and so VX), and the nuclear charge (Q2).

For more information see the booklet *Dynamic modelling system.*

Other programs

Several special purpose programs, including J. Campbell's 'Simulation of alpha scattering', and Software Associates 'ALPHA', plot the trajectory of an alpha particle near a nucleus.

There are others (for example R. Beare's 'Alphafoil') which simulate the Geiger and Marsden experiment. J. Harris's 'Particle scattering' focuses on modelling and contrasts the scattering produced by a hard sphere and by a force which obeys an inverse-square law.

SECTION F2
EXPONENTIAL DECAY

RADIOACTIVE DECAY AND HALF-LIFE

Experiments F6 and F7 enable students to follow the decay of a radioactive substance and to measure conveniently short half-lives. A further experiment, F8 on the throwing of dice illustrates a chance-controlled behaviour which has important similarities with the behaviour observed in a radioactive decay.

Questions 15 to *18* are concerned with decay and half-life.

EXPERIMENT
F6 Decay and recovery of protactinium

ITEM NO.	ITEM
130/1	scaler
130/3	GM tube holder
130/5	thin end-window GM tube
1155	small polythene bottle
1156	uranyl(VI) nitrate
1156	concentrated hydrochloric acid
1156	2-methylbutan-3-one (isopropyl methyl ketone)
503–6	retort stand base, rod, boss, and clamp
507	stopwatch or clock
	tray lined with absorbent paper

Details for students appear in the laboratory notes. These assume that the solutions are provided made up and sealed in the polythene bottle.

Contents of the bottle

The plastic bottle, nearly full, contains equal volumes of 2-methylbutan-3-one and acidified uranyl(VI) nitrate solution. Each layer must be as deep as the width of the GM tube window. Each 10 cm^3 of the acidified solution is made from 1 g of uranyl(VI) nitrate dissolved in 3 cm^3 of water, to which 7 cm^3 of concentrated hydrochloric acid is then added. See *School Science Review*, **45**, 157, pp. 597–605 (1964); Revised Nuffield Physics *Teachers' guide Year 5*, page 143. Chemical eye protection should be used when making up the solution, and the bottle should be labelled TOXIC, CORROSIVE, with the appropriate hazard symbols. The bottle may be a cheap, chain-store article, if the cork washer and plastic cap are protected by a sheet of thin polythene screwed under the cap. This also prevents leakage. A better-quality bottle, with a snap-on polythene cap, will leak less easily, but the walls must be thin or many beta particles will be stopped by them and the count rates will be undesirably low. Most glass bottles are too thick, and polystyrene ones are attacked by the contents.

In this experiment, ^{234}Pa is extracted from a solution containing its parent ^{234}Th and 'grandparent' ^{238}U, with which it is in equilibrium. It is extracted by the 2-methylbutan-3-one and the decay of the ^{234}Pa can be followed. It is also possible to follow the recovery of ^{234}Pa in the aqueous solution from which some was removed.

The decay scheme is:

	^{238}U	$\xrightarrow{\alpha}$	^{234}Th	$\xrightarrow{\beta}$	^{234}Pa	$\xrightarrow{\beta}$	^{234}U
Half-life	4.5×10^9 years		24.1 days	(low-energy)	72 seconds	(2.32 MeV)	2.5×10^5 years

The organic reagent removes about 95 per cent of the protactinium, some uranium, but no thorium from the aqueous solution of the uranium salt. The counter does not detect either the alpha particles from the ^{238}U or the weak beta particles (≈ 0.2 MeV) from ^{234}Th, but records only the energetic beta radiation from ^{234}Pa.

The experiment should be performed within or directly above the tray with its absorbent lining in case of spillage.

It is advisable to check the activity of the solution each year; some teachers report that more consistent results are obtained if fresh solution is used each year.

Before starting, teachers should ensure that students know the significance of background counts.

The bottle is shaken for 10 to 15 seconds and placed beside the GM tube so that the window of the tube is opposite the top half of the bottle (figure F9). As soon as the layers have separated, the scaler is started and a 10-second count taken every half minute.

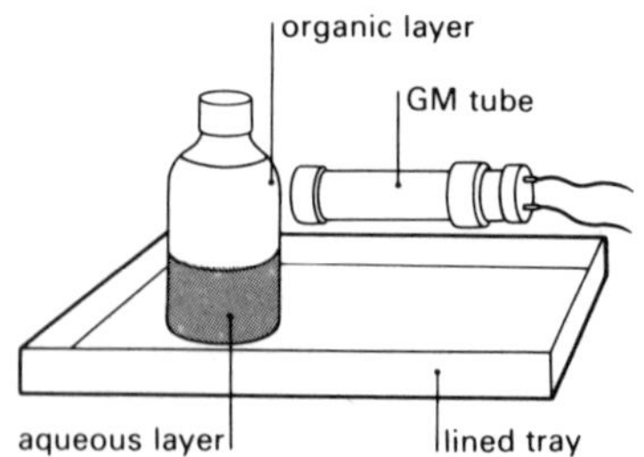

Figure F9
The decay of protactinium.

To observe the recovery, repeat with the GM tube window opposite the lower aqueous layer. To avoid any risk of contaminating the GM tube, it should not be allowed to come into contact with the bottle.

EXPERIMENT

F7 The decay of radon and the determination of its half-life

ITEM NO.	ITEM
1516	picoammeter (see below)
1108	ionization chamber
1066	radon generator
1033	2 cell holders with four cells
1000	leads

Details for students appear in their laboratory notes.

Picoammeter A current of about 10^{-11} A is to be measured in this experiment. As in experiment F2, this can be done by using an electrometer (very high impedance voltmeter) to measure the p.d. across an appropriate high resistance – in this case $10^{11}\ \Omega$.

Some ionization chambers are supplied both with solid lids and with gauze-covered lids: the solid lid is required here.

The ionization chamber has a probe rod running up its centre. This rod is connected to the electrometer input. The 12 V supply is connected between the wall of the ionization chamber and the earthed electrometer input.

For safety, the chamber should be connected to the radon generator by two tubes, so that air circulates between them and neither radon nor powder is puffed into the room when the bottle is squeezed.

Check the generator bottle each year before students handle it as some bottles have been known to split and to spill their contents.

If the bottle does split and powder is spilled, the area should be *gently* ventilated to remove the radon gas. A person wearing disposable plastic gloves, eye protection, laboratory coat, and (if possible) a disposable face mask, can then safely collect the bulk of the powder using two sheets of paper, and transfer it to a bottle for future use. The remaining powder can then be wiped up with damp tissues or filter paper. The contaminated tissues, disposable gloves, and mask should be sealed in a polythene bag and placed with other solid refuse.

Any bottles over ten years old should be *discarded*, and the powder transferred to a new bottle.

The decay scheme is:

$$^{232}\mathrm{Th} \xrightarrow{\alpha} {}^{228}\mathrm{Ra} \xrightarrow{\beta} {}^{228}\mathrm{Ac} \xrightarrow{\beta} {}^{228}\mathrm{Th} \xrightarrow{\alpha} {}^{224}\mathrm{Ra} \xrightarrow{\alpha} {}^{220}\mathrm{Rn} \xrightarrow{\alpha} {}^{216}\mathrm{Po}$$

	^{232}Th	^{228}Ra	^{228}Ac	^{228}Th	^{224}Ra	^{220}Rn	^{216}Po
Half-life	1.4×10^{10} years	6.7 years	1.1 hours	1.91 years	3.6 days	52 seconds	0.16 second

Note that the ^{220}Rn decays into ^{216}Po, which itself decays, also by emitting an alpha particle, with a very short half-life. This and the other decay products do not affect the result appreciably since they have half-lives which are very different from that of ^{220}Rn.

The thorium hydroxide is kept inside a soft polythene bottle fitted with a filter in the neck, or in a dustproof bag within the bottle, and the radon builds up to its equilibrium concentration inside. One gentle squeeze is usually enough to pass sufficient radon into the ionization chamber to give an ionization current of 10^{-11} A. This current diminishes exponentially with time as the radon decays. Meanwhile the concentration is building up again in the bottle, taking several minutes to regain equilibrium.

The experiment tells the same story as experiment F6, though in this case, the recovery of radon is not easy to follow. The experiments together have the value of emphasizing that the exponential decay pattern is common to all radioactive decays, while the rate of decay differs from material to material.

Having completed these experiments, students should know:

i that background counts need to be considered when determining count rates;
ii the general shape of the decay curve and of the growth or recovery curve;
iii how to determine half-life from a graph of activity against time;
iv that the decay of a radioactive isotope is part of a chain of events;
v that, when activity is low, unreliable results may be obtained because of the random nature of radioactivity decay.

Television or video

'The determination of radioactive half-life' (programme 6 in the Granada TV series *Experiment: physics*) could supplement experiments F6 and F7. It concerns the decay of an isotope of indium, and involves students in taking measurements from a filmed experiment.

EXPERIMENT
F8 A radioactive decay analogue

ITEM NO.	ITEM
1155	100 dice
	graph paper
	tray (*optional*)

Details appear in the students' laboratory notes, together with an explanation of why the experiment may be thought worth while.

The dice need to be kept in a box, so that they may be shaken around before being cast out onto the table or tray. Each time those that fall with, say, five pips uppermost are removed, being supposed to have 'decayed' in the 'time interval' represented by one throw of the 'active' dice. Those left are then returned to the box for another throw. Up to ten throws may be needed before the supply of dice is almost exhausted. The 'half-life' of a die is between three and four throws. Graphs of the number of dice remaining and of the number of dice eliminated are plotted against 'time', that is, against the number of throws.

Figure F10 illustrates the results that may be expected, idealized a good deal. Fluctuations are substantial. They can be reduced by repeating the exercise several times and averaging.

Students should be asked to check to see whether the graph is exponential in form, to estimate and to calculate the 'half-life'.

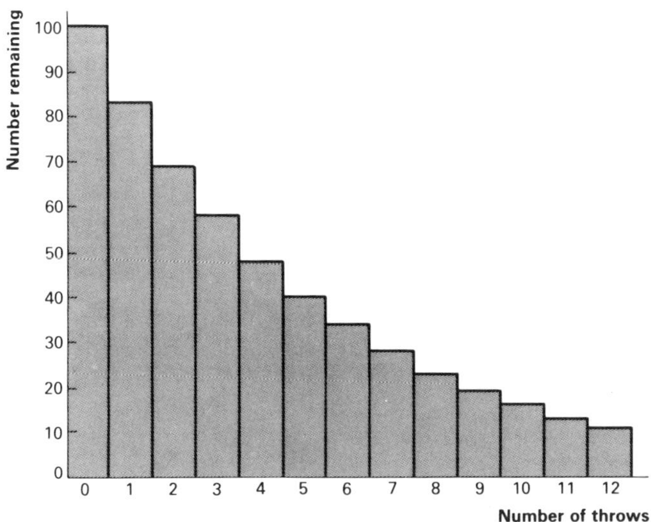

Figure F10
Decay of dice.

Our analogue has the advantage that dice are strongly linked in a student's mind with random events. Indeed, 'like the throw of a die' is what 'random' means to most people. Other analogues, such as those using ball-bearing balls rolling down a slope, give smoother exponential curves, but it is usually less clear how a *constant* chance of decay is built into the analogue.

Devices employing water running out of tubes are *not* recommended because they involve no random element.

Having completed this experiment, the student should know:

i that there are similarities between radioactive decay and the dice-throwing game, suggesting that radioactive decay is also a chance process;

ii that given a constant chance of decay, the variation of decay rate with time is exponential in form; and

iii that, when large numbers are involved, chance events lead to smooth decay rate–time graphs.

Randomness

The concept of randomness occurs again later in the course, especially in Unit K, 'Energy and entropy', and also in Unit L, 'Waves, particles, and atoms'.

Questions

Question 14 is about random events; questions 15 to 18 about radioactive decay, background count, and half-life.

Radioactive decay – a chance process

Experiments F6 and F7 give graphs which do not have their points lying on a near-perfect line. In this they differ markedly from the similar-shaped graph obtained for the discharge of a capacitor. The graph of experiment F8 shows the same variation. That experiment also reveals that the moment in time of the decay of any one die cannot be predicted. It appears that chance is an important factor in this process.

We might ask: 'What is the chance of a six falling uppermost when a six-sided die is thrown?' (1/6). 'Of a four, or a five?' (1/6).
'What is the chance of any one side falling uppermost when a ten-sided die is thrown?' (1/10).
'If a number of dice is thrown and all the sixes removed, what proportion is left?' (5/6). 'What proportion will be left after two such throws?' (5/6 × 5/6).
'How many such throws will halve the number of dice?' 'How many will reduce the number surviving to a quarter?'

Throws	Proportion remaining	Number remaining, starting with 100
1	5/6 = 0.833	83
2	5/6 × 5/6 = 0.694	69
3	5/6 × 5/6 × 5/6 = 0.579	58
4	5/6 × 5/6 × 5/6 × 5/6 = 0.482	48 (just under half)
5	5/6 × 5/6 × 5/6 × 5/6 × 5/6 = 0.402	40
6	0.335	33
7	0.279	28
8	0.232	23

Table F3

The proportion left after two 'half-lives' is *not* zero and this provides an opportunity to to clarify the meaning of the term.

These points may be plotted on to the graph obtained by the students in experiment F8. This will emphasize how the reality (with small numbers) differs from the theoretical.

'If one of the dice were marked and observed during the dice-throwing experiment, when would it decay?' (You can't tell; it could decay at the first or any other throw.) All that can be said is that the proportion that is left after each throw is approximately 5/6 and that the proportion that has decayed is about 1/6. This could be tried experimentally by marking one of the dice if necessary.

The chance of decay

In one minute, about half the ^{220}Rn nuclei decay in experiment F7. In one second, a constant small fraction of them decay (*not* 1/120). The rate of decay is the number of nuclei decaying per second.

For dice, at each throw, a constant small fraction of the total tends to 'decay'. The 'rate of decay' has no meaning for dice, unless one imagines throws to represent time intervals. At one throw of the dice, if there are N dice thrown and ΔN of them 'decay', it is to be expected that

$$\Delta N = -\frac{1}{6}N$$

the negative sign showing that the 'decayed' dice are removed, decreasing N.

Nobody, so far as we know, casts dice to decide the fate of a nucleus. There is no 'succession of throws', at each of which a nucleus may, or may not, decay. But the experiments on radioactive decay do suggest that a constant fraction of the nuclei present decay in any fixed time interval. That suggests writing $\lambda \Delta t$ for the chance that a nucleus will decay in a time interval Δt. The decay constant, λ, is the chance of decay per unit time (unit s^{-1}). Then if there are N atoms at the start of an interval Δt, on average $N\lambda\Delta t$ of them will decay in that interval. $\lambda\Delta t$ replaces the constant 1/6 appropriate to dice. Thus

$$\Delta N = -\lambda \Delta t N$$

or

$$\Delta N/\Delta t = -\lambda N$$

This is like an equation for the discharge of a capacitor (Unit B, 'Currents, circuits, and charge'):

$$\Delta Q/\Delta t = -Q/RC$$

The time interval must be small enough for N (or Q) not to change much within Δt; that is, $\Delta N/N$ ($\Delta Q/Q$) must be small. Even so, the equation does not deserve the full equals sign; $\approx$ might be fairer. (Of course, the decay of charge from a capacitor is not subject to the statistical variations which are so obvious in the case of radioactive decay.)

Like the decay of charge on a capacitor, the equation $\Delta N/\Delta t = -\lambda N$ leads to a dwindling away at a decreasing rate, as a numerical example now shows.

Using the equation $\Delta N/\Delta t = -\lambda N$ to produce a decay curve

The numerical example which follows takes a straightforward case of decay and produces graphs of the number of atoms remaining against time, and of the activity (the number of atoms decaying per second) against time. This latter graph is important because most measurements taken from radioactive materials are in terms of activity rather than numbers of atoms remaining.

Students may be given the initial data, shown how the table of results might be arranged, and how the first lines are worked out. They should then be able to complete the table for themselves. This might provide an opportunity to discuss the use of calculators in this and other work.

Let there be $N = 10\,000$ nuclei as yet unchanged into another type of nucleus and let $\lambda = 0.1\ \mathrm{s}^{-1}$. If $\Delta t = 2$ s, the number decaying in the first 2 s interval is 2000, leaving 8000 unchanged. In the next time interval of 2 s the number decaying is smaller, but it is the same fraction of the remaining 8000; that is, 1600 decay. Table F4 shows the example worked out for several successive intervals of 2 s.

Time in seconds	Number at beginning	Number decaying in 2 seconds	Number left after interval	Rate of decay $\Delta N/\Delta t$ per second
0	10000	2000	8000	1000
2	8000	1600	6400	800
4	6400	1280	5120	640
6	5120	1024	4096	512
8	4096	819	3277	410
10	3277	655	2622	328
12	2622	524	2098	262
14	2098	420	1678	210
16	1678	335	1343	168
18	1343	269	1074	134
20	1074	215	859	108
22	859	172	687	86
24	687	137	550	69

Table F4
Numerical example of decay.

It would be better to use a smaller time interval than 2 s, but the calculation would become tedious. Note that random fluctuations are ignored, though one could expect the 2000 decays in the first interval to fluctuate by perhaps 50.

The two graphs (of number of atoms remaining against time, figure F11; and of activity against time, figure F12) have a very similar shape. Indeed, the values of $\Delta N/\Delta t$ are each 0.1 of the corresponding values of N, and if the scale for $\Delta N/\Delta t$ is ten times larger than that for N, the graphs are identical. But, *of course* they are the same; that is what the equation says:

$$\Delta N/\Delta t = -0.1N$$

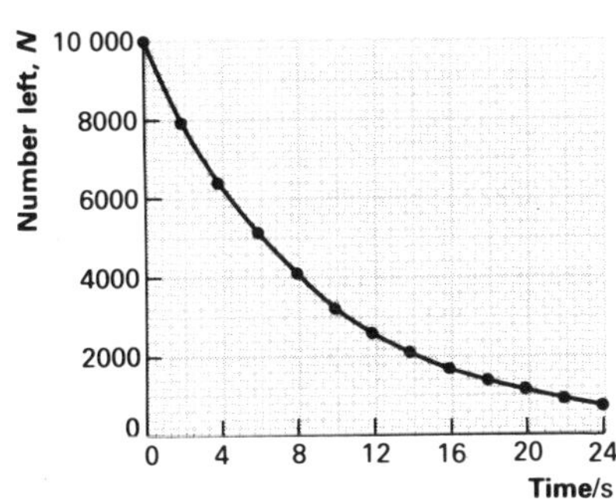

Figure F11
N against t for calculated decay of 10 000 atoms.

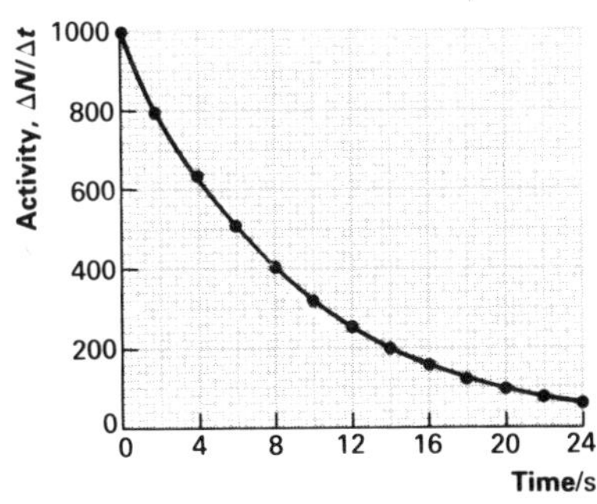

Figure F12
$\Delta N/\Delta t$ against t for a calculated decay of 10 000 atoms.

Subsequent discussion might consider whether or not a real decay curve has this shape.

Two properties of the curve can be checked:

i constant half-life (by testing two or three portions of the curve), and
ii constant ratio (by testing whether the activity falls by the same ratio in equal intervals of time).

For example, if we take a time interval of 4 s:

the activity at 0 s = 1000
ratio 1000/640 = 1.6
at 4 s = 640
ratio 640/410 = 1.6
at 8 s = 410
ratio 410/260 = 1.6
at 12 s = 260

These tests should be applied to the graphs obtained in experiments F6 and F7. Alternatively, the results from those experiments could be scaled to those of the activity curve in the example. If the curves fit they have the same shape and the same properties.

Students should also be encouraged to plot a graph of $\ln N$ against t. This is a straight line with a positive slope for exponential growth and a negative slope for exponential decay.

Use of calculators

In the example discussed above the choice of $\lambda = 0.1$ leads to very simple arithmetic. It may be worth asking students to use their calculators to show that the general form of the curves is the same for any value of λ, and that the larger λ is, the more steeply the curves slope.

Dynamic modelling system

The dynamic modelling system can of course be used to plot or tabulate solutions to $\Delta N/\Delta t = -\lambda N$, and makes it easy to explore the effect of varying λ, Δt, and the initial value of N. A suitable set of equations is:

```
DN = -L*N*DT
N = N+DN
T = T+DT
```

Initial values must be provided for N, L, T, and DT.

These equations express the fact that in each time interval (DT) a constant fraction of all the nuclei present will decay.

An interested student might be asked to write a set of equations expressing the fact that the chance of any *individual nucleus* decaying in one time interval is constant (*i.e.*, to write equations that model the dice analogue of radioactive decay). A possible set of equations is:

```
RANDOMIZE
D = 0
FOR I = 1 TO N
IF RND(1)<L THEN D = D+1
NEXT I
N = N-D
T = T+1
```

The variables N (number of nuclei), L (chance of decay), and T (time) must be given initial values.

It is easy to show how the decay curve becomes smoother if a higher initial value for N is used. (But note that the program gets quite slow for N > 1000, because so many random numbers have to be generated and tested.)

For more information see the booklet *Dynamic modelling system.*

The activity of radioactive sources

The activity of a radioactive source is quoted in *becquerel*, where 1 becquerel (Bq) is an activity of one disintegration per second. The earlier unit known as the curie (Ci) was the activity of one gram of radium (3.7×10^{10} disintegrations per second). Sources in use in schools may well have their activities given in μCi and will require re-labelling. The activity of a 5 μCi source is 185×10^3 Bq.

Questions

Questions 19 to 23 involve testing for exponential growth and decay in a variety of situations.

Summary

The students should now know that:

1 Radioactive decay is a chance process.

2 The chance of decay is constant with time.

3 The half-life of a radioactive isotope is the time for half of the nuclei present to decay.

4 The properties of random decay are best displayed if large numbers of events are involved.

5 The chance of a nucleus decaying in 1 second is known as the decay constant (λ). This is independent of temperature, pressure, and other physical conditions.

6 The equation $\Delta N/\Delta t = -\lambda N$ applies to the process of random decay.

7 The experimental data for radioactive decay and the graphs derived from them display the constant ratio property and give a value for half-life. This provides a way of recognizing this type of decay.

8 The unit in which activity is measured is the becquerel (Bq).

THE EXPONENTIAL FUNCTION

The form of the curve for the decay of radioactive nuclei or for the decay of charge on a capacitor shows that the changes are exponential; that is, that the rate of change of something is proportional to how much of that something there presently happens to be. In the two cases referred to, the change was a decay. But the description can apply to such other changes as the growth of bacteria or the spread of an epidemic. These concern growth, not decay, and it is growth we are going to look at in our search for a mathematical equation for exponential change.

The problem is to find a solution for the differential equation used in the numerical example. This yielded a graph of the equation $\Delta N/\Delta t = -0.1N$. 'What would the graph look like if the constant were to be changed from -0.1 to -1.0?' (The general shape would be the same, but the slope would be steeper.)

'What would a graph of $\Delta N/\Delta t = +0.1N$ look like?' (It would grow and grow rather than dwindle away. The bigger N became, the faster it would grow.)

To find a mathematical solution it is best to keep things as simple as possible. This is why we have changed the constant to $+1.0$. We are therefore looking for a solution of the equation $\Delta N/\Delta t = +1.0N$.

Mathematics – the exponential

This is one of the occasions on which we suggest that mathematical ideas can and should be taught within physics teaching. There are several things to be learned here. The suggested approach will, we hope, reveal more of what kind of thing a differential equation is, and what is involved in finding solutions of such equations. This is worth while because of the great importance of differential equations in all sorts of applied and pure science. In the process, the idea of a derivative should also become clearer.

Further, the function studied here – the exponential – is of wide interest and importance. Indeed, if one allows complex exponents, a large fraction of the functions of interest in science can be cast into exponential form. We do not suggest going so far, but the understanding of the exponential function with real exponents may well help later learning in this area.

Within the course, the exponential function has a significant role. It may already have been mentioned in connection with capacitor discharge. There are more examples of exponential growth in Unit G, 'Energy sources'. The exponential also appears in the Boltzmann factor $e^{-E/kT}$ in Unit K, 'Energy and entropy'.

We suggest a graph-drawing approach which is supported by a series of learning questions (24 to 27) in the *Students' guide*. Although not the only possible approach, this provides a detailed numerical look at 'how the thing goes'. It will enable students to work by themselves with little assistance. It may even be that no direct teaching, apart from a concluding summary, will be needed. What follows here is a summary of the development in the *Students' guide*, 'About exponential changes' (pages 394 to 399).

The graphical approach to the exponential function

The approach is in two steps. First the form $N = a^t$ is seen to have the necessary property of increasing by equal multiples when t increases in equal steps. This can be brought out with a numerical example or, more formally, in the following way.

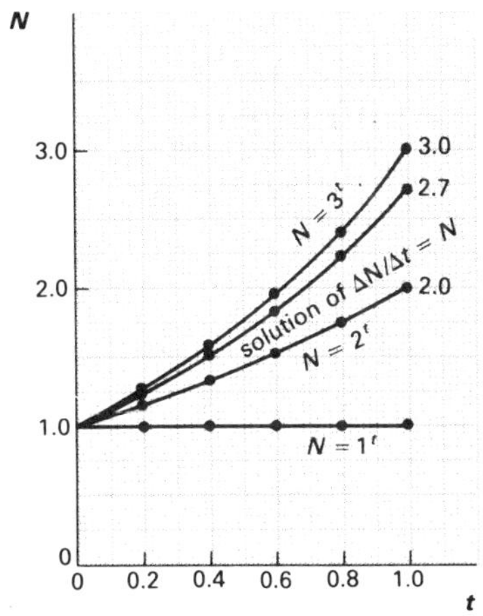

Figure F13
Plots of $N = 1^t$, $N = 2^t$, $N = 3^t$, and solution of $\Delta N/\Delta t = (1.0)N$.

It is convenient to plot graphs of $N = 1^t$, $N = 2^t$, and $N = 3^t$ with values of t from 0 to 1, in 0.2 steps, using either the y^x key on a calculator or tables of logarithms (see figure F13). The graphs can be inspected for a constant-ratio property – do, say, 2^0, $2^{0.2}$, $2^{0.4}$, $2^{0.6}$, etc., increase in constant ratio?

The link with the addition of exponents upon multiplication can be brought out: what, for example, is the value of $2^{3.2}$ and of $2^3 \times 2^{0.2}$?

The values can be found either from the graph or by using the y^x key of a calculator.

More formally, the exponents of the series a^t, a^{2t}, a^{3t}, etc., increase in equal steps t. But $a^{2t} = a^t \times a^t$; $a^{3t} = a^t \times a^t \times a^t = a^{2t} \times a^t$. Each term is larger than the one before by the factor a^t.

Construction of a solution to $\Delta N/\Delta t = 1.0N$

On the same axes as the graphs above, a step-by-step graphical solution to the equation $\Delta N/\Delta t = 1.0N$ can be constructed. Taking time intervals Δt of 0.1, and starting with $N = 1.0$, the solution is constructed by drawing segments of the graph one after the other as short, straight lines with slopes given by the equation. Figure F14 illustrates the process. The technique is just like that used in Unit B, 'Currents, circuits, and charge', and in Unit D, 'Oscillations and waves' for the discharge of a capacitor and for plotting displacement as a function of time.

In the simplest method, shown in figure F14, the slope of the next segment is decided by the previous value of N. This value is always too low, so the curve is not so steep as it ought to be. The proper value of N to use would be that in the middle of the segment to be drawn, but that value is as yet unknown. This simple method gives a graph which, if carefully drawn, will rise to $N = 2.59$ instead of $N = e = 2.718 \ldots$ at $t = 1$.

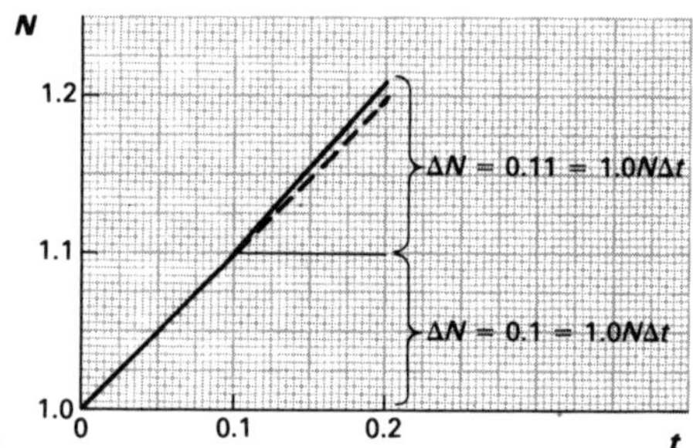

Figure F14
First two steps of a graphical solution of $\Delta N/\Delta t = (1.0)N$.

We see then, that at $t = 1.0$, the solution rises to a value of nearly 2.7, where $N = 2^t$ has risen to the value of 2.0 and $N = 3^t$ to the value 3.0. It may now seem reasonable to propose the form $N = 2.7^t$ for the graphical solution, and write this as $N = e^t$, where e is a number close to 2.7, that can be found more accurately by more accurate versions of the graphical solution process, or its equivalent.

To show that $N = e^{\lambda t}$ may be the form of a solution of $\Delta N/\Delta t = \lambda N$, a graphical solution of $\Delta N/\Delta t = 0.7N$ might be constructed. It is close to the curve $N=2^t$. Calculators, or the first graphical solution, show that $e^{0.7}$ is nearly 2.0. Thus when $\lambda=0.7$, the solution seems to be $N=(e^{\lambda})^t$, or $N=e^{\lambda t}$.

Summary

The growth equation $\Delta N/\Delta t=\lambda N$ has a solution of the form $N=a^t$, where a is some number. This form has the required property of yielding a 'constant-ratio' curve; one which increases N by a fixed factor when t increases in equal steps.

When $\lambda=1.0$ and $N=1.0$ at $t=0$, the value of a seems to be near 2.7, more accurately 2.718..., where this is the never-ending number e. This number e is the value of a for which the slope of $N=a^t$ is equal to 1.0 at $t=0$, $N=1.0$.

The graphical solution provides a rough estimate of e. More accurate methods are available, but all amount to doing arithmetic to find the value of N at $t=1.0$.

It might seem that to cope with $\Delta N/\Delta t=\lambda N$, one would need a table of values of the number a in $N=a^t$ appropriate to each value of λ. This is not so. The required value of a is just e^{λ}, and the solution (for $N=1.0$ at $t=0$) is $N=e^{\lambda t}$. A negative sign takes care of the case of decay. If $\Delta N/\Delta t=-\lambda N$, the solution is $N=e^{-\lambda t}$. This curve falls and falls, being the same as $N=1/e^{\lambda t}$.

In general, N is not 1.0 at $t=0$. This turns out to be easy to cope with. If $N=N_0$ at $t=1$, then N/N_0 is 1.0 at that time. If $\Delta N/\Delta t=-\lambda N$, then also $\Delta(N/N_0)/\Delta t=-\lambda(N/N_0)$. Here N has been scaled down by the factor N_0. So $N/N_0=e^{-\lambda t}$ is the solution. This may also be written $N=N_0\,e^{-\lambda t}$.

Testing curves for the exponential property

The arguments suggested above have stressed the 'equal-ratio' property that distinguishes exponential changes from others. The possible use of logarithms in the plotting of $N=a^t$ should suggest testing for exponential change of a quantity by plotting its logarithm against time (or whatever other variable is concerned). Students should try one or two such tests on data provided in, for example, questions 21 to 23. There will be further opportunity in Unit G, 'Energy sources'.

HALF-LIFE AND THE DECAY CONSTANT

The half-life of a radioactive isotope and the decay constant (λ) for that isotope are simply related.

When $N=\frac{1}{2}N_0$, $t=t_{\frac{1}{2}}$ (the half-life).

We may write $\frac{1}{2}N_0=N_0e^{-\lambda t_{\frac{1}{2}}}$

or $2=e^{\lambda t_{\frac{1}{2}}}$

Whence $\ln 2=\lambda t_{\frac{1}{2}}$

So that $t_{\frac{1}{2}}=\ln 2/\lambda=0.693/\lambda$

Question 28 uses the concepts of decay constant and half-life.

RADIOACTIVE DECAY AND RECOVERY

Figure F15 shows the familiar decay curve for a radioactive substance in which the number of atoms is plotted against time. At $t=0$, the number of atoms of isotope X is N_0. As X decays, the daughter isotope Y grows. 'If Y is stable (*i.e.*, does not decay) what will the growth curve look like?' (The broken line in figure F15.) 'Suppose there were no atoms of Y to start with. How many will there be after a very long time?' (Nearly N_0). 'How fast will the number of atoms of Y rise to begin with?' (As fast as the number of atoms of X falls.)

If the daughter Y is stable, the number of atoms of Y present is equal to the number of atoms of X which have decayed. This is (N_0-N). Since

$$N=N_0e^{-\lambda t}, \qquad (N_0-N)=N_0-N_0e^{-\lambda t}$$
$$=N_0(1-e^{-\lambda t})$$

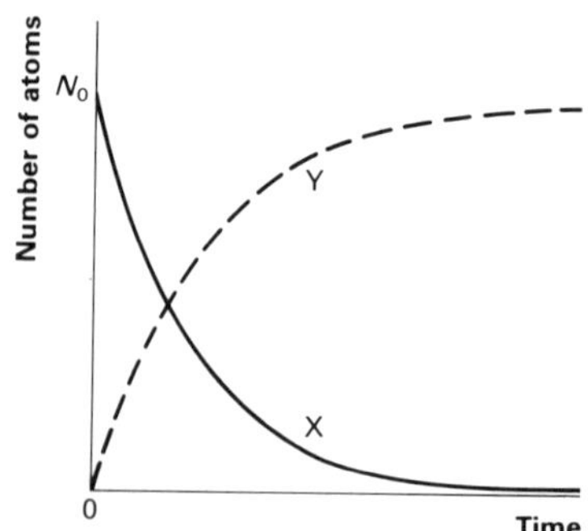

Figure F15

Now suppose that Y is also radioactive. Students should be able to explain why the number of atoms of Y falls exponentially (broken line in figure F16) beyond t_x where there are hardly any atoms of X left. (Almost no more Y atoms are being produced.) Initially, the number of atoms of Y rises almost as fast the number of X atoms falls, since there are very few Y atoms yet, and the number decaying is small, if the half-life of Y is not too short.

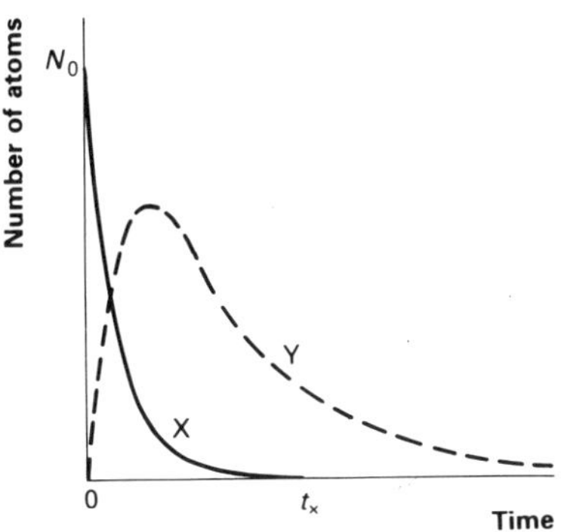

Figure F16
The parent only is present when $t = 0$. The daughter Y decays with a half-life five times that of the parent X.

In experiment F6 the decay of ^{234}Pa is observed in the organic layer, which does not contain any of protactinium's parent ^{234}Th. In the aqueous layer there will be virtually no protactinium immediately after shaking the bottle and allowing the two layers to separate. The decay of ^{234}Th in the aqueous layer produces protactinium. Because of its relatively long half-life (about three weeks) the amount of parent ^{234}Th present will not change appreciably during the course of the experiment; neither, therefore will the rate of production of the daughter ^{234}Pa. But protactinium's half-life is much shorter, and the amount present grows only until as many ^{234}Pa nuclei decay in a second as are formed in a second. Thus the recovery curve for ^{234}Pa in the aqueous layer is like the dashed curve in figure F15 – though for different reasons. After a few minutes the amount of ^{234}Pa present does not change appreciably. (It can be shown that the growth curve has the same 'half-life' as the decay curve for ^{234}Pa.)

Note also that the ^{234}Th comes from ^{238}U, with a half-life of over 10^9 years. The sample is more than a few weeks old, so the amount of ^{234}Th has also stabilized. The number of protactinium decays recorded in a second, from the solution containing also ^{238}U and ^{234}Th, when this is steady, is equal to the number of uranium decays in a second.

Dynamic modelling system

The use of the dynamic modelling system suggested on page 378 to show the decay of radioactive nuclei could easily be extended to show the growth of daughter nuclei, and, if they are radioactive, their decay as well.

Summary

Students should now know that:

1 The growth curve of a daughter product which does not itself decay is given by $N = N_0(1 - e^{-\lambda t})$.

2 If the daughter product does decay, the shape of the growth–decay curve will depend on the relative half-lives of the daughter and parent nuclei.

3 After the passage of a sufficient period of time, dependent upon the half-lives of the elements in the series, the series will attain equilibrium.

SECTION F3
THE NUCLEUS

This Section raises a series of questions, and provides partial answers. The aim is to give students some understanding of the nucleus and sufficient knowledge to follow the discussion of nuclear power and reactors in Unit G, 'Energy sources'. Teachers may be tempted to provide far more than is suggested here, but this should not be done at the expense of other parts of the course. On the other hand, students may wish to follow up some of the unanswered questions themselves.

Links with chemistry teaching about models of atoms

Some of the content of this Section will also be taught in chemistry (for example Revised Nuffield Advanced Chemistry, Topic 4). This is a good place to bring out the overlap in interests of the two subjects, and perhaps also their differing uses of the same ideas. Chemists will be more inclined to use the ideas to cope with a wide variety of problems, such as those of bonding. This may be reflected, in teaching, in a tendency of chemists to present the ideas as briefly as possible, so as to have time to go on and use them. Physicists may be more interested in using the ideas to probe further into the nature of atoms.

Neither party ought to cast doubt on the value of the other's activities. But chemists should be aware of the danger of accepting as a 'fact' some model or picture of an atom which is in need of further critical examination and development. And physicists should beware of having too limited a perspective; of failing to see the broad relevance of an atomic model for other people's problems.

These considerations suggest that this is a very good stage for a little examination, amongst students who are studying both physics and chemistry, of questions about which of the things being said are facts, which are fair inferences from some model, and which go beyond what a particular model or evidence can support.

The Rutherford model – questions raised

Once Rutherford's model had been established it provided answers to some of the questions which could not be answered by previous models. However, it did not provide all of the answers and it raised some more questions. The evidence collected is in line with the atom having a small, massive, charged nucleus. But at this point it offers no indication of how the nucleus is made or if it has constituent parts. And it offers no explanation as to the place of electrons in the atom, other than to suggest that they exist in the space surrounding the nucleus.

To try to answer some of these questions it is necessary to consider the elements. The fact that each element is different suggests that the atoms of any one element are all the same, but different in some way from the atoms of every other element. If an atom is made up of a nucleus plus electrons, perhaps the differences between atoms can be understood in terms of how they are built up from these parts.

Students should be familiar with the chemical concept of periodicity and the Periodic Table – whether or not they are taking chemistry at A-level. A serial number, Z, can be assigned to each element according to its position in the Periodic Table. Because, at this stage, we have no evidence about the significance of Z – other than that it is a serial number – it is convenient to use its old name 'atomic number' for the time being. While Z goes up regularly in steps of one unit, the mass number A rises a little irregularly, but roughly in steps of two. This suggests that there is more to the nucleus than might be apparent at first sight. As the nucleus is charged, does the charge rise with Z or with A? Are the chemical properties dependent upon the charge?

The charge on the nucleus from the scattering experiment

The alpha scattering experiment can be used to measure how much charge the nucleus has, although the experiment is difficult to do, and the calculation is not easy. The bigger the charge on the nucleus, the larger the angle through which it turns an alpha particle coming at it with a certain 'aiming error' (impact parameter). For an impact parameter p, the angle of deflection ϕ is given by $\cot \phi/2 = 2p/b$, where b depends on the charge on the nucleus and the alpha particle's kinetic energy (see, for example, Nuffield Advanced Physics, *Teachers' guide Unit 5*, Atomic Structure. 1st edn. Longman, 1971. Appendix A). Of course p cannot be controlled in an experiment with alpha particles, and theoretical analysis leads to a more complicated formula for the fractional number scattered at any angle when alpha particles are fired at a metal foil containing many nuclei. By counting just how many particles actually are swung through a certain angle (figure F17), it is possible to tell roughly how many electron charges the nucleus has.

The result of the theoretical analysis is that the fractional number detected at an angle ϕ is given by the equation

$$\frac{q^2 Q^2 nt \text{ (detector area)}}{(4\pi\varepsilon_0 \times \frac{1}{2}mu^2)^2 \;\; 16R^2 \sin^4(\phi/2)}$$

where

q = charge on alpha particle
Q = charge on nucleus
n = number of nuclei per unit volume
t = thickness of the foil
m = mass of alpha particle
u = initial velocity of alpha particle
R = distance from foil to the detector

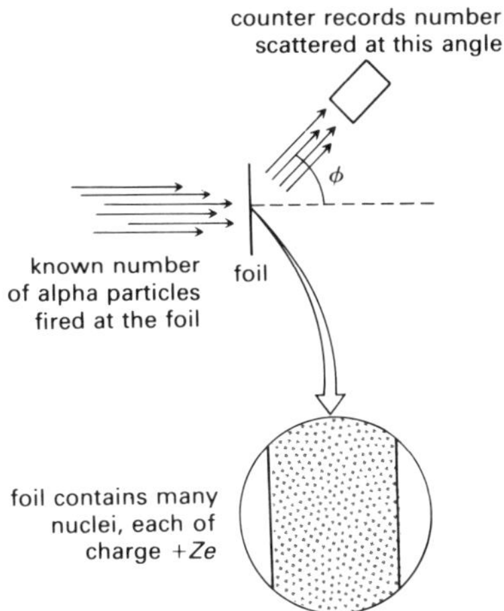

Figure F17
Measurement of Z from alpha particle scattering.

Students may be shown this result, though there is no intention that they should either know the formula or follow the steps in the derivation (which is given in full in Appendix A of Nuffield Advanced Physics, *Teachers' guide Unit 5*). Figure F18 is an attempt to show what went into the argument. It may be possible in this way to show students that such a derivation is possible and the steps that go into it, without laying out the argument in detail. It may help to emphasize that a gulf sometimes has to be bridged before experimental evidence can be interpreted in terms of a physical model.

The equation indicates that the number of alpha particles detected depends on the square of the charge on the nucleus. Thus the number of electronic charges in a nucleus can be determined. The results obtained are shown in table F5. The first result for gold is that from Geiger and Marsden (1913); the other results are from Chadwick (1920).

Element	Mass number	Atomic number	Z from scattering
gold	197	79	99 ± 20
gold	197	79	77.4 ± 1.0
silver	108	47	46.3 ± 0.7
copper	63.5	29	29.3 ± 0.5

Table F5
Mass number, atomic number, and experimentally determined value of Z.

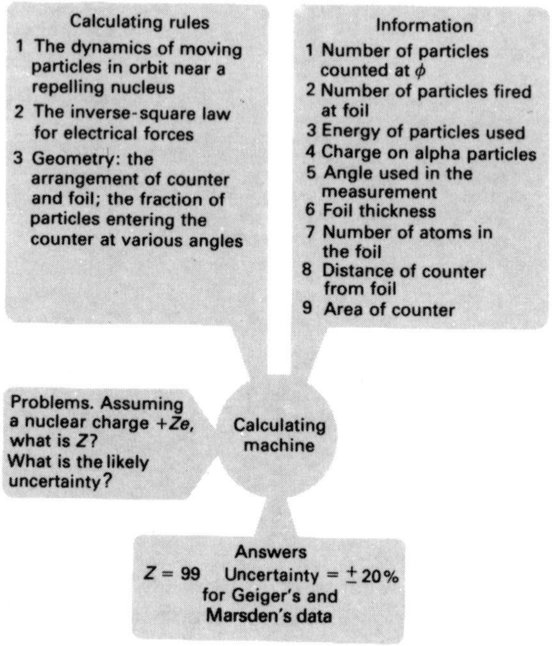

Calculating rules
1 The dynamics of moving particles in orbit near a repelling nucleus
2 The inverse-square law for electrical forces
3 Geometry: the arrangement of counter and foil; the fraction of particles entering the counter at various angles

Information
1 Number of particles counted at ϕ
2 Number of particles fired at foil
3 Energy of particles used
4 Charge on alpha particles
5 Angle used in the measurement
6 Foil thickness
7 Number of atoms in the foil
8 Distance of counter from foil
9 Area of counter

Problems. Assuming a nuclear charge $+Ze$, what is Z? What is the likely uncertainty?

Calculating machine

Answers
$Z = 99$ Uncertainty $= \pm 20\%$ for Geiger's and Marsden's data

Figure F18

These results show that the charge on the nucleus is equal to the Periodic Table serial number, *i.e.*, the atomic number, Z, and not the mass number, A.

The model has been used here to answer an important question; but in so doing many other questions have been raised. The model may need modification in the light of the experimental evidence which has been generated and the new questions which now need answering.

Background for teachers: evidence for nuclear charge as the periodic serial number

There are several pieces of evidence which are all too difficult to present in detail to students at this stage.

1 The absolute and relative amounts of alpha scattering. The theory shows that the scattering depends on Z^2, and Geiger and Marsden compared the amount of scattering from different elements. See pages 174–7 of WRIGHT *Classical scientific papers – physics*, or page 160, CONN and TURNER, *The evolution of the nuclear atom.* Because mass number A is nearly equal to twice the atomic number Z, this test does not show whether the charge is equal to Z or to A, but only that it varies in proportion to either. But the absolute proportion of particles scattered at some angle does give a measure of Z. See *Classical scientific papers – physics*, p. 180, or *The evolution of the nuclear atom*, p. 162.

Some years later Chadwick made more precise measurements than Geiger and Marsden (see table F5).

2 Evidence from X-ray wavelengths. Moseley, in two papers reprinted in *Classical scientific papers – physics*, reports the remarkable work in which he measured X-ray wavelengths for many elements. Even without a theory it can be seen that the measurements support Z rather than A as a fundamental measure of the number of charged building-bricks in the atom. In 1913, of course, Bohr's theory had recently been published, and so Moseley was able to use it to suggest a plot of $\sqrt{\text{frequency}}$ against atomic number, and to obtain straight-line graphs.

3 Amount of X-ray scattering. Barkla showed that the amount of scattering of X-rays, which was thought to depend on how many electrons the atoms had, was consistent with the number of electrons per atom being about half the mass number, A. If atoms are neutral, this also supports a nuclear charge of Z rather than A.

4 The wavelengths of X-ray absorption edges. When X-rays of varying wavelengths pass through materials, each element has a sharp increase of absorption at a characteristic wavelength, when inner electrons absorb energy. In effect, this evidence bears the same relation to Moseley's as the absorption spectrum of, say, sodium bears to its emission spectrum.

Some unanswered questions

The picture that has emerged of an atom containing a small, massive, charged nucleus raises several questions:

1 How is a nucleus of mass number A and charge Z made up?
2 What happens to a nucleus that decays by emitting alpha, beta, or gamma rays?
3 What can be said about the number, energy, and arrangement of the electrons in an atom?
4 Are there electrons in the nucleus?

How is a nucleus of mass number *A* and charge *Z* made up?

The questions which now arise are:

i Which part of the nucleus carries the positive charge?
ii What is the rest of the nucleus made from?
iii What keeps the particles within the nucleus confined in such a small space?

The following are suggestions as to the way in which the nucleus might be constructed:

a A protons, $A - Z$ electrons
b Z protons, $A - Z$ something else.

'Do these agree with the known facts?' (Consider both charge and mass.)

The possibility of a neutral particle – perhaps a proton–electron combination – was suggested by Rutherford. 'Why would it be hard to detect?' (No charge, therefore little ionization). From 1914 to 1932, some people thought that there might be such neutrons, but no one observed them. Then Chadwick, who had spent much time looking for neutrons, found them. It was known that when alpha particles bombarded beryllium, a very penetrating radiation was emitted – figure F19(a). Chadwick found that if this radiation fell on material which contains plenty of hydrogen (for example wax), fast protons (hydrogen nuclei) emerged – figure F19(b).

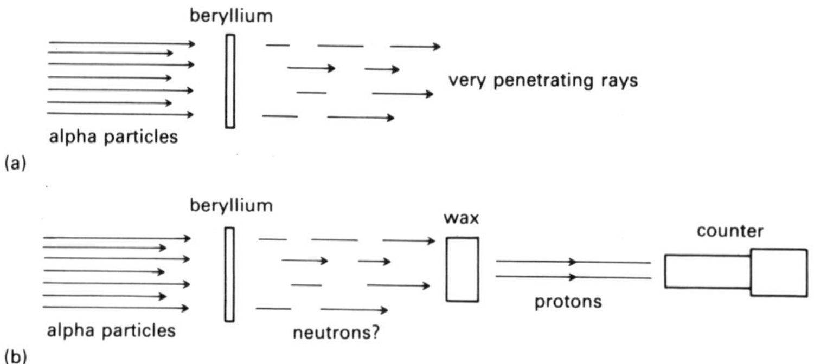

Figure F19
Production and identification of neutrons.

From the speed of the protons, Chadwick used considerations of dynamics to obtain the mass that neutrons would have to have, if they were the penetrating radiation. He showed that the mass of a neutron would be about that of a proton. Other arguments eliminated the other obvious possibility – that the radiation was gamma rays. (A reference back to the discussion of momentum and energy transfer in collisions in Unit A, 'Materials and mechanics', would be appropriate here.)

Thus the neutron has charge zero and mass approximately the same as a proton. This mass is approximately 1800 times the mass of an electron. The discovery of the neutron adds much in favour of a model of the nucleus made up of Z protons and $(A - Z)$ neutrons. And it is now appropriate to use the modern terms for Z and A: proton number and nucleon number.

Radioisotopes and their uses

Since its discovery in 1932 the neutron has played an increasingly important role in pure and applied science. An important example is the use of neutrons from reactors (or 'piles') to produce radioactive isotopes which are used in a wide variety of fields. Many examples are given in the article 'Radioisotopes' in the Reader *Particles, imaging, and nuclei.*

Students are expected to be familiar with the conventional representation, for example ${}^{12}_{6}C$, ${}^{14}_{6}C$.

Questions

Question 29 introduces the unified atomic mass unit u.
Questions 30 to 33 are about isotopes and their uses.

What happens to a nucleus that decays by emitting alpha, beta, or gamma rays?

In this course we consider only alpha decay, beta decay, and gamma emission. Students should know the changes in proton number and nucleon number involved in alpha and beta decay:

$${}^{A}_{Z}\mathrm{X} \rightarrow {}^{A-4}_{Z-2}\mathrm{Y} + {}^{4}_{2}\alpha$$

$${}^{A}_{Z}\mathrm{X} \rightarrow {}_{Z+1}{}^{A}\mathrm{Y} + {}_{-1}^{0}\beta$$

They should also know that gamma radiation is not particulate, but is the very short-wavelength electromagnetic radiation emitted when a nucleus adjusts its energy levels. It may help to compare this radiation with the light emitted when an atom adjusts its electronic energy levels. There is no change in Z or A when a gamma ray is emitted.

Optical spectroscopy gives information about atomic energy levels, as will be seen in Unit L, 'Waves, particles, and atoms'. Similarly, nuclear spectroscopy gives information about nuclear energy levels. But these are not part of the course.

The nature of gamma emission may be easier to explain after the discussion of ionization energy and nuclear binding energy later in this Section.

Questions

Questions 39 to 41 are about radioactive transformations.

What can be said about the number, energy, and arrangement of electrons in an atom?

Our picture of the atom is now one with a nucleus of Z protons and $A-Z$ neutrons, with Z electrons outside the nucleus. So far, little information about the electrons has been gathered and the high-energy alpha particles do not seem to have told us much about them.

Information about the nucleus came from bombarding it with an alpha particle. The mass of the alpha particle was comparable with the mass of the target nucleus, and this suggests that to gain information about electrons they should be bombarded by other particles of comparable mass, that is, other electrons.

The energies of the bombarding electrons need only be in the region of 10–20 eV and this represents a considerable difference from the MeV alpha particles used previously.

Are there electrons in the nucleus?

In Unit L, 'Waves, particles, and atoms', it will be shown that free electrons cannot exist inside a nucleus. To cram them down into so small a space requires so short a de Broglie wavelength, and, from $\lambda = h/mv$, so large a momentum, that the electrons would fly out again at once. At this stage, it may be best to say that theory gives reasons for rejecting a protons-plus-free-electrons picture of the nucleus, and that a sketch of the theory will come later.

What happens when atoms are bombarded by low-energy electrons?

Students should know that the radioactive sources they have been using cause ionization, and that this is how the radiation is detected (*e.g.* in cloud chambers and in Geiger–Müller tubes). They should also know that air can be ionized by bombardment with much lower-energy electrons; for example, from a hot wire, or by ions from a flame. See, for example, Revised Nuffield Physics *Teachers' guide Year 5*, Demonstration 80. A coil of wire (say a tight coil made from about 0.3 m of 0.4 mm nichrome) heated by passing a current (say from the l.t. variable voltage supply) will also discharge the electroscope.

The energy required to remove an electron from an atom can be determined by controlling the energy of the electrons bombarding the atoms of a gas at a low pressure; the p.d. accelerating the electrons being increased until their energy is just sufficient to cause ionization.

Such experiments can be done in specially constructed apparatus, or in gas-filled 'valves' (thyratrons). The latter are becoming difficult to obtain: they are obsolete in electronics, having been replaced by solid-state devices.

OPTIONAL DEMONSTRATION

F9 Ionization by electron collision

a Using thyratrons

i Xenon
ii Argon

ITEM NO.	ITEM
1049	thyratrons and thyratron base
59	l.t. variable voltage supply
1064	low voltage smoothing unit
	either
1005	multirange meter, 25 V range
	or
1507	voltmeter, 100 V
27	transformer
1040	clip component holder
1000	leads
	i xenon thyratron EN91 or 2021
1507	milliammeter, 10 mA
	either
1151	protective resistor, 1000 Ω
	or
1017	resistance substitution box
	ii argon thyratron 884
1507	milliammeter, 100 mA
1151	protective resistor, 100 Ω

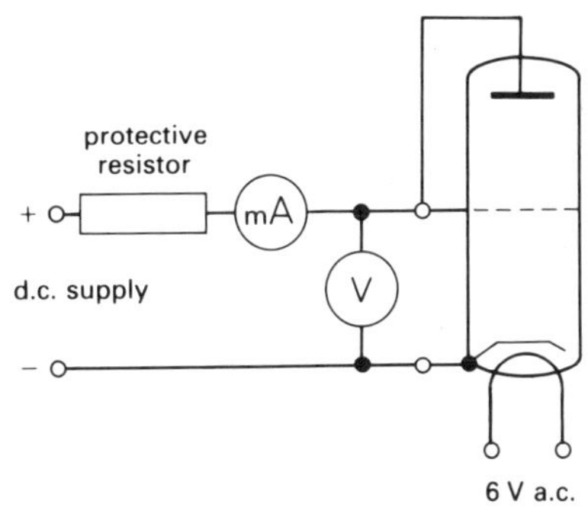

Figure F20
Thyratron circuit.

The grid and anode are connected together, the tube being used as a diode. The protective resistor is essential to prevent damage to the tubes.

Electrons from a hot filament can be accelerated by a potential difference, and the larger the p.d., the higher is the kinetic energy of the electrons. A diode containing gas as well as a hot filament would pass a current of electrons and then, if ions were formed by collision between electrons and gas atoms, there would be an extra ion current. So the experiment is to raise the p.d. across the tube slowly, looking for a rise in current due to the extra ions.

Figure F21 shows the result. At some fairly definite p.d., the current suddenly rises sharply. The gas atoms are ionized by collisions with electrons in the tube only when the electrons have at least the energy corresponding to the p.d. at which the current first starts to climb more steeply.

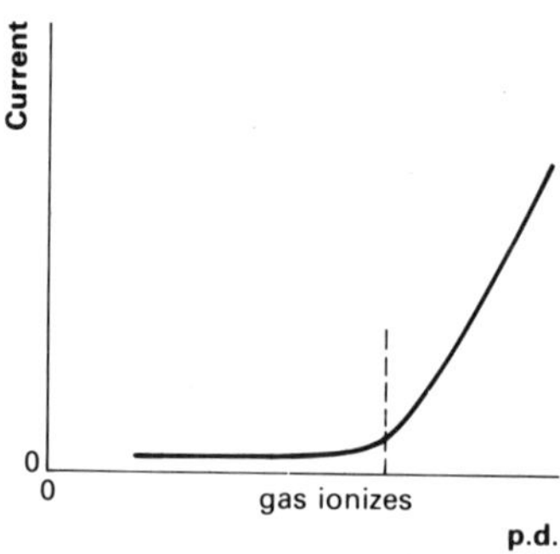

Figure F21
Graph of current against accelerating p.d. for a thyratron.

The gases ionize at different electron energies, the critical p.d.s being about 12.1 V for xenon and 15.7 V for argon.

For each gas there is some one electron energy below which the gas atoms are not ionized. Too feeble a collision will not chip an electron off an atom.

It may be desirable to have a vacuum diode available so that the quite different current–p.d. curve obtained with no gas in the tube may be shown as a contrast.

b Using gas-filled demonstration tubes

Either the gas-filled triode (Teltron TEL 532) or the critical potentials tube (TEL 533) could be used. Both contain helium at low pressure.

Follow the manufacturer's instructions as to appropriate power supply, meters, etc. If the potentially dangerous h.t. supply is used, connections to it must be made with leads having *shrouded* 4 mm plugs (*e.g.*, RS Components stock number 489–037).

i Gas-filled triode
When the filament is heated the current should show a sudden increase when the accelerating p.d. across the tube reaches about 25 V.

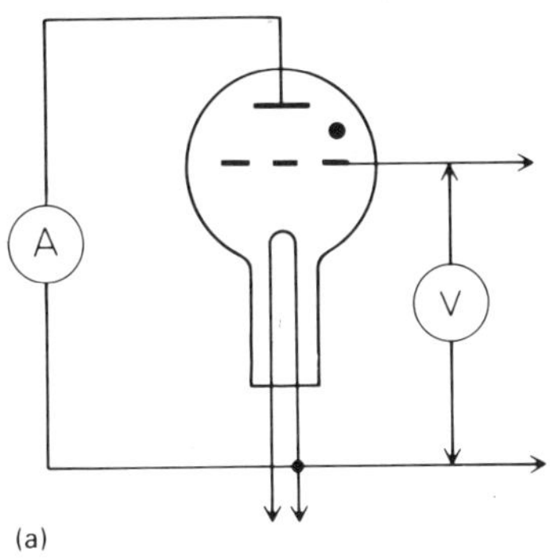

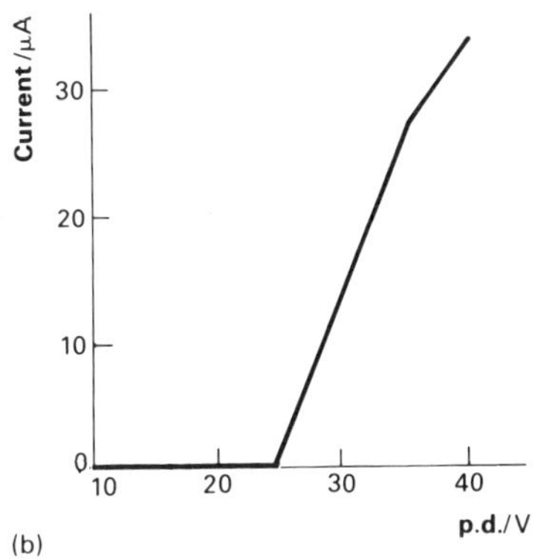

Figure F22
Gas-filled triode:
(a) circuit;
(b) current–p.d. graph.
Based on: Data sheet 532. *Teltron Ltd.*

ii Critical potentials tube
This tube also gives evidence for excitation at electron energies below the ionization potential. The important outcome now is the large effect at ionization. (See figure F23.)

The ionization potential of helium is 24.6 V. Again it may be desirable to show the current–p.d. characteristic of a vacuum tube (*e.g.*, planar diode TEL 520) to emphasize the role of the low-pressure gas in these demonstrations.

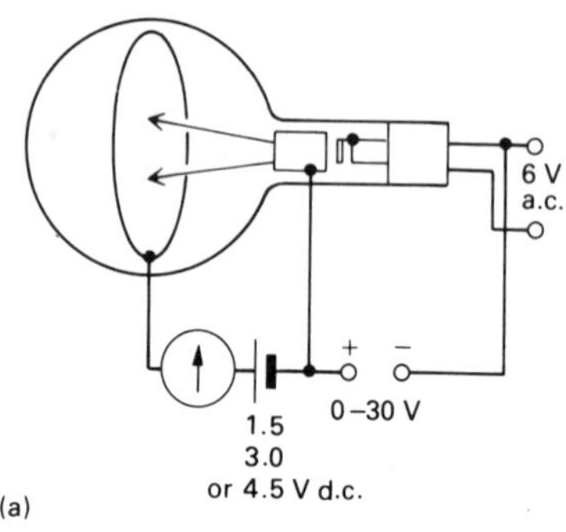

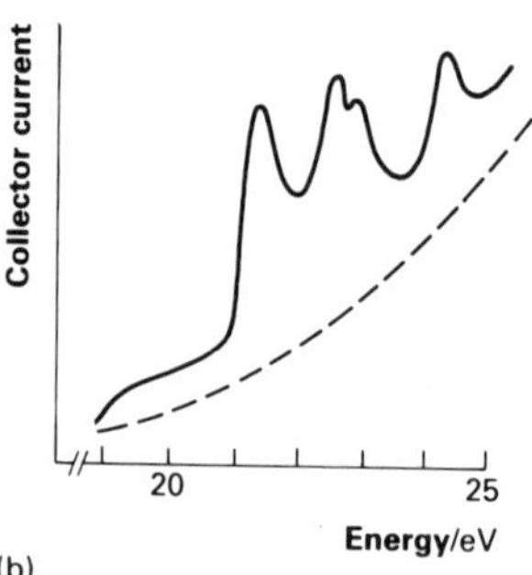

Figure F23
Critical potentials tube:
(a) circuit;
(b) collector current and electron energy.
Based on: Data sheet 533. *Teltron Ltd.*

Television or video

The programme 'Measurement of ionization potential' made by Granada TV as part of the series *Experiment: physics* could be used at this point as an alternative to the demonstrations suggested above.

Chemistry

Ionization is an important matter for chemists; see for example, REVISED NUFFIELD ADVANCED CHEMISTRY, Topic 4.

Ionization and electronic binding energies

Experiments with gases show that the gas atoms are ionized at a particular accelerating p.d., releasing more electrons into the tube, and this is shown by the rise in current. These extra electrons are 'liberated' from the atom and the energy required to liberate them is equal to the accelerating p.d. × charge on the electron. For an electron in a xenon atom this energy is 12.1 eV and for an electron in an argon atom it is 15.7 eV. Electrons are not released from the gas atoms below these energies.

A free electron – one which has been released from the atom but has no kinetic energy – has zero energy. So the energy of an electron bound to the atom will be less than this, that is, it will be negative. The idea should be familiar from Unit E, 'Field and potential'. Thus the electron in the xenon atom has an energy of −12.1 eV, and this is referred to as the *binding energy* for the electron. (Negative because it needs energy to bring it up to zero.) Similarly, for argon the binding energy for the electron is −15.7 eV. But it is more conventional to use the term *ionization energy*. This is the energy which must be *added* to the system to free the electron, and so it is a *positive* quantity.

Questions

Questions 34 and 35 give practice in using the electronvolt as a unit of energy, and conversion from electronvolts to joules.

More precise evidence about ionization energies, and also about excitation energies, is given by spectroscopy. The subject will be taken up again in Unit L, 'Waves, particles, and atoms'.

Each element has a different ionization energy for the first electron and this value can be plotted against Z, the atomic number. The graph is printed in the *Students' guide*, page 368.

Points that might be brought out from the graph

1 Ionization energies are in the range 4 to 24 eV and may typically be described as of the order of 10 eV.

2 The least reactive elements have large ionization energies and the most reactive have small ionization energies.

3 Each element has its own different ionization energy.

4 Next to each noble gas (He, Ne, Ar, ...) is a highly reactive element (Li, Na, K, ...).

Questions

Questions 36 and 37 are about ionization energies and the sizes of atoms and ions.

The fact that an energy of about 10 eV is needed to remove an electron from an atom allows a rough estimate to be made of the atom's size (*i.e.* the distance of the electron from the nucleus). The result, about 10^{-10} m, agrees well with evidence for the size of an atom from Unit A, 'Materials and mechanics' and shows that the electrons lie well outside the nucleus. The estimate is rough because it ignores the electrons' kinetic energy.

Question 38 asks students to make this estimation.

Conclusion

Electrons are bound to the atom and have typical binding energies of -10 eV. The electrons are held in the atom because of the electric attraction between the positively charged nucleus and the negatively charged electron.

HOW ARE NEUTRONS AND PROTONS HELD IN THE NUCLEUS?

The aims of this part of the course are modest. The strong nuclear force is touched on briefly and some discussion of nuclear binding energies is an essential prelude to any consideration of nuclear power. The equation $\Delta E = c^2 \Delta m$ is introduced in an *ad hoc* manner and without justification. It is the basis for our calculations of nuclear binding energies.

As protons are positively charged, the electric force between protons in a nucleus will be a repulsive one. Whatever attractive force holds the nuclear particles together must be strong enough to overcome the repulsive force between protons. (A simple calculation shows that the gravitational force between two protons is too weak by a factor of about 10^{-36}.) 'Strong nuclear force' is an appropriate term.

Question 42 is concerned with this calculation.

Other information about the strong nuclear force and its range comes from scattering experiments and from knowledge of nuclear binding energies.

In Rutherford scattering experiments 5 MeV alpha particles might typically be used. Such alpha particles can approach to within about 5×10^{-14} m of the nucleus (question 11). The results of such scattering experiments are well accounted for by Coulomb's Law: there is no evidence for another force between nuclear particles at this distance. On the other hand, later experiments with energetic protons show that a new force begins to affect them at a distance of 2 or 3×10^{-15} m from the nucleus.

Information about nuclear radii comes from electron scattering experiments (electron energies of several hundred MeV are used). It turns out that the density of nuclear matter is more or less constant, which means that each individual nuclear particle occupies about the same space whether it is in a light or in a massive nucleus. This evidence all suggests that the force between nucleons is strongly attractive at distances of 1 or 2×10^{-15} m, but falls off rapidly beyond this. It is a 'short range' force. To prevent the collapse of nuclear material the force must become repulsive at less than 1×10^{-15} m.

The graph of total nuclear binding energy against nucleon number (*Students' guide*, figure F10) shows that the more nucleons there are in the nucleus, the more negative the binding energy becomes: more energy is needed to take apart a big nucleus than a small one (just as more energy is needed to evaporate a large drop of water than a small one). The fact that the graph is more or less linear is further evidence for the short-range nature of the strong nuclear force. (If every nucleon interacted with every other nucleon in the nucleus we would expect the binding energy to depend approximately on $(\text{number of nucleons})^2$, whereas, in fact it seems as if each one only interacts with a small number of near neighbours.)

Background for teachers

Although we do not expect students to know more about the strong nuclear force than that it is strong, independent of charge, and short range, teachers may find it useful to have more information themselves. GRIFFITH and TEBBUTT *Notes for guidance for the nuclear physics option* is a useful source.

But the total binding energy does not change quite linearly with A, and this has very important consequences. Figure F11 of the *Students' guide* (page 370) shows the probably more familiar graph of average binding energy per nucleon against A. The significance of the negative sign should be emphasized throughout: energy is required to release a nucleon from the nucleus against the attractive nuclear force.

Points to be brought out from the curve

1 For most values of A the average binding energy per nucleon is about -8 MeV, which is the slope of the graph of binding energy against nucleon number.

2 The lowest value, *i.e.* the most stable nucleus, is at ^{56}Fe.

3 Some nuclei, most notably ^{4}He, lie in pronounced troughs in the curve. This suggests that this nucleus is particularly stable and offers an explanation of the importance of the alpha particle in radioactive decay.

Question 43 is based on the curve of average nuclear binding energy per nucleon.

Background for teachers on nuclear binding energies

Teachers may know that the binding energy of a nucleus is fairly well accounted for by the 'semi-empirical formula':

$$-\mathrm{BE} = aA - bA^{2/3} - c(Z^2/A^{1/3}) - d(N-Z)^2/A$$

where a, b, c, and d are all positive constants.

The first term simply expresses the fact that the more nucleons there are in the nucleus, the more tightly bound they are; the second is a 'surface energy term' – the surface area depends on $A^{2/3}$, and nucleons at the surface are less tightly bound; the third term expresses the Coulomb repulsion between the protons. The fact that stable nuclei tend to have neutron number $N \approx Z$ is accounted for by the last term. See for example GRIFFITH and TEBBUTT *Notes for guidance for the nuclear physics option.*

Note that many sources treat binding energy as a *positive* quantity. To be consistent with other parts of this course we regard it as negative.

How do we know these values for nuclear binding energy?

By this time we hope that students are asking how physicists know the values of nuclear binding energies. We can offer no theoretical basis for $\Delta E = c^2 \Delta m$ and its significance is often woefully misunderstood. If it were not the basis for the calculation of nuclear binding energies, it would not be part of this course.

It is wrong to talk about mass being converted into energy. Einstein's achievement (it is part of his special relativity theory of 1905) was in uniting the two apparently unrelated conservation principles: conservation of mass and conservation of energy. This can be observed when an electron is accelerated to a speed approaching that of light. It is continually gaining energy by being in an electric field. Since it cannot travel faster than light, the 'ultimate speed', its speed cannot go on increasing indefinitely: its mass increases, and can rise without limit. This mass increase has to be allowed for in the design of particle accelerators. Einstein told us that mass and energy are measurements of the same thing. $\Delta E = c^2 \Delta m$ tells us the exchange rate between, say, kilograms and joules. $\Delta E = c^2 \Delta m$ applies not only to moving bodies and kinetic energy, but to all forms of energy.

It would be wrong to let students go away thinking that the equation $\Delta E = c^2 \Delta m$ applies only to nuclear physics. It is always true: the mass of any body or system of bodies is a measure of its (their) energy. So, for example, the mass of a charged dry cell is larger than the mass of the same cell when discharged; the mass of a spring increases when it is stretched; the mass of a heap of gunpowder is more than the mass of the reaction products (at the same temperature); the mass of the Earth is *less* than the sum of the masses of its individual atoms when they are far apart (because the total has less energy than the isolated constituent parts).

Questions

Questions 44 and 45 give practice in using $\Delta E = c^2 \Delta m$ in non-nuclear problems.

Reading

We offer no derivation of $\Delta E = c^2 \Delta m$. More important than trying to follow a derivation is an appreciation of what the relationship says and does not say. Among references which teachers may find useful are:

ROGERS *Physics for the inquiring mind.* Chapter 26.
DAVIES *The forces of nature.* Pages 31–33.
CARO, McDONNELL, and SPICER *Modern physics.* Section 6.5.

Calculation of nuclear binding energy: an example

The rest masses of the neutron, the proton, and the electron are known with considerable accuracy. The values, in unified atomic mass units, are:

mass of proton, $m_p = 1.007\,3$ u

mass of neutron, $m_n = 1.008\,7$ u

mass of electron, $m_e = 0.000\,5$ u

(1 u $= 1.66 \times 10^{-27}$ kg, one-twelfth of the mass of an atom of ^{12}C)

The total mass of the constituent parts of the atom $^{A}_{Z}$X is

$Z m_p + (A - Z) m_n + Z m_e$

The results of the calculation are always found to be *greater* than the accurately measured mass of the atom (*e.g.*, from a mass spectrometer). For example, the mass of an atom of ^{4}He is 4.002 6 u, whereas the total mass of two protons plus two neutrons plus two electrons is 4.033 0 u. There is a mass loss of 0.030 4 u or $0.050\,5 \times 10^{-27}$ kg. From $\Delta E = c^2 \Delta m$ this corresponds to an energy loss of 4.54×10^{-12} J or 28.4 MeV. Thus the nuclear binding energy of ^{4}He is -28.4 MeV, corresponding to an average binding energy per nucleon of -7.1 MeV, as shown in figure F11 of the *Students' guide*.

Questions

Question 46 asks students to go through this calculation for ^{4}He.
Question 29 gives practice with the unified atomic mass unit, u.

The direct conversion between u and MeV is 1 u = 931 MeV.
A book of data such as Tennent, *Science data book*, or the Revised Nuffield *Book of data* is invaluable for providing data for such calculations.

INDUCED FISSION

The average binding energy per nucleon is less negative for the massive nuclei (*e.g.* U) than for nuclei in the middle of the range (*e.g.* Fe). If the nucleons in one of the massive nuclei could be redistributed among two or more less massive nuclei then some energy would be released in the process. Although fission is energetically possible, it rarely happens spontaneously. But it can be induced, as for example when ^{235}U captures a 'slow' neutron* to form an excited ^{236}U nucleus which then splits up:

$$^{235}\text{U} + \text{n} \rightarrow {}^{236}\text{U} \rightarrow \text{X} + \text{Y} + x\text{n} + \text{Q}$$

In this reaction X and Y are the fission fragments: Ba and Kr are typical examples but there are many possibilities. Fission is usually asymmetric, that is, X and Y are different. Two or three neutrons are released, which can in turn trigger the fission of further ^{235}U nuclei in a chain reaction. Q represents the energy released; it is of the order of 200 MeV per nucleus split (which is equivalent to 8×10^{13} J per kg of ^{235}U). Most of this energy appears as kinetic energy of the fission fragments.

FUSION

If two nuclei from the lighter end of the curve of average binding energy per nucleon could combine into a more massive nucleus, then some energy should be released as the product nucleus will be lower on the curve. (The extreme conditions required for fusion are the subject of one of the 'Energy options' in Unit G, 'Energy sources'.)

For example, the fusion reaction

$$^{3}\text{H} + {}^{2}\text{H} \rightarrow {}^{4}\text{He} + {}^{1}\text{n} + \text{Q}$$

gives a net energy release of 17.5 MeV (2.8×10^{-12} J). Although this is

*A slow (or thermal) neutron is one with energy equal to the average thermal energy of the atoms of the medium (*i.e.*, about 0.025 eV at 20 °C).

less than the 200 MeV released by fission of a uranium nucleus, the energy release per unit mass of 'fuel' should be considered.

Fission and fusion are dealt with in more detail in Unit G, 'Energy sources', which includes questions dealing with the energy released.

Reading

The background article 'Our nuclear history' in the Reader *Particles, imaging, and nuclei* discusses fusion as the Sun's source of energy.

Summary

Using the graph of average binding energy per nucleon against A it is possible to calculate the energy which could be released in fission and fusion processes.

THE FOUR INTERACTIONS

Much has been omitted (models of the nucleus, energetics of alpha and beta decay, other modes of decay) and many questions left unanswered (which nuclei will be unstable and how will they decay...). But there is no time for more about the nucleus in this course. Instead one might end the Unit by offering students a very broad view and telling them that they have now met three of the four forces or interactions of physics: gravitational, electromagnetic, and strong nuclear. (The fourth, the weak nuclear interaction, governs beta decay.) One of the main goals of theoretical physics is unification. Magnetism and electricity were united by the work of Faraday and Maxwell, and finally Einstein, who showed how the magnetic effects which arise when charges are in motion can be explained by relativity theory. Recent experiments with high-energy particles (1982) support a theory which brings together the electrical and weak nuclear interaction. The aim of grand unification theories (GUTS) is to bring all four interactions together.

Further reading

PROJECT PHYSICS Text Unit 6, *The nucleus*, especially chapters 23 (Probing the nucleus) and 24 (Nuclear energy; nuclear forces) is a very good source for students who want to read more about these topics.

The article 'The particles and forces of nature' in the Reader *Particles, imaging, and nuclei* deals with the four forces and with attempts to produce a unified theory.

The article 'Lasers probe the atomic nucleus' in the Reader *Particles, imaging, and nuclei* describes a current area in nuclear research.

Rutherford Appleton Laboratory Monographs are short pamphlets dealing, mostly, with particle physics. Among those so far published which have some relevance to this Unit are *Atoms, particles, leptons and quarks, Experimental particle physics, The universe and Man.*

Unit G
ENERGY SOURCES

Maurice Tebbutt
Faculty of Education, University of Birmingham

Contributors
Michael Carrick, North Bromsgrove High School;
Bev Cox, Streetly School, Sutton Coldfield;
John Gardner, Shenley Court School, Birmingham;
Tom Gregory, Codsall School, Staffordshire;
John Minister,Bishop Walsh School, Sutton Coldfield;
Richard Spiby, King Edward VI Camp Hill School for Boys, Birmingham

Suggested time allocation: three weeks

PLAN OF THE UNIT

Section G1
The background to energy supply and demand

Prior knowledge		Topic		Leads to
GCSE	►	**Conversion and conservation of energy**		
GCSE	►	**Efficiency** **Second Law of Thermodynamics**	►	Unit K, 'Energy and entropy'
exponentials (Unit F)	►	**Patterns in supply and demand for fuels**		

Section G2
Power from nuclear fission

Prior knowledge		Topic		Leads to
nuclear masses (Unit F)	►	**Origin of the energy**		
electrical potentials (Unit E)	►	**Obtaining a chain reaction**	►	nuclear power stations: Section G4
collisions (Unit A)	►	**Moderation**		

Section G3
Conserving fuel in the home

electrical resistance (Unit B) ► **Thermal transfer of energy and thermal resistance**

Surface resistance

Cavity resistance

Double glazing, cavities, and cavity insulation ► domestic energy conservation: Section G4

Section G4
Energy options

Nuclear power

Solar energy

Fossil fuels

Other sources of internal energy

Water power

Conservation

Storage

INTRODUCTION

This Unit provides students with the physics background to energy supply and demand. There is no shortage of material on this topic. Some material, like *Science in society* and *Siscon in schools*, has been written for schools, and some is intended for a more general audience. While there is bound to be a little overlap between this Unit and existing material, our intention is to provide a somewhat different focus – specifically, the particular contribution that physics can make to the debate on energy supply and management. Thus the emphasis is on the basic physics which applies to many aspects of energy sources and the numerical data which are involved.

It is not possible, however, to ignore areas such as economics in which physicists are normally not involved. Here, more variables are involved, they are less controllable, and the situations are altogether more problematic than is usually the case in physics.

The content of the Unit is, therefore, somewhat unusual in that it ventures into other disciplines. The style adopted in this Unit is also different from that of most others in the course, since it must reflect the uncertainties inherent in much of the content.

Energy supply is a continuing problem likely to be subject to rapid change. This has both advantages and disadvantages. The Unit's concern with a 'live' issue should help motivation, but students often object if materials seem out of date. However, data can be kept reasonably up to date by revising the background book *Energy sources: data, references, and readings*, and the questions which are asked will continue to focus on important aspects of the problem.

THE STRUCTURE OF THE UNIT

Section G1: The background to energy supply and demand

Basic ideas

Conversion and conservation of energy

Efficiency and the need for fuels

Second Law of Thermodynamics

Personal energy

Patterns of growth in fuel consumption

Resources and their lifetimes

Detail

Contribution of different fuels

Balance of supply/demand for different nations

Difference in *per capita* consumption for different nations

What fuels are used for

Characteristics of fuels

Undesirable consequences of fuel use

Factors influencing further growth

Section G2: Power from nuclear fission

Energy from fission

- calculation from nuclear masses
- origin from potential

Triggered by extra neutron

Chain reaction

Critical mass

Nuclear cross-section

Isotopic composition of uranium

Neutron energy required for fission

Moderation and control

Enrichment

Section G3: Conserving fuel in the home

Analogy between thermal flow of energy and electric current

Thermal resistance

Thermal resistance coefficient

Surface resistance

Factors affecting surface resistance

Calculating temperatures in thermal transfer systems

Heating buildings:
—thermal lag
—ventilation

Section G4: Energy options

Nuclear power

A Nuclear power stations
B Nuclear fusion
C Nuclear safety

Solar energy

D Passive systems
E Active systems
F Photovoltaics
G Biomass
H Wind energy

Fossil fuels

I Power stations
J Alternative sources of fossil fuels

Other internal energy sources

K Geothermal
L Heat pumps
M Combustion of refuse

Water power

N Hydro-power
O Wave energy
P Tidal energy

Conservation

Q Domestic
R Industrial
S Transport

Storage

T Storage

Table G1

It is difficult to describe the material in the Unit in an entirely linear sequence. Table G1 outlines the content and shows some of its structure. The more detailed comments below should be read in conjunction with this table.

All students should study Sections G1, G2, and G3. It has been assumed that Sections G1 and G2 will be taught using the questions in the *Students' guide* as seems appropriate. Section G1 provides the basis for most of the work of the Unit. It is in two parts: 'Basic ideas' and 'Detail'. The principles of nuclear fission are dealt with in Section G2. This is a compulsory section since any student of A-level physics in the latter part of the twentieth century must surely come to grips with these principles. No value judgment about nuclear power is implied in this, and the application of these principles to electric power generation takes its place with other energy options in Section G4.

The core syllabus for physics at A-level includes thermal conduction. Section G3 sets this in the practical context of conserving fuel in the home. While this Section could be taught in class, it has been written so that the students could complete it on their own, perhaps as homework.

Each student will then study one or two of the energy options in Section G4. This will provide practice in the technique of reading to find information and reporting to the rest of the group, and also ensure that a reasonably balanced view is obtained of this very broad area. Because some students find reading for learning difficult, some guidance is given in the *Students' guide*.

THE PLACE OF THE UNIT IN THE COURSE

There are a number of reasons why this Unit may be appropriately placed towards the end of the first year of the course. Since it deals with a particularly relevant topic it can be regarded as an intermediate end-point. The Unit can also be taught piecemeal, which may be useful at the latter end of the summer term when work in schools tends to be disrupted. It could, for example, be taught in parallel with other Units, either as a whole or in part; or Section G4 might be scheduled for homework or as a vacation task. This last suggestion has *not* been assumed in estimating the time to be spent on the Unit.

The Unit involves practice or extension of the ideas of flow of charge met first in Unit B, 'Currents, circuits, and charge', exponential change (Unit B and Unit F, 'Radioactivity and the nuclear atom'), and nuclear structure (Unit F); and it provides a basis for work on internal energy met again in Unit K, 'Energy and entropy'.

Time allocation

The topic 'Energy sources' raises enough issues for a complete course. But within the context of an A-level physics course three weeks seems an appropriate allocation.

It could be divided like this:

Section G1 The background to energy supply and demand. Rather more than one week + homework.

Section G2 Power from nuclear fission. About half a week + homework.

Section G3 Conserving fuel in the home. No more than one double period + more than the Section's share of homework time.

Section G4 Energy options. Less than half a week + homework.

Reporting back. Half a week.

These suggestions are very approximate because the flexibility of the Unit will allow the proportion of homework time to be increased if this is useful. The allocation for Section G3 assumes that students will do most of the work on their own as suggested above. Alternatively, more class-teaching time will need to be found.

On the other hand, if students were to use the summer holiday for the reading on energy options, some class time could be saved.

Both *Students'* and *Teachers' guides* for Section G1 are rather longer in proportion to the whole Unit than the Section time allocation would justify. This is because a number of ideas which are likely to be unfamiliar to students and teachers are introduced; Sections G2 and G3, however, deal with more familiar topics and are relatively short.

BIBLIOGRAPHY

Teachers may need two different kinds of information to help in teaching the Unit – either general or specific. There are so many sources of general material that the problem is likely to be that of making a choice. Paperback books are likely contenders on grounds of cost and of being up-to-date, though even this cannot be guaranteed.

Useful ones to start on might be:

RAMAGE *Energy: a guidebook.*
McMULLEN, MORGAN, and MURRAY *Energy resources.*
CHAPMAN *Fuel's paradise.*
FOLEY *The energy question.*

Specific information may also be needed on nuclear power, thermal conduction, and the energy options.

Nuclear power is covered in the first two references and also in:

GRIFFITH and TEBBUTT *Notes for guidance for the nuclear physics option.*

Information on the particular approach to thermal conduction used in this Unit is not easy to obtain in one source, but is to be found, in part, in some books on building technology or heating system design, such as:

BASSETT and PRITCHARD *Heating.*

This book's approach to the problem of heating buildings contains a great deal of physics and it would be a very useful source of some of the material in Section G3. A search of the appropriate section of a school, college, or local public library may, however, be more productive of sources of similar information than an attempt to obtain any particular book that might be recommended here.

Specific articles on the 'Energy options' are included in the background books *Energy sources: data, references, and readings* and *Energy options: a reader.* The former also contains specific references and guidance for both teachers and students in locating suitable material.

Full bibliographic information about the sources listed above is given on page 505 of this *Teachers' guide.*

TEACHING ENERGY

Energy is rightly regarded as an important idea. In physics it is a highly abstract concept which has little connection with direct experience and which must draw on a number of contributory concepts if it is to be understood. 'Energy' is also used in everyday language, and is something that the general public is increasingly aware of. This is partly to do with its links with fuel (from which it is usually not distinguished) and fuel supplies; partly because of its everyday use; and partly because of the increasing use of energy as a unifying concept in elementary science courses.

The problems produced by introducing the concept at an early stage in school curricula have been dealt with in detail in the following:

WARREN *The teaching of physics.*
WARREN 'The nature of energy', *Eur. J. Sci. Educ.*
WARREN 'Energy and its carriers: a critical analysis', *Phys. Educ.*

Attempting to teach the concept of energy before it can be built on a framework of supporting concepts may run the risk of allowing (even encouraging) misconceptions and the need to unlearn these later. Alternatively, leaving the teaching of the concept until it can be taught logically would deny most of the population access to it. There is no doubt that energy should form part of an A-level course in physics, as should, we believe, a wider discussion of the issues concerned with energy supply, demand, and management. Whatever approach is used, clearly those students who take this course should be helped to have as clear an understanding of the concept as possible. Teachers may find the references listed above useful here.

The volume of discussion about the teaching of energy, including the teaching of 'heat', suggests a distinct lack of agreement about how this should be done. In addition to the references above teachers may wish to consult the following in order to appreciate the detailed arguments:

OGBORN 'Dialogue concerning two old sciences'.
SCHMID 'Energy and its carriers'.
SUMMERS 'Teaching heat: an analysis of misconceptions'.
WARREN 'The teaching of the concept of heat'.

We cannot reiterate all the arguments here but we offer some suggestions which may be helpful in dealing with various points, some arising in the references listed and some arising only in this Unit.

Energy is an abstract but very useful concept

When any well-defined system changes from one definite state to another, its energy changes by the same amount, no matter how the change of state came about. We can say how to measure the change of energy by choosing one particular way of changing the state: most commonly we keep it thermally isolated (adiabatic change) and then the energy change is the work performed. Strictly, we can only speak of energy *changes*, but it is useful to speak of the energy of a system, as long as we do not alter the implied zero value. Because energy is conserved it is a useful book-keeping concept enabling us to keep track of energy transfers from one part of the system to another.

An important part of the energy of a system is its *internal energy*. For example, a bullet has kinetic energy because it is moving; it has potential energy because it is above the surface of the Earth; and it has internal energy because it is hot. (The internal energy is simply the sum of the kinetic and potential energies of all the individual particles of the bullet.) If the bullet is suddenly stopped it loses kinetic energy. Some of this kinetic energy is transferred to internal energy of the bullet and the bullet gets hotter. As internal energy is transferred to the surroundings again the bullet cools.

Mistakes can arise from not considering the *whole* system. Thus to speak of the 'chemical energy of a lump of coal' is to ignore the oxygen which is needed to burn it. While it would be cumbersome to say 'the chemical energy of the fuel-plus-oxygen system' every time, the point should not be lost sight of.

Energy is not an objective substance, nor is it a fluid

The uncritical use of the idea of energy flow could imply some notion that energy is a substance; of course it is not. Nevertheless there are some circumstances in which it is useful and sensible to think of energy 'flowing'. We do not speak of energy flowing from potential to kinetic as a ball falls. But it is reasonable to think of energy flowing between parts of a system. If one part loses energy and another gains it, through some interface, it makes sense to think of a flow across that interface. Thus we do think of energy from a radiator entering a room, and then leaking out through walls and windows to the outside. For the idea of flow to be useful the situation has to be one in which energy is lost by one part of a system to an immediately adjacent one.

Thus the idea of flow is a useful generalizable concept, and there are other areas of physics where it is used, although no objective fluid substance actually flows. Examples in this course occur in Unit B, 'Currents, circuits, and charge' and Unit H, 'Magnetic fields and a.c.' For example, flux (whose very name means flow), is a useful concept to physicists and engineers in electromagnetism.

See also the discussion of flow in the article 'Electromechanical similarities' in the Reader *Physics in engineering and technology*.

Forms of energy

In earlier courses phrases such as 'chemical energy' or 'electrical energy' will have been used, as well as potential energy and kinetic energy. One difficulty with 'chemical energy' has already been mentioned. Another is apparent from a consideration of the meaning of the term. Consider a chemical reaction, for example burning a fuel in air, and heating the surroundings. In such a chemical reaction there is a difference between the potential energy of the bonds in the reactant molecules and those in the product molecules. In the example chosen this potential energy decreases, the internal energy of the surroundings and the product molecules (*i.e.*, their kinetic energy and the potential energy due to forces *between* molecules) increases. Similarly, a careful analysis of other terms, for example 'nuclear energy', reveals that what is involved can be described in terms of changes in potential and kinetic energy and, ultimately, internal energy. While the notion of many different forms of energy has its uses, it may be worth while, from time to time, to raise the issue of their reducibility to potential and kinetic energy.

'Heat' is not a 'substance' but a process

We commonly speak of the heat in a body, meaning its internal energy, but scientific usage of the word 'heat' has changed. A simple way to look at the matter is to regard the statement of the first law of thermodynamics

$$\Delta U = \Delta Q + \Delta W$$

as saying that the internal energy, ΔU, can be changed in two ways: by working, ΔW, and by thermal transfer of energy across a temperature difference, ΔQ. Calculations of energy transferred by either of these processes are changes in internal energy, so it is wrong to use 'work' and 'heat' to mean amounts of energy *in* something. Our solution is to use the terms 'working' or 'doing work', and 'heating' or 'thermal transfer' to describe processes and to avoid the use of 'heat' as a noun.

This still requires care since 'heating' has a range of colloquial meanings, including 'making hotter', and also the process of thermal transfer of energy in situations when there is no temperature rise, such as melting a block of ice, or maintaining the temperature of a house. There is, of course, in all these cases, thermal transfer of energy due to a difference in temperature. For these reasons it is clearer to use 'thermal transfer' for all cases where energy flow is the result of a temperature difference; but 'heating' may be less clumsy and will be satisfactory as long as its full meaning is appreciated.

Language difficulties

Energy, heat, and work all have colloquial meanings which can easily interfere with the precise meanings appropriate to physics, and some perseverance is needed to instil the latter. Much of the discussion above relates to this point, and students may also be told that what they have been used to calling 'heat' in everyday conversation is more properly called 'internal energy' in scientific speech and writing.

Hidden difficulties

An example is question 1 of this Unit which is intended as revision of a familiar situation, leading to the conservation of energy and extension to the need for fuel. While at first sight it may seem an innocent question of a familiar type, a closer examination brings up some difficulties. For example, once the steady state is reached, the potential energy of the (mass + Earth) system is decreasing, but the rotational kinetic energy of the generator is constant. Does it make sense to speak of the former being transformed into the latter? This issue has been raised in the question, but it is left to teachers to emphasize the question's main purpose rather than raising too many doubts and difficulties which may destroy confidence.

KEEPING UP-TO-DATE

Some of the information in this Unit (for example figure G4 in the *Students' guide*) is inevitably out-of-date even before it is published. We hope that the Department of Energy will continue to publish that figure, in updated form, and that teachers will be able to obtain supplies sufficiently regularly to keep these data in the *Students' guide* up to date. Other information will need to be brought up-to-date too, and it is intended to revise the background book *Energy sources: data, references, and readings* from time to time to make sure it contains current information and statistics, and a list of references to new books, articles, and teaching aids.

As time goes on schools will no doubt build up a library of relevant books and articles. Since some suitable examples will be expensive and perhaps difficult for schools to come by, most of *Energy sources: data, references, and readings*, and the whole of *Energy options: a reader* contain a selection of articles which will help to make a start. When *Energy sources: data, references, and readings* is republished we hope that as well as current data it will contain new articles. It is not intended to republish the Reader *Energy options.*

SECTION G1

THE BACKGROUND TO ENERGY SUPPLY AND DEMAND

The scope of the topic and its treatment

As will be seen from table G1 the Unit as a whole, and this Section in particular, involve the bringing together of many ideas. Although some may appear to be familiar to students, many may prove difficult. While there is little conventional experimental work in the Unit the approach throughout has been to apply the techniques of physics to an important practical problem. Like most practical problems, the problem of energy supply and demand has many ramifications, and it will not be possible to explore all the implications. It will also be necessary to treat some questions rather lightly, while maintaining their seriousness of purpose. They are intended to raise issues, rather than to explore them in depth. Where it has been possible to draw attention to complications without making questions too cumbersome this has been done, but it has not always been possible. Every student need not answer every question, and teachers are encouraged to omit or share out introductory and practice questions as the needs and strengths of their students dictate.

In this Section a number of topics are introduced fairly briefly in the first part and then taken up in more detail in the second part – the spiral approach.

Energy is conserved; Sankey diagrams

Students are reminded that the conservation of energy is a basic principle of physics; the need for fuels arises because energy spreads out and becomes less useful (question 1). Sankey diagrams are a convenient way of representing energy transfers. The idea of flow is useful in this context, but must not be used uncritically.

Efficiency; personal energy

The key concept of this part is efficiency. It is applied to various energy converters, including the human body, in order to pave the way for such later work as the thermodynamic limitation on efficiency in this Unit and in Unit K, 'Energy and entropy'; the need for fuels to replace the energy which has become 'spread out' and hence less useful; and the dependence on fossil fuels of the developed countries in particular.

Questions

Questions 3, 4, and 5 all provide opportunities for students to perform simple calculations and to become familiar with units. In addition, questions 3 and 5 aim to familiarize students with the amounts of energy involved in physical activities and in domestic appliances. The idea of 'energy slave', introduced in question 5, is a useful way of highlighting our dependence on fuels while providing a different slant on the purposes indicated above.

Teachers (though probably not students) may be surprised to find that the mechanical efficiency of the body is as high as about 35 per cent (question 3). If it were treated as a simple heat engine operating between 37 °C and, say, 15 °C, its efficiency would be $[(310\text{–}288)/310] \times 100\,\% \approx 7\,\%$. The body, which is an *open system*, is much more efficient than an ideal heat engine working between these temperatures.

Fuels: energy converters; primary, end-use, and functional energy; energy units

Energy is conserved but the supply of fuel is finite. It is therefore important to consider the efficiency of a variety of energy transfer processes. Table G1 in the *Students' guide* shows that efficiency may range from 90 per cent for processes which involve mechanical change, but do not involve thermal flow of energy from hot to cold (for example, turbine, electric motor, or generator) down to as low as 10 per cent for a steam engine. This last case is an example of those processes which inevitably involve thermal flow of energy from hot to cold, which put limits on the proportion of the total energy which can be transferred to do useful work for us. But if our aim is to heat a building or some water, then the process of burning fuel may be very efficient. Note that the figures given in table G1 in the *Students' guide* for the efficiencies of various boilers are only less than 100 per cent because not all the energy is transferred to internal energy of *that part of the system we are interested in*; some heats the flue gases and is lost up the chimney. But given that all the energy released in the chemical reaction when fuel is burned is transferred to internal energy the process could be regarded as 100 per cent efficient.

(The formula, efficiency $=(T_1 - T_2)/T_1$ is simply quoted. It will be discussed in Unit K, 'Energy and entropy'.)

Home experiment

Home experiment GH1, The cost of a 'cuppa', in which students compare the amount of fuel used, the cost, and the efficiency of using electricity and gas to boil water, is relevant at this point.

Learning the language and reading the signs

The last part of this Unit (Section G4) is designed to provide students with practice in reading and reporting back. Therefore some of the emphasis in this first Section is on introducing them to terms and units which may be used in articles, and giving practice in interpreting data published in the form of graphs, tables, or diagrams.

Units of measurement are a major problem, as the *Students' guide* explains. This Unit does not use SI exclusively. Articles on energy do not tend to use SI and students might be discouraged from reading if they suddenly came across the plethora of units which are used. Instead, an attempt has been made to explain the systems which are used and to provide a conversion table. (Table G4 in the *Students guide.*)

A second difficulty with articles about energy is the use of terms which are not precisely defined and which are used in different ways on different occasions. To specify a definition and stick to it would have been an easy route to consistency within this Unit but again, would not have helped students in reading articles. The work on primary and secondary energy and fuels, end-use, and functional energy (and question 5 on energy slaves) is intended to be helpful in itself, and also to make the point that many similar and loosely defined terms, which it has not been possible to consider in such detail, are used in this topic.

Questions

Question 4 gives practice with various energy units. Question 6 is about efficiency; question 7 is about primary, end-use, and functional energy: both are intended to provide practice in interpreting data presented in complex diagrammatic forms. For this reason an official publication has been used unmodified. Some parts of question 6 have been deliberately framed so that they cannot be answered precisely from the diagram and some parts of the diagram are not entirely clear. The intention is for students to interpret the information as well as they can, or to find supplementary data if necessary.

Changes in fuel supply; making predictions; is there a crisis?

The intention of this part is to revise and extend students' knowledge of exponential change in determining whether growth in fuel consumption is exponential or not. They should also see the implications of two simple growth patterns (exponential growth or steady consumption) for the lifetime of the World's probable or proved reserves of fossil fuels.

Many factors make it difficult to obtain even an elementary view of present and future fuel consumption patterns. Some of these are mentioned in the *Students' guide*, but this topic is particularly dependent on variations in the data. For instance, total World fuel consumption was effectively constant for the three years up to 1983. Some commentators suggest that this was due only to economic recession and that demand will eventually resume exponential growth as it did after the 1973 Middle East war. Others suggest that consumption will actually fall from now on.

This part of the *Students' guide* has been written to try to avoid the need for major changes in the teaching whatever pattern of consumption becomes established. Since the emphasis is not on students predicting future consumption in detail, but rather on seeing what current patterns are and might imply for the future, only a limited amount of up-dating of factual data, rather than of detailed predictions, should be necessary.

Questions

Question 8 asks students to determine whether World population growth is exponential. The doubling-time for population growth is used in question 17.

Question 9 makes the point that, during exponential growth, as much resource is used in each doubling-time as has been used in the whole time hitherto. Hence the resource is used up at an increasing rate.

Questions 10 and 11 are revision questions on the nature of exponentials and may be omitted if not required. Question 11 establishes the relation $t_D = 0.693/g$, where t_D is the doubling-time and g is the growth constant. This relation should already be familiar, in terms of half-life and decay constant from Unit F, 'Radioactivity and the nuclear atom'.

Question 12 introduces terms to do with reserves and involves working with tables of data to calculate lifetimes of resources assuming constant consumption.

Question 13 is similar, but more complicated, since it assumes continued growth in consumption at the average growth rate of the last ten years. This question is much more manageable if the 'Dynamic modelling system' is available. If students have had some experience of this already they should be well able to set up the appropriate equations. Questions 14 and 15 consider what happens to growth as resources become more difficult to obtain. Question 14 develops the bacterial colony analogy of question 9, while question 15 is concerned with fuels. The 'Dynamic modelling system' is useful here, too, but the equations are rather harder to set up.

Useful sets of equations are:

For exponential growth (question 13)

Initial values are needed for LC, G, DT, R, T

```
DC = C*G*DT
C = C+DC
R = R-C*DT
T = T+DT
```

For consumption with limited resources (question 15)

Initial values are needed for C, G, DT, R, A, T

```
DC = C*G*DT*(R-2*A)/R
C = C+DC
A = A+C*DT
T = T+DT
```

Notes

1 The factor (R−2*A)/R is arranged to cause the growth in consumption (rate) to become negative when half the total resource has been used up. This fraction is entirely hypothetical and the effect of other fractions could be investigated.

2 An initial value will need to be set for A. This can be done by examining historical consumption figures or from a knowledge of current consumption and average doubling-times.

For each set of equations the meaning of the symbols is:

C current consumption (rate)
DC increment in C
A total resource consumed to date
R current or total value of the resource
G rate of growth of consumption
T time
DT time increment

Question 16 may appeal to the students' sense of humour, but we hope that its serious purpose – highlighting the difficulties of prediction – will be apparent, and that it will not prove offensive.

Questions 17 and 18 are ultimately to do with food production. Question 17 concentrates on the availability of land for food production if population growth continues exponentially, and is intended to draw attention to this in preparation for the next topic. Food production in the developed world is heavily dependent on fossil fuels – partly for machinery and partly for artificial fertilizers. Question 18 concentrates on this.

MORE ABOUT FUEL USE

The second part of this Section provides more detail about fuel use and the growth of fuel demand and hence develops some of the ideas in the first part. The Section has been arranged like this because introducing all the detail at once might make the material difficult to comprehend. Although the text only scratches the surface it should enable students to read articles on energy options with reasonable understanding.

Distribution

The first part of this Section examined fuel use on a World scale. This part of the Section makes a more detailed examination.

One way is to examine the large and increasing contributions made to global fuel supplies by fossil fuels.

A global perspective has been adopted to try to counter the unduly insular attitudes which are sometimes taken. However, national energy policies will be determined in part by the degree of each country's own self-sufficiency in fuels.

There remains the large disparity in *per capita* fuel use between the developed and the less developed countries. The link between fuel use and industrial activity is often demonstrated by diagrams, such as figure G12 (*Students' guide*) which shows a substantial correlation between *per capita* fuel use and gross domestic product *per capita*. Some commentators object to such diagrams on the grounds that they show bias in using parameters which more properly represent the activity of industrial societies than non-industrial ones. To resolve this and many similar issues would require more time and expertise than is available, but raising the issues may encourage students to be generally more critical of these and similar data.

In spite of the link with industrial activity, a substantial proportion of the U.K.'s fuel use is in fact devoted to space heating, resulting partly from our northern geographical location. This provides a justification, in addition to the core syllabus requirement, for the approach taken to thermal conduction in Section G3; and also needs to be borne in mind for a number of the energy options which feature in Section G4.

Some characteristics of fuels

This part prepares the way more overtly for the energy options by adding another dimension to the increasingly complex picture of fuel supply. By now students will have realized that it may be difficult enough to satisfy future demand for fuel. Other characteristics, such as energy density, transportability, time, and feasibility will make a difficult task worse.

Undesirable consequences of fuel use

It would be difficult to produce such a relatively short Unit on energy sources, which would be universally accepted as balanced. While it would, therefore, be wrong to neglect any mention of the undesirable consequences of fuel use, to do much more than mention them presents problems. The difficulty is well illustrated by the example of thermal pollution. The details of the Earth's energy balance are quite complex because of the variety of causes of reflection, absorption and re-radiation, together with absorption of energy in biomass. A reasonable treatment of radiation would have required the introduction of Stefan's Law. Only then can the effect of fuel consumption on the Earth's temperature be predicted. Students with a good mathematical background would cope adequately with the derivation and use of the relation

$$\frac{\Delta T}{T} = \frac{1}{4}\frac{\Delta E}{E}$$

but this would be an added complication for others. It is for these reasons that only a little has been made of thermal pollution and similar topics. However the points mentioned above may find some application later, since most of the energy options in Section G4 involve a study item concerned with environmental factors. It should be emphasized at this point that thermal pollution results not only from combustion of fuels but from any form of energy 'consumption'.

Similarly, detailed treatment of risks is beyond the scope of this Unit but discussions of *Students' guide* figures G17 and G18 will raise the issues sufficiently. Figure G17 shows how the frequency of accidents of various kinds in the U.S.A. involving more than some number of fatalities X, varies with X. Thus an explosion causing more than 1000 fatalities has a frequency of 10^{-2} per year; that is, on average one such accident might be expected every 100 years. Figure G18 uses some similar data to point up an apparently similar slope to all the graphs. A major difficulty with all such graphs is the scarcity of data relating to the energy industry. A number of points about risks are made in the *Students' guide* which may help discussion. A related point which is not made in the *Students' guide* is that all risks apply to the future, and that we should think about the welfare of the generations to come. There are risks for these generations associated with our use of nuclear power, for instance. But if we choose to use up a substantial proportion of the oil or coal resources now, then future generations will have to live with the consequences of a decision they have not been party to.

Questions

Some of these questions are intended to provide the basis for discussion. It is not necessary for students to do every question, as long as they meet the issues in discussion. Thus, questions 19, 20, and 22 could be shared out with each student doing only one question, unless there is time to spare. The discussion will then be concerned with the following:

Question 19 discusses the effect on the energy policy of some nations of imbalance between fuel supply and demand.

Question 20 is intended to encourage students to examine figure G12 in the *Students' guide* closely and to provide more opportunities for calculation.

Question 22 shows that substantial savings by the developed world will buy some time but not very much. The fact that savings might be made in fuel use for space heating provides some justification for Section G3.

All students ought to do questions 21, 23 and either 24 or 25 since they provide more opportunities for calculation and also emphasize energy density, time, and transportability. Question 25 makes use of the Gas Laws. Question 26 gives practice in estimation.

MORE DETAIL ABOUT GROWTH

The strategy here is to reflect the increasing complexity of the picture of fuel supply and demand, having established some basic ideas in the earlier part of the Section. There is not enough time to develop the picture very far but the implications of two 'scenarios' can be examined.

One of these is of continuing exponential growth of population. If *per capita* fuel consumption remains the same, growth of fuel consumption will clearly be exponential. Question 27 deals with this.

The second 'scenario' assumes that the *per capita* fuel consumption of every citizen of the World rises to somewhere near the current levels in the developed nations. Even if population does not change, demand for fuel will increase to calculable extents as shown in question 28.

Question 29 deals with a situation which is closer to what is likely to happen – rising population together with increasing expectations. This combination predicts a further reduction in fossil fuel resources. Question 30 provides a contrast to this by indicating the vast resources of energy available from the Sun.

Questions

Questions 27, 28, and 30 form a group which could be shared among students. Question 29 depends on both 27 and 28 and could be used in class to extend the discussion of those questions.

There are, of course, many other factors which could affect future fuel consumption (see the references in the background book *Energy sources: data, references and readings*). For example, growth of population might follow a more complicated pattern than it has been possible to assume in this Unit. If World population reaches a peak before the middle of the twenty-first century, then long-term predictions based on assumptions of exponential growth will obviously be incorrect. The implications for the growth which will occur in the medium term will, however, be clear, and students should be able to understand the more complicated predictions.

Table G6 in the *Students' guide* may help to provide a focus for discussion to round off this Section and look forward to the remaining Sections in this Unit. It shows the variety of factors which influence 'energy planning'. Leave students to study it now, and come back to it later, after only a brief introduction.

SECTION G2

POWER FROM NUCLEAR FISSION

This Section links closely with Section F3, 'The nucleus'. Examination of figure G19 in the *Students' guide* (average binding energy per nucleon against nucleon number) to see the possibility of energy release on fission provides an overlap between the Units and should help students to make a rapid transition. The quantity of energy which is released on fission can be calculated using nuclear masses. What is not obvious is that this is something of a circular process since most nuclear masses were determined by measurements of the energies involved in nuclear reactions. Moreover, the physical source of the energy is not apparent from such a calculation. It is for this reason that question 31 requires a parallel calculation of the initial potential energy of the fission fragments which is transformed to kinetic energy as the charged fragments repel each other.

MAKING FISSION WORK

It is one thing to obtain energy from the fission of a single nucleus, but quite another to extract power on a commercial scale. To do this a number of conditions must be satisfied simultaneously, and it is this complexity, together with the difficulty of some of the ideas involved, which may cause problems.

Critical mass

The twin needs for many fissions per second, and for neutrons to trigger the fissions, are met when a nuclear chain reaction is established. In turn this requires a critical mass of fuel.

Nuclear cross-section

The idea of nuclear cross-section is introduced to explain first that fission occurs either with energetic neutrons and ^{238}U (utilized in fast reactors – see Section G4, part A, Nuclear power stations), or with ^{235}U and neutrons which have very low kinetic energies – so-called 'thermal' neutrons (utilized in thermal reactors, the subject of the remainder of this Section).

Absorption of neutrons, predominantly by ^{238}U, competes with the fission reaction and this is the second feature illustrated by means of nuclear cross-sections. The third element of this rather complicated picture is the very small proportion of ^{235}U in either natural or artificially-enriched uranium metal. Students are taken through this argument step by step by means of two learning questions, 34 and 40.

Moderation and control

The process of moderation involves instructive applications of dynamics and may be useful as revision of earlier work.

Control of the chain reaction also makes use of the concept of nuclear cross-section but requires little extension of the work already done on moderation.

THE SCOPE OF THIS SECTION

As in many parts of this Unit, it has been necessary to omit detail. One such point is that control would be very difficult to achieve if all the neutrons were emitted at the instant of fission. In fact, enough of them are sufficiently delayed to make control possible.

Similarly, although the deliberate enrichment of the uranium, and the need for reprocessing to remove fission products which may absorb neutrons, are mentioned in the *Students' guide*, the increasing amount of fission which is produced in ^{239}Pu, resulting from

$$^{238}_{92}U + ^{1}_{0}n \rightarrow ^{239}_{92}U \rightarrow ^{239}_{94}Pu + 2\,^{0}_{-1}e$$

is not. These points are made to emphasize that this Section is only intended to provide a brief treatment of nuclear energy: there are opportunities for students to examine nuclear energy in more detail, as well as other energy options, in Section G4.

Questions

Question 31 compares the physical origin of the energy from fission with its mere calculation using nuclear masses.

Questions 32 and 34 are learning questions: question 32 tries to give an idea of the meaning of nuclear cross-section and how it might be measured, while question 34 uses data on cross-sections to determine the neutron energy required for most effective reactor operation. Question 33 is about neutron energies.

Questions 35 to 39 are to do with moderation and control. Question 35 has links with Unit F, 'Radioactivity and the nuclear atom'; questions 33 and 36 to 38 use ideas from Unit A, 'Materials and mechanics'. All students should do questions 37 and 39 but questions 36 and 38 might be better reserved for students with good mathematical skills.

Question 40 involves calculations on a power reactor.

SECTION G3

CONSERVING FUEL IN THE HOME

The word 'conserving' in the title of this Section is used in the everyday sense of 'not wasting' and therefore refers to fuels rather than to energy. Since about 40 per cent of the U.K.'s fuel consumption is for space heating, there is obvious potential for making savings. It can be argued with some force that using fuel more effectively is equivalent to discovering a new source of fuel equal in size to the quantity of fuel saved. In this Section thermal conduction is treated in the practical context of home heating.

The Section also provides opportunities to revise ideas of flow which were introduced in Unit B, 'Currents, circuits, and charge', by making use of the analogy between thermal and electrical conduction. Electrical conduction can be said to involve a current (or flow of charge) which arises when a potential difference is established across a length of conductor (that is, when an electric field exists). Analogously thermal conduction involves a flow of internal energy which arises when a temperature difference is established across a length of conductor.

The features of the analogy are as shown in table G2.

Feature of the analogy	Electrical conduction	Thermal conduction
Variable 1	potential difference V(volt)	temperature difference $\Delta\theta$ (kelvin)
Variable 2	current I(ampere)	rate of flow of energy ϕ(watt)
Relationship: (variable 1) = constant × (variable 2)	$V = IR$	$\Delta\theta = \phi\mathscr{R}$
Proportionality constant	Electrical resistance, R ($\mathrm{V\,A^{-1}}$ or Ω)	Thermal resistance, $\mathscr{R}$ ($\mathrm{K\,W^{-1}}$)
Determinants of resistance	$R = \rho\dfrac{l}{A}$	$\mathscr{R} = \dfrac{1}{k}\dfrac{l}{A}$*
	ρ is resistivity of the material (Ωm)	k is thermal conductivity of the material ($\mathrm{W\,m^{-1}\,K^{-1}}$)

*Note this is only true in practice if the length l is small compared with the area A.

Table G2
Analogy between electrical and thermal conduction.

It will be clear that the equations which describe the two situations must be similar in form since they describe the essential feature of the two systems – that of flow. Other common features are the ideas of resistance and conductivity.

But of course the systems are not identical. The units obviously cannot be the same. Moreover the things that are said to flow (charge and energy) are not only not the same but are not of the same kind. It is worth emphasizing the point about models and flow which was made on page 415 of this *Teachers' guide*.

Teaching strategies

The *Students' guide* for this Section places the questions within the text rather than in one block at the end of the Unit. This is done to allow teachers to choose between (at least) two strategies for teaching the material.

Strategy A
The Section may be taught like any other, using the material in the *Students' guide* to support the teaching in class and using questions in class or for homework as seems appropriate. If this strategy is adopted the following example serves only as a brief guide to teachers of the main points of the Section.

Strategy B
The example which follows may be used in class to introduce the main ideas of the Section while accepting that students will not understand the work in detail. The students may then work through the text in the *Students' guide* on their own, either in class or at home, in order to develop their understanding.

Stage 1 Introducing the equations:

$$\mathcal{R}=\frac{l}{kA} \qquad \text{and} \qquad \Delta\theta=\phi\mathcal{R}$$

The equations will be approached by analogy with electric current, but since the students will subsequently be working through the *Students' guide* there is no need to treat the analogy in great detail at this stage.

Stage 2 Calculation on a single-glazed window produces a surprising result
Consider a room at 20 °C when the outside temperature is 0 °C.

Calculate the rate of flow of energy through a single-glazed window of the dimensions 2.0 m × 1.0 m, and 6 mm thick, where the thermal conductivity of glass, $k = 1.0\ \text{W m}^{-1}\ \text{K}^{-1}$.

Thermal resistance of the glass

$$\mathscr{R} = \frac{l}{kA} = \frac{6 \times 10^{-3}\ \text{m}}{1\ \text{W m}^{-1}\ \text{K}^{-1} \times 2\ \text{m} \times 1\ \text{m}} = 3 \times 10^{-3}\ \text{K W}^{-1}$$

$$\phi = \frac{\Delta\theta}{\mathscr{R}} = \frac{20\ \text{K}}{3 \times 10^{-3}\ \text{K W}^{-1}} = 6.7\ \text{kW}$$

This is clearly much too high.

Stage 3 Introducing surface resistance

Examination of the equations shows that the calculated value of $\mathscr{R}$ must be too low.

We have assumed that the window glass separates a body of air at a uniform temperature, θ_1, from another body of air, also at uniform temperature, θ_2, outside. This is too simple a picture. The glass surfaces are not at θ_1 and θ_2. The situation is rather complex but can be treated as a boundary or surface layer of air on each side of the glass. Energy from the bulk of air in the room must pass through these two layers as well as the glass. They present extra thermal resistances 'in series' with the glass and there is a temperature difference across each layer. As the calculation below will show, the 'surface resistances' are actually greater than the thermal resistance of the glass, and most of the temperature drop between interior and exterior is across the surface layers, not across the glass.

These surface layers are of variable thickness and have diffuse boundaries. However, for practical purposes, it is possible to regard the layers as presenting ordinary thermal resistances, providing that the conditions are similar to those of a typical domestic situation, for example, having limited temperature differences.

Resistances of unit area of conductor are referred to in this Section as *resistance coefficients* (but they are also known as unit thermal resistances).

Typical values of surface resistance coefficients (*i.e.*, surface resistances for one square metre) are

$0.13\ \text{m}^2\ \text{K W}^{-1}$ for internal surfaces

$0.06\ \text{m}^2\ \text{K W}^{-1}$ for external surfaces.

The power loss through the window can now be recalculated.
For 1 m² of glass

$$\mathcal{R} = \frac{l}{kA} = \frac{6 \times 10^{-3}\,\text{m}}{1\,\text{W m}^{-1}\,\text{K}^{-1} \times 1\,\text{m} \times 1\,\text{m}} = 6 \times 10^{-3}\,\text{K W}^{-1}$$

This resistance and the internal and external surface resistances act in series and their values therefore add to give the total thermal resistance of one square metre of window as

$$0.06 + 0.006 + 0.13 = 0.196\,\text{K W}^{-1}$$

and the power loss through each square metre is

$$\phi = \frac{\Delta\theta}{\mathcal{R}} = \frac{20\,\text{K}}{0.196\,\text{K W}^{-1}} \approx 102\,\text{W}$$

Hence the total power loss through the whole window with area 2 m² is 2 × 102 W = 204 W. This is much more reasonable.

Home experiment

Home experiment GH2, Surface layer, is relevant at this point.

Stage 4 Using the analogy to calculate temperatures

An electrical analogue of one square metre of window is shown in figure G1.

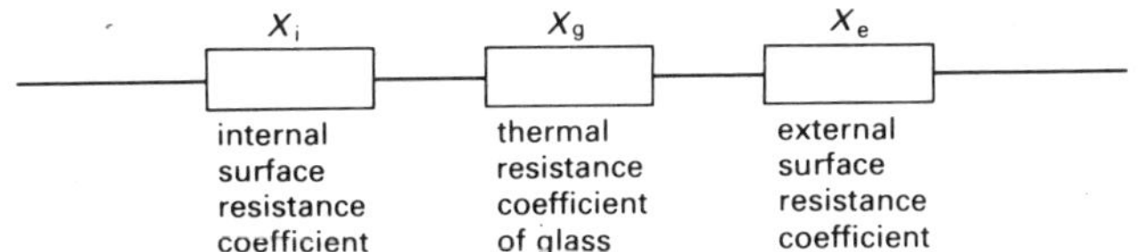

Figure G1

The resistances act in series and have a total temperature difference of 20 °C across them when the thermal energy flow is 102 W. A calculation analogous to the electrical one gives temperature differences proportional to the resistances

i.e. $$\frac{\Delta\theta_e}{0.06} = \frac{\Delta\theta_g}{0.006} = \frac{\Delta\theta_i}{0.13} = \frac{20}{0.196}$$

giving
$$\Delta\theta_e = 6.1\,°\text{C}$$
$$\Delta\theta_g = 0.6\,°\text{C}$$
$$\Delta\theta_i = 13.3\,°\text{C}$$

The reason for the error in the original calculation of power loss is now clear: the temperature difference across the glass is much less than 20 °C.

Stage 5 Double glazing
Now suppose that the window is double-glazed with a cavity 20 mm wide, filled with air which is assumed to act just as a normal conductor, having conductivity $k = 0.024\ \mathrm{W\,m^{-1}\,K^{-1}}$. (This is, in fact, an invalid assumption, as is shown later.)

The resistance coefficient of the cavity is

$$\frac{l}{kA} = \frac{0.020\ \mathrm{m}}{0.024\ \mathrm{W\ m^{-1}\ K^{-1}} \times 1\ \mathrm{m^2}} = 0.83\ \mathrm{m^2\ K\ W^{-1}}$$

The total resistance coefficient of the window (cavity plus two pieces of glass plus external and internal surfaces) is therefore

$$0.83 + 2(0.006) + 0.06 + 0.13 = 1.032\ \mathrm{m^2\ K\ W^{-1}}$$

The power loss per square metre is

$$\frac{\Delta\theta}{\mathscr{R}} = \frac{20\ \mathrm{K}}{1.032\ \mathrm{K\ W^{-1}}} = 19.4\ \mathrm{W}$$

and hence for the $2\ \mathrm{m^2}$ window it is 39 W.

Even double-glazing firms do not claim an 80 per cent reduction! In fact, the cavity does not act as a simple slab of conductor. Nor does it act simply as two internal surfaces, since radiation and convection have a substantial influence as the *Students' guide* shows.

For the moment it is enough for students to know that a cavity can be represented by yet another resistance – a cavity resistance. The cavity resistance coefficient for this cavity is $2.1 \times 10^{-1}\ \mathrm{m^2\ K\ W^{-1}}$.

The resistance coefficient for the window is now

$$0.13 + 2(0.006) + 0.21 + 0.06 = 0.41\ \mathrm{m^2\ K\ W^{-1}}$$

and the power loss per square metre $= 20\ \mathrm{K}/0.41\ \mathrm{K\ W^{-1}} = 49\ \mathrm{W}$.

Hence, for the whole window the power loss is an altogether more reasonable 98 W.

After this introduction students should now be able to work on the *Students' guide* which deals with this same material but with the following slight changes:

a there is more material in the *Students' guide*, for example, work on ventilation;

b there is more detail, for example spelling out the effects of convection and radiation on energy flow through a cavity;

c the treatment is slightly changed in order to be a little different from this example *e.g.*, the use of the relation $\mathscr{R} = X/A$ where X is the thermal resistance coefficient of a material, that is to say, the thermal resistance of unit area. (We choose to use X rather than the more familiar U-value often used by heating engineers, so that the discussion can be in terms of resistance which is an idea students are used to. The thermal resistance coefficient, X, is in fact the reciprocal of the U-value. X is, in fact, sometimes used to compare the effectiveness of different kinds of roof insulation, for example.)

Questions

Questions 41 to 45 deal with the electrical analogy and give practice in calculating thermal resistance and power loss.

Questions 46 to 48 deal with thermal resistance coefficient X – the thermal resistance of unit area. The relationships $\mathcal{R} = X/A$ and $X = l/k$ are introduced. It is important for students to grasp that the thermal resistance of a large area of wall, window, etc., is less than that of a small one.

Questions 49 and 50 are about surface resistance and surface resistance coefficient.

Questions 51 to 57 deal with cavities: cavity walls and double glazing.

Finally, questions 58 to 60 are about the overall heating of buildings: new factors including thermal capacity of the building, ventilation, and the power output of the occupants of a room are introduced.

SECTION G4
ENERGY OPTIONS

In this Section students study a selection of energy options. Some of these options are actually energy sources while others are major factors associated with energy sources, such as conservation and storage.

Students should learn about these topics by reading and help others by reporting their findings. Organizing their knowledge in order to report back effectively will, of course, also help them to learn.

In order to help students to get started some 'study items' (mostly questions, but some instructions, or hints) have been included for each option. Although we have made these 'study items' reasonably comprehensive (so that conscientious students should be successful), students should be encouraged to go beyond them.

The background books *Energy sources: data, references, and readings* and *Energy options: a reader* contain some 'starter' articles and references to suitable sources of information. Many books on energy topics are now available, but a sufficiently large library, even of paperback books, will be expensive and take time to build up. Spreading this part of the course over a few weeks, perhaps while other Units are being taught, will give students the opportunity to use public libraries (even if they are initially reluctant to do so).

Some suggestions are made in the *Students' guide* about reporting back (verbal report; written report; slides or other illustrations; poster session). These will be familiar to teachers, but are mentioned since students may be working on their own, perhaps during the holidays, and written help may be useful. The suggestions will, of course, have to be interpreted for the circumstances of each school.

A simulation, such as the 'Power Station Game', either the original IEE version or the shorter one from *Science in society* could form a good finale to the Unit (details of both of these are in the background book *Energy sources: data, references, and readings*). The time needed could be saved by reporting back in the form of duplicated notes or a poster session.

Seven energy options are listed in the *Students' guide*; each is divided into a number of topics.

The individual topics are of different lengths and the amount of work involved is unpredictable; a good student will find a large amount of information on an apparently 'thin' topic. Two topics might be the norm, but teachers will be able to give advice about the number to be investigated, in the light of their knowledge of their students and the topics which are being considered.

The energy options (and topics) are:
Nuclear power (nuclear power stations; nuclear fusion; nuclear safety).
Solar energy (passive solar devices; active solar devices; photovoltaic devices; biomass; wind energy).
Fossil fuels (power stations; alternative sources of fossil fuels).
Other sources of internal energy (geothermal; heat pumps; combustion of refuse).
Water power (hydro-power; wave energy; tidal energy).
Conservation (domestic; industrial; transport).
Storage

ANSWERS TO QUESTIONS

UNIT A
Materials and mechanics

1 The graph is shown in figure Q1.

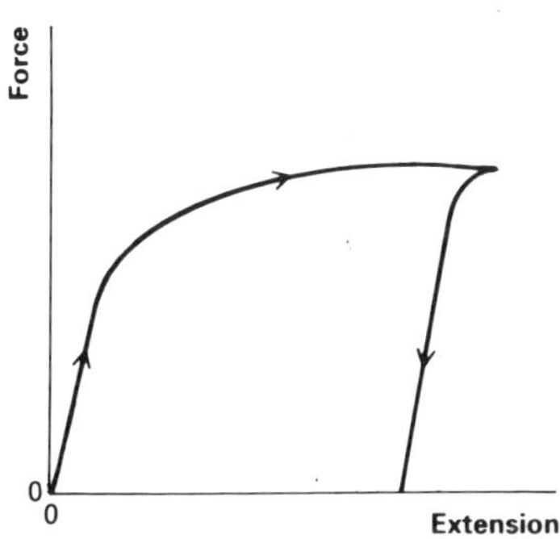

Figure Q1

2 Question **1** asked for a translation from words to a graph: this one asks students to reverse the process. Being able to 'read' a graph is an important skill to acquire. The four substances are: polythene strip, copper or aluminium wire, rubber cord, and steel or cast iron wire (this last one is too curved to be a thin glass rod).

3 'Tough' is the opposite of 'brittle'; 'plastic' and 'ductile' are almost the same; 'strong' refers to a high breaking stress. (See table Q1.)

4a *ii*
b The person will oscillate up and down (might even bounce). The oscillations are likely to be heavily damped and die away quickly. The final equilibrium position will be as in part **a**.

5a $100\,\mathrm{N\,m^{-1}}$
b 0.02 m; $50\,\mathrm{N\,m^{-1}}$
c 0.005 m; $200\,\mathrm{N\,m^{-1}}$
d Halved; doubled.

6a $10^6\,\mathrm{N\,m^{-2}}$
b 0.33
c 10 N. This is very thin wire — about 0.12 mm in diameter.
d About 3.6 mm for a one tonne car.
e 0.02 m
f Assuming that the rubber band is 1 mm × 3 mm when stretched to three times its original length (which it isn't), the force is 6 N.
g The interpretation is wrong in the sense that a steel wire would not survive such a strain.

7a The force is doubled. $F = 0.8$ N.
b*i* 0.16 N. *ii* Over 10 N. It is not likely that Hooke's Law is applicable over such a range.
c Doing the straight-forward calculation would suggest a mass of 2000 kg. But the wire is unlikely to stand this sort of stress. So the answer is 'There's no knowing'!

	ductile	elastic	plastic	strong	tough	brittle
rubber	✗	✓	✗	✓	✓	✗
copper wire	✓	for tiny strains	✓	fairly	fairly	✗
biscuit	✗	✗	✗	✗	✗	✓
glass	✗	✓	✗	✓	✗	✓ (with cracks)

etc.

Table Q1

8a Glass is stiffer than Perspex; it requires more force per unit cross-sectional area to stretch it the same amount.
b It is easier to flex wood across the grain – in the low Young modulus direction.
c Wood.

9a 10^{11} to 10^{12} $\mathrm{N\,m^{-2}}$
b 100 to 1000 times greater.
c 3.4 mm

10a $\mathrm{M\,L\,T^{-2}}$
b $\mathrm{M\,T^{-2}}$
c $\mathrm{T^{-1}}$
d $\mathrm{T^{-1}}$ *i.e.*, frequency.
e $1/2\pi$ is purely a number. It does not represent a mass, length, or any other physical quantity, as k does.
f [stress] $= \mathrm{M\,L^{-1}\,T^{-2}}$; strain is simply the ratio of two lengths, so it has no dimensions; [the Young modulus] $= \mathrm{M\,L^{-1}\,T^{-2}}$

11a 100 N
b 10 N
c 50 N
d 10 J
e $\frac{1}{2}Fx$
f Less. The area under the force–extension graph (*i.e.* between the curve and the extension axis) is less than it would be if the force were proportional to extension.

12a No. The tension in the elastic cord is continually decreasing.
b About $0.4\,\mathrm{m\,s^{-1}}$, kinetic energy about 0.08 J.
c Using $\frac{1}{2}Fx$, the energy stored is about 0.11 J.
d The calculation of energy stored gives a larger value than the actual kinetic energy found, partly because some energy is dissipated in friction and partly because, for rubber, the force is not proportional to the extension.
e, f If the force, F, were proportional to extension, x, so that $F = kx$, then the energy $\frac{1}{2}Fx$ would be $\frac{1}{2}kx^2$. The graph of potential energy against extension would be a parabola.

13a 1 N
b*i* 2 N to the left.
ii $2.5\ \mathrm{m\ s^{-2}}$ to the left.
c*i* 0.2 J
ii 0.25 J
d 0.05 J

14a Tensile stress $= F/A$
b Tensile strain $= x$/original length
c $\frac{1}{2}$ (stress × strain)
d Rubber.
e Steel springs; yew bows; tendon: shock absorber for body or athletic ability; rubber: elastic uses.

15 $v \approx 8\,\mathrm{m\,s^{-1}}$; the end of the wire is likely to travel at least twice as fast, *i.e.*, about $20\,\mathrm{m\,s^{-1}}$.

17a*i* About 0.1 J.
ii About 0.07 J.
b The graphs are shown in figure Q2.

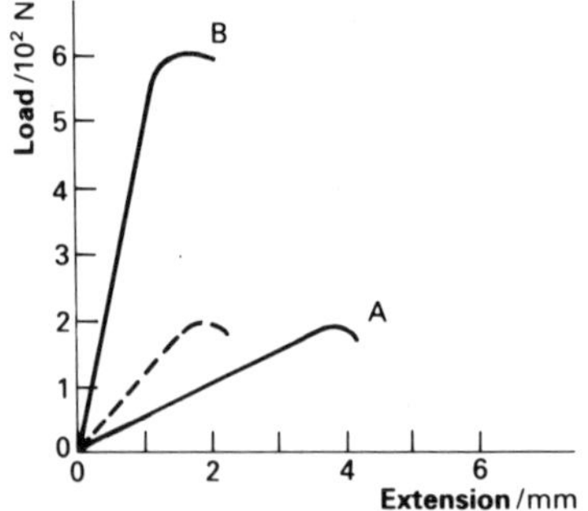

Figure Q2

c The longer cable can stretch more than a shorter one before breaking, and can store more energy. The longer one is therefore less likely to snap when there are sudden jerks.

18b $2.4 \times 10^{-9}\,\mathrm{m}$
c 10 %; 1 %; 30 %
d The layer might be more than one molecule thick.
e About 10.
f About 10^{19}.

19a 6000:1
b 5×10^{-9} m
c Smaller.

20a Na: 22.99 g; Cl: 35.45 g.
b NaCl: 58.44 g.
c 6×10^{23}
d The charge per mole of electrons (or any other singly charged particles, for example hydrogen ions) is a universal constant called the *Faraday constant*, $F = 9.65 \times 10^4$ C mol^{-1}. Electrolysis measurement of F and X-ray diffraction methods of finding L are used to find an accurate value of e, the electronic charge.
e 3×10^{-5} m^3 mol^{-1}

21a 9.4×10^{24}
b 8.4×10^{28}; 8.4×10^{22}
c 1.2×10^{-29} m^3; 2.3×10^{-10} m
d 1.04×10^{-5} m^3 mol^{-1};
1.00×10^{-5} m^3 mol^{-1} Atoms are about the same size despite the large difference in mass.

22a Taking one grain as a cube of side 10^{-5} m its volume will be 10^{-15} m^3 and its mass about 2×10^{-12} kg. So 1 mole of grains is 6×10^{12} kg.
b About 10^{17} atoms, *i.e.*, 10^{11} million.
c A tyre lasts about 3×10^4 km during which about 0.01 m thickness wears away. This gives an answer of several thousand kilometres.

23a 3
b Each bubble attracts other bubbles, and the result is that they cluster as closely as possible. The bubbles are round, and attract other bubbles equally in any direction, whereas, if the forces tended to act in certain special directions, some arrangement like **1** or **4** might result.
c By analogy, it might be true that the attractive forces between atoms in such metals were equal in all directions.
d Because the sizes of the Na^+ and Cl^- ions are very different.

24a 7.12×10^{-6} m^3
b*i* 8.68×10^{-30} m^3
ii 8.20×10^{23} mol^{-1}
iii This is an upper limit for L because the spherical copper atoms cannot fill all the space in a crystal of copper.
c*i* 16.6×10^{-30} m^3
ii 4.30×10^{23} mol^{-1}
iii This is a low value for L because the assumption of a 'square' array over-estimates the volume taken up by each atom.
d 6.15×10^{23} mol^{-1}

25a There are no forces between the spheres until they touch. The repulsive force then becomes very, very large; there is a discontinuity.
b r_0 is the diameter of either sphere.
c Suppose the balls obey Hooke's Law with a large force constant (the larger the force constant, the steeper the line). (See figure Q3.)

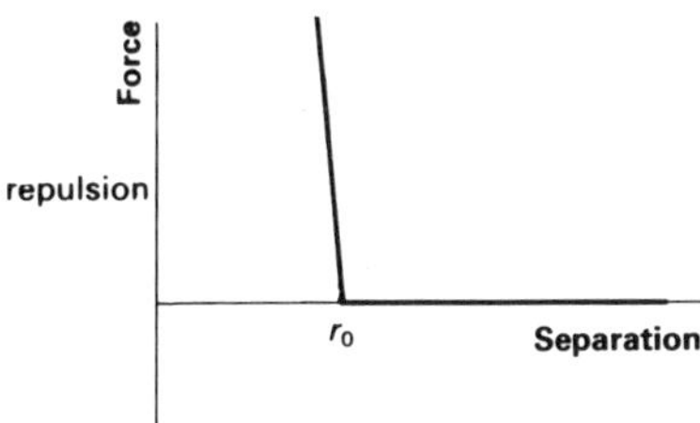

Figure Q3

d The potential energy curve will be parabolic. (See figure Q4.)

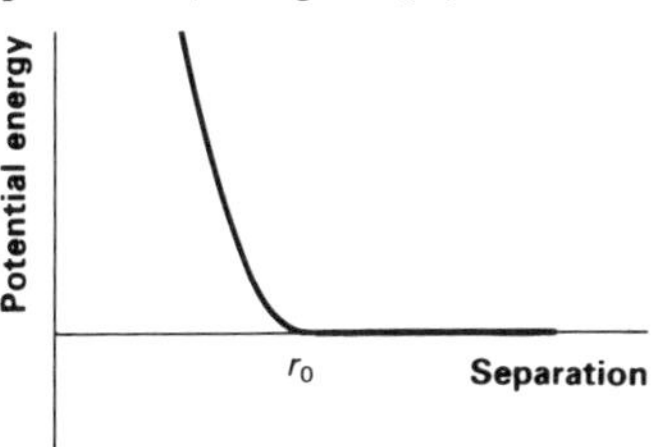

Figure Q4

26a, b, c The graphs are shown in figure Q5.
d 0A is the equilibrium separation. 0B indicates the maximum separation between atoms before the substance yields.
e 0A is about 2.5×10^{-10} m.

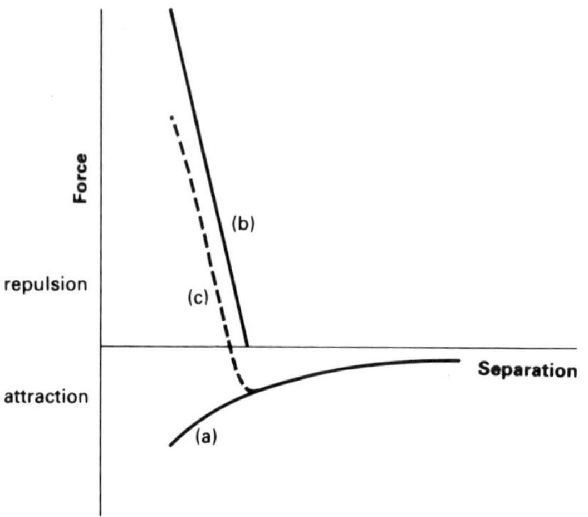

Figure Q5

27a The potential energy of the two atoms decreases as they move together with a net attractive force. 0Q gives the mean equilibrium separation, r_0. As the repulsive force is predominant for closer distances than r_0 the potential energy rises. It becomes positive for separations less than 0P.
b 0Q on the potential energy–separation graph = 0A on the force–separation graph.
c The sum of potential and kinetic energies must be constant if there is no external source or sink of energy. So as the kinetic energy increases the potential energy will decrease, and vice versa. The potential energy will vary by ΔE as the atoms oscillate between separations X and Y (see figure Q6).

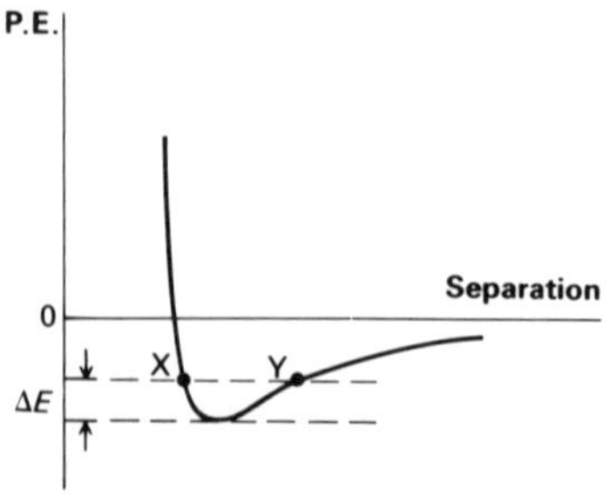

Figure Q6

d The depth of the potential well at Q gives the binding energy, E (see figure Q7).

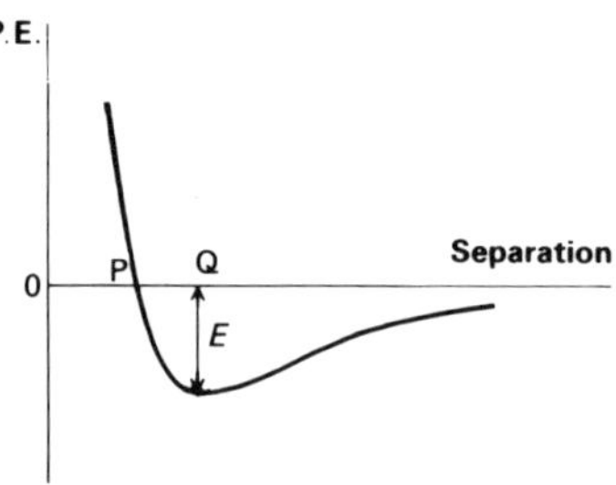

Figure Q7

28a A
b B
c B
d B

29a 2 N
b $400\,\mathrm{N\,m^{-1}}$

30a nr_0
b $n\Delta r$
c $\Delta r/r_0$
d $\mathrm{N\,m^{-1}}$
e $mk\Delta r$
f $1/r_0^2$
g $k\Delta r/r_0^2$
h k/r_0
i $60\,\mathrm{N\,m^{-1}}$

31a $x\sqrt{n}$
b xn
c About 9; since strain is $(xn - x\sqrt{n})/x\sqrt{n}$.

32 The Summary for Section A2 gives a brief description of these terms under the heading 'Structure of solids'. Several text books have good explanations too.

33a Metals bend because the crystal grains can slide over each other. The size of the grains is important in determining the mechanical properties of a material. There are dislocations too.
b Toffee will shatter if it is cold. Glasses do not have a sharp melting-point (unlike crystalline solids) but just become more viscous as they cool, hence the disorderliness and lack of regular crystal structure.
c Unlike many crystalline solids, glass-like substances do not have a sudden change of density when they melt, nor do they absorb energy (latent heat).
d Plastic flow occurs in metals where only one atom has to move at a time, *i.e.*, where there are dislocations in the grains which can move easily. (See question **34**.)

34 '.. in the next picture the dislocation has moved on by one atom. (See figure Q8.) In each succeeding picture the dislocation moves on by one atom until it reaches the edge.'

35 See, for example, Akrill, Bennet, and Millar *Physics*, Chapter 5.

36a A
b B
c A is high carbon steel; B is mild steel.
d The introduction of foreign atoms makes the material harder; there is less chance of slip. Many foreign atoms will cause the substance to become hard and brittle – no slip possible (A). With fewer, there is still some hardening and also slip, so some ductile effect (B).

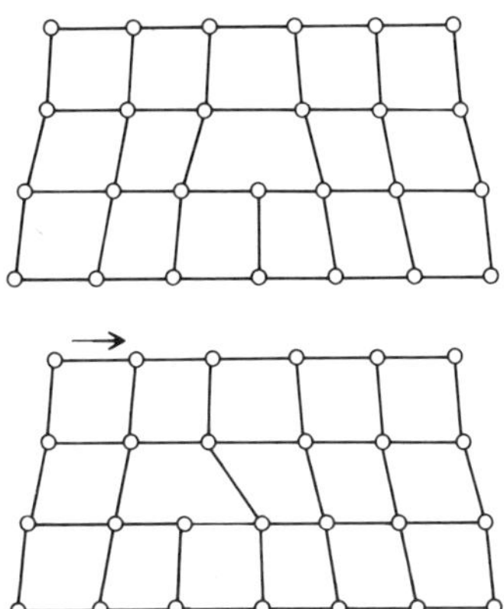

Figure Q8

37a The diagrams are shown in figure Q9.

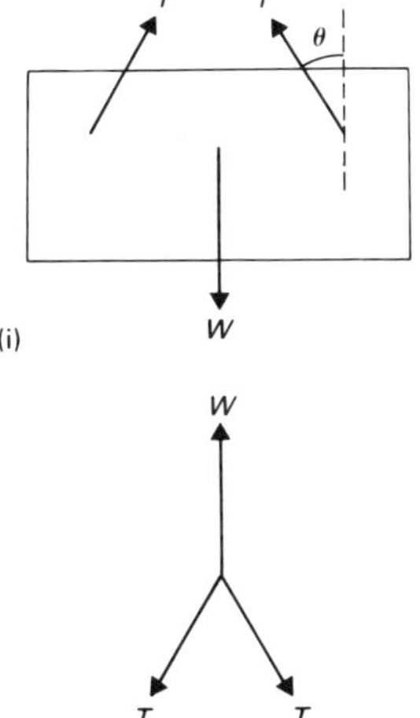

Figure Q9
(i) Forces on the picture.
(ii) Forces on the hook.

b The tension $T = W/2\cos\theta$. As $\theta \to 0$ so $T \to W/2$. The tension is less in a long cord than a short one.

38a $F = N \sin\theta$
b $W = N \cos\theta$
c $F = W \tan\theta$
d $F \cos\theta = W \sin\theta$

39a The force resulting from his weight and the pull of the rope acts perpendicularly to the rock face through his feet. It is balanced by the normal force exerted by the rock on him.
b 500 N; 500 N.
c In equilibrium, the three forces must still pass through a point. If he leans forwards the contact force will increase and the tension decrease. Students should use the angles from a space diagram to draw a force diagram to show this. The contact force *on* the rock is made up of a normal force and a frictional force parallel to the rock (and acting down the slope).

40a Hinge 2 is the top one.
b See figure Q10.

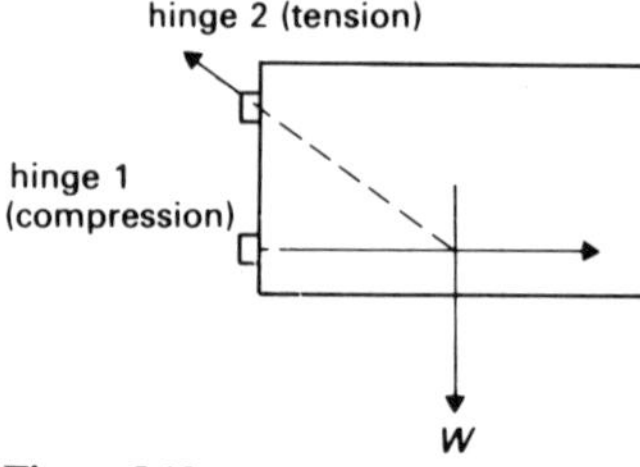

Figure Q10

c Force at hinge 2 = 1414 N, at 45° to vertical. Force at hinge 1 = 1000 N, horizontal.

41 Forces are 9.7 kN and 12.3 kN at left- and righthand supports respectively.

42a *mg*
b Greater.
i $T = mg/\cos\theta$
ii $F_e = mg\tan\theta$
c For small angles $\tan\theta \approx \theta$, and $\theta \propto d$.

43a*i* 400 N
ii 346 N
iii 346 N to the right.
b*i* 200 N
ii 173 N
iii 173 N to the right; 100 N upwards.
c*i* 600 N
ii 520 N
iii 520 N to the right; 100 N upwards.
d*i* 795 N
ii 521 N
iii 521 N to the right; 200 N *downwards.*

44a*i* The strut is in compression.
ii The tie is in tension.
iii Compression force in strut is 14 kN; tension force in tie is 10 kN.
b See figure Q11.

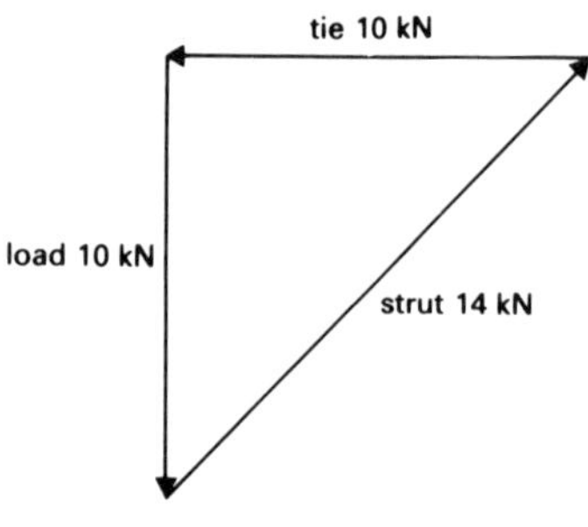

Figure Q11

c Strut.

45a BC and CD are struts (compression); AB and AC are ties (tension).
b BC and AC.
c They prevent the structure bending.
d Ties: BC, CD, AC, CE.
Struts: AB, DE, BD.

46 See, for example, Gordon *The new science of strong materials* for information on cracks, structure of wood, and composite materials.

47a The weight of the earth above the pipe forces it into an oval shape. The breaks occur where the greatest (tension) stresses are.
b X is at a point of maximum compression rather than tension. Concrete is strong in compression and weak in tension.
c The concrete surface is under compression from the plastic. So when the pipe is buried the concrete remains in reduced compression or only low tension. The plastic cover also has to stretch.

d Concrete is porous, the plastic prevents leakage. Surface damage is also stopped.

48a 4:1
b 1:1

49a Glass fibre has as great a tensile strength as steel. Concrete and wood are very bulky materials to hold large loads.
b Steel: elastic, flexible, high Young modulus, slightly malleable and ductile, brittle fracture; can be machined, melts.
Glass fibre: will give in matrix only slightly, rigid, no grain, composite material; cast in mould.
Concrete: weak matrix, brittle, strong in compression, composite material; cast in mould.
Wood: grained, strength with grain, weak in compression (empty cells collapse). What effect do cracks have? Fatigue?
c Think about: volume available, use, availability, appearance, corrosion, inflammable, insulating/conducting, ease and cost of manufacture in appropriate shape, installation cost, etc.

50a $8\,\mathrm{m\,s^{-1}}$
b $6\,\mathrm{m\,s^{-1}}$
c 1.4 J
d $1.4\,\mathrm{kg\,m\,s^{-1}}$
e 35 N

51a The centre of mass of the system of man and truck never moves as there are no unbalanced external forces. While the man runs, the truck moves to keep the system's centre of mass in the same place. When the man stops, the truck does also.
b $mv/(M+m)$
c 0
d $mvt/(M+m)$

52a 6000 N s
b $6000\,\mathrm{kg\,m\,s^{-1}}$
c About $10\,\mathrm{m\,s^{-1}}$.
d The force at any instant is proportional to the slope of the graph of velocity against time. The graph is shown in figure Q12 (i).

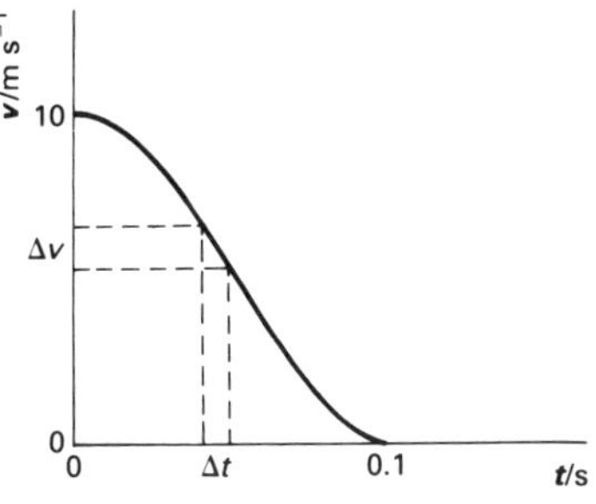

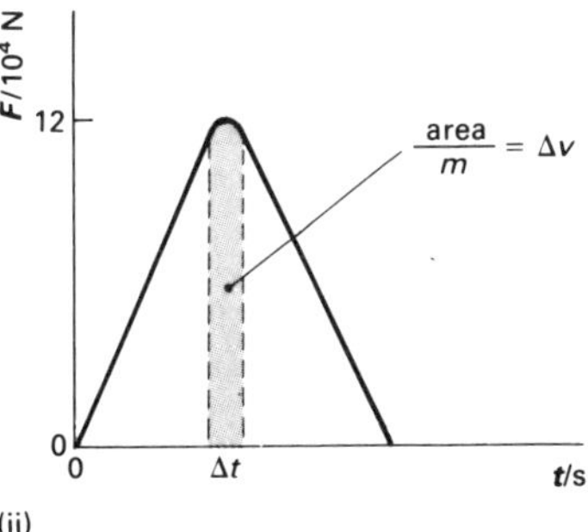

Figure Q12

53a $+0.08\,\mathrm{kg\,m\,s^{-1}}$; $-0.12\,\mathrm{kg\,m\,s^{-1}}$ (positive to right).
b $-0.04\,\mathrm{kg\,m\,s^{-1}}$
c $0.096\,\mathrm{kg\,m\,s^{-1}}$ in magnitude.
d 0.096 N s in magnitude.
e 0.96 N in magnitude.
f $-0.08\,\mathrm{m\,s^{-1}}$
g 0.0165 J; 0.024 J
h 0.0016 J
i Thermal energy, sound, and perhaps permanent deformation of vehicles.

54a The Earth.
b The pull of the Earth on the book, the push of the table on the book. The force pairs of Newton's Third Law are the two gravitational forces (Earth on book, book on Earth), and the two contact forces (book on table, table on book). Newton's Third Law does not require that the two forces acting on the book be equal. If they were not the book would accelerate (Newton 1).
c The Earth moves up towards the body with equal and opposite momentum.

d The train pushes on the Earth which moves 'backwards' as the train moves 'forwards'. The Earth and train make the closed system.

The ratio of masses ensures that the effect on the Earth's motion is negligible in both cases.

56d Electron loses 4×10^{-4} of its kinetic energy; energy would have to be measured to better than 0.04 %.

57a 7499:7501; 0.03 %; yes, the approximation is justified.
b $400\,\text{m s}^{-1}$; $402\,\text{m s}^{-1}$. The kinetic energy of the molecules will increase, *i.e.*, the gas will become hotter.

58 One is a nearly elastic collision, the other perfectly inelastic. The areas and times of impact are very different so the forces involved are very different even if the impulses are similar.

60a $2.89 \times 10^{-2}\,\text{kg mol}^{-1}$
b $8.29\,\text{J mol}^{-1}\,\text{K}^{-1}$; a more accurate value is $8.31\,\text{J mol}^{-1}\,\text{K}^{-1}$.
c 2.08×10^{25} molecules.
d $3 \times 10^{-9}\,\text{m}$; about 10 times diameter of an atom.
e $1.38 \times 10^{-23}\,\text{J K}^{-1}$

61a 42 moles
b 7.0 kg
c $6.0\,\text{m}^3$

62a $c_x/2l$
b $2mc_x$
c mc_x^2/l
d mc_x^2/l
e $Nm\overline{c_x^2}/l$
f $\frac{1}{3}Nm\overline{c^2}/l^3$
g $\rho = Nm/l^3$; $\frac{1}{3}\rho\overline{c^2}$
h $\overline{c^2} = 2.5 \times 10^5\,\text{m}^2\,\text{s}^{-2}$; $c_{\text{r.m.s.}} = 500\,\text{m s}^{-1}$

63a $350\,\text{m s}^{-1}$
b $17\,\text{m s}^{-1}$ to left.
c $3.8 \times 10^{-21}\,\text{J}$
d $390\,\text{m s}^{-1}$
e $390\,\text{m s}^{-1}$. Root mean square speed is the name given to the speed calculated in part **d**. For the large number of particles in a gas the r.m.s. speed is 1.09 times the mean speed.
f $8.5 \times 10^{-25}\,\text{kg m s}^{-1}$ to the left.
g Parts **b** and **f** only.

65a $5.65 \times 10^{-21}\,\text{J}$
b $4.71 \times 10^{-26}\,\text{kg}$
c $5.65 \times 10^{-21}\,\text{J}$
d $1300\,\text{m s}^{-1}$

66a Using $\frac{1}{2}MV^2 = \frac{1}{2}mv^2$ gives $400\,\text{m s}^{-1}$ (to one significant figure, which is all that the data given justifies).
b Using $c_{\text{r.m.s.}} = (3p/\rho)^{\frac{1}{2}}$ gives the same value.

67a $210\,\text{m s}^{-1}$
b 2.4 ms
c 110 km; 1.2×10^{12}
d $9.2 \times 10^{-8}\,\text{m}$ *i.e.*, approximately $10^{-7}\,\text{m}$.
e $1.5 \times 10^{-10}\,\text{m}$
f Yes. The assumption for the theoretical calculation is that all molecules are of the same diameter. This is approximately true.

68a $1.7 \times 10^6\,\text{s}$ (nearly three weeks).
b Convection currents in a room cause circulation as will any draughts.

69a pA
b $pA\Delta l$
c $p\Delta V$
d $p\Delta V$
e No frictional losses at piston; change happens slowly.
f When air is compressed the work done on it increases its internal energy; when it expands it does work, its internal energy is reduced.

UNIT B
Currents, circuits, and charge

1a 0.16 C
b 10^{18} particles.
c 6×10^{22} particles per metre length.
d $1.67 \times 10^{-5}\,\text{m}$
e $1.0 \times 10^{-6}\,\text{m s}^{-1}$

f The algebraic answers to **a–e** become: It, It/Q, nA, It/QnA, I/QnA

The speed at which anything moves obviously does not depend on the length of time for which its motion has been observed.

g Reducing n increases v; fewer particles have to move faster.

Reducing Q raises v; each carries less charge, so they must move faster to carry the same total charge in a given time.

A molar solution is strong, but can be made, and contains 6×10^{23} solute molecules in one litre, and so 6×10^{26} molecules in one cubic metre. If every molecule provides an ion, there will be 6×10^{26} ions in each cubic metre. Water molecules are not by any means all dissociated into ions in very pure water while, say, paraffin would contain almost no ions.

2a Unless there is some evidence that the junction acquires a net charge, all charge flowing in flows out.

b No.

c Yes.

d Not if both are copper.

e See answer to **a**.

3a A possible estimate can be reached by considering the time for which the cell could light a torch bulb (say 5 hours) carrying a current of, say, 300 mA.

b Estimate mass and cruising velocity and calculate $\frac{1}{2}mv^2$. An allowance for inefficiency is necessary (typically the overall efficiency of conversion of internal energy into mechanical energy is about 30 per cent). An all-electric train will have a much higher efficiency *by itself*, but the overall inefficiency is transferred back to the power station.

c With no losses, about 110 seconds.

d*i* 1000 A *ii* 100 A *iii* $(10)^2 = 100$ times greater. *iv* 100 times longer.

4 B

5a An estimate of mass equivalent to 1 kg of water and a temperature rise of 80 K gives 4.7 A (assuming a 240 V supply).

b*i* $0.076\,\mathrm{kg\,s^{-1}}$. A lower limit.

ii If the water inlet temperature drops the output temperature can only be maintained by either reducing the flow rate or increasing the power.

Likewise a drop in flow rate requires a reduction in power for the same temperature rise.

Inlet temperature and flow sensors can thus in principle initiate control actions. A full control strategy would have to consider the upper limits imposed on power and flow rate.

6a If a p.d. is in some way equivalent to pressure in a liquid then the units of each ought to show some parallel. The equivalent of joule/coulomb would be joule/(metre)^3.

$1\,\mathrm{J\,m^{-3}} = 1(\mathrm{N\,m})\,\mathrm{m^{-3}} = 1\,\mathrm{N\,m^{-2}}$

i.e., 1 pascal, the pressure unit.

If a charge, Q, moves through a p.d. of V, the energy transformed is VQ. Similarly the work done when a volume of fluid, V, is driven through a pressure difference p is pV.

b Some points to bring out:

Apart from overheads, payment is for fuel used in power stations. Kirchhoff's First Law has to hold.

The 'return path' for water flow into a house tends to be ignored: there are of course many, and we think of water as being 'consumed'.

At any instant the current flow into a house is equal to the current leaving – whether it is a.c. or d.c. is thus irrelevant. The *energy* flow is related to the outward *and* return conductors independent of current direction – it is directed *out* of the power station. (An interesting parallel is with a motor driving some machinery by means of a pulley on a drive belt.)

7 The resistance of (a) is larger than the resistance of (b) and both seem to be linear components, obeying Ohm's Law. (c) might be a diode, as its resistance changes when the p.d. is reversed. (d) could be a filament lamp, for its resistance increases as the current increases. Initially, (d) conducts better than any of the others. One could also compare the forward and reverse resistances of (c) with the resistances of (a) and (b).

8 **B**

9a Only graph (b) obeys Ohm's Law as stated. Graph (a) is linear, but there is current when there is zero applied p.d. (perhaps the component included a source of p.d.). (c) shows a constant current, not a constant rate of change of current with change of p.d. Graph (d) is non-linear, though it might be said to obey Ohm's Law for p.d.s in one direction only.
b Yes. A graph of current against p.d. is a straight line through the origin. Alternatively, equal increases in current go with equal increases in voltage. The resistance at 2.0 mA, and at all other currents given, is $14.5 \times 10^3\,\Omega$.
c The answers here are partly matters of opinion. Certainly Ohm's Law is not applicable to all objects that conduct, but it does compactly describe the behaviour of some. Suggestion **3** avoids trouble by defining it away; sometimes physicists do this, but in this particular case they usually speak of 'ohmic' or 'linear' materials. **4** is wrong, for Ohm's Law requires the current–voltage graph to pass through the origin and a small section of the curve, though nearly straight, will not project back to the origin. Whether Ohm's Law is important and general enough to count as a law is a matter of opinion. It does not rank with those 'Laws of Physics' to which we know no exception (give an example).

10a 2 A
b 4 A
c 6 A
d 2 Ω
e The key step is the application of Kirchhoff's First Law to the junction of R_1 and R_2, giving $\frac{V}{R}=\frac{V}{R_1}+\frac{V}{R_2}$

11a*i* The resistance of A increases with applied p.d.
ii The resistance of B is constant.
b For 2 A, $V_A = 5$ V and $V_B = 10$ V so the combined resistance is $\frac{(5+10)\,\text{V}}{2\,\text{A}} = 7.5\,\Omega$.
c If they are in parallel, the p.d. is the same across each. A p.d. of 5 V produces $I_A = 2$ A and $I_B = 1$ A, *i.e.*, a total current of 3 A.
d The resistance of a parallel combination is less than the resistance of the smallest component. The parallel combination of A and B has a minimum resistance when that of A is as low as possible, *i.e.*, at a low value of p.d.

12b The *proportional* change in resistance of the whole circuit which the rheostat can make is insufficient to bring about the current change necessary for this variation in p.d. across the lamp.

13a Their resistivities rise in constant ratio.
b The logarithm to base ten of the resistivity.
c More than 10^{20} km, over ten million light years; further from us than the nearest galaxy.

14 Notice that the question says *might* apply, not *does* apply.
a X will if any do; one might argue for Y on the grounds that even a tiny current could represent a flow of electrons.
b Z
c Y or Z because resistivity falls with rise in temperature.
d X because resistivity rises with temperature.
e Y, identifying X with metals and Z with insulators.
f Z
g Y, excluding graphite.
h Z
i X

15a 1.6 A
b 2*R*

16b The coefficients of $(\delta l/l)$ and $(\delta r/r)$, the *proportional* changes in length and radius, in the expression for $\delta R/R$, are the same as the *powers* of l and r in the formula for R. This result is general and is particularly useful in thinking about accuracy of measuring and experimental uncertainty.
c 8.3 Ω
d The change in resistance is proportional to strain.

17a $I/A = V/AR$
b $I/A = V/\rho l$
c $5.9 \times 10^7\ \mathrm{A\,m^{-2}}$; current in a $1\ \mathrm{mm^2}$ wire is 59 A.
d $v = 3.7 \times 10^{26}/n$
e 8.5×10^{28} atoms, or electrons, per cubic metre.
f About $4\ \mathrm{mm\,s^{-1}}$.

18a A is connected electrically to B, B is not connected to D (unless by a very high resistance). A may or may not be connected to C (or D).
b The effective resistance between A and B has 1.5 V across it when a current of (4.5/1000) A passes, and is thus just over 330 Ω.
c The effective resistance between A and B is just over 300 Ω, in agreement with **a** and **b**.
d If there were resistors connected between A and C or between B and D, the meter would read more than 18 mA (but **a** suggested that B and D were not connected).
e If A and C were connected, the meter would give a reading when the switch was open. So they are not. The reading of 18 mA when the switch is closed confirms earlier suggestions that the circuit is as in figure Q13.

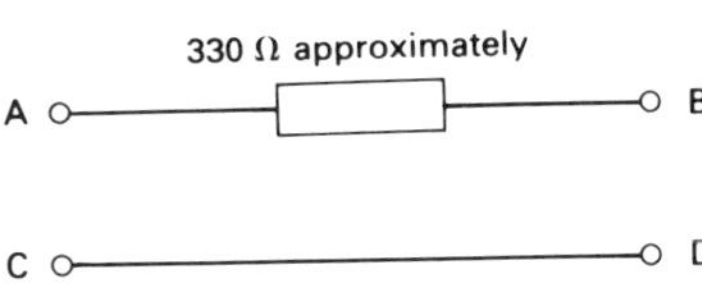

Figure Q13

19a (a)
b (b)

20a 11 000 Ω
b 0.27 mA
c Three-quarters of the length of CD.
d The lamp resistance (15 Ω working) is so low compared with the resistance of the part of the potentiometer across which it is connected that it effectively short-circuits the output, producing very nearly zero p.d. across it.

21 **B**

22 1 Ω. Students should find, and perhaps be able to prove, that the power delivered has a maximum value when R is 1 Ω, that is, the resistance of the external circuit is equal to the internal resistance of the cell.

23a **B**
b **B**

24 **E**

25 **C**

26 Briefly:
a X and Y are at the same potential.
b $8 \times 10^{-6}\ \Omega\,\mathrm{m}$
c From Y to X, since the potential of X is now more negative.

Note: Fuller answers would of course be expected in an examination.

27a From the sliding contact into the meter.
b $I_1 - I_2$
c Three; only two are needed.
d 83 μA

28a The current is zero at first; rises sharply to 2 A; remains steady for ten hours; drops sharply to zero at 6 p.m.
b 2 C
c 7.2×10^4 C
d The charge passed is represented by the shaded area in figure Q14.
e The charge passed is still shown by the area between the graph and the time axis.
f About 5×10^4 C.

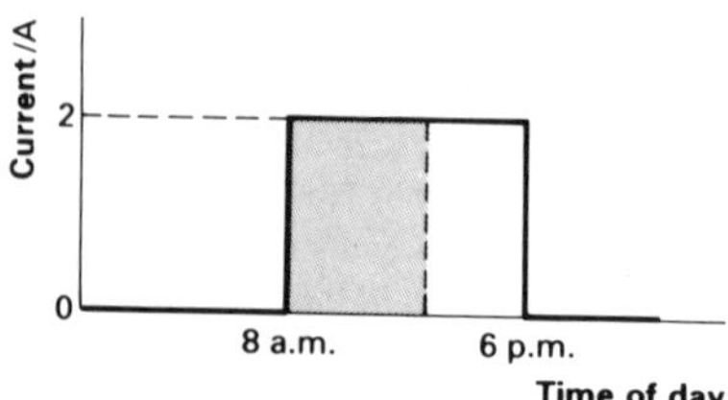

Figure Q14

29a The same as meter 1.
b They move 5 divisions to the left, and then return to zero.
c None!
d Charging, anti-clockwise; discharging, clockwise.
e When the capacitor is being charged the equality of meter readings 1 and 2 shows that as much charge flows onto one plate as flows off the other.

30a As in figure Q15.

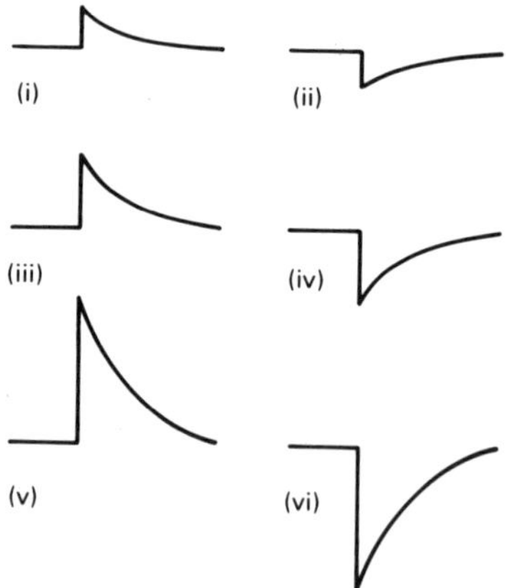

Figure Q15

b*i–iv* As for **a***i*. *v* As for **a***vi*.
c Equal changes in p.d. produce equal flows of charge (identical traces).

31a 1 mA
b 0.1 C
c 0.01 C
d 0.01 F
e Zero (and indeed at every other instant during the process the total net charge on the plates is zero).

32a 10^{-3} C
b 6.25×10^{15}
c The resistance would have to be decreased at a steady rate.
d 10^{-4} A
e See figure Q16.

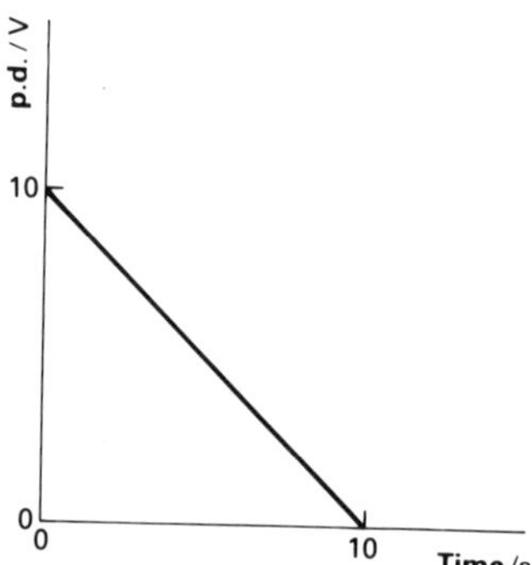

Figure Q16

33a 1 mA
b About 1.5×10^{-4} C.
c 15 μF
d As in figure Q17.

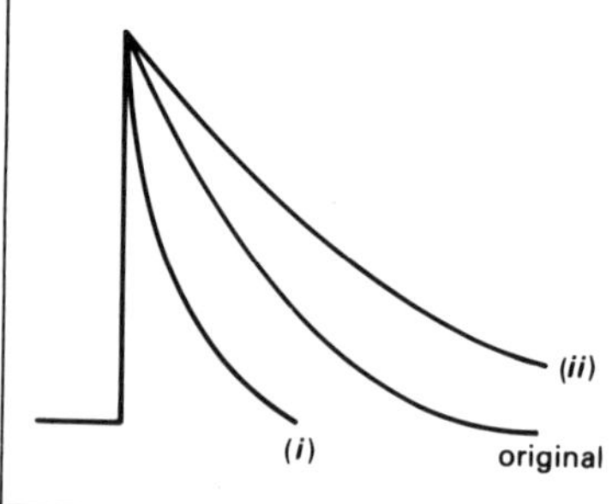

Figure Q17

34a 100 μC
b 200 μC
c 10 V
d 10 V
e 300 μC
f 30 μF
g Yes!

35a*i* No *ii* No
b No, from **a**. (Or consider the isolated part of the circuit consisting of the

righthand plate of C_1 and the lefthand plate of C_2.)
c Q/C_1
d Q/C_2
f $C=Q/V$

36a 3 μF
b 6 μF

37a 0.01 F
b 0.02 C
c 0.38 J. The taller shaded strip in figure Q18.

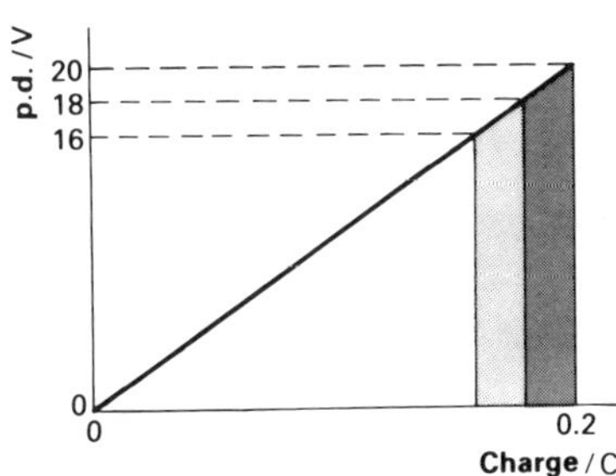

Figure Q18

d 0.34 J
e The two shaded strips in figure Q18.
f 0.38 J + 0.34 J + ... + 0.20 J = 2 J
g The total area of the triangle under the graph ($\frac{1}{2}$ × base × height) represents the energy transformed.

38a 4.5 J
b 9 ms

39a 2.1 kV
b Some points to consider:
The physical size of the bank of capacitors (the plates have to be far apart to withstand the high p.d.).
The energy will be delivered as a pulse.
Recharging arrangements.
Safety.

40a When the diaphragm bulges, there is as much more water on one side as there is less on the other. When a capacitor is charged, there is as much extra positive charge on one plate as there is missing on the other. If the pump were removed and a pipe substituted, the extra water would flow round the circuit. If a capacitor is connected to a wire, the extra charge on one plate flows round to the other. No water enters or leaves the system as a whole. No charge is created or destroyed in a capacitor circuit. The analogy is rather good.
b Yes.
c Make the diaphragm easier to stretch.
d Pressure differences across tank, corresponding to p.d., and amount of extra water in one side (and equal amount less in the other side) corresponding to charge.
e The capacitor conducts slightly through the insulating layer between the plates.

41a Identical to figure B89(b) in the *Students' guide*, but negative current.
b The currents will have initial values half that of **a** (50 μA) but will last for twice the time (time constant doubled).

42a 100 kΩ (about 40 seconds for the current to fall to $1/e$ of its starting value).
b 5 V
c Over the 80 seconds the area represents about 1700 μC. Using $Q=CV$ gives 2000 μC.
d The current is three times larger at any instant of time, so the time taken to decay to any given proportion of the initial value is unchanged.

43a See figure Q19.

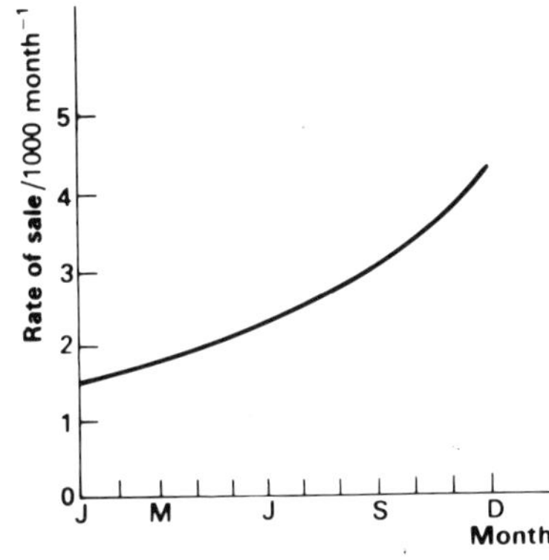

Figure Q19

b 31 310 (This can also be represented as the area under the graph.)
c The ratio is constant at about 1.08 except for the last two months.
d Christmas?
e The constant factor of 1.08 for most of the year indicates a constant *proportional* growth rate of 8 per cent per month. The growth is therefore exponential, *i.e.* the rate of increase is proportional to the total number sold. There is unlikely to be one single reason why this should be so; students should list some factors which affect demand.
f Approximately 47 months from January, assuming the growth rate for the first 10 months. Presumably the growth rate would have flattened out before this time as the market approached saturation.

44 One might use a circuit with a capacitor discharging through a variable resistor, with the discharge current holding the relay on while it exceeds 0.1 mA. The time constant of this circuit determines how long the relay is closed and hence for how long the lamp lights. The variable resistor could be calibrated for the required lighting time. Students are invited to sketch the curve of discharge current against time for these approximate values: resistance 10 kΩ, capacitance 700 μF, charging p.d. 10 V.

45a About 6×10^{15} electrons.
b About 6×10^{9} electrons.
c About 6×10^{3} electrons.
d The current is too high for the effect of individual charge carriers to show up. Even in as short a time as 10^{-12} s several thousand electrons hit the screen. Perhaps at very much lower currents (of the order of 10^{-6} or 10^{-9} A) the arrival of individual electrons on the screen might be observable. At very low currents, or over very short time intervals one might expect to see some random variation in the rate of arrival of electrons at the screen.

46a $6 \times 10^{5}\ \mathrm{m\,s^{-1}}$ approximately.
b $60 \times 10^{5}\ \mathrm{m\,s^{-1}}$ approximately.

47 The current is the ratio of power to potential difference. An estimate of 10 W power gives a current of 5×10^{-4} A. This is the same as 3×10^{15} electrons per second which, on an area of $10^{5}\ \mathrm{mm}^2$, gives a density of 3×10^{10} electrons per mm^2. Notice that the number of electrons per square millimetre of screen must be large otherwise it might be possible to see random fluctuations in the brightness as the electrons arrived randomly.

48a QV
b Fd
c $QV = Fd$. If $F = W$, then $Q = Wd/V$.
d See table Q2.

V **p.d. to balance drop/V**	Q **Charge on drop/C**	n **Multiple**
470	4.75×10^{-19}	3
820	3.18×10^{-19}	2
230	6.44×10^{-19}	4
770	1.64×10^{-19}	1
1030	1.57×10^{-19}	1
395	7.83×10^{-19}	5

Table Q2

e Basic charge e is about 1.6×10^{-19} C.
f The values of Q and n in the table lead to the following values for e:
1.58×10^{-19} C, 1.59×10^{-19} C,
1.61×10^{-19} C, 1.64×10^{-19} C,
1.57×10^{-19} C, 1.56×10^{-19} C. Given that the true value of e is 1.60×10^{-19} C, the uncertainty in these calculated values is about 3 per cent, suggesting that the *sum* of the percentage uncertainties in V, W, and d is also about 3 per cent. From the results, V, which is usually given to two significant figures only, seems to have the largest uncertainty, and d the least.

UNIT C
Digital electronic systems

1a Suppose the numbers are 1342 and 5768. Select the end digits of each number (2 and 8). Add them. The sum is 10. Store the digit 0 as the first digit of the answer, and carry the digit 1 to the next step. Select the next two digits (4 and 6). Add them, and the 1 carried over, getting 11. Store 1 and carry 1. Then continue. A better system would have a stopping device that turned the adding process off when all of the incoming digits (carry as well) became zero. Without it a mindless computer would go on for ever, producing the answer

......000000000000000000000000000007110

b When the clock reaches a preset time it switches on the power to a heater immersed in cold water. When the water boils, the heater is switched off. The 'water boiling' signal also operates a device that pours the water onto the tea leaves. After a suitable delay for the tea to infuse, a 'tea ready' alarm is rung. The human part of the system then adds milk and sugar.
c Figures Q20 to Q22 show systems, or parts of systems, to do the suggested jobs.

2 Figure Q23 shows the series of events involved when listening to a song on a record player.

4 The power supply provides the energy required. For example, a logic gate may be arranged to light a light-emitting diode (L.E.D.). In a sense the gate directs the energy to the L.E.D. under certain conditions. The energy comes from the power supply. All the systems in question **1c** have a power source. The system controls the use of that energy.

5 There are only two digits in the binary system: 0 and 1. It is easy to define two voltage levels – high and low; supply voltage and zero volts – representing the digits 0 and 1. The binary system readily lends itself to electronic systems. Defining intermediate levels to represent more digits is much more complicated.

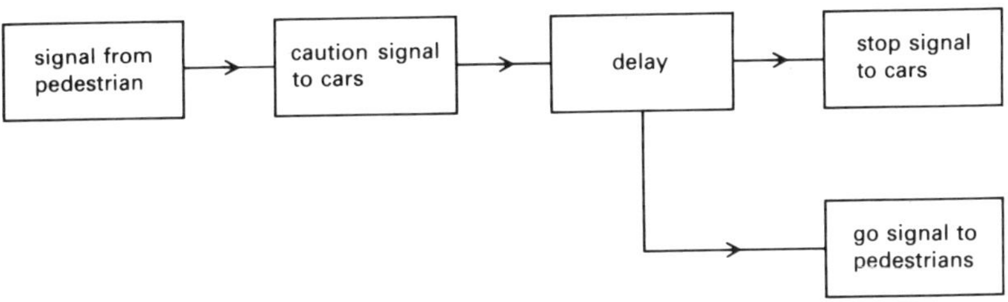

Figure Q20
Part of a system for controlling cars. Other parts should be added to operate go and stop signals for cars and pedestrians.

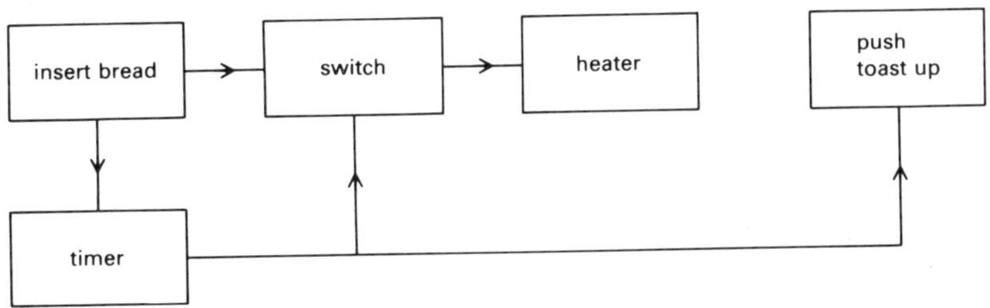

Figure Q21
A system for a 'pop-up' toaster.

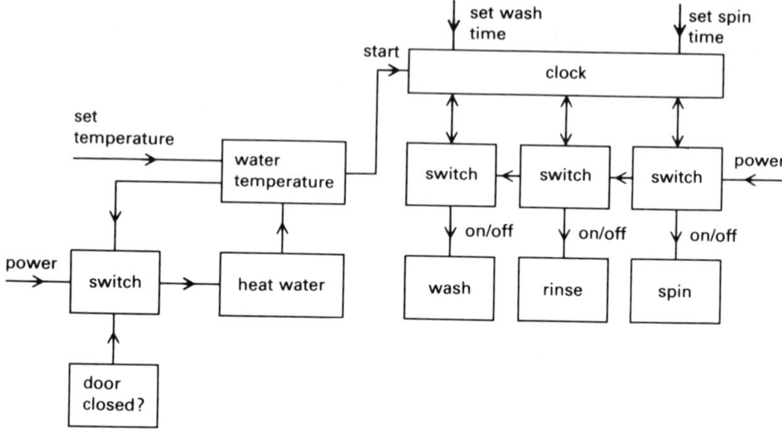

Figure Q22
A system for an automatic washing machine.

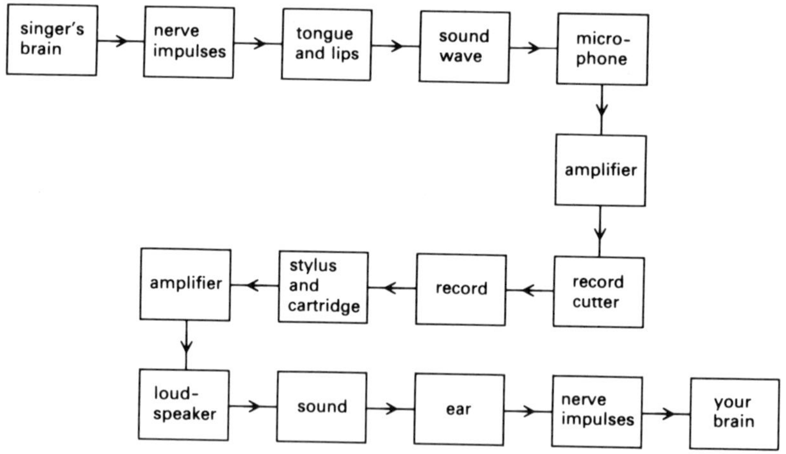

Figure Q23

6*i* A simple on–off system, information easily transmitted by a digital signal.
ii Temperature is continuously variable, so it is difficult to use a simple digital code (although not impossible, as may be seen later).
iii Depends on the number of directions. Two would be easy: more directions would require the use of more than one binary digit. For example, four directions could be coded using two 'bits'. If the directions are north, east, south, and west, then north could be 00, east 01, south 10, and west 11.
iv The magnitude of an electric current is like temperature: continuously variable. However, direction is likely in most cases to be easier to transmit with a binary code.
v The number of pages in this book can be expressed as a binary number which can be transmitted as a sequence of 1s and 0s. For example, if there are 76 pages: decimal 76 is binary 1001100.

vi The area of the page may also be expressed as a number and so can be expressed in binary form.
vii Again, this is a number and can be transmitted in digital form.

7a Crossroads, such as the one in figure Q24, require two sets of lights to control the traffic. Set A controls vehicles approaching from north and south; set B controls vehicles from east and west. Each set has 3 different lights (red, amber, and green) making 6 different lights to control altogether. Since each light is either on or off the information could be conveyed with six bits.

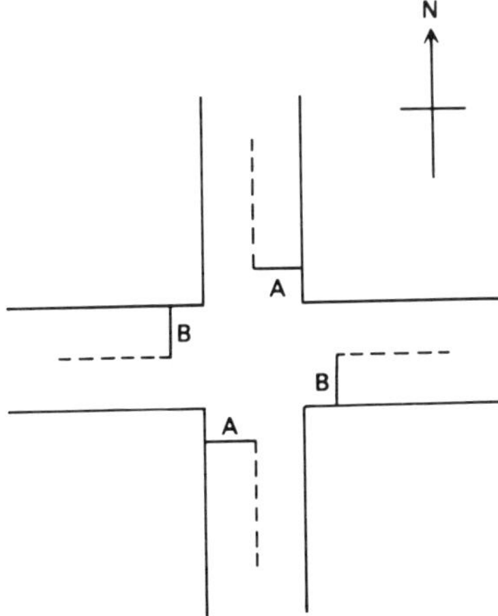

Figure Q24

However, not all possible combinations of lights are used. In fact, only the four shown below are used:

Set A	*Set B*
red	green
red and amber	amber
green	red
amber	red and amber

The computer need only convey which of these four combinations of lights is required, if some sort of decoder is used at the lights. These four states could be communicated from the computer to the lights with only 2 bits, representing the binary numbers from 00 to 11.

b $1024 = 2^{10}$. The term 'kilo-' is used because 1024 is roughly 1000.
c One byte is 8 bits. Each bit can be 1 or 0. This gives 2^8 (=256) possible combinations. So 256 different characters can be coded and communicated.
d It is usual to use one byte for each character. Fifty-two different characters are needed to represent the alphabet in upper and lower case form. There are also numbers and other characters: full stops, commas, brackets, dashes, etc. A code is also required for spaces and 'carriage returns' at the end of lines. (For more information about the ASCII code, see, for example, Horowitz and Hill *The art of electronic* section 10.17)

8 Circuits (a) and (b) are both inverters.
(c) This circuit functions as an inverter.
(d) In this circuit the output is always low.
(e) In this circuit the output is always high.
(f) This circuit functions as an inverter.

9 The circuit is an AND gate.

inputs A	B	C	D	output E
0	0	1	1	0
1	0	0	1	0
0	1	1	0	0
1	1	0	0	1

Figure Q25

10a This circuit is a NOR gate.

inputs A	B	C	D	output E
0	0	1	1	1
1	0	0	1	0
0	1	1	0	0
1	1	0	0	0

Figure Q26

b This circuit is an Exclusive OR gate.

inputs A	B	C	D	E	output F
0	0	1	1	1	0
1	0	1	0	1	1
0	1	1	1	0	1
1	1	0	1	1	0

Figure Q27

c This circuit is a three-input NOR gate.

inputs A	B	C	D	E	output F
0	0	0	1	1	1
1	0	0	0	1	0
0	1	0	0	0	0
0	0	1	1	0	0
1	1	0	0	0	0
1	0	1	0	0	0
0	1	1	0	0	0
1	1	1	0	0	0

Figure Q28

d This circuit is a three-input AND gate.

inputs A	B	C	D	E	output F
0	0	0	1	1	0
1	0	0	1	1	0
0	1	0	1	1	0
0	0	1	1	1	0
1	1	0	0	1	0
1	0	1	1	1	0
0	1	1	1	0	0
1	1	1	0	0	1

Figure Q29

11 The solution to this problem is called a Parity gate (figure Q30). It is the inverse of an Exclusive OR gate (see question **10b**).

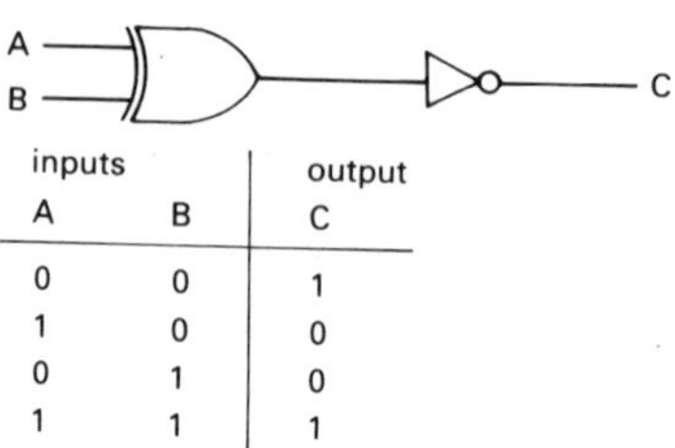

inputs A	B	output C
0	0	1
1	0	0
0	1	0
1	1	1

Figure Q30
A Parity gate.

12 The signals are shown in figure Q31.

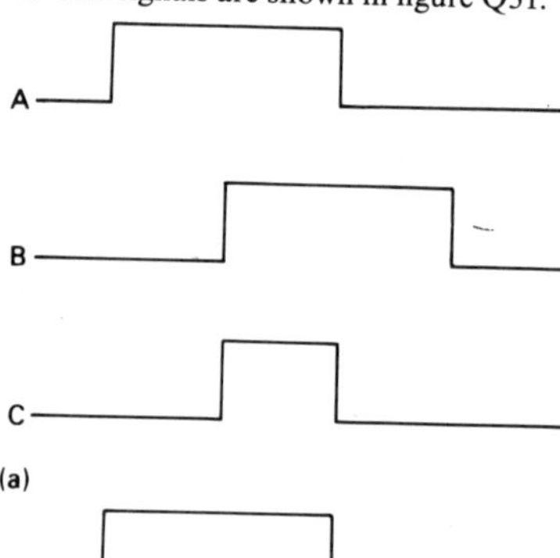

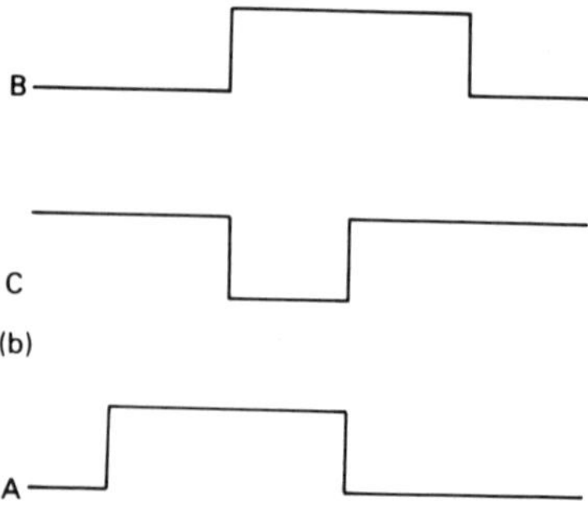

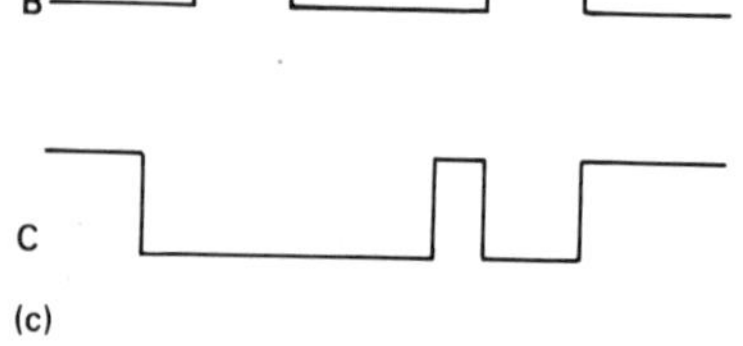

Figure Q31

13 If the system is thought of as a box with four inputs and three outputs, then truth tables can be constructed to satisfy the conditions specified in the question. For example, the red L.E.D. is controlled by sensors A and B in the way that the truth table in figure Q32 shows.

A	B	output red
1	1	1
1	0	0
0	1	0
0	0	0

Figure Q32

There are two other truth tables for the green and the yellow lights. The circuits in figure Q33 satisfy the truth tables.

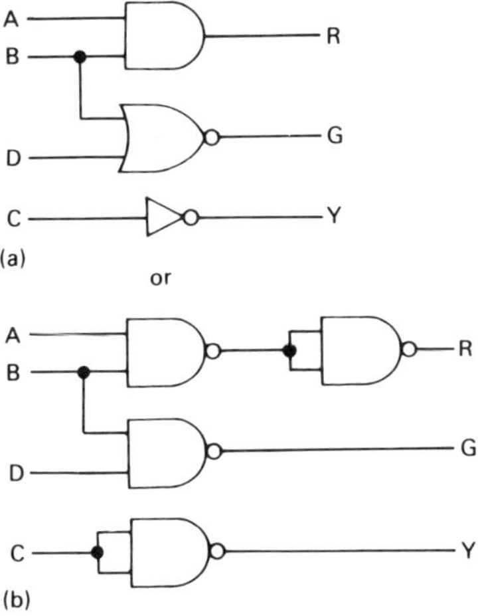

Figure Q33

14 This circuit will have three inputs and one output. Without showing the whole truth table, the necessary combination of inputs to make the output high is shown in figure Q34(a).

The circuit in figure Q34(b) satisfies this condition. Any other combination of inputs gives a low output.

A	B	C	output
1	0	0	1

(a)

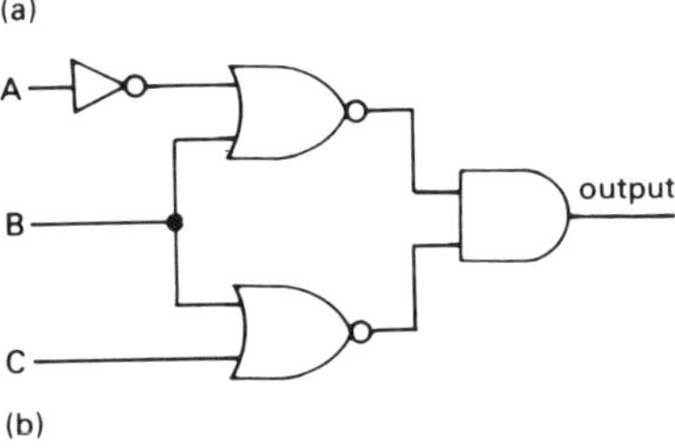

(b)

Figure Q34

15 The circuit in figure Q35(a) is suitable. Both switches must be marked 'Yes' in the V_s position. The truth table is shown in figure Q35(b).

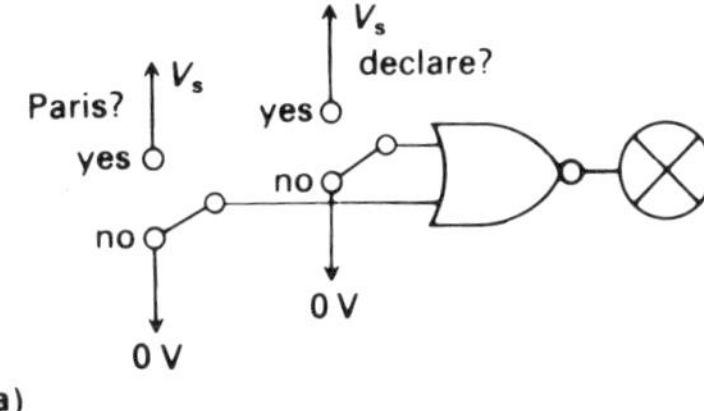

(a)

inputs: I have something to declare	inputs: I have come from Paris	output: you may proceed
no	no	yes
no	yes	no
yes	no	no
yes	yes	no

(b)

Figure Q35

If passengers have nothing to declare and have not come from Paris they may proceed without customs examination.

16 The circuit in figure Q36 shows how a third switch can be added so the 'proceed' lamp only lights when passengers have passed it.

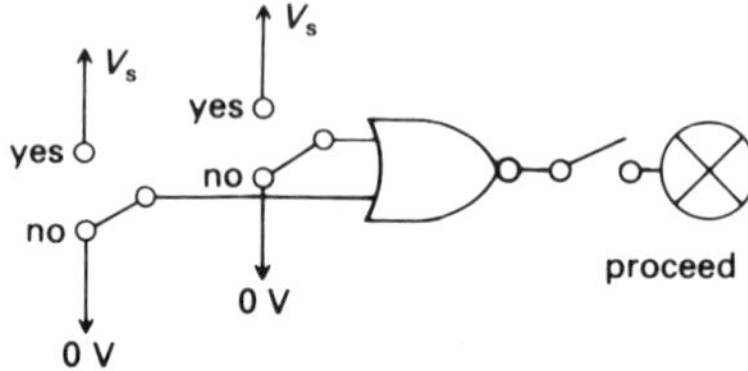

Figure Q36

17 The truth table is that of the Exclusive OR gate, although the second input is now labelled 'control'. Examination of this truth table shows that the 'input' is passed unchanged to the 'output' if the 'control' is 0. If 'control' is 1, then the 'output' is the inverse of the 'input'.

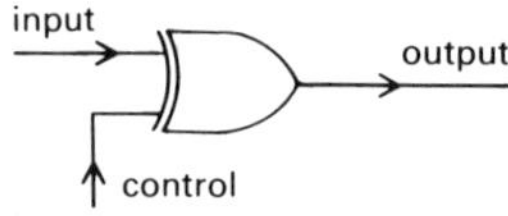

input	control	output
0	0	0
0	1	1
1	0	1
1	1	0

Figure Q37

18a Note that the actual value of the variable used depends on the computer used. Some machines (for example, Apple and Sinclair) should be used with input values of 0 or 1. Other machines (for example, Research Machines, B.B.C.) require values of 0 or −1.

b Program for Exclusive OR gate (question **10b**):

```
 10 INPUT A
 20 INPUT B
 30 LET G = A AND B
 40 LET C = NOT G
 50 LET H = A AND C
 60 LET D = NOT H
 70 LET J = C AND B
 80 LET E = NOT J
 90 LET K = D AND E
100 LET F = NOT K
110 PRINT F
120 GOTO 10
```

Program for three-input NOR gate (question **10c**):

```
 10 INPUT A
 20 INPUT B
 30 INPUT C
 40 LET G = A OR B
 50 LET H = B OR C
 60 LET D = NOT G
 70 LET E = NOT H
 80 LET F = D AND E
 90 PRINT F
100 GOTO 10
```

Program for three-input AND gate (question **10d**):

```
 10 INPUT A
 20 INPUT B
 30 INPUT C
 40 LET G = A AND B
 50 LET H = B AND C
 60 LET D = NOT G
 70 LET E = NOT H
 80 LET J = D OR E
 90 LET F = NOT J
100 PRINT F
110 GOTO 10
```

19a 5 volts. All the kits will work from a 5 V supply.
b When the output of one gate is connected to the input of another, the 'output low' of the first gate must be less than or equal to the 'input low' of the second gate, if the logic is to be preserved.

For the same reason, 'output voltage "high"' is higher than 'input voltage "high"'.

c Basic unit and TTL gates have operating limits that are defined by fixed power supply limits. Examine the characteristic curves in figure C13 (*Students' guide* page 167).

CMOS gates have a very sharp transition from high to low, but the level of this depends on the supply voltage. Since the supply voltage (V_s) can vary over a wide range, it is best to express the limits in terms of V_s.

20a The two bare wires in figure C70 in the *Students' guide* would be arranged so that rain water would short them. This would cause the input of the gate to go high and the indicator to light.

b The circuit in figure C71 in the *Students' guide* would not light the indicator brightly as soon as rain shorts the two wires. A large current is needed to light the lamp but only a small current is needed to control the logic gate.

c Pure water does not conduct electricity.

d A pump will require more power than the logic gates can control directly. The pump can be controlled by a relay energized by the gate (figure Q38).

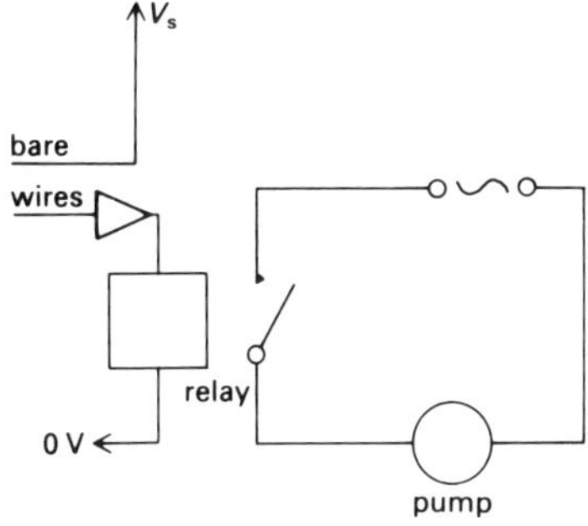

Figure Q38

21a Figure C72(b) in the *Students' guide*. As the resistance of the thermistor drops so does the potential difference across it and therefore the input to the inverter goes low. So the output of the inverter becomes high, lighting the warning lamp.

The circuit shown in figure C72(a) in the *Students' guide* will warn against low temperatures.

b The circuit in Figure C73 in the *Students' guide* would be used to warn against decreases or increases in temperature. The indicator, which in this circuit is on when conditions are safe, will go out if either A or B goes high. The resistors are chosen so that the inputs are low when the temperature is within the safe range. If an OR gate were used instead, the indicator would be off when both inputs were low. It would then light when either input went high, that is, when the temperature went out of the safe range.

22a 5 V

b 800 Ω

c 1250 Ω

d Preferred values are
1, 1.2, 1.5 ... kΩ (5 and 10 %);
1, 1.1, 1.2, 1.3 ... kΩ (2 % and 1 %).

e The gate switches when the p.d. between 0 V and the input to the gate is 5 V. So the p.d. across R_2 is 5 V and the p.d. across the L.D.R. is 4 V. I_1, the current through the L.D.R.

$$= \frac{4\,\text{V}}{1\,\text{k}\Omega} = 4\,\text{mA}$$

So $I_2 = I_1 - 1\,\text{mA} = 3\,\text{mA}$

and $R_2 = \dfrac{5\,\text{V}}{3\,\text{mA}} = 1\tfrac{2}{3}\,\text{k}\Omega = 1.7\,\text{k}\Omega$, to 2 s.f.

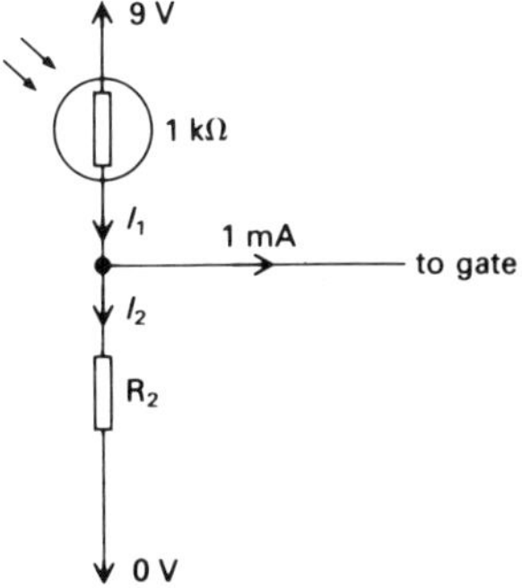

Figure Q39

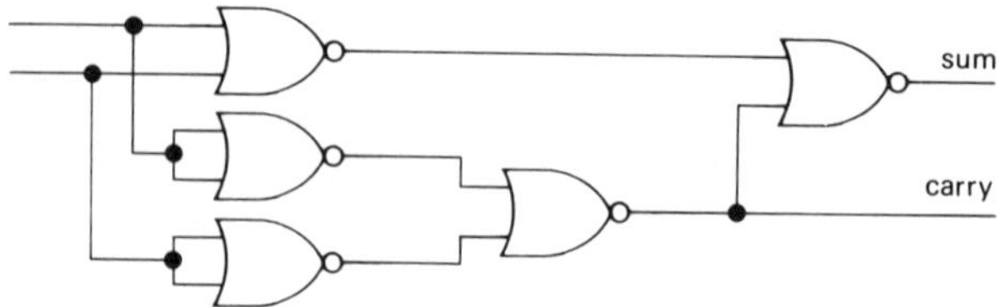

Figure Q40
A half adder circuit using NOR gates.

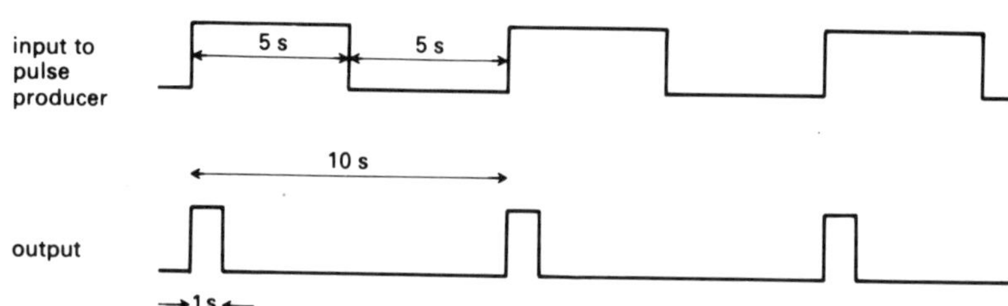

Figure Q41

So to allow for current drawn by the gate itself R_2 must be larger. (It's like connecting a low resistance voltmeter in a circuit. It actually reduces the p.d. measured.) If the circuit uses a variable resistor that can be increased to 2 or 3 times the calculated value, then problems like this can be overcome. (In fact 1 mA is huge for a CMOS gate – almost unheard of; it is even large for the base current of a transistor.)
f It is easy to adjust R_2 to compensate for the loading of the potential divider by the gate; or to make the indicator come at different levels of illumination.

23 See pages 168–9 of the *Students' guide*.

24a Figure Q40 shows a half adder using NOR gates.
b See *Students' guide* figure C17 for full adder from half adders.
c See figure C17 in the *Students' guide*. Only if A = 1 and B = 1 is the carry from the first half adder 1. But then the sum from the first half adder is 0, so there can be no carry from the second half adder.

25a A pulse producer – see figure C19 in the *Students' guide*.
b Input 1 goes from low to high; output 1 goes low, stays there for a while, and then returns to high. When output 1 goes high this triggers circuit 2, whose output then goes low, stays there for a while, and then goes high again.
c The process would continue indefinitely – just like an astable circuit.

26a Use a pulse producer which is triggered by a voltage rising from zero to V_s. Adjust the RC value of the pulse producer so that it produces a pulse of 1 second duration. Figure Q41 shows the relationship between the input and output.
b Assuming the pulse producer used in **a** produces a positive pulse, all we need do is add an inverter at the output. (Of course, if the circuit in **a** produces a negative going pulse, the inverter will be needed in part **a** and not in part **b**.)

27 The output D is high when either input A **or** input B are high.

A	B	C	D
0	0	1	0
0	1	0	1
1	0	0	1
1	1	0	1

Figure Q42

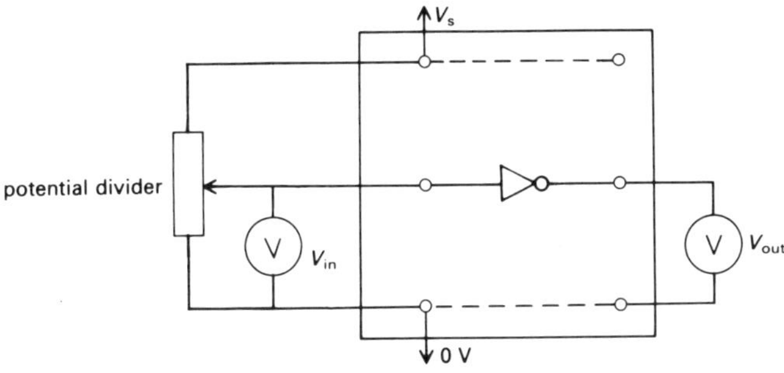

Figure Q43

28a C is high.
b D is low; D keeps A low.
c D is high; feedback keeps A high.
d D goes high and feedback makes A go high as well. The circuit has changed its state.
e A can only stay low when both A and B are low. Both must be made low.
f It has two stable states; with D (and A) low or high.

29a The output of the second also goes from low to high.
b Nothing happens to it next. It stays high. The output from the second keeps it high.

30 The shop runs out of chocolate sooner. This is also positive feedback.

31a The switches in *Students' guide* figure C79 keep the input to the non-inverting gate low. If any one is opened the gate input will go high and the bell will sound.
b The alarm system can not be disconnected by cutting the wires to the switches.
c When a switch in *Students' guide* figure C79 is opened the bell sounds. The bell can be easily turned off again by closing the switch. In *Students' guide* figure C80 the output of the OR gate is fed back to one input. When a switch is opened the output of the gate goes high and stays high even if the switch is closed again.

32 B

33 B

34a The additional circuit components are shown in figure Q43.
b The graphs are shown in figure Q44 (below and on the next page).

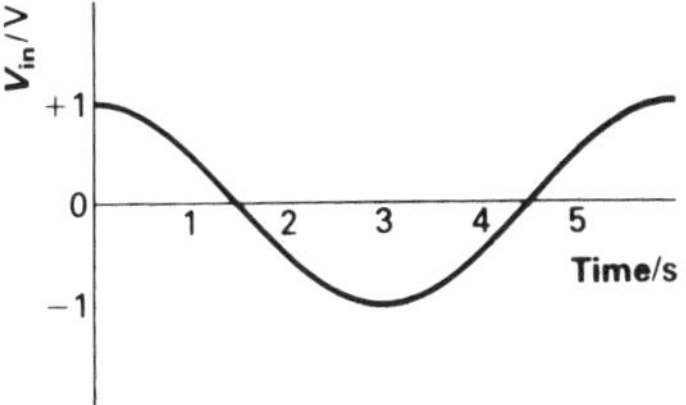

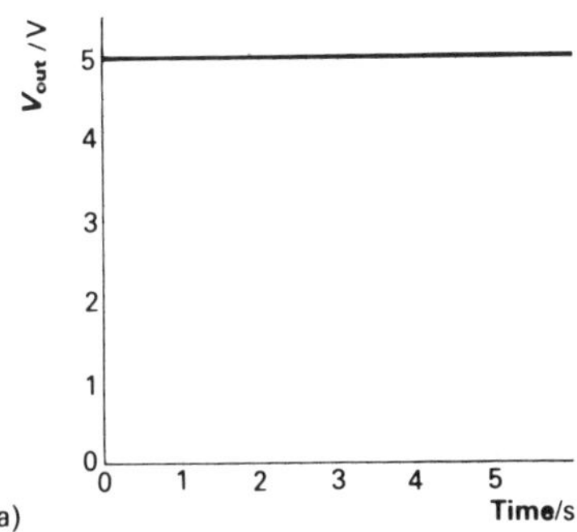

(a)

Figure Q44 (part)

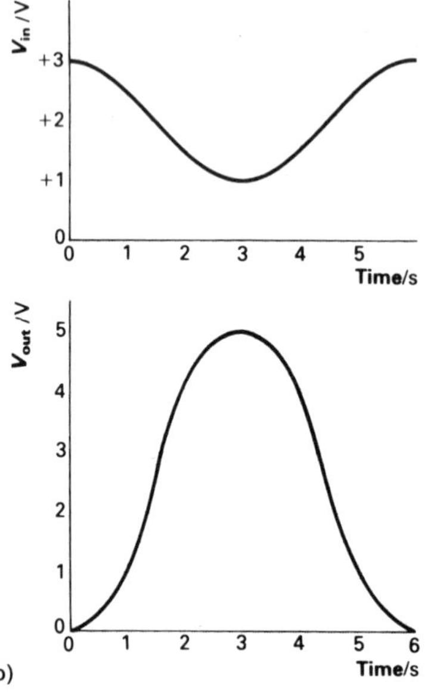

Figure Q 44 (part)

UNIT D
Oscillations and waves

1a It is probable that the car body has initially been set into oscillation, and the energy of oscillation is gradually being reduced due to damping. (This particular car has rather worn shock absorbers – the energy should disappear in only one or two cycles.) The energy is being distributed among the many atoms making up the surroundings of the car body. The reverse process – concentration of energy from many individual atoms into oscillations of one object – never happens.

But the motion shown could be reversed if the car body were being given a properly timed series of pushes by an external agent – that is, being set into forced oscillation.

b Something like the graphs in figure Q45 should appear.

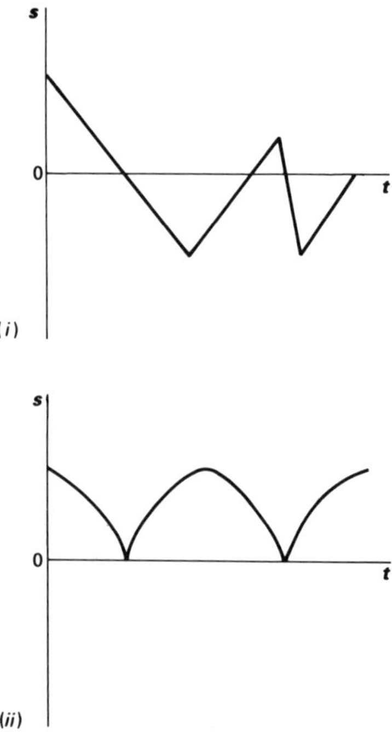

Figure Q 45 (part)

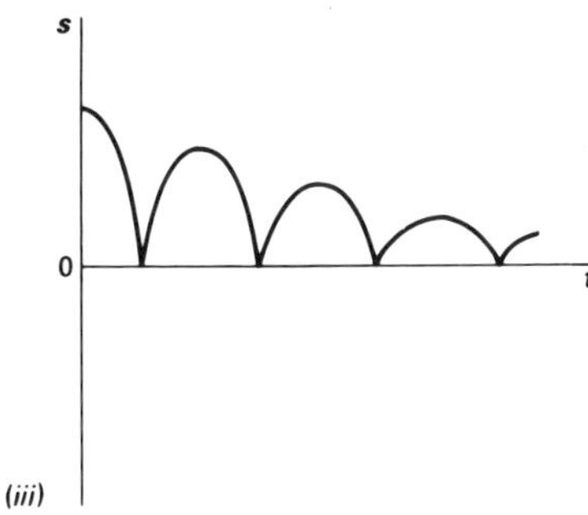

Figure Q 45 (part)

2a 1.0 m
b*i* $\frac{1}{4}$ cycle
ii 0; -1.0 m
c*i* -0.5 m
ii 1.5 s, 4.5 s, 7.5 s, etc.

3 The story of Harrison and his chronometers is told in a booklet obtainable from the National Maritime Museum.

The point of carrying a clock for navigation is to compare the time of noon (say) at the ship's unknown longitude with that, indicated by the clock, at some other definite longitude, which has gradually come to be that of Greenwich. If, for instance, local noon is six hours earlier or later than at Greenwich, the ship is one-quarter of the way around the Earth from Greenwich. At the Equator, a time difference of one minute corresponds to a distance in longitude of nearly thirty kilometres, so an accurate clock is needed.

The way to test the clock is to determine, by making astronomical observations, the longitude of the place reached (Jamaica), and to calculate the expected time of noon compared with that at Greenwich. The clock, if it is accurate, will still show Greenwich time, and its reading can be compared with the calculated time.

4a *1* To test for regularity, it would compare the number of dots punched out in fixed intervals of time as indicated by its own clock, and do so at several times of day.

2 To test for accuracy, it would need to count the number of dots produced over a long period as measured by its own clock, and see if there were, say, 5000 dots for every 2000 pips. There would be no point in doing so if the test for regularity had shown the 'dot-clock' to be irregular, compared with the 'pip-clock'.
b It is very unlikely that the clock can be accurate over long periods if it is not regular – that is, if it disagrees even with itself. But it can be regular, yet not punch dots at exactly $\frac{1}{5}$ second intervals.
c Only by *deciding* that some clock or other is going to be treated as a regular time marker. This standard time marker cannot be tested, though the choice would fall on one of a number of such clocks that show each other to be regular, and for which, if possible, there are theoretical reasons for thinking that the rate will not be affected by most kinds of laboratory disturbances. Thus, atomic clocks are preferred to clocks based on the oscillations of some lump of matter.

If the laboratory decides to trust its pip-clock, the issue will have been settled by decision, not by trial. The inventor could claim that the decision should have gone his way. Indeed, atomic clocks at present in different countries do show tiny disagreements, and there is no way to tell which one is 'right'.

5 This is discussed in the reading 'Quartz and atomic clocks' on pages 237–9 of the *Students' guide*.

6 2 radians; 115°

7a*i* $r\theta$
ii $r \sin \theta$
b $\sin \theta \approx \theta$ if θ is small.

8

θ/degrees	20	10	5	2	1	0.5	0.1
θ/radians	0.3491	0.1745	0.08727	0.03491	0.017453	0.0087266	0.00174533
$\sin\theta$	0.3420	0.1736	0.08716	0.03490	0.017452	0.0087265	0.00174533
$\cos\theta$	0.9397	0.9848	0.9962	0.9994	0.9998	1.0000	1.0000
$\tan\theta$	0.3640	0.1763	0.08749	0.03492	0.017455	0.0087269	0.00174533

Table Q3

a*i* 0.5 %
ii 0.006 %
iii 0
b*i* $\cos\theta \approx 1$
ii $\tan\theta \approx \theta$

9 $6.7\ \mathrm{m\,s^{-1}}$

10a

	Velocity	**Acceleration**
Q	zero (but about to move rapidly down)	large, downward
R	large, upward	zero
S	zero	large, downward

Table Q4

b*i* Tension in the Slinky (T) – a large tension will mean that for a given displacement there will be a large restoring force, accelerating displaced coils back towards their mean position more quickly.
ii Mass per unit length of the Slinky (μ) – a given restoring force acting on a large mass, will accelerate displaced coils more slowly towards their mean position.
In fact

$$c = \sqrt{\frac{T}{\mu}}$$

11 Both. Any parts of the spring in motion possess K.E.; and any parts in tension or compression compared with their rest state possess P.E. The superposition to give no net compression or expansion is instantaneous: at that moment all the energy is K.E.

12 See figure Q46.

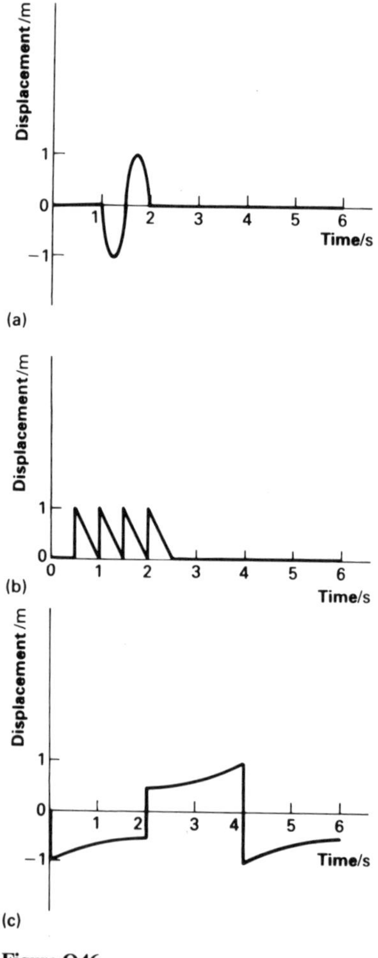

Figure Q46

13a*i* Measure the frequency directly (measure the time for, say, 10 waves to pass a point using a stopwatch).
ii Find the magnitude using $a = \omega^2 A = 4\pi^2 f^2 A$.
At P the acceleration is downwards; at Q it is to the left.
b See figure Q47.

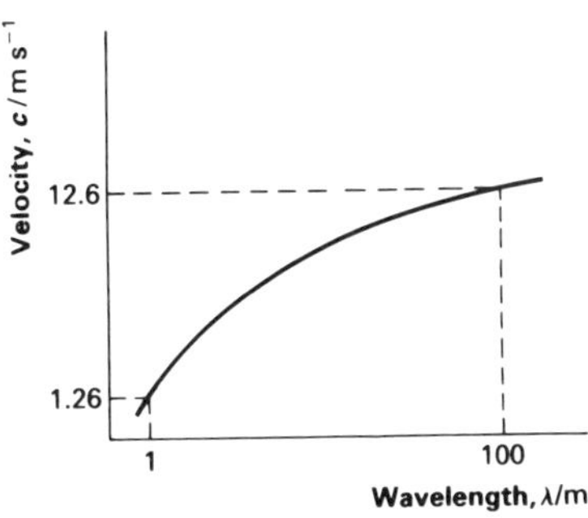

(i) For a straight line, plot c^2 against λ

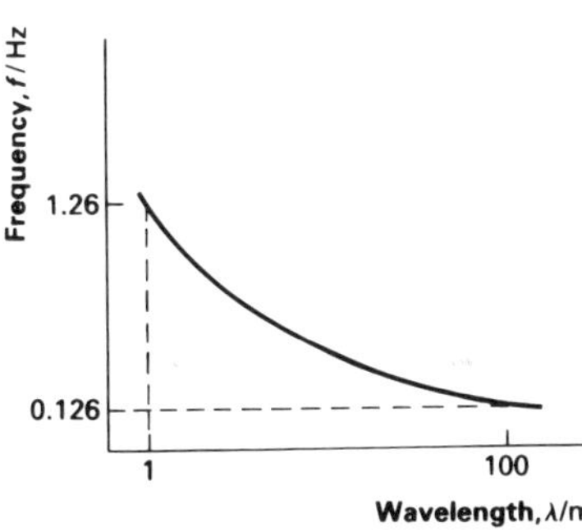

(ii) For a straight line, plot f^2 against $1/\lambda$

Figure Q47

c*i* All waves in this range of wavelengths will be felt. (Minimum speed needed by waves to reach ship $= 1.16\,\text{m}\,\text{s}^{-1}$.)
ii $7.9\,\text{m}\,\text{s}^{-1}$. The boat will move in a looped path forwards, with $v_1 = 9.9\,\text{m}\,\text{s}^{-1}$, and $v_2 = 5.9\,\text{m}\,\text{s}^{-1}$. The frequency of the loops is 0.106 Hz.
iii $6.3\,\text{m}\,\text{s}^{-2}$; maximum force $\approx \frac{2}{3}$ weight.
d $1.26\,\text{m}\,\text{s}^{-1}$; $31.6\,\text{m}\,\text{s}^{-1}$.

14a $\lambda = S_2C - S_1C$.
b*i* The pattern spreads out.
ii and *iii* The pattern closes up (distances like AB become less).
iv Minima at A, C; maxima at B, D.
v The pattern changes with time. At any instant, an interference pattern would be visible over the whole surface, but it isn't stationary. At any one point, there are beats: each point is alternately at a maximum then at a minimum.

c The amplitude would only be zero at a minimum if the amplitudes of the disturbances arriving from S_1 and S_2 were equal. The wave energy is spreading out from the sources, so the amplitude of an individual wave decreases with distance travelled. Since in general waves from S_1 and S_2 will have travelled different distances to reach a minimum point, they will not have equal amplitudes.

15 The path difference has increased by an odd number of half-wavelengths. So λ could be 2 m, or $\frac{2}{3}$ m, or $\frac{2}{5}$ m, or ...

16 See figure Q48.

17 There are many answers. Wavelength measurements usually involve using the fact that waves superpose to give a large or small resultant effect, depending on whether the path difference is an even or an odd number of half-wavelengths.

Two-source experiments are possible in practice with waves of reasonable wavelength, for the sources need not then be very close together. But the sources must emit in phase. For light, two-source experiments are still possible, but the two sources must be imitated by splitting the light from one narrow source.

Diffraction gratings can be used, and are especially suitable for visible light and for X-rays. The grating spacing must not be very much larger than one wavelength if the diffraction angle is to be reasonably large. For X-rays, the grating is tilted so that the X-rays graze its surface, and the spacing looks small to the X-rays.

In the microwave and v.h.f. region (wavelength from 0.01 m to 1 m roughly), simple arrangements of reflectors which introduce a path difference can be used. A problem here is to have big enough reflectors, for a reflector must be bigger than one wavelength in linear dimensions to reflect an appreciable amount of wave energy.

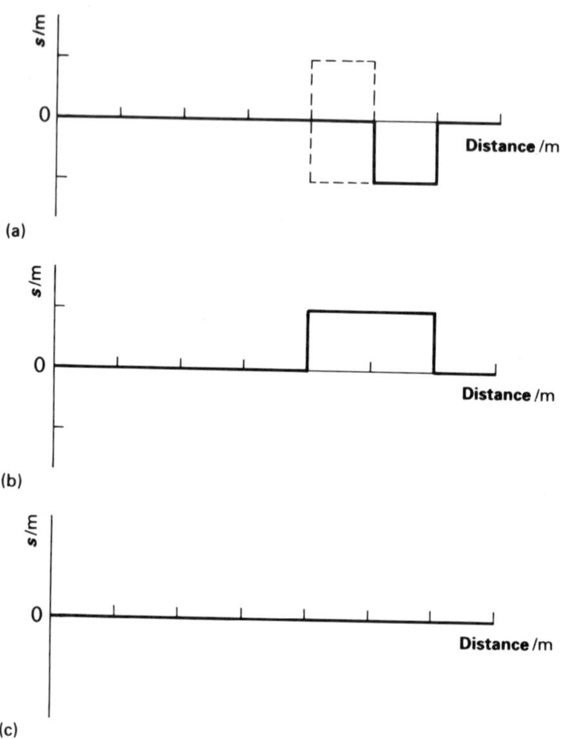

Figure Q48

18a 120 mm
b As S_2 is moving further from T, it is reflecting a decreasing amount of wave energy; and as R is also now further away, a smaller proportion of this amount reaches R. Hence the two superposing signals will no longer have equal amplitudes, which they must have to cancel out exactly.

19 D

20a Out of phase.
b In phase.
c The light following paths A and C superposes destructively – thus reducing the intensity reflected. B and D, however, give constructive superposition, increasing the intensity transmitted.
d There will be relatively little light intensity reflected in the middle of the visible band, and relatively more at the ends – that is, blue and red. These combine to make the reflected light appear purple.

21 78 m

22 In general, superposition occurs between light reflected at the top and the bottom of the oil film. If there are no changes of phase, the path difference in figure Q49 is (QS + SU) – (PR + RT). If the film is very thin, this may equal one wavelength for one particular colour – in which case the observer will see that colour at that angle θ. At a slightly different angle, the path difference will alter, and he will see a different colour.

23a Doubled.
b Doubled.
c Doubled.
d Same time.

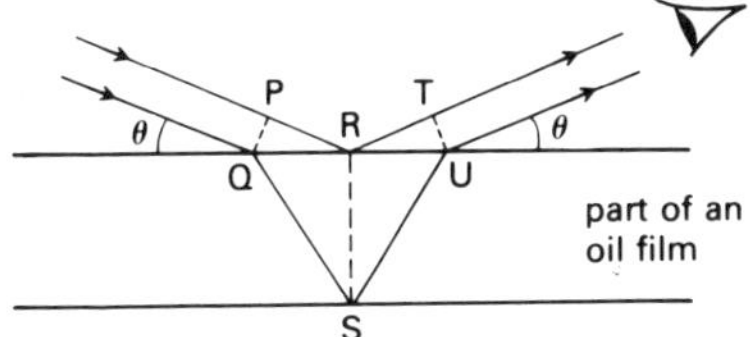

Figure Q49

24a Stiffer.
b Doubled.
c Acceleration in (b) is four times that in (a).
d k in (b) is four times k in (a).
e $T^2 \propto \frac{1}{k}$.
f Decrease the mass.
g Mass would increase four times.
h $T^2 \propto m$

25 (a) corresponds to (b). (b) corresponds to (a). (c) corresponds to (c).

26a B C A D
b Same magnitude, opposite direction.
c Both zero.
d 0, D and F.
e 0, D and F.
f 0B = $T/4$, 0D = $T/2$, 0F = T, BE = $T/2$, DF = $T/2$.

27 (b) will oscillate. (Do not confuse the 'weightlessness' of the mass with lack of inertia. Its mass is the same as on the Earth, and it behaves as an oscillator with the same period, neglecting any relativistic effects.) Students can discuss the others.

28a 0.32 s
b 20 rad s^{-1}
c From diagram, *i* $\omega = 1100\,^\circ\,s^{-1}$
ii $\omega = 19$ rad s^{-1}.

29a −1 m s^{-2}
b −0.05 m s^{-1}
c −0.05 m s^{-1}
d −0.005 m
e 0.095 m
f −0.95 m s^{-2}
g −0.095 m s^{-1}
h −0.145 m s^{-1}
Table Q5 shows the other values.
i The graphs are shown in figure Q50.
j 2.0 s. It compares well (1.99 s).
k These answers would all be doubled; the period would be shorter.
l These answers would all be halved; the period would be longer.

31 The description is plausible; but the lack of definition in the photograph, and the associated uncertainty in plotting points on a graph, make it difficult to tell for certain.

32a $2x\rho gA$
b The restoring force is proportional to the displacement x.
c*i* Acceleration $= -\frac{2xg}{l}$
ii Angular velocity $= \sqrt{\frac{2g}{l}}$
iii Periodic time $= \pi\sqrt{\frac{2l}{g}}$

t/s	s/m	a/m s^{-2}	v/m s^{-1}	Δs/m
0	0.1	−1	0	
0.05			−0.05	−0.005
0.10	0.095	−0.95		
0.15			−0.145	−0.0145
0.20	0.0805	−0.805		
0.25			−0.226	−0.0226
0.30	0.0580	−0.580		
0.35			−0.283	−0.0283
0.40	0.0296	−0.296		
0.45			−0.313	−0.0313
0.50	−0.0017			

Table Q5

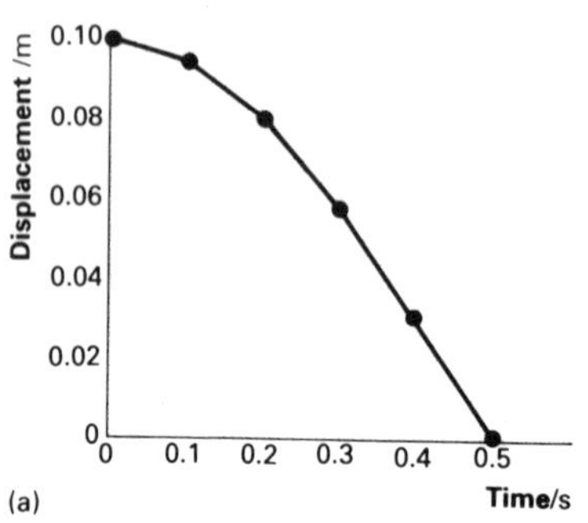

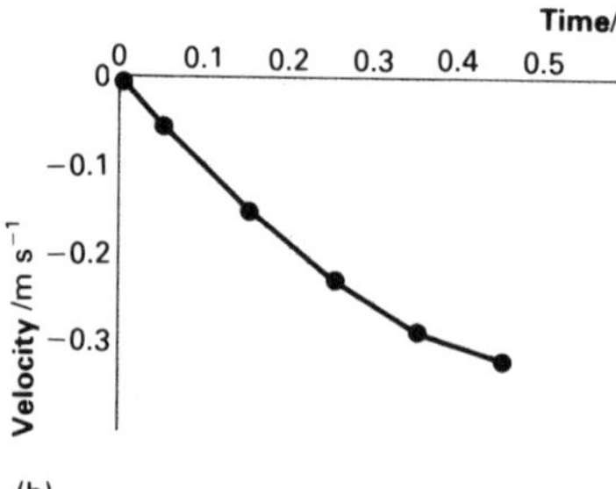

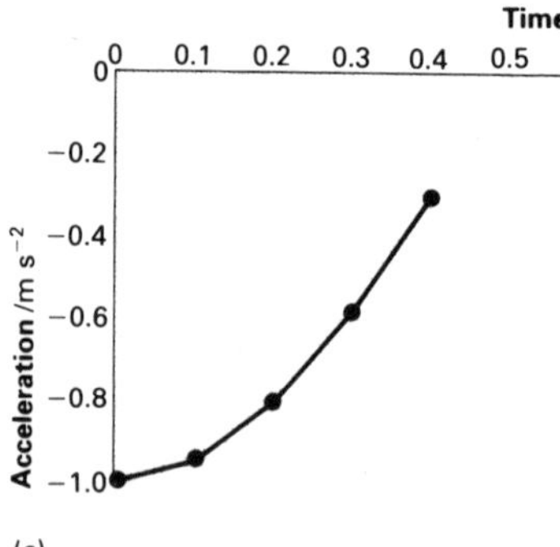

Figure Q50

33 $2\pi\sqrt{\frac{hd}{\rho g}} = 2\pi\sqrt{\frac{L}{g}}$

34 It must be doubled.

35a $0.52\,\text{rad h}^{-1}$
b $d = 5\cos(0.52 \times t)$ metres. (t measured in hours from high tide.)
c 1.14 p.m.

36a About $\frac{1}{20}$.
b $400\,\text{N m}^{-1}$
ci $2.8 \times 10^{-20}\,\text{J}$ *ii* 0.18 eV

37a 0.1 m
b $10\,\text{rad s}^{-1}$
c 0.16 s
d $1\,\text{m s}^{-1}$

38 **A**

39 **B**

40a See figure Q51.

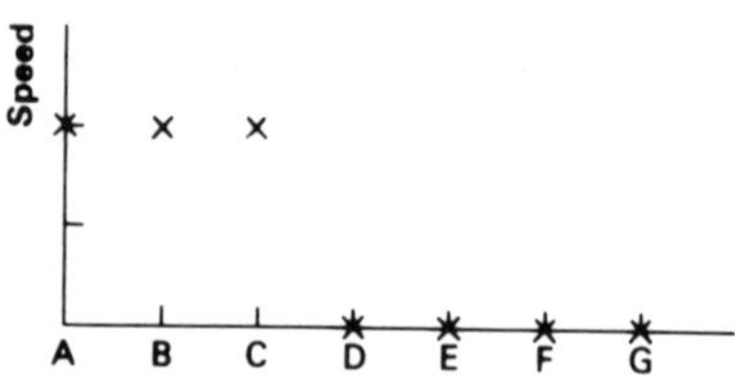

Figure Q51

b Spring constant k, mass of each trolley m, length of each section.
c One possible point is that we believe metal atoms are arranged in regular lines (Unit A); another point from Unit A is the macroscopic elasticity of metals, suggesting the existence of spring-like bonds between atoms.
d The model has only a single line of trolleys and springs, whereas a bar must have a three-dimensional array of atoms and bonds; real metals have dislocations and grain boundaries in their structures; and there is no feature of the metal directly analogous to the frictional damping of the trolleys.

41a $8 \times 10^{12}\,\text{m s}^{-1}$
b $5600\,\text{m s}^{-1}$
c Compares quite well (within about 10 %).

42a ρx^3
b $\sqrt{E/\rho}$
c $5100\,\text{m s}^{-1}$
d The Young modulus for aluminium must be less than that for steel by the

same factor as the density of aluminium is less than that of steel.

43–45 and **47–52** These questions involve explaining ideas about resonance, and making estimates based on students' practical experience.

46a Both sets of ions would experience forces, Na^+ to the right and Cl^- to the left.
b 3.8×10^{-26} kg
c 1.2×10^{13} Hz (effective $k = 200\ \text{N m}^{-1}$).
d*i* 2.6×10^{-5} m
ii Infra-red.

53a The mass has constant velocity; therefore no unbalanced forces act upon it. The springs will therefore be stretched symmetrically on both sides of it.
b The mass is accelerating to the left; therefore there must be an unbalanced force acting upon it to the left. This must be caused by the lefthand spring being more stretched, and the righthand spring being less stretched. Thus the mass is displaced to the right.
c $\dfrac{ma}{k}$
d See figure Q52
e Zero friction would be a bad idea, because the mass would always oscillate, and never settle to a steady reading.
f If Q is the quality factor of the oscillator, then the oscillations will die away in time $\approx QT$. For a usable instrument, $QT \ll t$. Since $Q \geqslant 1$, we want $T \ll t$. (If $Q < 1$, damping is heavy and the instrument may not give a reading at all.)
g 0.02 m
h k/m would be $1\ \text{s}^{-2}$. The maximum displacement would need to be 2 m, requiring an instrument 4 m long – impractical. The springs would be long, but would require $k = 0.1\ \text{N m}^{-1}$ if $m = 0.1$ kg, say; these springs are very soft, and would sag.

54a 10^7 cycles
b 6×10^{14} Hz
c 1.7×10^{-8} s
d 5 m

55a,b,c In each case a low Q is desirable; the more important factor, though, in **a** and **b** is to ensure that the natural oscillation frequency is well away from any frequency at which the system may receive a periodic driving force.
d,e High.
f Low. In this case the natural frequency of the arm should be near to 50 Hz.

56a See figure Q53.
b*i* 4 m   *ii* 2 m   *iii* 4/3 m
c $60\ \text{m s}^{-1}$
d The system responds with large amplitude only at special resonant frequencies. The mass-on-spring has only one resonant frequency, however, whereas the string has a series of resonant frequencies: its fundamental frequency,

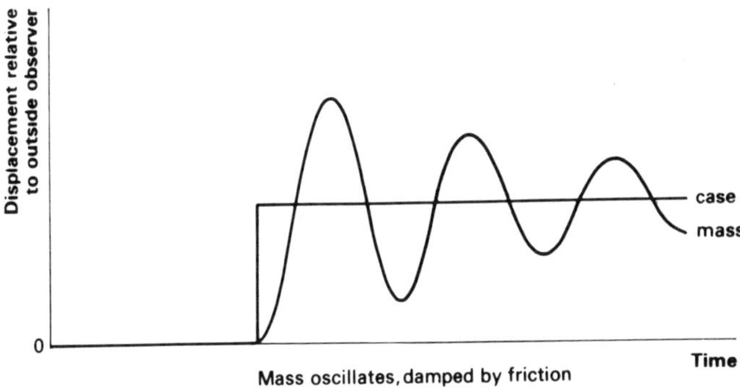

Figure Q52

and multiples of this. The resonance of the string depends on waves arriving back from the far end in phase with the new waves being generated: and this can happen when the path of the waves (4 m) is any whole number of wavelengths.

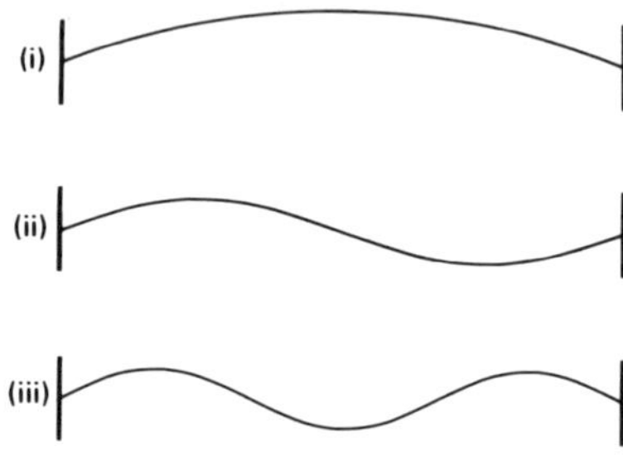

Figure Q53

57 192 Hz, 320 Hz.

58 C

59a $f=\dfrac{1}{2L}\sqrt{\dfrac{T}{A\rho}}$
b $f=650$ Hz
$L=0.5\,\mathrm{m}$
$A=0.1\times10^{-6}\,\mathrm{m}^2$
$\rho=6000\,\mathrm{kg\,m}^{-3}$
c $T=4l^2A\rho f^2\approx300$ N
d A and ρ, in **b**, are estimated to 1 significant figure: so the answer can only be quoted to 1 significant figure. Even this one figure (3) depends heavily on the estimated quantities: really only the order of magnitude is reliable.

UNIT E
Field and potential

1a To the right. The charges on the plates cause a force to be exerted in this direction on the positively-charged ball.
b The ball becomes negatively charged. It moves to the left.
c Negative.
d Negative.
e Anticlockwise.
f Clockwise.
g So that charge can flow on and off.
h The p.d., V, of the supply, and the separation, d, of the plates. The frequency rises with increasing V or decreasing d.
i Charge is displaced over the conducting surface of the ball; the larger force on the near side of the ball causes it to be attracted to the plate, whereupon it acquires the same charge as the plate and the process continues as before.

2 At the smaller spacing with the same p.d. the field strength (volts per metre) is larger, and the foil hangs at a larger angle.

The field strengths are $62.5\,\mathrm{kV\,m}^{-1}$ and $125\,\mathrm{kV\,m}^{-1}$.

To get the same field strength at 40 mm separation, reduce the p.d. to 2500 V.

3a Fd
b QV
c $F/Q=V/d$
d $\mathrm{V\,m}^{-1}$
e*i* joules/coulomb
ii newtons × metres
iii $\dfrac{\mathrm{V}}{\mathrm{m}}=\dfrac{\mathrm{J}}{\mathrm{C\,m}}=\dfrac{\mathrm{N\,m}}{\mathrm{C\,m}}=\dfrac{\mathrm{N}}{\mathrm{C}}$
f*i* $10^4\,\mathrm{V\,m}^{-1}$
ii 10^{-9} C
iii 10^4 N

4a Use $E=V/d$. $V\approx2000$ V.
b Briefly, the higher the pressure the shorter the distance a particle moves before a collision with another. If in that distance the occasional ion or electron can acquire enough energy (about 30 electronvolts or about 5×10^{-18} J) from the p.d. it moves through, it may ionize the molecule it hits, and the ion or electron so freed may make another in the same way, leading to an 'avalanche' of ions, and a spark. At higher pressures, then, a greater p.d. is needed to start a spark, for the ions have to acquire a fixed energy in a shorter distance. The question suggests that the sparking p.d. in a car engine exceeds 2 kV, for the pressure in the cylinder is several times atmospheric pressure.

5 The completed diagrams are shown in figure Q54.

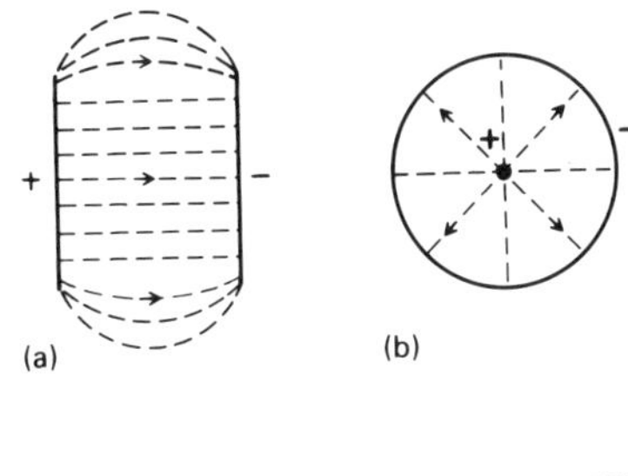

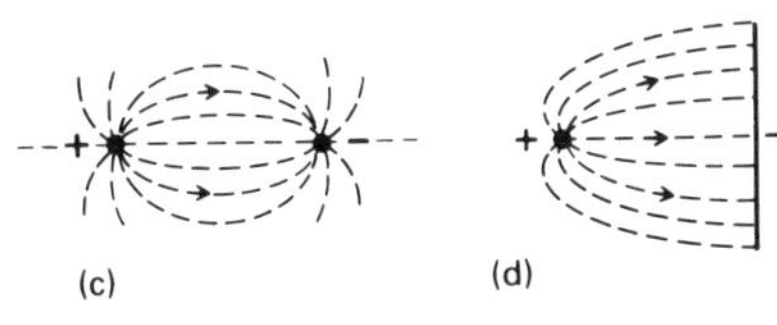

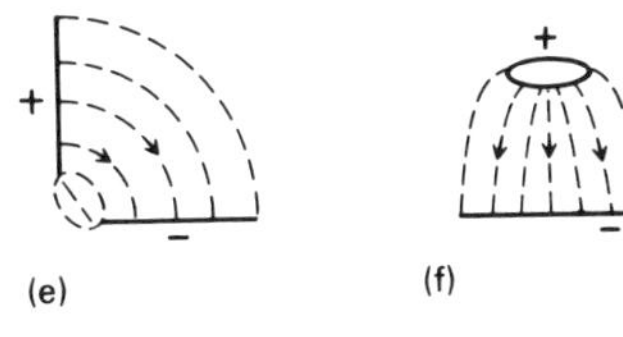

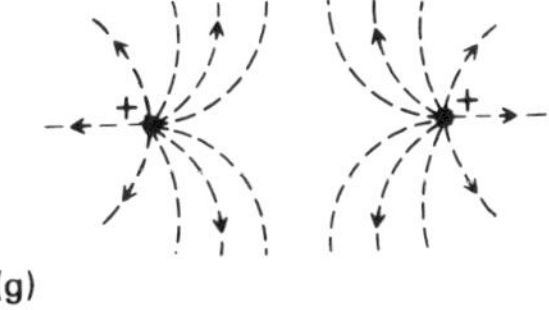

Figure Q54

6a The diagram is shown in figure Q55.

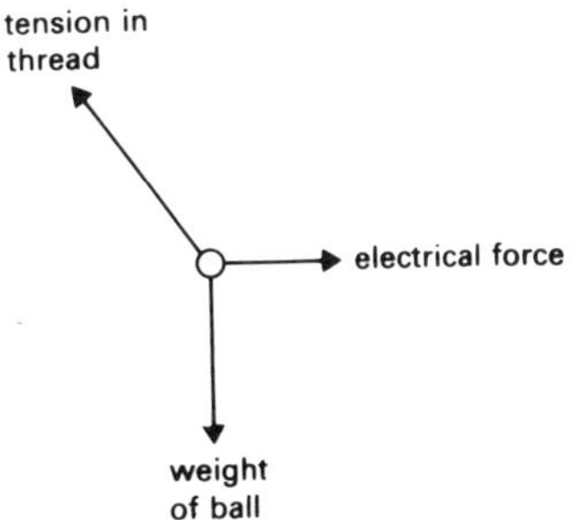

Figure Q55

b 10^{-3} N (equal to the weight of the ball).
c 10^{-8} C. Negative.

7a A large charge would affect the charges on the plates, completely altering the field between them. A small charge, however, would give such a small force that it could not be measured.
b With no flame, the probe may acquire an induced charge and thus affect the field around it, altering the potential at the tip. The flame contains ions which discharge the tip of the probe so that there is no potential difference between it and its surroundings. The electroscope gives little or no indication (unless a charged plate is touched) without the flame.
c Any source of ions would do, for example, a radioactive source giving out alpha particles. These are in fact used in probes in balloons in the upper atmosphere.
d Set up as described, it will acquire positive charge.
e Towards the electroscope.
f If the charge does not leak off the electroscope, then this current will drop to zero. In practice, it settles at a small steady value, enough to keep the electroscope 'topped up'.

8a Positive.
b To the left.
c Negative.
d Increase.
e See figure Q56.

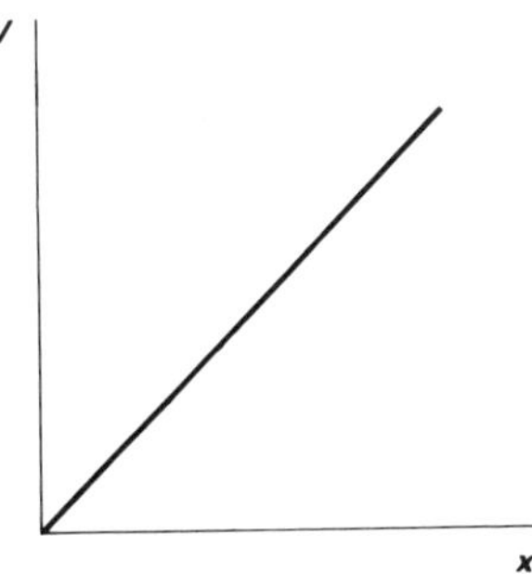

Figure Q56

f Positive.

g $E = -\dfrac{\Delta V}{\Delta x}$ or

'field strength = – potential gradient'.

9a The graph is shown in figure Q57.

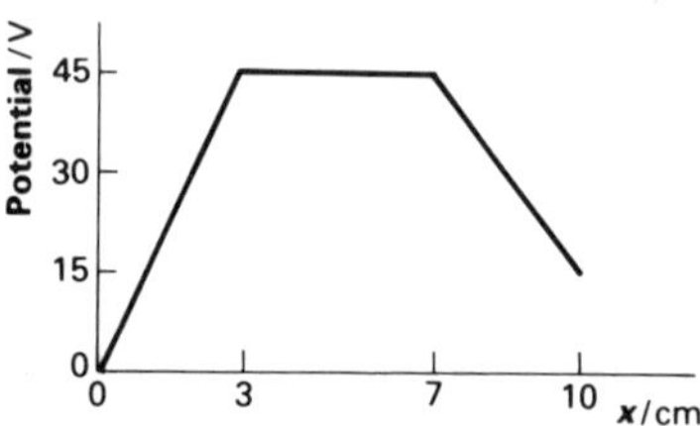

Figure Q57

b For CG_1 it is $-1500\ \text{V m}^{-1}$; for G_1G_2 it is $0\ \text{V m}^{-1}$; and for G_2A it is $+1000\ \text{V m}^{-1}$.

c Between C and G_1 there is constant acceleration up to maximum energy of 45 eV. Between G_1 and G_2 there is constant speed. Between G_2 and A there is constant deceleration down to energy of 15 eV.

d Since it has only 10 eV it cannot reach A, so will 'fall back' towards G_2.

10 The gradient gives the magnitude of the electric field strength (figure Q58).

11 Discharge is most likely to take place around sharp points where the electric field strength is greatest.

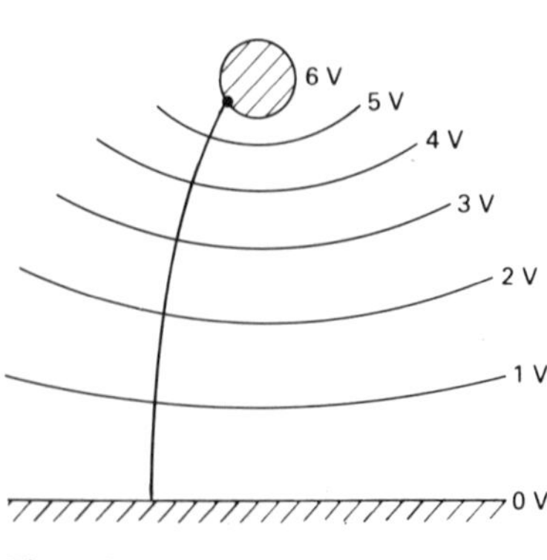

Figure Q58

a will be most dangerous, then **c** (the tree conducts, especially if wet, and charge will flow if it is struck). The horse is 'earthed' by its metal shoes and the rider forms a fairly sharp 'point', so would be vulnerable. The rubber tyres of a bicycle, *if dry*, might offer a little more insulation, though the safest places would be **d** or **b**, as there can be no electric field *inside* a closed conductor containing no charge.

12a 10^{-11} F.

b Approximately 10^{-8} F.

c Approximately 0.1 per cent.

d The effective capacitance is now about 2×10^{-11} F. Only 50% of the charge would be transferred; nevertheless, the p.d. across the voltmeter would be about 500 V, which could cause serious damage if an electrometer was being used as the high resistance voltmeter.

13a The statement is correct: $Q = 10\ \mu\text{C}$, so $I = 0.5$ mA at 50 Hz.

b 10^{-9} F.

14a $10^3\ \text{V m}^{-1}$.

b $10^{-8}\ \text{C m}^{-2}$

c 6×10^8

d 10^{-16} N

e 2×10^{-7} m

f 10^{10} atoms

15a When it is rolled up, otherwise B and D will touch and short circuit.

b About 100 m.

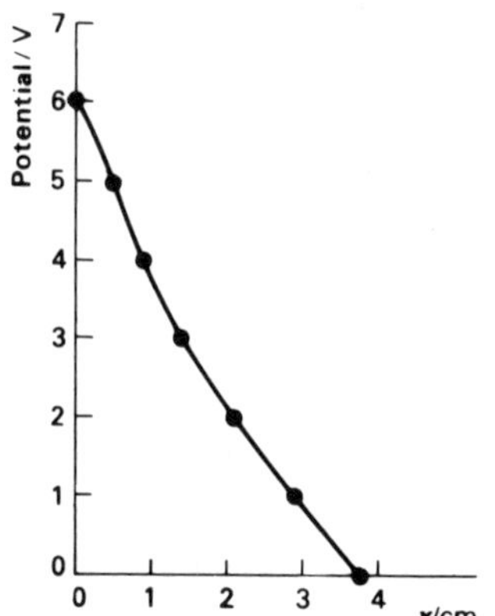

c 2

d It would be halved, *i.e.*, about 50 m.

e The cross-sectional area can be considered as πr^2, or as length (50 m) $\times$ thickness of sandwich (0.2 mm). This gives a diameter of about 0.15 m, rather larger than one might want of a 1 μF capacitor. In practice paper of this thickness would not be used.

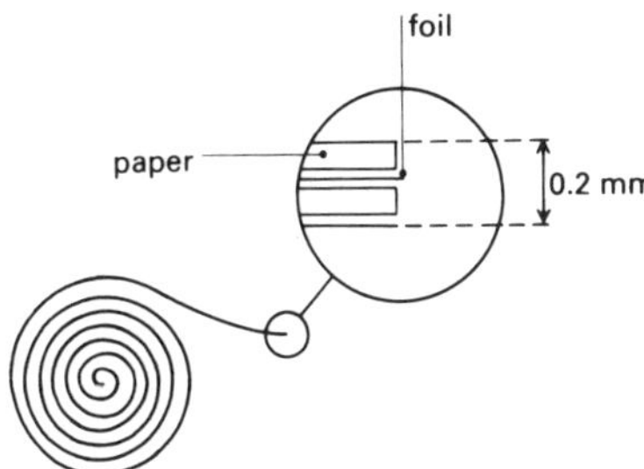

Figure Q59

16a*i* halved *ii* same *iii* halved *iv* halved *v* halved.

b*i* halved *ii* same *iii* doubled *iv* doubled *v* same.

c In **a** energy is transformed and heats the battery as half the charge flows 'the wrong way' through the battery. In **b** energy has to be supplied to move the plates apart.

17 Between 10^{-14} and $10^{-15}\,\Omega^{-1}\,m^{-1}$.

The area and distance from the bench are not required.

18a*i* Increases, since $C = \varepsilon_0 A/d$.

ii Decreases, since $V = Q/C$: Q is the same, C is bigger.

b Charge would flow onto the capacitor plates until a p.d. of 300 V was restored.

c If the plate separation is small, then small changes in it will produce relatively large changes in capacitance.

d 0.01 s

e*i* The p.d. will fall below 300 V (as there will be insufficient time for charge to flow and restore the p.d. to 300 V).

ii The p.d. will remain at about 300 V (since there is now plenty of time for charge to flow and maintain the p.d. at 300 V).

f*i* For slow vibrations, there is always time for charge to flow and the p.d. of 300 V to be maintained; therefore the p.d. across BC varies little for 50 Hz vibrations but may vary much more for 10 kHz vibrations.

ii The time constant could be increased (for example, by decreasing d and thereby increasing C) so that even 50 Hz appeared to be a fairly 'high' frequency.

19a $4\,m\,s^{-1}$

b $4\,m\,s^{-1}$

c*i* 8 J *ii* 8 J *iii* 640 J

d Approximately 1 m apart for 10 J intervals of energy; 100 m apart for 1000 J intervals.

e Contour lines, as on a map, joining points at the same potential energy.

20a*i* 50 000 J *ii* 200 000 J *iii* −50 000 J (*i.e.*, it *loses* P.E.)

b*i* 50 J *ii* 200 J *iii* −50 J

c*i* $50\,J\,kg^{-1}$ *ii* $200\,J\,kg^{-1}$ *iii* $-50\,J\,kg^{-1}$

d*i* $150\,J\,kg^{-1}$ *ii* $-100\,J\,kg^{-1}$

e 120 000 J

21a 10 m

b*i* 50 N *ii* 50 N

c*i* 1500 J *ii* 500 J

d 1000 J

e*i* $28\,m\,s^{-1}$ *ii* $32\,m\,s^{-1}$

22a 101.5 mm

b $6.66 \times 10^{-11}\,N\,m^2\,kg^{-2}$

23a The Earth's gravitational pull slows it down since it acts almost in the opposite direction to the motion.

b $7.65 \times 10^{-3}\,m\,s^{-2}$

c $7.65 \times 10^{-3}\,N\,kg^{-1}$

d $7.9 \times 10^{-3}\,N\,kg^{-1}$

e $2.2 \times 10^{-4}\,N\,kg^{-1}$. Its contribution is beginning to become significant, although still small ($\approx 3\,\%$ of the Earth's field strength).

24a

Mean distance $r/10^6$ m	Mean acceleration g/m s^{-2}
27.7	−0.453
55.4	−0.122
96.5	−0.042
170.4	−0.013

Table Q6

b Numerical values of the gravitational field strength are identical to those of the acceleration; units are N kg^{-1}.
c See figure Q60.

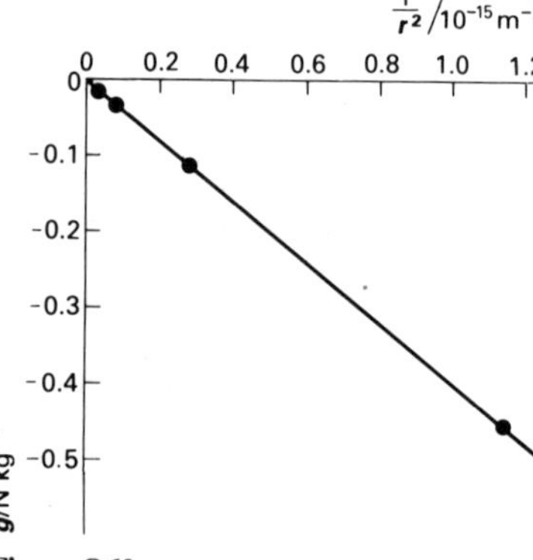

Figure Q60

25a One step. Assuming that g is constant over such a large interval leads to a poor estimate of the area under the graph.
b 1000 steps. This yields the best estimate of ΔV_g.
c One could use an even smaller step size.
d The program would take a very long time to run.
e 2 %
f Nearest the Earth, since this is where g varies most.

26a See figure Q61.

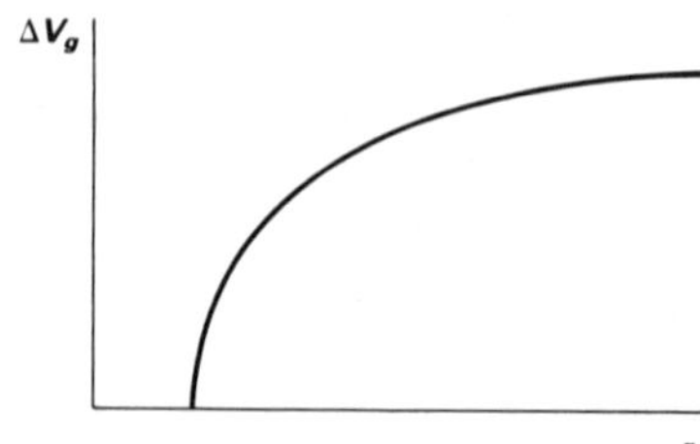

Figure Q61

b*i* 42.5×10^9 J
ii 10.6×10^9 J
iii 2×10^6 J
c 62.5 MJ
d*i* It will reach a distance of about 32×10^6 m from the centre of the Earth before falling back towards it.
ii It will escape completely from the Earth, its kinetic energy tending towards 7.5 MJ when it is a long distance from the Earth.
iii It will just escape from the Earth (assuming no other forces act on it), its kinetic energy tending towards zero as its separation from Earth increases.

27a*i* See figure Q62.

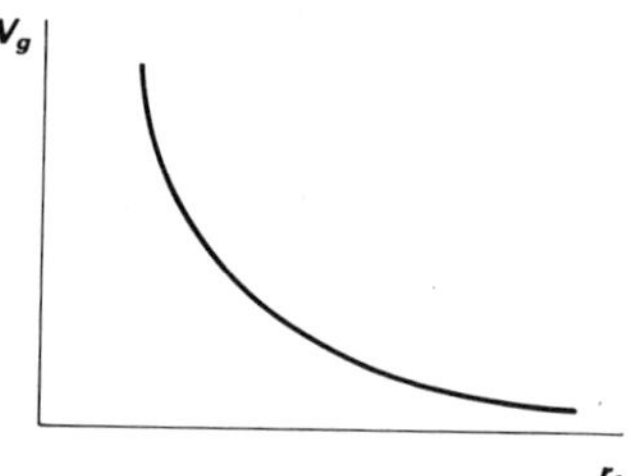

Figure Q62

ii The field strength is decreasing as r_1 increases, so the energy required to move one kilogram a given distance also decreases.
iii Students' suggestions might include $\Delta V_g \propto 1/r_1$; $\Delta V_g \propto 1/r_1{}^2$; $\Delta V_g \propto e^{-r_1}$
But note that whereas $\Delta V_g = 0$ when $r_1 = 50 \times 10^6$ m, ΔV_g will never become zero in any of the above relationships, however large r_1 becomes.
b*i* See figure Q63.

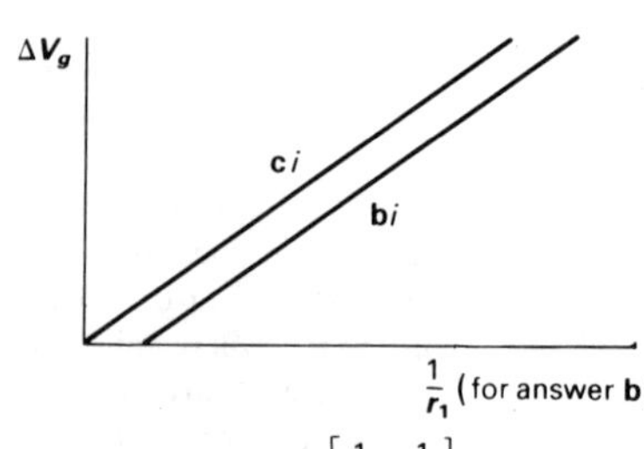

Figure Q63

ii Gradient $\approx 4.0 \times 10^{14}$ N m^2 kg^{-1}.

iii Intercept on $\frac{1}{r_1}$ axis is very close to the value of $\frac{1}{r_2}$, *i.e.*, 2×10^{-8} m^{-1}.

iv $\Delta V_g = 4 \times 10^{14} \times \left(\frac{1}{r_1} - \frac{1}{r_2}\right)$.

c*i* See figure Q63.

d It would have had the same gradient but a smaller intercept.

28a Zero

b No

c ΔV_g now represents the amount of energy required to remove 1 kilogram to an infinite distance from the Earth, or 'as far away as one would wish'.

d Gravitational potential energy is lost.

e Negative

f -62.5×10^6 J kg^{-1}

g See figure Q64.

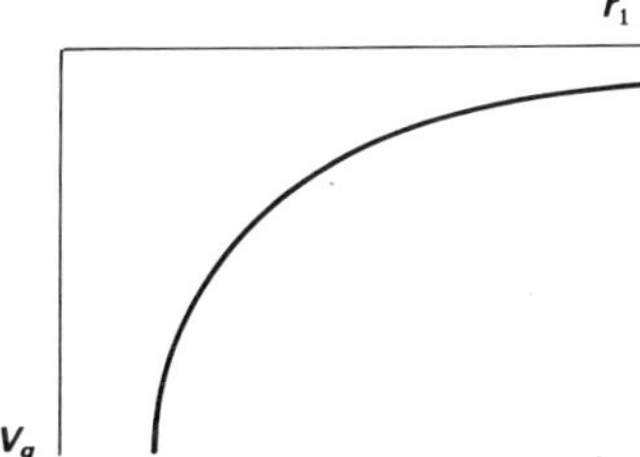

Figure Q64

29a $\Delta V_g \approx \frac{4 \times 10^{14}}{r_1}$

b $V_g \approx -\frac{4 \times 10^{14}}{r_1}$

c $GM = 3.98 \times 10^{14}$ N m^2 kg^{-1}. The two expressions are equal.

d 63×10^6 J kg^{-1}, using the approximate value for GM.

62.5×10^6 J kg^{-1}, using the more accurate value.

30a They are equal in magnitude and opposite in sign, assuming that no other forces act.

b -34.41×10^6 J kg^{-1}

c 34.41×10^6 J kg^{-1}

d Zero.

e 35.33×10^6 J kg^{-1}

f See figure Q65.

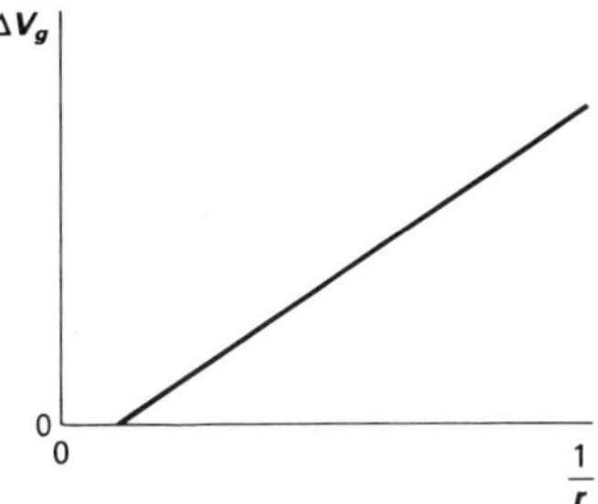

Figure Q65

g The gradient is very close to 3.98×10^{14} N m^2 kg^{-1}, the value of GM_E.

h The intercept is approximately 2×10^{-9}, giving a value of r_0 of 5×10^8 m.

i Zero.

j $\Delta V_g = \frac{GM_E}{r}$

31a 10^5 J

b 10 N. This is approximately the force at the Earth's surface, though it decreases with height.

c 61.5×10^6 J

d Because the force is much less than 10 N for most of the distance.

e 62.5×10^6 J

f 11 200 m s^{-1}

g The energy needed is $\frac{GMm}{r} = \frac{1}{2}mv^2$.

m cancels so a larger mass will require the same *velocity*, though of course it has greater *energy*.

h See figure Q66.

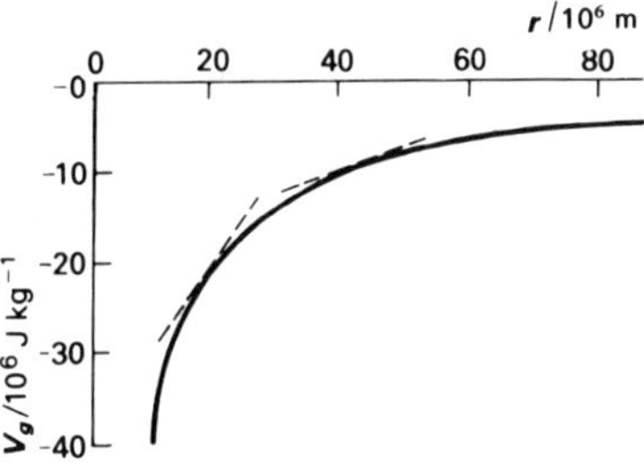

Figure Q66

i $1.0\,\mathrm{N\,kg^{-1}}$; $0.25\,\mathrm{N\,kg^{-1}}$.
j They represent the magnitudes of the field strengths at those points.
k The ratio is 4:1. Since the ratio of the distances is 1:2, this is quite consistent with an inverse-square field.

32a $-62.5 \times 10^6\,\mathrm{J\,kg^{-1}}$; $-2.8 \times 10^6\,\mathrm{J\,kg^{-1}}$
b $59.7 \times 10^6\,\mathrm{J\,kg^{-1}}$
c $597 \times 10^9\,\mathrm{J}$
d*i* $-62.5 \times 10^6\,\mathrm{J\,kg^{-1}}$
ii $-4.02 \times 10^6\,\mathrm{J\,kg^{-1}}$
iii $-1.36 \times 10^6\,\mathrm{J\,kg^{-1}}$
iv $-1.32 \times 10^6\,\mathrm{J\,kg^{-1}}$
v $-1.35 \times 10^6\,\mathrm{J\,kg^{-1}}$
vi $-3.94 \times 10^6\,\mathrm{J\,kg^{-1}}$
e–g See figure Q67.
h $612 \times 10^9\,\mathrm{J}$

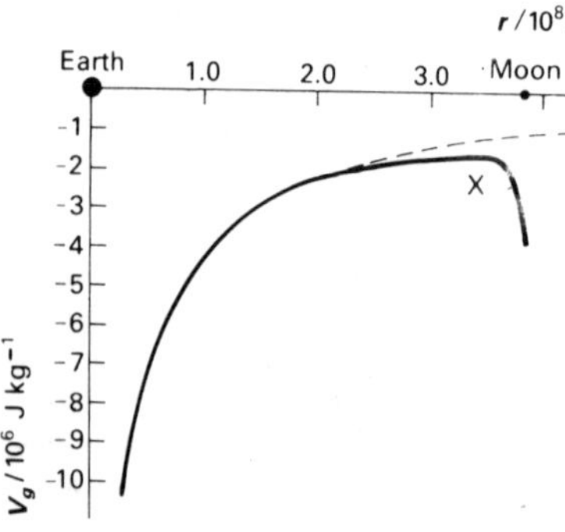

Figure Q67

i Extra energy will be gained in 'falling' from X towards the Moon. This must be transformed using reverse thrust. Also energy will have been 'lost' as the spacecraft is heated in passing through the Earth's atmosphere; and much of the fuel itself has been transported some distance from the Earth, which requires further energy.
j $2400\,\mathrm{m\,s^{-1}}$. It needs only sufficient energy to reach X, which is only about two-thirds of that required to escape completely. After X it will 'fall' towards Earth.
k Many molecules will have speeds in excess of the escape velocity (which is the same for all masses) and so will escape. However many molecules might have been present, a good proportion will always escape, so any atmosphere will soon dwindle to zero.

33a $r_1\theta$
b $\pi r_1^2\theta^2$
c $\sigma\pi r_1^2\theta^2$
d,e $G\dfrac{(\sigma\pi r_1^2\theta^2)}{r_1^2} = G\sigma\pi\theta^2$
f The result is of the same magnitude.
g Zero. The contributions are equal in size but opposite in direction.
h Zero. There is no field anywhere inside the hollow sphere.
i No.

34a Zero
b $-\dfrac{GM}{r^2}$
c $\frac{4}{3}\pi r^3\rho$
d $-\frac{4}{3}\pi\rho G r$
e See figure Q68.

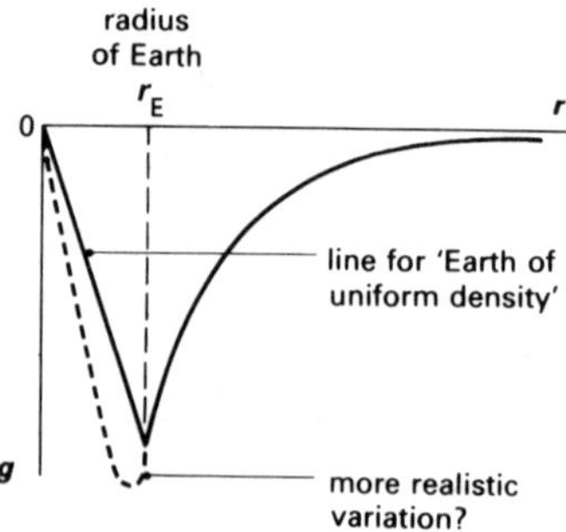

Figure Q68

35 $a = -\frac{4}{3}\pi\rho G r$
b The equation for acceleration is of the form $a = -\omega^2 s$, where s represents the displacement from the centre and $\omega^2 = \frac{4}{3}\pi\rho G$, which is constant. Therefore the motion is simple harmonic.
c None.
d More.
e It is lower (*i.e.* more negative). The potential difference between the centre and the surface of the Earth can be

calculated by finding the area under the g against r graph (*e.g.*, the answer to question **34e**) between these two points. The potential at the centre is lower than that at the surface by this amount.

f $\sin\phi = \frac{PC}{OP} = \frac{x}{r}$

Component of g towards C is $g \sin\phi = g\frac{x}{r}$

$= -\frac{4}{3}\pi\rho G r \frac{x}{r}$

$= -\frac{4}{3}\pi\rho G x.$

So acceleration $\propto$ −displacement from centre ⇒ the motion is S.H.M.

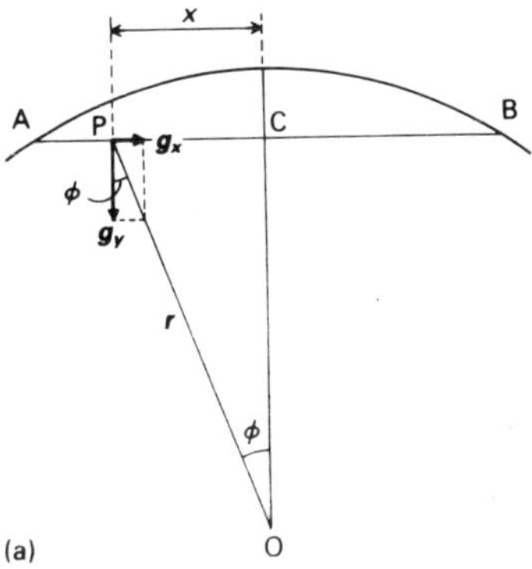

Figure Q69

g None.

h See figure Q69(b). This would be the kind of slope in which one would expect an object to perform S.H.M. It traces out the route between A and B which requires the minimum energy. In designing an underground railway, engineers might take this into consideration, as a route like that above requires the least acceleration and braking between stations and therefore uses least fuel.

36a The ball continues in a straight line (towards Q) at constant velocity because no overall force acts on it.

b An overall force (the pull of the string) now acts on the ball causing it to accelerate towards O. This acceleration is sufficient to maintain circular motion about O.

c Explanations should refer to the centripetal force required to make the contents of the dryer move in a circle. This is provided for the clothes by the drum but not for the water (at least not where there are holes in the drum) so that water which has already started moving (being carried along with the clothes by friction) continues in a straight line making its way out of the drum at a tangent. (Viewed from the inside the effect is of water being 'flung outwards'.)

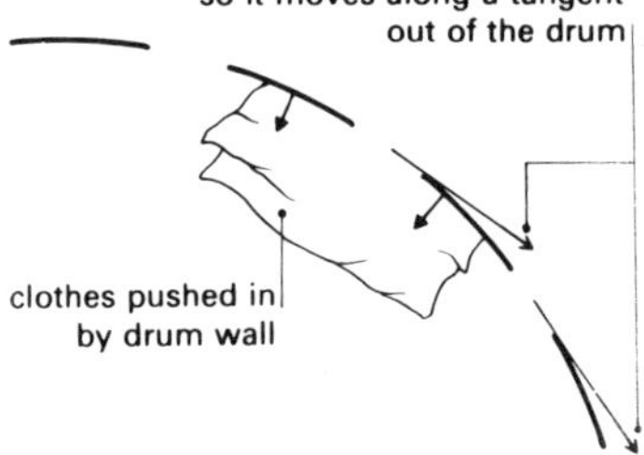

Figure Q70

37a*i* $\Delta t = \frac{1}{2}T$ *ii* $\Delta v = 2v$

iii $a = \frac{2v}{\frac{1}{2}T} = \frac{4v}{T}$ *iv* $T = \frac{2\pi r}{v}$

v $a = \frac{4}{2\pi} \times \frac{v^2}{r} \approx 0.64\frac{v^2}{r}$

b*i* $\Delta t = \frac{1}{4}T$ *ii* $\Delta v = \sqrt{2}v$ *iii* $a \approx 0.90\frac{v^2}{r}$

c*i* $\Delta t = \frac{1}{6}T$ *ii* $\Delta v = v$ *iii* $a \approx 0.95\frac{v^2}{r}$

d*i* $\Delta t = \frac{\theta}{2\pi}T$ *ii* $\Delta v \approx PQ = v\theta$

iii $a = \frac{v^2}{r}$ *iv* Acceleration is at right angles to motion.

e Centripetal acceleration $= \frac{v^2}{r}$

f Centripetal force, $F = \frac{mv^2}{r}$

g A negative sign, $F = -\frac{mv^2}{r}$

38a $\frac{2\pi r}{T}$

b $-\frac{v^2}{r}$

c $-\frac{4\pi^2}{T^2}r$

d $-\frac{GM}{r^2}$

e $\left(\frac{GMT^2}{4\pi^2}\right)^{\frac{1}{3}}$. Only one particular value of r gives the correct period.

f 36×10^6 m above the Earth's surface. $53 \times 10^6\,\mathrm{J\,kg^{-1}}$.

g Because nowhere else could the satellite stay directly above a point and move in a circular orbit *with the centre of the Earth as its centre.*

h Shorter.

i $v = \left(\frac{GM}{r}\right)^{\frac{1}{2}}$. Faster.

j Briefly, energy is transformed during the impact so the satellite loses energy and cannot remain in its present orbit. In dropping to a lower orbit, it speeds up, as potential energy is now transformed into kinetic.

k A satellite in 'geostationary' orbit above the Equator can 'see' nearly a third of the Earth's surface and so can be used for telecommunications over a very extensive area. However, with nearly 200 communications satellites in this orbit, space is becoming a little crowded, especially in the popular region over the Atlantic. (Signals from nearby satellites must be modified to prevent 'interference'.)

Satellites in low orbits lose energy quickly as they encounter the upper levels of the Earth's atmosphere. Sometimes their short life is an acceptable price to pay for the better 'view' they afford, especially if they are designed for monitoring or military purposes. The more stable orbits further out tend to accommodate satellites serving more civilian purposes such as weather forecasting. Many satellites engaged in scientific research have extremely eccentric elliptical orbits which enable them to penetrate wide regions of the magnetosphere. All in all the first 25 years of space exploration have seen over 3000 satellites go into orbit.

39a $E_p = -\frac{GMm}{r}$

b $-\frac{mv^2}{r}$

c $-\frac{GMm}{r^2}$

d $E_k = \frac{1}{2}\frac{GMm}{r}$

e $E = -\frac{1}{2}\frac{GMm}{r}$. Total energy negative means that the satellite is bound in the Earth's gravitational field.

f $\sqrt{2}$. This would double the kinetic energy making the total energy zero.

40a r is the (average) radius of the orbit, T its period.

b $\frac{v^2}{r}$

c $\frac{2\pi r}{T}$

d $\frac{\left(\frac{2\pi r}{T}\right)^2}{r} = \frac{4\pi^2 r}{T^2}$

e $\frac{r^3}{T^2}$ is constant so the acceleration is proportional to $\frac{1}{r^2}$.

f The force is also inverse-square.

41a Twice the distance.

b Twice the height.

c The same.

d The same.

e Four times the area.

f One-quarter.

g One-ninth.

h It is a constant (called the 'flux' of the light).

i In a parallel beam of light both intensity and area remain constant with distance and so does the flux.

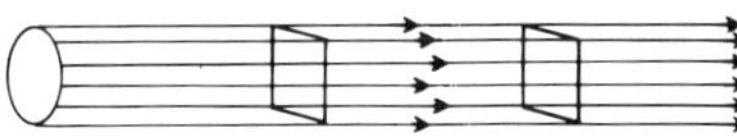

Figure Q71

But wherever light converges to or diverges from a focal point, the change in intensity with distance must be inverse-square if flux is to remain constant.

The 'conservation of flux' is a key concept in more advanced physics and the idea also applies to another type of field – magnetism.

42a Approximately, F is proportional to d for small angles. The deflection is doubled.
b The points on a graph of d against $1/r^2$ lie near to a straight line.
c $0.01\ \mu F = 10^{-8}$ F. At 0.5 V the charge on it is 0.5×10^{-8} C. This is nearly the charge that was on the ball. The fluctuations in charge could explain why the points plotted in **b** are scattered on either side of a straight line.
d,e The force constant is of the order of $10^{10}\ N\ m^2\ C^{-2}$, and certainly lies between 5×10^9 and $5 \times 10^{10}\ N\ m^2\ C^{-2}$.

43a 10^{-4} N
b 10^{-9} C
c 10^{-11} A
d 10^3 V
e $10^{14}\ \Omega$

44a See figure Q72.

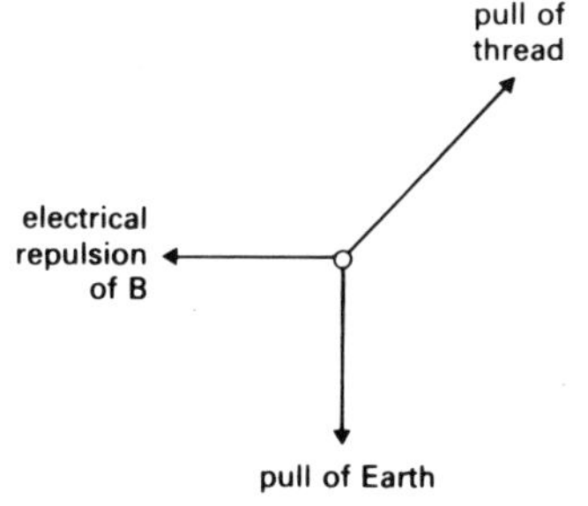

Figure Q72

b 6×10^{-5} N
c 8×10^{-9} C

45a $\frac{1}{4\pi\varepsilon_0}\frac{Q}{r^2}$
b $2r$
c $4Q$
d $\frac{1}{4\pi\varepsilon_0}\frac{4Q}{(2r)^2} = \frac{1}{4\pi\varepsilon_0}\frac{Q}{r^2}$
e*i* Upwards.
ii Small.
iii Approximately uniform.
f Approximately inverse-square.

46a 1.5×10^5 V
b 2.5×10^{-6} C
c $1.0 \times 10^6\ V\ m^{-1}$
d 0.05 m, assuming a roughly uniform field between the sphere and the hand.

47a $3 \times 10^{20}\ N\ C^{-1}$
b 1×10^2 N
c 7×10^6 V
d 2×10^{-12} J
e 12 MeV, assuming all the electrical potential energy is transformed into kinetic energy.

48a $5.8 \times 10^{11}\ N\ C^{-1}$
b 9.2×10^{-8} N
c 28.8 V
d -28.8 eV
e Zero.
f Zero.
g 57.6 V
h -57.6 eV
i 14.4 eV
j -43.2 eV
k The ion has negative potential energy and so is a bound system; the two protons without the electron would fly apart. This very simple view of a 'covalent' bond has the electron 'shared' by both protons. It is not a complete view, as the electron has kinetic energy as well, for instance, but it gives results of the right order of magnitude.

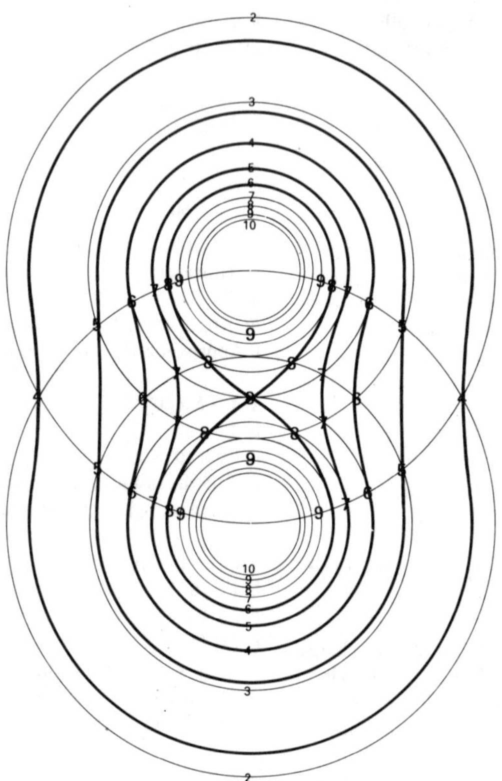

Figure Q73

49a Students may produce a pattern something like that shown in figure Q73.
b See figure Q74.
c The field patterns should be similar to those obtained between two 'point' electrodes in experiment E3 with semolina particles.

50a*i* 270 V *ii* 150 mm
iii 1.5×10^{-8} C
b Less than 900 V. Mutual repulsion between the charges causes them to be displaced somewhat toward the outer sides of the spheres, thereby increasing the effective separation of the charges and thus reducing the potential.

51a 900 V
b 1200 V at X, 1080 V at Y; by simple addition of the potentials due to A and B.
c Yes, because there is a potential gradient across it.
d Charge would flow towards Y until the potential gradient fell to zero.
e No.
f The surface of a conductor must always be an equipotential, even if some parts of the surface are charged.

52a 10^{-47} N
b 10^{40}
c Both the electric force and gravity obey inverse-square laws so one force falls off with distance at the same rate as the other.
d Approximately 10^{36}.
e No.
f It falls off very sharply indeed and is quite insignificant outside the nucleus.

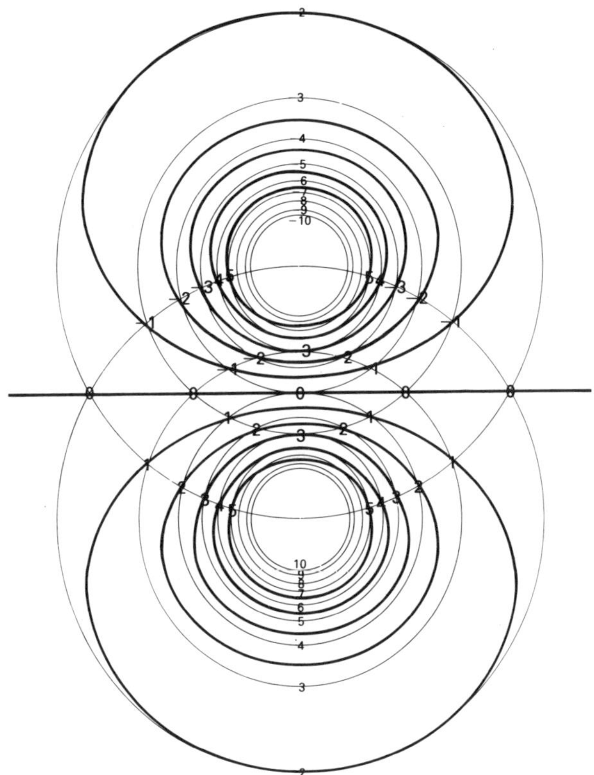

Figure Q74

53a $1.4 \times 10^{11}\,\mathrm{N\,C^{-1}}$ (= 1 unit)
b $\frac{1}{100}$ unit
c $-\frac{1}{9}$ unit at C; $-\frac{1}{144}$ unit at D
d $\frac{8}{9}$ unit at C; $\frac{44}{14\,400}$ unit at D
e $\frac{11}{3200} \approx \frac{1}{290}$
f More rapidly (particularly if one considers that the distance of D from the middle of the dipole is only just over five times that of C). From a good distance the field strengths due to A and B almost cancel; indeed the dipole 'looks' almost like a neutral object.
g The array is neutral when 'viewed' from a distance and it is only very near any individual charge that the local field dominates significantly.

54 Lack of space does not permit a full answer here. (Examination question.)

55a Electrical/gravitational field strength is the force acting on unit charge/mass.
b By taking the negative of the gradient of the graph.
c By computing the area under the graph between the two points.
d By multiplying by the mass or charge.
e The negative of the force.
f The force is always attractive, so the force vector is directed *inwards*, in the opposite direction to the separation or displacement vector.
g This is taken care of by the charges which may be positive or negative, giving both attractive (negative) and repulsive (positive) forces.

h Because the potential energy of a mass at all points is less than that at the agreed zero of potential (at 'infinity').
i A positive charge would have positive potential energy and a negative one negative potential energy.

UNIT F
Radioactivity and the nuclear atom

1a One method would be to use three distinct detectors: a spark counter which only responds to α-particles, a thin end-window GM tube which responds to beta and gamma radiation, and a thicker tube which responds to gamma radiation only.

If a thin end-window GM tube is the only detector available, then note the effect of a range of absorbers (thin paper for alpha radiation, a few millimetres of aluminium for beta radiation, leaving gamma radiation which it is very difficult to absorb).

Magnetic deflection experiments are not very satisfactory; a field powerful enough to produce appreciable bending of an alpha particle track will have such a major effect on the track of the beta particles that they will be sent back to the source.
b Some energy is transferred to atoms during elastic collisions and some in exciting and in ionizing atoms.
c*i* The magnetic field is too weak to produce measurable bending of the path of the alpha particles.
ii Use a very strong magnetic field and work *in vacuo*.

2a $10^6\,\mathrm{V\,m^{-1}}$
b $1.6 \times 10^{-13}\,\mathrm{N}$
c $1.6 \times 10^{16}\,\mathrm{m\,s^{-2}}$
d $3.3 \times 10^{-10}\,\mathrm{s}$
e $9 \times 10^{-4}\,\mathrm{m} \approx 1\,\mathrm{mm}$
f If $L = l$, the deflection y will be more than twice as big as s, because the straight path from the edge of the plates to the screen cuts the undeflected path through the plates (if projected back) somewhere between the plates, not at the far end of the plates. An estimate of 3 or 4 mm for y might be fair.

3a $2 \times 10^8\,\mathrm{s^{-1}}$
b 5×10^4
c 1.5 MeV

4a Towards the end of a track the particle has lost a lot of energy and is travelling more slowly. It is therefore spending more time in the vicinity of the atoms of the air; these atoms are more likely to be ionized.
b The number of ion-pairs is given by the area below the graph from the distance 50 mm back to the origin. One rough estimate put the number of ion pairs at about 200 000.
c The total energy of such an alpha particle is around 6 MeV.

5a $10^9\,\mathrm{J}$
b The total energy is larger, $5 \times 10^9\,\mathrm{J}$.
c A minimum estimate.
d 10^{16}
e $10^{-2}\,\mathrm{J\,s^{-1}}$
f About 2×10^{-12}.
g Of the order $5 \times 10^{11}\,\mathrm{s}$. Note that this is something of an overestimate, the half-life being 1600 years ($5 \times 10^{10}\,\mathrm{s}$). But the errors involved in the approximations made could easily amount to a factor of ten.

6a $8.7 \times 10^{-19}\,\mathrm{J}$
b $8 \times 10^{-13}\,\mathrm{J}$
c $2.4 \times 10^{12}\,\mathrm{J}$
d The energy released in the radioactive decay of radium is greater than that released on oxidation by a factor of over a million.
e Even with 1 mole of radium (226 g – a very large sample), the energy is dissipated over thousands of years. In the first 1600 years, 0.5 mole will have decayed.

7 Students' laboratory notes for this Unit are prefaced by a note about the precautions which should be observed when handling radioactive substances. Make sure that students read this.

A full answer to the question is given in the *Students' guide* page 485.

8 197 g of gold contain 6×10^{23} atoms and occupy $\frac{197 \times 10^{-3}}{19.3 \times 10^{3}}\,\text{m}^3$.

1 atom of gold occupies $1/(6 \times 10^{23})$ of this space, which is $17 \times 10^{-30}\,\text{m}^3$.

The cube root of this gives an estimate of the diameter of a gold atom ($\approx 2.6 \times 10^{-10}$ m).

9 If the gold atom is about 2.6×10^{-10} m across, the number of layers n in a foil 6×10^{-7} m thick is 2.3×10^3 approximately.

If d is the diameter of a nucleus, then

$$\frac{d^2}{(2.6 \times 10^{-10})^2} \approx \frac{1}{8000n}$$

$$d \approx 6 \times 10^{-14}\,\text{m}$$

The roughness of the data does not justify an accurate calculation, for example allowing for the way gold atoms pack together. The important thing is the order of magnitude of the answer, some 10^4 times smaller than the diameter of the atom.

10a The incident particle will stop. The other particle moves on in the same direction at $10^6\,\text{m s}^{-1}$.

b Momentum is conserved in any collision. After the collision the velocity of the alpha particle is $0.6 \times 10^6\,\text{m s}^{-1}$ in the original direction; the velocity of the proton is $1.6 \times 10^6\,\text{m s}^{-1}$, also in the original direction.

The mathematics produces another solution, in which the alpha particle goes straight on without change of speed and the proton stays at rest. That is, the alpha particle misses the proton! The algebra wasn't told that they must collide!

c The gold nucleus, mass number 197, is much more massive than the alpha particle and its velocity after impact can only be very small. Even if the momentum change of the alpha particle and the nucleus is as large as possible ($2mv$), the *velocity* of the nucleus is still small.

11a 8×10^{-13} J

b As the alpha particle approaches the nucleus, this kinetic energy is converted to electrical potential energy. When the speed of the particle is zero, this conversion is complete and the energy stored in the system is 8×10^{-13} J. The potential energy is given by

$$E_p = \frac{(2e)(79e)}{4\pi\varepsilon_0 r}$$

on the assumption that the gold nucleus acquires very little kinetic energy.

c $r = 4.6 \times 10^{-14}$ m

d The electric potential at this distance is

$$V = \frac{79e}{4\pi\varepsilon_0 r} = 2.5 \times 10^6\,\text{V}.$$

e Since the alpha particle carries charge $+2e$, its original kinetic energy in eV will be double this number, *i.e.*, 5.0×10^6 MeV.

f The electric field at this distance

$$E = \frac{V}{r} = \frac{2.5 \times 10^6\,\text{V}}{4.6 \times 10^{-14}\,\text{m}} = 5.4 \times 10^{19}\,\text{V m}^{-1}.$$

The force on a charge of $2e$ is about 17 N.

This is about the same as the weight of a mass of 1.5 kg – an enormous force to find exerted on a single atomic particle. The book was the best guess.

12a An attractive force which decreases with distance from A.

b A repulsive force which decreases with distance from A.

c No significant force at the distances shown.

d A repulsive force which suddenly affects P at a small distance from A.

13a The alpha particles are being repelled by the nucleus N.

b The force on B will be one-quarter of that on A.

c The forces are along the lines NA and NB.

d See figure Q75.

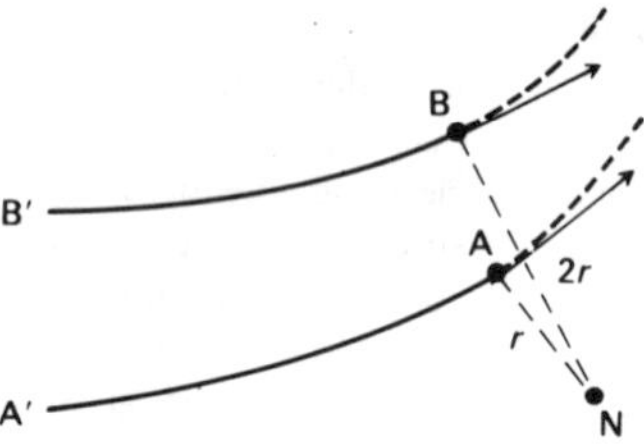

Figure Q75

e The velocity at A will be less than that at A′ as some of the kinetic energy of the alpha particle has been converted to potential energy.

f See figure Q76. At X the potential energy will be a maximum and the speed of the alpha particle a minimum.

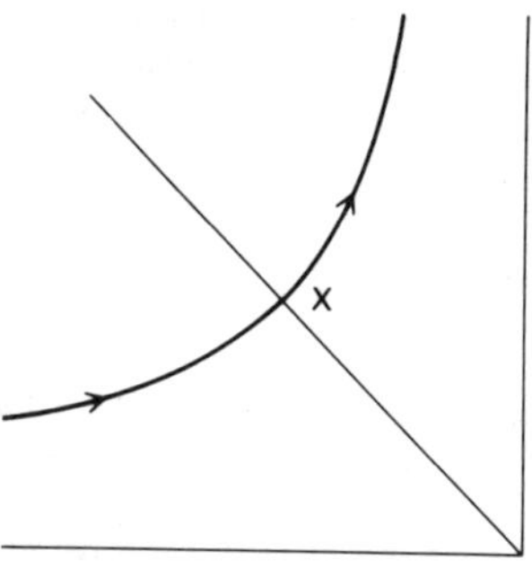

Figure Q76

g The potential energy at B is half that at A, so that the speed at B will be greater than that at A. Less kinetic energy has been transformed to potential energy.

h No. The force between the particles is always along the line joining them and so there is no force deflecting the alpha particles out of the original plane surface.

14a*i* Not very good. It describes roughly what happens, but may give the wrong impression about why. The particles come in by chance, and if the chance of arrival stays the same, over a long time there will be a steady average rate.

ii We tried to make this a correct description.

iii Wrong. The time between arrivals may have any value, but not all intervals are equally likely. The average interval is most likely; much longer or shorter intervals are less likely.

b To say that in the kinetic model the molecules of a gas have random motion implies that a molecule may be moving in any direction at any instant (not one of which is preferred). Molecular speeds are distributed randomly about a mean value; this mean value is the most likely to be found whilst far higher or far smaller speeds are unlikely.

15a In the same short time, say one second, more sodium atoms than strontium atoms will disintegrate, because the sodium has a shorter half-life.

b See figure Q77.

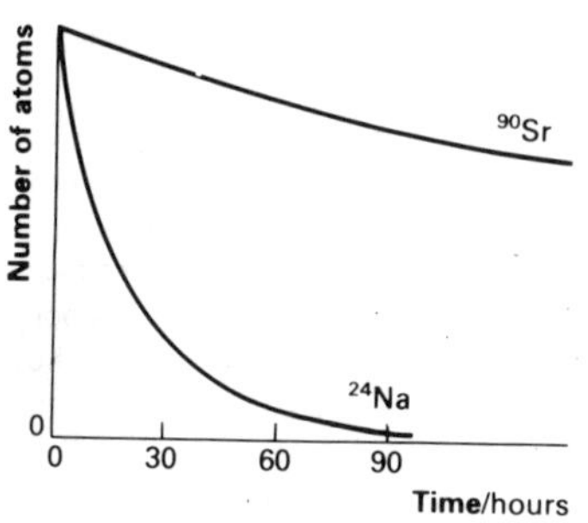

Figure Q77

c 27 years is about 1.6×10^4 times longer than 15 hours, so the sodium will initially be about this many times more active than the strontium. In 15 hours the activity of the sodium will fall to one-half; in twice as long to one-quarter; and so on. It will have fallen by a factor 1.6×10^4 after n periods of 15 hours, where

$2^n = 1.6 \times 10^4$, and therefore

$n \approx 14$

Fourteen times fifteen hours is about 9 days; much less than the 27 year half-life of the strontium, so the strontium activity will not have diminished very much at all. So 9 days may be a fair estimate of the time taken for the two activities to become equal.

16 A common answer is that a radioactive sample has an infinite lifetime 'in theory but not in practice'. Although not absolutely wrong, this seems to us to be a bit feeble.

The smooth mathematical model of exponential decay does not exactly fit the behaviour of radioactive atoms. As one finds by experiment, the rate of decay of a sample fluctuates considerably. The average rate of decay is close to the value to be expected from the mathematical model, but need not be equal to it.

When the sample is reduced to only a few atoms, the smooth exponential model is a poor fit. Yet it is not enough to say that 'the theory breaks down'. The smooth exponential decay is a consequence of supposing that there is a constant chance of decay for each atom in each time interval, and also that there are very many atoms. When there are not many atoms left, the smooth decay is not to be expected. But so far as is known, the chance of decay remains constant. When one atom remains, that atom may decay in the next second, or in the next hundred years. It is possible to say how long it will last on average: that is, how long one atom will last in many trials. But it is not possible to say how long one particular atom will last.

17 The background count is likely to be 119/4 counts per minute $\approx$ 30 counts per minute. After a further 30 s the total count is likely to be 135.

18a The background (30 counts per minute) should be deducted from each of the counts per minute tabulated.
b The half-life is about 80 minutes.
c The count rate would not change significantly if the counting were continued for a longer period; very little of the radioactive substance remains after 8 hours, anyway.
d No; random fluctuations within the background count are probably responsible for the increase.
e The experiment could be repeated and readings could be taken at intervals closer than one hour during the first few hours.

19a Both are exponential. The lengths of the straws decrease in a fixed proportion.
b Plot the natural logarithm of y against x; or check whether y changes by a constant factor as x increases in equal steps.

20a The data do fit the model; the constant ratio over the first six hours averages 0.61, and the graph of ln (activity) against time is a straight line.
b Since $N = N_0 e^{-\lambda t}$, $\ln N = \ln N_0 - \lambda t$, where λ is the gradient of the graph of ln (activity) against time and λ is the decay constant. The slope and therefore the decay constant is $-0.0084\ \text{min}^{-1}$.

Since $t_{\frac{1}{2}} = \dfrac{0.693}{\lambda}$, $t_{\frac{1}{2}}$ is 83 minutes.

21a Although the graph of ln (capacity in GW) against time does not follow a straight line, there are periods when it does so to a good accuracy. From 1951 to 1958 it was very nearly straight (and the growth exponential), and also from 1964 to 1970.
b The growth constant diminished in about 1962. During the 1970s growth ceased and the capacity remained approximately constant.
c The 'doubling times' were
i about 10 years and
ii about 11 years.

22 A logarithmic plot shows that from 1956 to 1962 the rise is closely exponential.

23 The slope of the plot of ln (population) against time has increased since the period 1650–1750, when the growth was closely exponential. (See figure Q78.)

24a The change in the number of families with video recorders.
b The rate at which the number of such families changes.

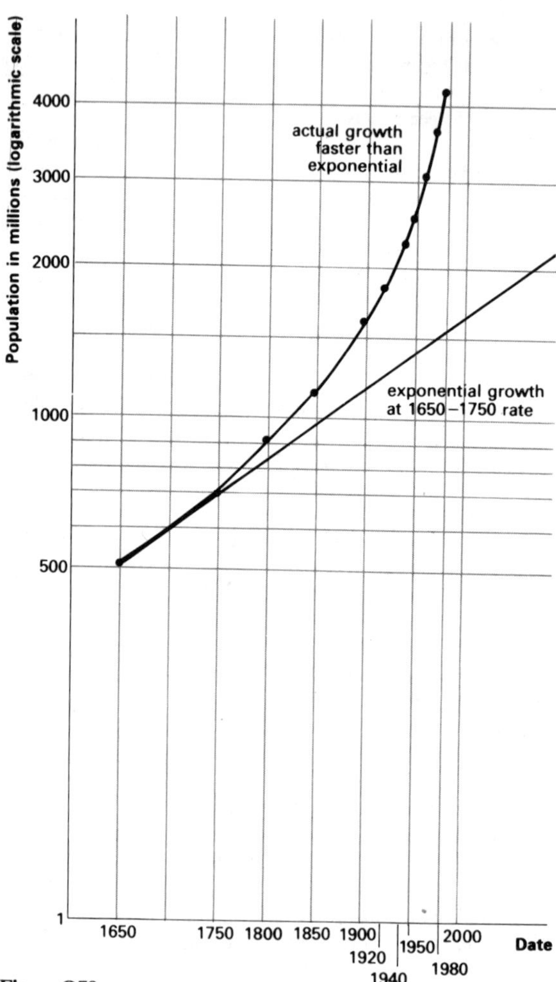

Figure Q78

c The model $\Delta N/\Delta t = kN$ suggests that the rate at which the number of families with video recorders grows is directly proportional to the number of families already having video recorders. This would give an exponential growth. In real life this could be the result of 'keeping up with the Joneses'.
d There could be a limit on the supply of money available, costs might rise unexpectedly, or the supply of the equipment might become limited. Such changes would reduce the value of k after a time.
e The existence of television could increase the value of k.

f See figure Q79.

25a Press, in order, the keys 2, x^y, 7, = (or the keys 2, logarithm, ×, 7, =, 10^x; or multiply the logarithm of 2 by 7 and find the number whose logarithm is this product).
b $N = 1.414$
c, d The graphs start at $N = 1$ because, at $t = 0$, $N = a^0 = 1$. (See figure Q80).
e At each step of 0.2 in t, the value of N is multiplied by the same factor, which is $2^{0.2}$, or 1.15.
f $2^{0.2}$ is 1.15; $2^{3.2}$ is 9.20.

g In the series a^t, a^{2t}, a^{3t}, etc., each exponent is equal to the previous one with the addition of t. Adding t to the exponent of a means multiplying by a^t; similarly, adding x to the exponent of a means multiplying by a^x.

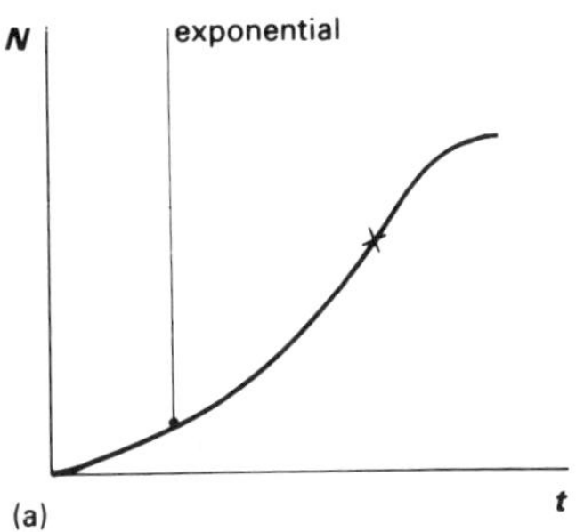

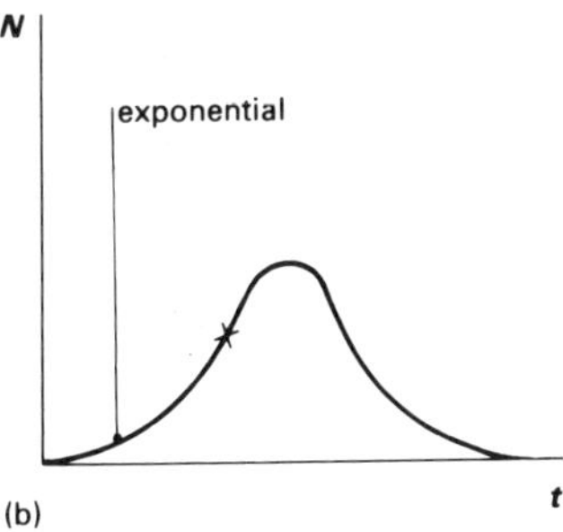

Figure Q79

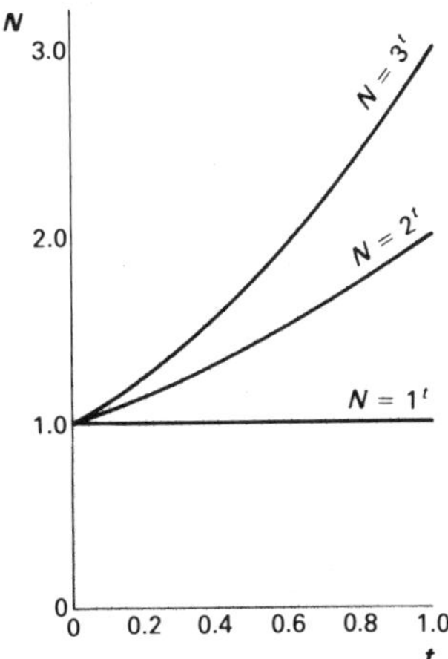

Figure Q80
Plots of $N=1^t$, $N=2^t$, and $N=3^t$.

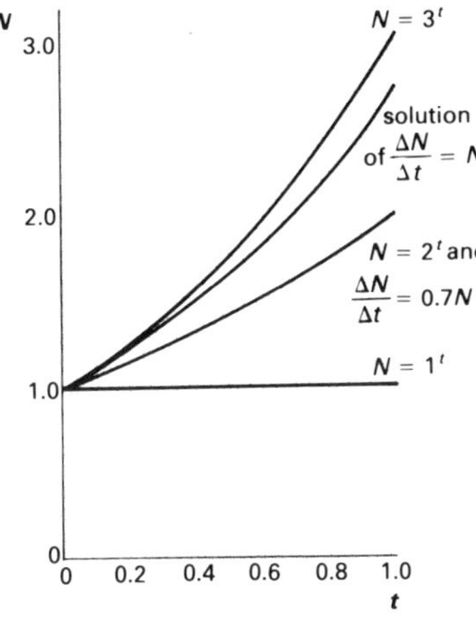

Figure Q81

26a, b See figure Q81.
a ΔN is a little larger in the second 0.1 s interval because the starting value, N, is now larger (1.1 instead of 1.0).
b Growth is slower because the constant ratio is now 0.7, not 1.0.
c If $N=2^t$, then $N=2$ at $t=1$.
d $N=a$ at $t=1$.
e For $\Delta N/\Delta t=(1.0)N$, $N\approx 2.7$ at $t=1$, so $a\approx 2.7$. For $\Delta N/\Delta t=(0.7)N$, $N\approx 2.0$ at $t=1$, so $a\approx 2.0$.

27a $e^{0.7}=2.01\approx 2.0$.
b Since $(e^{0.7})^t=e^{0.7t}$, the best speculation is $a=e^k$.

28a The steps are:
i Calculate N, the number of ^{40}K atoms in 0.6×10^{-6} g; use $dN/dt=-\lambda N$ to calculate λ.
ii Use $t_{\frac{1}{2}}=0.693/\lambda$ to calculate $t_{\frac{1}{2}}$ (in *seconds*).
b Total mass of ^{40}K originally present was 4.8×10^{-6} g. The amount remaining, 0.6×10^{-6} g, is 1/8 of this. $8=2^3$, so it takes three half-lives for the amount of ^{40}K to decrease to 1/8 of its initial value.
c Background count is likely to be very much greater than $0.16\,s^{-1}$.

29a 1.993×10^{-26} kg
b 1.661×10^{-27} kg
c 1.007 u; 1.008 u
d 0.000 548 u

30 The mass of heavy hydrogen is twice that of 'normal' hydrogen. As the two, in a mixture, are at the same temperature, they should have the same average kinetic energy, according to the kinetic theory of gases. The average velocity of heavy hydrogen molecules will therefore be less than that of normal hydrogen. Thus, more normal hydrogen will evaporate than heavy hydrogen.

The method will be less effective with neon because the velocities will differ by a smaller factor.

31 The data suggest that, while the average molar mass of lead from a variety of sources is around 207 g mol^{-1}, the lead in particular minerals may have a molar mass which is more or less than this. In particular, it is possible that the lead in thorite comes in part from the decay of thorium. A glance at the decay series of ^{232}Th (see REVISED NUFFIELD ADVANCED SCIENCE *Book of data* page 29) shows that the series ends with ^{208}Pb. On the other hand, lead from the decay of ^{238}U is the isotope ^{206}Pb. It seems probable that the lead in the first four minerals mentioned comes at least in part from the decay of uranium and not from the decay of thorium to any appreciable extent. Pitchblende was the uranium mineral studied by Becquerel and by the Curies, who extracted radium from it.

32a One gram of carbon contains 0.5×10^{23} atoms, since 12 g contain 6×10^{23} atoms. Of these, about 0.5×10^{13} atoms are the isotope ^{14}C. In one second, 4×10^{-12} of the ^{14}C atoms will decay, so one may expect around 20 decays per second, an activity of 20 Bq.
b 11 000 years is close to two half-lives, so the activity might be around 5 Bq from one gram.
c The rate of decay is small, and the smaller it is, the harder it will be to measure the rate accurately unless one is prepared to measure it over a very long period. Even with counting times of about a day, the ^{14}C method is subject to pretty large errors for times above about two half-lives, and is not of great assistance much beyond three half-lives, say 20 000 years.

34 Mass of alpha particle $\approx 7 \times 10^{-27}$ kg. Speed $\approx 7 \times 10^{6}$ m s^{-1}. Note that to use the equation, kinetic energy $= \frac{1}{2}mv^2$, with m in kilograms, the energy must be in joules, not electronvolts.

35a 1.6×10^{-18} J
b*i* 0.8×10^{-18} J *ii* 3.2×10^{-18} J
c*i* 1.3×10^{6} m s^{-1} *ii* 1.9×10^{6} m s^{-1}
iii 2.6×10^{6} m s^{-1}

36a These are the inert gases, which rarely form compounds with other elements. They have high ionization energies. It is not easy to remove an electron from one of these atoms.
b These are the alkali metals, all very reactive, readily forming singly charged positive ions. In a crystal of sodium chloride, the sodium atoms are all ionized, and the crystal is a vast assembly of Na^+ and Cl^- ions. All the alkali metals have low ionization energies, indicating the ease with which an atom loses an electron.
c Neon has a large ionization energy; sodium, with one more electron, has a low ionization energy; it seems that the 'last' electron in a sodium atom is rather loosely held. (In fact, it takes 47 eV to remove a second electron from sodium, so it is only the 'last' one which is nearly free.)
d Yes, each comes first on the rise following a trough. They are members of another family with chemical similarities.
e If francium is the element one before radium, it must be an alkali metal, like Li, Na, K, Rb, Cs. Their ionization energies are all seen to be low, from 3 to 5 eV. The value drops slowly as one goes along the

list, so francium will probably have an ionization energy nearer the lower end of this range.

37a An electron a long way from a nucleus will be weakly bound to it, as electrical forces diminish with distance. If at least one electron is a long way out from the nucleus, the atom will be big, and will have a low ionization energy. (The fact that the energy needed to remove a second electron from Na is 47.3 eV, while that to remove the first is only 5.1 eV, suggests that only one electron is a long way out in Na.)
b The ion Na^+ is a good deal smaller than the atom Na, suggesting that just one electron in Na, and in other similar elements, is a long way out. If so, the energy needed to remove this electron will be small, as the data in figure F31 of the *Students' guide* show it to be.
c The simple general rule is that all atoms are about the same size, whatever the number Z of electrons bound to a nucleus of charge Z. While Z varies from 1 to 90, the radius goes up and down around 1.5×10^{-10} m. The smallest radius is 0.3×10^{-10} m; the largest shown is 2.7×10^{-10} m, a factor of only 9. This is surprising. Despite the extra attraction of a strongly charged nucleus, the electrons of a massive atom occupy much the same space as those of a light one. Or, to put it another way, ninety electrons pack into the same space as three.

38a -2.18×10^{-18} J
b $-e^2/4\pi\varepsilon_0 r$
c 1.0×10^{-10} m
d The *total* energy of the electron is -2.18×10^{-18} J. Since it has some kinetic energy, which must be positive, the potential energy is more negative than -2.18×10^{-18}, *i.e.*, the electron is closer to the proton than this calculation suggests. (In fact, as will be seen in Unit L, 'Waves, particles, and atoms', the electron is not precisely located at one distance from the proton. But this calculation gives the correct order of magnitude for the proton–electron distance.)

39 X is $^{207}_{82}Pb$.

40 A

41a $^{212}_{82}Pb \rightarrow ^{212}_{83}Bi + ^{0}_{-1}e$
b $^{212}_{83}Bi \rightarrow ^{212}_{84}Po + ^{0}_{-1}e$
c $^{212}_{84}Po \rightarrow ^{208}_{82}Pb + ^{4}_{2}He$

42a F_e is repulsive if Q_1 and Q_2 are both positive or both negative; F_g is always attractive.
b $F_e/F_g = Q_1 Q_2/4\pi\varepsilon_0 \, G m_1 m_2$
c Both forces vary in the same way with distance (inverse-square law), so their ratio is the same at all distances.
d 1.2×10^{36}
e No; it is much smaller than the repulsive electrical force.
f Nuclei of neighbouring atoms would be attracted to each other: they would not stay about 10^{-10} m apart.

43a ^{12}C is the more stable.
b ^{2}H, ^{6}Li, ^{4}He, ^{235}U, ^{12}C, ^{208}Pb, ^{56}Fe.
c Helium is at a minimum in the curve. The average binding energy per nucleon is lower for some heavier nuclei (*e.g.*, ^{12}C, ^{16}O), but to form nuclei like ^{6}Li which are likely to be involved in intermediate steps in the build-up of heavier nuclei requires an input of energy.
d C, O, Fe.

Students can read more about the evolution of the elements in 'Our nuclear history' in the Reader *Particles, imaging, and nuclei.*

44a 1050 kJ or 1.05 MJ.
b 1.2×10^{-11} kg
c About one part in 4×10^{11}.

45a 0.001 kg (1 gram).
b 0.5 J (taking $k = 100$ N m^{-1}, and extension = 0.1 m).
c 6×10^{-18} kg
d 6×10^{-13} per cent.

46a $2m_p + 2m_n + 2m_e = 4.0330$ u.

The mass of an atom of $^{4}_{2}He$ is 0.0304 u *less* than the mass of its parts.

b*i* -4.54×10^{-12} J *ii* -28.4 MeV
c -7.1 MeV
(Note that because the information refers to a neutral helium atom – nucleus plus electrons – the binding energy calculated includes the binding energy of the electrons as well as that of the nucleons. However, since electrons are so much more weakly bound than nucleons, the value obtained is still a good estimate for the average binding energy per nucleon.)

UNIT G
Energy sources

1a*i* The potential energy of the mass is the work which would be done by the gravitational field if the mass were allowed to fall to some reference point (often the surface of the Earth). The mass needs the presence of the Earth in order to have potential energy, which is therefore a property of the Earth–mass system.
ii The gravitational field acting on the mass transfers energy to the generator by working. The mass loses energy and the generator gains energy (to an equal extent in an ideal system).
iii This means that a quantity of energy (15 J in this example) would be transformed if the generator were brought to rest.
b Boxes should be inserted at every stage of conversion. They all represent internal energy of the surroundings.
c The boxes should have solid boundaries since they represent continuing 'losses' to the system.
d It depends what you mean by the system. If this is taken to be the 'Universe' then the energy is constant, though clearly it is not so if the mass–generator–lamp arrangement is regarded as the system. In this latter case energy is lost from the system. Because the 'losses' are only losses in this limited sense, quotation marks are used.

2 The Sankey diagram is shown in figure Q82.

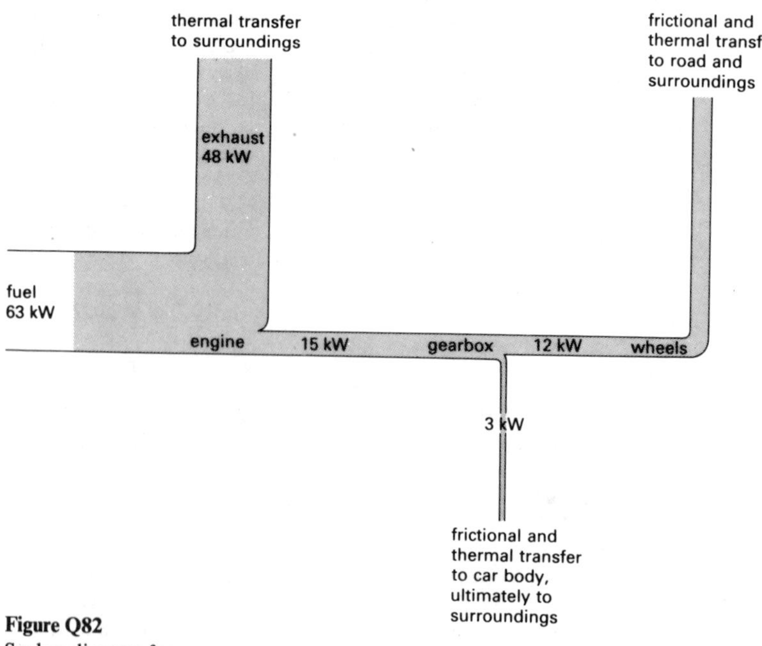

Figure Q82
Sankey diagram for a car.

3a 10–100 W depending on the activity and the length of time for which it is sustained.
b*i* 1–10 MJ approximately depending on the answer to **a**.
ii The cost of approximately one-third to three units. The minimal diet would be about $\frac{2}{3}$ kg of rice whose cost would be considerably more than that of the electricity. A more balanced diet would be even more expensive.
iii About 25 per cent (range 10–100 %).
iv About 115 W. Power from 500 people $\approx$ 60 kW. This seems a lot – but it is difficult to say whether it is enough to heat a theatre without other information. Heating would be required for raising the temperature to comfortable levels but thereafter the 'body heat' of the audience might be enough. The latter part of Section G3 deals with heating buildings.
c*i* The central parts of figure Q83 cannot easily be quantified. For example, the proportion of the energy 'contained' in food which the body can actually convert depends on the composition of the food.
ii Reducing food intake will apparently reduce all the subsequent energy flows, including that for building body parts, *i.e.*, slimming occurs. In fact this may not happen at all, or only in the short term, since the metabolic rate may have reduced. Similarly, increasing the quantity of work which is done without changing the food input should lead to slimming since less energy is available to contribute to building body material.

Again, reducing the need for energy to heat the body means that more may be available to contribute to building body material. Thus battery farms are maintained at high enough temperatures for the animals not to need to use as much energy as normal to maintain body temperatures, in order to maximize their rate of growth.

4a 3.6 MJ
b About 35 MJ.
c About 29 × 10^9 J.
d 3.6 × 10^{12} J
e 10^3 J
f About 250 kJ.
g 2.9 × 10^{20} J
h 1.6 × 10^{-11} J
i 1060 J

5a This use of the term ignores the fact that the appliances will have different power ratings and many will also be used infrequently.

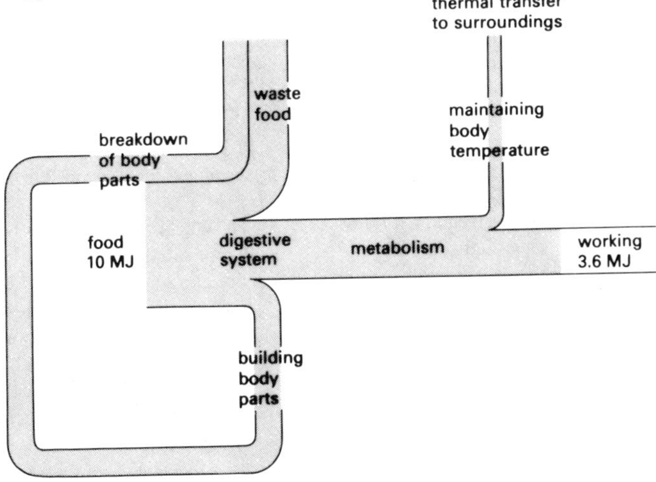

Figure Q83
Sankey diagram for a mammal.

b This version deals with part of the problem but does not take account of the time for which each appliance may be used.
c According to the Law of Conservation of Energy, energy cannot be 'consumed', if this is taken to mean 'used up'. But energy can be transferred from one part of the system (for example, the generating station) to another part (such as appliances) either by working or by thermal transfer.
d The value will usually be much bigger. This relates to the data in table G5, page 419 on the proportion of fuel used for space heating.

6a The fuel input (1983 figures in 10^9 therms) is 18.8 from coal, 1.9 from oil, 0.1 from gas, and 4.7 from nuclear and hydro. This last figure is conventionally taken to be the quantity of coal or oil which would produce the same quantity of electrical energy as is actually produced by nuclear or hydroelectric power stations. Thus the total input is 25.5.

The electrical output seems to be labelled as 7.5, but it is not clear whether the small unlabelled loss is included in this value or not. Adding the final energy use figures provides a cross-check, giving a total of 7.4.

Since the main loss from power stations is labelled as 17.4, the unlabelled loss seems to be about 0.6, and seems to be a distribution loss. Thus the actual output from the power stations seems to be about 8.1. The operating efficiency is therefore about 32 %.
b For the reason given in **a** it is not possible to say exactly, but the distribution losses are small – assumed to be 0.67×10^9 therms in **a**. Hence distribution efficiency is about 93 %.
c The overall efficiency is the product of the operating efficiency and the distribution efficiency – about 29 %.
d The total of the crude oil figures is 61 (in units of 10^9 therms). Some of this (29.6) is exported or bunkered, so 31.4 actually goes to refineries.

The output of refined fuel is (clockwise from the top) $(1.9+22.4+2.3+3.4+7.2) = 37.2$, of which 6.1 was imported already refined, and 2.3 represents losses. Hence the output from the refineries is 28.8 and the efficiency is $\frac{28.8}{31.4} \times 100 = 92\,\%$.

Note that the total input and output figures do not balance but the difference is slight (0.3).
e About 91 %.
f Roughly 3.5 times as much as the useful output.

7a Each kW h of energy delivered electrically to the final user can produce 1 kW h of heating. Table G1 in the *Students' guide* shows that gas boilers are about 75 % efficient. Hence $1\frac{1}{3}$ kW h of end-use energy are needed for each 1 kW h of heating. For equal heating costs electricity can be about 1.3 times as expensive as gas.
b 1 kW h of heating by gas requires $1\frac{1}{3}$ kW h of end-use energy and hence a little more than this of primary fuel (question **6e**). 1 kW h of heating by electricity requires about

$\frac{100}{32} = 3.1$ kW h of primary fuel

(question **6a**).

Hence, overall, using electricity for heating requires

$\frac{3.1}{1.3} = 2.4$ times as much primary fuel as

using gas for the same purpose.

8a World population has been growing exponentially since about 1950 with a doubling-time of about 37 years.
b Between 1925 and 1981 World fuel consumption increased by a factor of about 7, whereas the population grew by a factor of about $2\frac{1}{2}$. Clearly the fuel consumption *per capita* increased during this period.

9a The range of values is such that the graphs can only be plotted over about 12 hours, from 0–12 or from 12–24 hours. Breeding rate is taken to be the number of new bacteria at the end of each hour, rather than the ratio of new bacteria to old ones (see figure Q84).

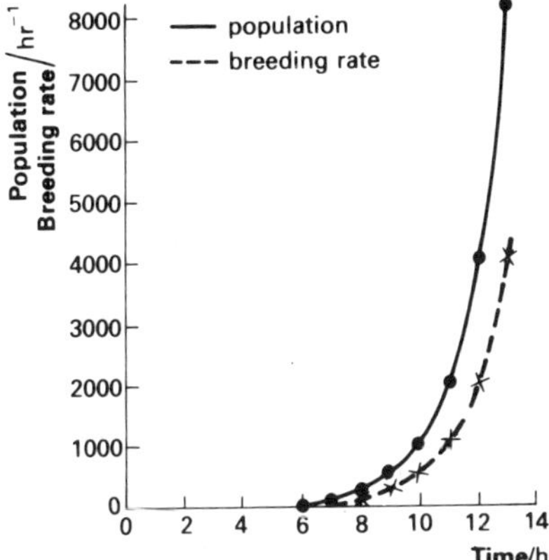

Figure Q84
Population and breeding rate of bacteria.

b One hour before it is full, *i.e.*, after 23 hours.
c Probably not until $\frac{1}{8}$ or $\frac{1}{4}$ of the nutrient is used up. This means at 21 or 22 hours, *i.e.*, with only two or three hours to go before all the resource is used.
d 1 jar; 2 jars.
e It would only give one more hour's life.

10 Doubling-time is 35 months, to the nearest month.

11 $t_D = 0.693/g$

12 Coal 226 years; oil, including shale, tarsands, etc., 57 years or 29 years for oil only; natural gas 53 years.

It is not clear in table G10 in the *Students' guide* whether the reserves of oil and gas are proved reserves or recoverable reserves. If they represent proved reserves, the lifetimes would be less for the smaller proportion which can be recovered. On the other hand, oil and gas discoveries continue to be made which will increase the lifetimes. The main point is, however, that the lifetimes are very short, even assuming zero growth.

13a Oil consumption increased by 8.7 per cent over ten years from 1972 to 1982. Assuming a growth rate of 0.87 % *per annum* gives a doubling-time of $0.693/0.0087 \approx 80$ years.

Similarly, the doubling-time for coal consumption is 27 years and for gas also 27 years.
b Taking the growth rate as an average of ten years' growth smooths out the variations which occur over a shorter period. If, however, the growth is very rapid (as in the case of nuclear power from 1972 to 1982) the average value over ten years is not the same as the year-on-year value which applies throughout the period.

Thus an annual growth rate of 0.87 % leads to growth in ten years by a factor of $e^{(0.0087 \times 10)} = 1.091$, rather than 1.087. Thus, using the 'average' growth rate overestimates the growth. The difference here is quite small – the annual growth rate required to give the correct 10-year growth is 0.83 %. However, for nuclear power the 10-year growth was 464 %. The annual growth rate required to produce this is about 15.4 % rather than the average of 46.4 %.
c Using 1982 figures the lifetimes of the fuels are: oil – 26 years; coal – 76 years; natural gas – 34 years.
d The additional resources increase the lifetimes as follows: oil – 46 years (assuming the oil shale and tarsands reserves are all recoverable); coal – 152 years; natural gas – no change since no additional resources are listed.
e If the price of fuel rises, the figures for recoverable reserves also rise since reserves which could not previously be recovered economically now become viable.

14a The broken line shown in figure Q85.
b About 23 hours. If the impending lack of nutrient had had no effect on breeding rate at all then point B (half maximum population) would be reached one doubling-period before 24 hours, *i.e.*, at 23 hours. If the impending shortage had begun to have an effect then half maximum population would be reached more slowly, *i.e.*, in more than 23 hours.

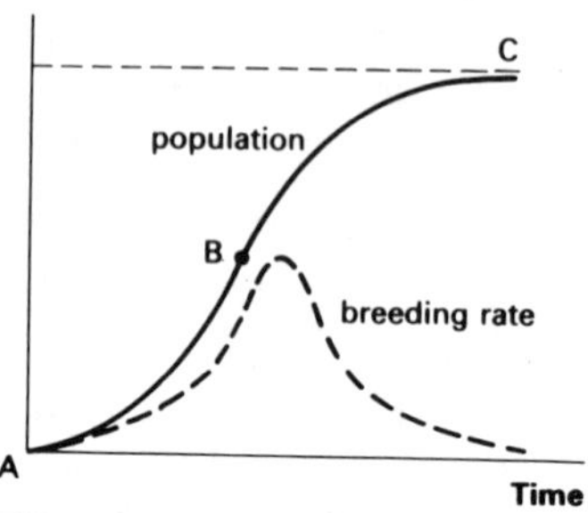

Figure Q85
Population and breeding rate of bacteria with limited resources.

c The time corresponding to point C might be said to be infinite since the population value will approach the maximum more and more slowly.

For practical purposes the time corresponding to C can be given a range of values depending on the form of the growth curve. If the curve is symmetrical, the time corresponding to C is twice that corresponding to B, *i.e.*, about 48 hours or so. Asymmetry in the growth curve could reduce or increase this value.

d The main effect in this analogy is that the organism's breeding rate would be reduced (or its death rate increased). The effect of decreasing fuel supplies would be to reduce the standard of living in various ways – some of them quite drastic – depending on the size of the reduction and what steps were taken to deal with the problem.

15a, b See figure Q86.

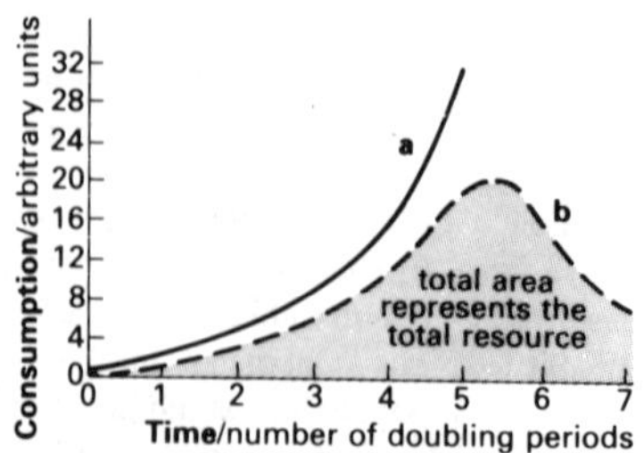

Figure Q86
Consumption of fuel.
a Constant growth rate.
b Effect of increasing difficulty of extraction.

c The area under the graph up to a particular time represents the resource used up to that time. The total area under the graph represents the total resource, which must be the same for all graphs, regardless of initial growth rate.

d If the maximum consumption is high the lifetime will be correspondingly low.

e In order to prolong the lifetime of fossil fuel resources we should start reducing consumption as soon as possible.

16 This question is intended to be speculative. Exponential growth could have occurred while London was small and rural. Food for horses could be obtained *in situ* or relatively easily. Greater urbanization would increase the difficulty of obtaining food for horses and would limit the number of horses which the city and its hinterland could support.

Alternatively, the limitation in numbers of horses could have been due to the development of new technology, for example, vehicles powered by steam.

Implications for the use of fuels could be either the need to match demand to resources or the possibility of developing a new technology such as fusion.

17a Area of Earth's surface $= 5.1 \times 10^8\,\text{km}^2$.

b Land area $= 1.5 \times 10^8\,\text{km}^2$.

c Area per person $= 0.03\,\text{km}^2$.

d Cultivated land per person $= 2.1 \times 10^{-3}\,\text{km}^2$. Grassland per person $= 8 \times 10^{-3}\,\text{km}^2$.

Apart from forests the rest is rocks, desert, tundra, snowcap, *etc.*

Some nations produce too much food and store or even destroy it for economic reasons.

e Doubling-time for population = 37 years

Hence 2018 is 1 doubling period after 1981. In 2018 the cultivated land area available per person would be $1.05 \times 10^{-3}\,\text{km}^2$, and in 2055 (after two doubling periods) $0.53 \times 10^{-3}\,\text{km}^2$ would be available.

18a Columns **1** and **3**.

The reasons for high productivity are complicated. They might include the ability to make use of powered machinery and fertilizers, but factors like the development of high-yielding strains of crops could also be important.

b Using columns **1** and **2**: input per hectare year for rice growing is 48 kW h; for grain growing, 8200 kW h; for battery farming, 4900 kW h.

All of these are consistent with **a**.

c The statement is reasonable. The implications of depleting fuel supplies are that less will be available for fertilizer production and for machines. Productivity will decrease and eventually people may starve.

19a The U.K. and U.S.S.R. have a slight overall surplus, North America has a slight deficit, and Japan has a large deficit.

b For water power and nuclear energy the figures for production and consumption are identical. Only one set of figures is published, presumably because storage of electrical energy on any reasonable scale is not possible. The figures for coal, natural gas, and oil can be processed as in part **a**. The calculations are simple but the real point of the question is dealt with in parts **c** and **d**.

c Short-term shortfalls can be dealt with by importing from nations which have surplus resources. The importing nations must have some means of earning the currency to buy the fuel and the fuel-exporting nations may have some degree of control over the economies of the importers (for example, the Organization of Petroleum Exporting Countries (OPEC) in the late 1970s).

d Nations like Japan, with continuing shortfalls, are likely to wish to reduce their dependence on imported fuels. They may have to develop alternatives much more purposefully than do nations with fossil fuel resources. To some extent, what happens to Japan today happens to the World tomorrow.

20a The energy equivalent of food for the average West European is 10 MJ per day (question **3**). This is equivalent to about 0.09 toe *per annum* The value for the less developed countries will be rather less than this.

b On the basis of burning 3 kg of wood per head per day the equivalent in terms of oil is about 0.2 toe per head per annum. The value will depend on the estimates made but precision is not required here.

The total of **a** and **b** is about 0.3 toe per head per year.

c If the diet is assumed to be rice, $\frac{2}{3}$ kg of rice would provide the daily requirement of 10 MJ (assuming, as usual, complete conversion). The annual requirement of 250 kg would have cost, say $100 in 1972 when the data for figure G12 were produced.

The effect of incorporating these figures into G12 would be to raise the points and move them to the right, but both changes are quite small.

21a*i* The area needed is 0.13 km^2.

The total area needed for the U.K. population is 1.8×10^6 km^2. This is greater than the area of the U.K. (approximately 0.25×10^6 km^2).

ii 0.007 miner for the needs of a family or about 100 000 miners for the whole population. About 200 000 miners were employed by the N.C.B. in 1983. Much of their production goes, of course, to power stations (see part **b** of this question) and either directly or indirectly to providing for the needs of industry – not the focus for this question.

iii Area of panel needed = 70 m^2 for a family, or about 1000 km^2 for the whole population. The feasibility depends on the mismatch of times of maximum supply and demand, and the consequent storage problem. The material resources needed for constructing solar panels on this scale are also a problem.

b Roughly three times as much resource would be needed.

22 Conservation measures could reduce space heating demands. (The asterisked categories in table G5 in the *Students'* *guide* are essentially space heating.) If these were to be halved, demands for transport reduced by 20 % and for fuel used in industrial production by 10 %, the total saving could be about 30 %.

A saving of 20 % in North America and Europe would be 10 % of the World's total. While this is not a vast saving it is worth while and might be influential in reducing expectations.

23a 0.04 %
b 4 %
c 5.5 %
d 7.5 %
e 0.25 %

24a A base load capacity of about 39 GW is required. This is the level for which area A = area B (figure Q87). In practice, taking an average of maximum and minimum power levels is reasonably satisfactory.

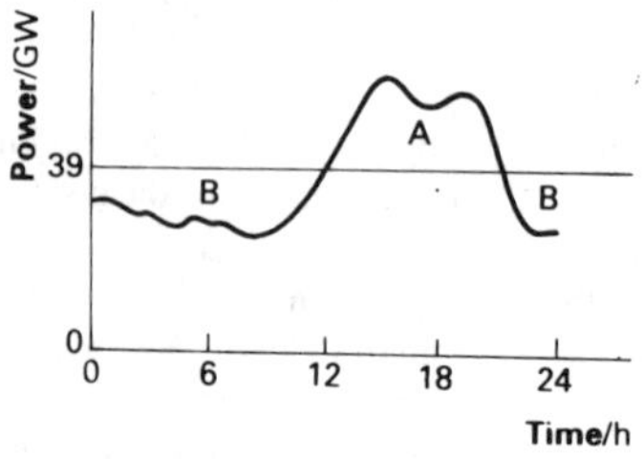

Figure Q87
Variation in power demand during 24 hours.

About 9 GW of hydroelectric power will also be needed.
b 9000 m^3 of water would be needed per second to provide the peak power output. The total energy required is about 98 GW h. This would be provided by the contents of a lake of area about 12 km^2 and 30 m depth falling through 100 m. This arrangement would be quite feasible.
c The simplest annual demand curve might be as shown in figure Q88.

The overall base load can be assumed to be 27 GW. The daily energy requirement is 12 × 24 GW h during the winter months, *i.e.*, a total of 12 × 24 × 180 GW h. This would require a total area of pumped storage reservoir of about 6000 km^2, *i.e.*, about 75 km square or 2000 km of flooded valley of average width 3 km. This does not seem to be feasible. A more realistic demand curve would make no essential difference to this estimation exercise.

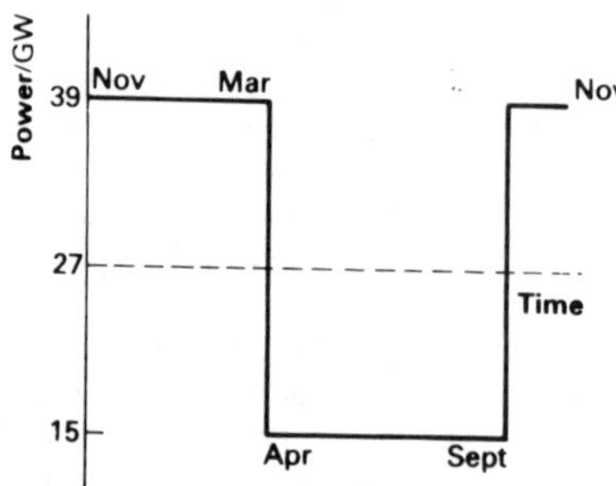

Figure Q88
Simplified annual demand curve.

25a Volume of the ring main = 56 000 m^3.
b The pressure falls.
c About 1.4 × 10^6 m^3.
d About 3.9 × 10^6 m^3.
e The increase in pressure which would result from storing all this gas in the main would be about 70 × 10^5 Pa (*i.e.* from 25 × 10^5 to 95 × 10^5 Pa) which takes the main pressure above its limit. Over half ($\frac{40}{70}$) would have to be stored elsewhere.

26 Some means must be found of estimating the fuel consumption of London. If fuel consumption is assumed to be proportional to population, then London consumes $\frac{1}{6}$ of the fuel used in the U.K. (see, for example, table G11 in the *Students' guide*), hence the rate of use is about 4 × 10^{10} W. The area of London is about 1800 km^2 hence fuel is consumed at a rate of about 25 W m^{-2}.

Insolation is about 175 W m^{-2}. Hence, overall fuel consumption per unit area is in excess of 10 % of the insolation, and this is likely to raise the temperature. In fact, temperatures in Greater London are generally greater than those of its surroundings.

27 For population, $t_D = 37$ years and $g = 1.9\%$.

For fuels, $t_D = 17$ years and $g = 4\%$ over much of this time.

Hence, growth in demand for fuels is not merely due to increasing population.

28 Raising global *per capita* fuel consumption to Western European levels would double World fuel consumption. If this occurred over 30 years the annual rate of growth for this reason alone would be 2.3 %.

29 From question **27** increasing expectations seem to account for a rate of growth of 2.1 %. Table G11 in the *Students' guide* shows much larger increases for less developed countries than for developed ones.

30 For the purpose of this question the Earth can be regarded as a flat disc which is always illuminated by the Sun. This takes care of the day–night and inclination variables and gives a value for the energy incident at the bottom of the atmosphere of 2.8×10^{24} J per year. The World's annual fuel consumption is about 2.8×10^{20} J. Thus the average energy delivered by the Sun in an hour is about the same as the Earth's fuel consumption in a year.

31a*i* No. The number of electrons is the same on each side.
ii Change in mass is 0.192 u or 3.18×10^{-28} kg.
iii 2.86×10^{-11} J
b*i* $\dfrac{Q_1Q_2}{4\pi\varepsilon_0 d}$
ii 4.61×10^{-11} J
iii The fission fragments are very close together at first, perhaps close enough for the nuclear force to have some effect. The fission fragments may be attracted initially as well as being subject to electrostatic repulsion. The kinetic energy of the fragments would then be less than that calculated on the basis of electrical potential energy.
c*i* 9×10^{13} J
ii About the same.
iii 143 kg
iv 143 tonnes
v About 30 times as much material needs to be used to provide the same amount of energy from coal as from uranium.

32a a/A
b $\dfrac{nAva}{A} = nav$
c NAd
d $nav \times NAd$
e $R = \dfrac{nav \times NAd}{A \times d} = nNav$
f $a = R/nNv$
g Measure the rate of interaction per second per unit volume, R; measure the neutron density (number of neutrons per unit volume), n; the neutron velocity, v; and the number of target nuclei per unit volume, N.

33 0.1 eV. This is a factor of about 10^7 less than the energy of a neutron released in fission of ^{238}U.

34a*i* and *ii* See table Q7.
iii See table Q8.
Chain reaction is most likely for energies <1 eV.

Isotope, interaction	**^{238}U fission**	**^{238}U scattering**	**^{235}U fission**
i $\sigma/10^{-28}\,m^2$	0.63	8	1.3
ii Probability of interaction for 2 MeV neutrons	63	800	1.3

Table Q7
Cross-sections and interaction probabilities for 2 MeV neutrons.

Energy	Cross-section/10^{-28} m^2		Probability/arbitrary units		Ratio $\frac{^{235}\text{U fission}}{^{238}\text{U capture}}$
	^{235}U fission	^{238}U capture	^{235}U fission	^{238}U capture	
10 keV	8	0.63	8	63	0.13
100 eV	40	16	40	160	0.25
1 eV	100	0.63	100	63	1.6
0.01 eV	950	4	950	400	2.4

Table Q8
Cross-sections and interaction probabilities for neutrons of various energies.

iv For 3 % enriched uranium the ratio of ^{238}U to ^{235}U is 33:1. For natural uranium the ratio is 143:1. Taking 100 instead of the more exact values makes no significant difference. It can increase the numbers in the last column in table Q8 by up to three times, or could decrease them at worst by 33 %.

b*i* Arrange the fuel in thin rods or pellets surrounded by a large quantity of moderator – enough to thermalize (slow down) the neutrons before they reach another fuel element, as in figure Q89.

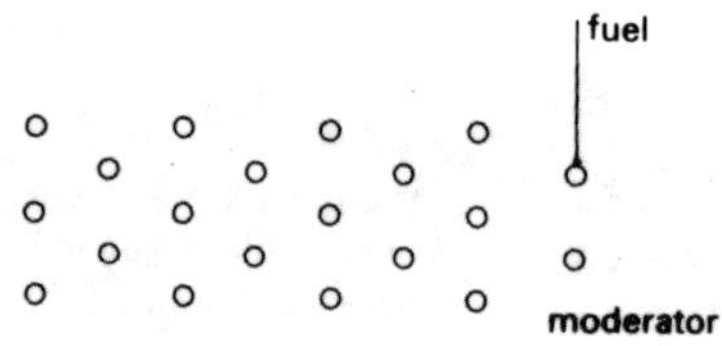

Figure Q89

ii There may be energy levels for neutrons in the ^{238}U nucleus. Each spike in the cross-section graph corresponds exactly with the difference in energy between two particular energy levels.

35a Kinetic energy of the fission fragments.

b The fission fragments cause a very large number of ionizing events and slow down rapidly. The ionized atoms eventually revert to the ground state. The energy released increases the internal energy of the material, heating it up.

c By ionizing atoms of air in the chamber. This mechanism cannot be used for neutrons since they are not charged and do not produce ionization.

d Collision with atoms (or nuclei).

36 Momentum is conserved:

$$mu = mv + MV \quad \textbf{[1]}$$

Kinetic energy is conserved in an elastic collision:

$$\tfrac{1}{2}mu^2 = \tfrac{1}{2}mv^2 + \tfrac{1}{2}MV^2 \quad \textbf{[2]}$$

Initial kinetic energy, $E_k = \frac{1}{2}mu^2$

Change in kinetic energy of A,
$\Delta E_k = \frac{1}{2}mu^2 - \frac{1}{2}mv^2 = \frac{1}{2}MV^2$ **[3]**

The rest is algebra; the key step is to eliminate v.

From **[3]**:

$$m(u^2 - v^2) = MV^2 \Rightarrow m(u+v)(u-v) = MV^2 \quad \textbf{[4]}$$

From **[1]**:

$$m(u-v) = MV$$

Substitute MV for $m(u-v)$ in equation **[4]**:

$$MV(u+v) = MV^2$$

$$\Rightarrow u + v = V$$

$$\Rightarrow v = V - u$$

Now substitute $(V-u)$ for v in equation **[1]**:

$$mu = m(V-u) + MV$$

$$\Rightarrow 2mu = (m+M)V$$

$$\Rightarrow V = \frac{2mu}{m+M}$$

So

$$\Delta E_k = \tfrac{1}{2}M\left(\frac{2mu}{m+M}\right)^2$$

$$= \frac{4mM}{(m+M)^2} \times \tfrac{1}{2}mu^2$$

$$= \frac{4mM}{(m+M)^2}E_k$$

37a $\Delta E_k/E_k = \dfrac{4mM}{(M+m)^2}$.

b For maximum value of $\Delta E_k/E_k$, M must be 1 u.

c See table Q9.

M/u	1	2	10	12	16	112	238
$\dfrac{\Delta E_k}{E_k}$	100	89	33	28	22	3.5	1.7
Relative value	$4\frac{1}{2}$	4	$1\frac{1}{2}$	$1\frac{1}{4}$	1	$\frac{1}{6}$	$\frac{1}{13}$
Element	H	D	B	C	O	Cd	U

Table Q9
Percentage energy loss for neutrons in head-on collision with certain nuclei.

d Hydrogen.

38 54. This is less than the actual number since the collisions will generally not be head-on in practice.

39a Low absorption (capture) cross-section. High scattering cross-section.

b See table Q10.

c On the basis of these figures the choice would be oxygen or deuterium. When account is also taken of relative energy loss the choice is confirmed and the choice of hydrogen on the basis of energy loss only is overidden.

d Boron, cadmium.

40a 3.1×10^{19}

b 1.2×10^{-5} kg

c Mass of ^{235}U used per year = 383 kg. Total mass is three times this, 1150 kg.

d 383 kg represents 2.2 % of the uranium present. The mass of the core is therefore 52×10^3 kg.

e The main reason is the low efficiency at which internal energy is converted to electrical energy.

41a Thermal flow of energy.

b Temperature difference.

c*i* C s^{-1}, A.

ii Temperature difference $\propto$ rate of thermal flow of energy; joule per second, watt; power.

d Measure the power transmitted through the conductor for different temperature differences. Problems are loss of energy from sides of conductor, and difficulties in measuring the power, etc.

e K W^{-1}

f Ω m

g W m^{-1} K^{-1}

42a 0.63 K W^{-1}

b 1.3 kW

43a 2.7×10^{-3} K W^{-1}

b 16×10^{-3} K W^{-1}

c*i* 1.3 kW

ii 7.5 kW

iii 8.8 kW

d*i* $I_T = I_1 + I_2$

ii The two calculations are analogous.

44a 19×10^{-3} K W^{-1}

b 570 W

45a 10.6 MW

b 170 W

Element	H	D	B	C	O	Cd	U
$\dfrac{\sigma_s}{\sigma_c}$	61.5	6396	4.7×10^{-3}	1397	13 925	2×10^{-3}	1.2
$\dfrac{\Delta E_k}{E_k} \times \dfrac{\sigma_s}{\sigma_c}$	277	25 600	7.2×10^{-3}	1850	13 925	0.3×10^{-3}	0.1

Table Q10
Cross-sections and relative energy loss for neutrons in collision with various nuclei.

46a $X/2$, $X/3$, X/A.
b $\mathscr{R} = X/A$
c $m^2\,K\,W^{-1}$
d $X = l/k$

47a $0.17\,m^2\,K\,W^{-1}$
b $0.13\,m^2\,K\,W^{-1}$
c The one which has the bigger resistance coefficient, *i.e.*, brick.

48a $2 \times 10^{-3}\,K\,W^{-1}$
b 10 kW
c Much too high.

49a Convection currents in the air and how easily these can flow; forced convection (winds) reduces the resistance of external surfaces.

50a Series
b $0.194\,m^2\,K\,W^{-1}$. The power loss is reduced to about one-fiftieth of its previous value.
c Outer surface of glass is at 6.2 °C. Inner surface of glass is at 6.6 °C.

51a, b The graph is shown in figure Q90.
c $0.26\,m^2\,K\,W^{-1}$. If the surface layers are assumed to be of zero thickness then the resistance coefficient of the cavity would be independent of thickness. In practice, the surface layers are of finite thickness so the resistance coefficient would be likely to increase with thickness until the cavity is two surface layers thick. Thereafter the resistance would be constant.
e A thin cavity behaves more like a slab of conductor than a thicker one. A faced horizontal cavity with downward flow of energy continues to behave like a slab of conductor as its thickness increases above 20 mm, whereas the behaviour of the other types of cavity is increasingly unlike that of a slab of conductor.
f Facing the cavity increases the resistance coefficient. If the effect were simply due to conduction, the metal foil would marginally decrease the resistance coefficient. The data suggest that radiation is also a mechanism in thermal transfer of energy. The effect is important – facing the surface increases the resistance coefficient by a factor of about three.
g Convection occurs when there is a possibility of upward movement of heated fluid, for example with hor↑ cavity. Compared with the hor↓ cavity the thermal resistance is approximately halved – hence convection is an important effect.

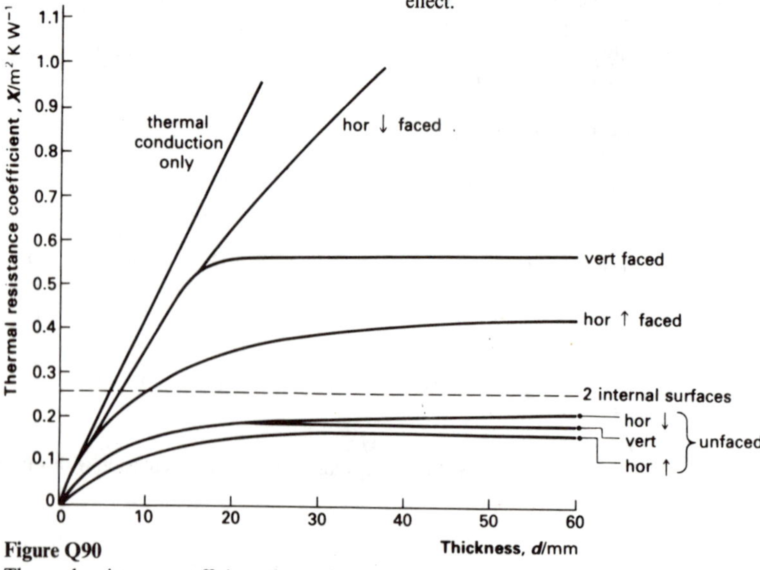

Figure Q90
Thermal resistance coefficients for various cavities.

h No, since none of the measured graphs coincides with the graph for two surfaces, except for a thickness up to about 8 mm. Perhaps this indicates that each surface layer has an effective thickness of about 4 mm for the conditions under which the measurements were taken.
i Measured values are always less than values calculated assuming that conduction is the only mechanism, often much less.
j Resistances need to be connected in parallel.

52b Power loss through the double-glazed window is about 120 W, *i.e.*, reduced to 0.6 of the single-glazed value.
c, d See figure Q91.
e No. The dependence of resistance on thickness is likely to be non-linear and hence the calculation is invalid. However, the non-linearity of resistance and, consequently, of temperature, has no effect on calculations which are solely concerned with the components in their entirety. We are using what amounts to a 'systems' approach to the problem.

53 A drawn curtain creates a cavity near the window.

54 The filling reduces the possibility of convection and hence increases the thermal resistance of the cavity.

55b A single-glazed window has $X = 0.194\,\mathrm{m^2\,K\,W^{-1}}$, while a double-glazed window has $X = 0.338\,\mathrm{m^2\,K\,W^{-1}}$.
c Wood is a poorer conductor than glass since a wood-framed window has a higher overall value of thermal resistance coefficient than that calculated for glass alone. The converse holds for a metal frame.
d Choose a single-glazed window with a relatively large area of metal frame, provided that when double-glazed the frame can be insulated to give a higher resistance coefficient than the double glazing.

56a See figure Q92.
b $0.114\,\mathrm{K\,W^{-1}}$
c 140 W
d 15.6 °C
e $0.074\,\mathrm{K\,W^{-1}}$
f 217 W
g 11.7 °C
h On the glass; it is colder.

57 Area of window $= 1.25\,\mathrm{m^2}$.

58a Reasonable values only need to be given, for example, temperature at B could be about 20 °C, and A could be between 5 and 10 °C.

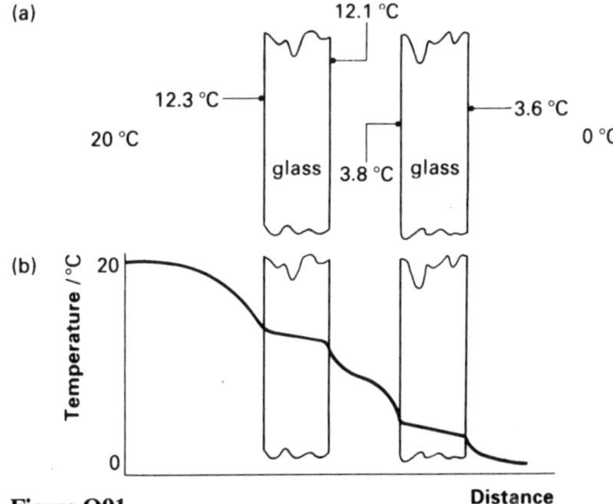

Figure Q91
(a) Temperature at surfaces in a double-glazed window.
(b) Temperature variation with distance near a double-glazed window.

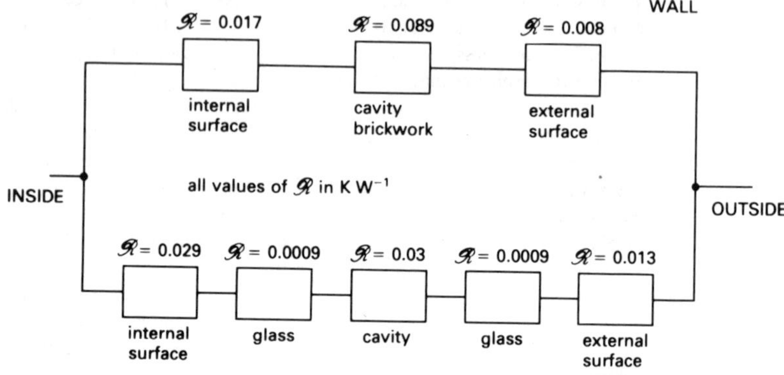

Figure Q92
Thermal resistances for wall with window.

b The rate of loss of energy depends on both the external temperature and the internal temperature. If the external temperature is 0 °C then the rate of loss of energy for an internal temperature of 20 °C is twice that for an internal temperature of 10° C or four times that for a temperature of 5 °C.
c Mass of building material, specific heating capacity of material, insulation value of the material.
d The heating system has to be operating for a long time before the building reaches its operating temperature. This is likely to require a large amount of fuel.
e Use a thermostat. The temperature would oscillate to some extent.
f, g See figure Q93.
h Light building materials of low specific heating capacity; high insulation values; minimum ventilation.

59 Power needed to heat the air = 1.3 kW. This is substantially greater than the loss through either single- or double-glazed windows in questions **50** and **52**.

60a For a group of 15 the power needed to heat the air is about 5.5 kW, assuming that the air has to be heated through 20 K.
b The power provided by the occupants is 1.5 kW, about 30 % of the power needed.

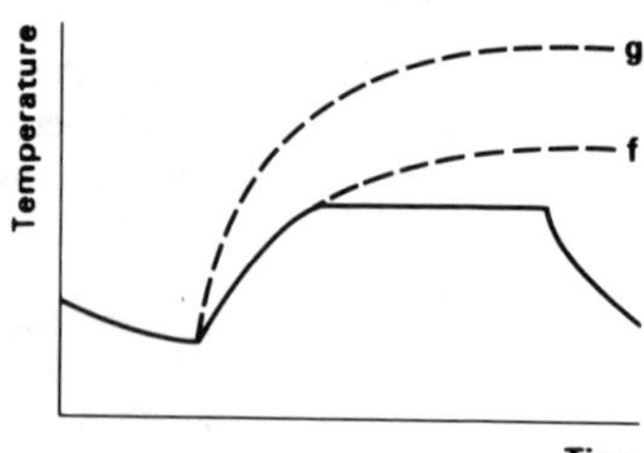

Figure Q93
The variation of temperature with time for a building **g** and for a lightly constructed hut **f**.

REFERENCE MATERIAL

TEXTBOOKS AND FURTHER READING

The students have a list of textbooks and other references on pages 491–2 of *Students' guide 1*. Here we give a list of books and references for teachers and the library.

Books useful throughout the course

AKRILL, T. B., BENNET, G. A. G., and MILLAR, C. J. *Physics*. Arnold, 1979.

BOLTON, W. *Patterns in physics*. McGraw-Hill, 1974.

DUNCAN, T. *Physics: a textbook for advanced level students*. Murray, 1982.

OR {
DUNCAN, T. *Advanced physics: materials and mechanics*. 2nd edn. Murray, 1981.
DUNCAN, T. *Advanced physics: fields, waves, and atoms*. 2nd edn. Murray, 1981.
}

FEYNMAN, R. P., LEIGHTON, R. B., and SANDS, M. The Feynman lectures on physics *Volume 1: Mainly mechanics, radiation, and heat*. Addison-Wesley, 1963.

NUFFIELD REVISED PHYSICS *Years 1 and 2, Year 3, Year 4*, and *Year 5*. Longman, 1978–80.

ROGERS, E. M. *Physics for the inquiring mind*. Oxford University Press, 1960.

WENHAM, E. J., DORLING, G. W., SNELL, J. A. N., and TAYLOR, B. *Physics: concepts and models*. 2nd edn. Addison-Wesley, 1984.

Unit A Materials and mechanics

ADKINS, C. J. *Thermal physics*. Hodder and Stoughton, 1976.

AKRILL, T. B. and MILLAR, C. J. *Mechanics, vibrations and waves*. Murray, 1974.

BLUNDELL, A., HAWKINS, R., and LUDDINGTON, D. Schools Council Modular Courses in Technology, *Structures*. Oliver and Boyd, 1981.

BOLTON, W. Study Topics in Physics, *Volume 1: Motion and force. Volume 2: Materials*. Butterworths, 1980.

BRONOWSKI, J. *The ascent of Man*. Futura, 1981. (First published by B.B.C., 1973.)

COLLIEU, A. McB. and POWNEY, D. J. *The mechanical and thermal properties of materials*. Arnold, 1973.

CRAC Hobsons Science Support Series, *Gases and gas laws*. Hobsons Press, 1983.

CRAC Hobsons Science Support Series, *Stress, strain and strength*. Hobsons Press, 1982.

FARRAR, R. A. Methuen Studies in Science, *The mechanical properties of materials*. Methuen, 1971. (Out of print.)

GORDON, J. E. *The new science of strong materials*. Pitman, 1979. (First published by Penguin, 1968.)

GORDON, J. E. *Structures*. Pitman, 1979. (First published by Penguin, 1978.)

HOCKEY, S. W. and MILLS, J. R. *Physics by experiment*. Wheaton, 1973.

LOFTAS, A. A. and GWYNNE, P. *Advances in materials science*. London University Press, 1967.

McSHEA, J. Schools Council Modular Courses in Technology, *Materials technology*. Oliver and Boyd, 1981.

MARTIN, J. W. and HULL, R. A. *Elementary science of metals*. Wykeham, 1975.

MARTIN, J. W. 'Strength of materials'. *Contemporary physics reprint*. Taylor and Francis, 1968.

SCIENTIFIC AMERICAN *Materials*. W. H. Freeman, 1967.

MOFFAT, W. G., PEARSALL, G. W., and WULFF, J. *The Structure and Properties of Materials*. Wiley, 1964.

NUFFIELD ADVANCED PHYSICS *Physics and the engineer*. Longman, 1973. (Out of print.)

NUFFIELD REVISED ADVANCED CHEMISTRY *Students book I* and *Teachers' guide I*. Longman, 1984.

NUFFIELD REVISED ADVANCED CHEMISTRY Special Studies *Metals as materials*. Longman, 1985.

NUFFIELD REVISED CHEMISTRY *Teachers' guide II*. Longman, 1978.

NUFFIELD REVISED CHEMISTRY *Handbook for pupils*. Longman, 1978.

OGBORN, J. *Molecules and motion.* Longman, 1973. (Out of print.)
OVERMAN, M. *Roads, bridges, and tunnels: modern approaches to road engineering.* Aldus, 1968.
PSSC *College Physics.* 5th edn. Heath, 1981.
ROSENBERG, H. M. *The solid state: an introduction to the physics of crystals.* Oxford University Press, 1979.
SCHOOLS COUNCIL Engineering Science Project, *The use of materials.* Macmillan, 1975. (Out of print.)
TABOR, D. *Gases, liquids, and solids.* Cambridge University Press, 1980.
WARREN, J. W. *Understanding force.* Murray, 1979.
WILLIAMS, D. J. *Force, matter, and energy.* Hodder and Stoughton, 1974.

The Institution of Metallurgists publishes several leaflets on metals and materials; particularly relevant to this Unit are:
Which materials?
Materials in Action Series: *Materials and the human body, Concorde*, and *Superlastic alloys.*
These can be obtained free from:
The Education Officer
The Institution of Metallurgists
PO Box 471
1 Carlton House Terrace
London SW1Y 5BE
Tel: 01-839 1963.

Books on theories and models

BONDI, H. Science Study Series No 12, *The Universe at large.* Heinemann, 1962. (Out of print.)
BRONOWSKI, J. *The common sense of science.* Heinemann, 1979.
POPPER, K. R. *Conjectures and refutations: growth of scientific knowledge.* 3rd edn. Routledge, 1969.
WATSON, J. D. *The double helix.* Penguin, 1970. (First published by Weidenfeld and Nicholson, 1968.)

Unit B Current, circuits, and charge

CRAC Hobsons Science Support Series, *Instrumentation systems* and *Stress, strain, and strength.* Hobsons Press, 1982.

Unit C Digital electronic circuits

A large and rapidly growing number of books is available covering the digital electronics in this Unit. We have found these listed below useful. Teachers may also find the books suggested for students useful (*Students' guide I* page 491). Teachers and students may find other books and magazines equally useful and should be encouraged to use them.

Suggestions for teaching

BEVIS, G. and TROTTER, M. (Eds) *Microelectronics: practical approaches for schools and colleges.* BP Educational Service.
GOUGH, C. E. *et al. Notes for guidance for the electronics option in Physics (Advanced).* Joint Matriculation Board. 1981.

Textbooks on electronics

GIBSON, J. R. *Electronic logic circuits.* 2nd edn. Arnold, 1983.
HOROWITZ, P. and HILL, W. *The art of electronics.* Cambridge University Press, 1980.
JONES, M. H. *A practical introduction to electronic circuits.* Cambridge University Press, 1977.
TOCCI, R. J. *Digital systems: principles and applications.* Prentice-Hall, 1980.

Computers

THOMPSON, D. L. *Inside the micro.* Unilab, 1982.

Designing circuits

LANCASTER, D. E. *CMOS cookbook.* SAMS, 1978.
LANCASTER, D. E. *TTL cookbook.* SAMS, 1978.
McWHORTER, G. *Understanding digital electronics.* Texas Instruments Inc., 1978.
MULLARD *CMOS digital integrated circuits,* Book 4, part 4. 1983.
TEXAS INSTRUMENTS *Designing with TTL integrated circuits.* McGraw-Hill, 1971.

Nerve impulses

ADRIAN, R. H. Carolina Biology Readers No. 67, *The nerve impulse.* 2nd edn.

Carolina Biological Supply Company, distributed by Packard Publishing, 1980.
HODGKIN, A. L. *The conduction of the nervous impulse.* Liverpool University Press, 1964.

Unit D Oscillations and waves

As well as the general list of textbooks on page 501, these are particularly useful for Unit D.

Reference to school experiments

ERICSON, T. J. 'Velocity of sound in a steel bar'. *School Science Review* **57**, 1976, page 741.
FORD, D. and SOPER, P. 'Velocity of sound in metallic rods'. *School Science Review*, **60**, 1978, page 114.
MACE, W. K. 'Speed of sound in a steel rod'. *School Science Review*, **61**, 1980, page 527.
MACE, W. K. 'Resonance experiments for pupils'. *School Science Review*, **61**, 1980, page 537.
SHIPSTONE, D. M. 'Experiments with travelling waves on water'. *School Science Review*, **50**, 1969, page 816.
CONTEMPORARY PHYSICS 'Sources of physics teaching, part 1'. *Contemporary Physics.* Taylor and Francis, 1968.

Books

AKRILL, T. B. and MILLAR, C. J. *Mechanics, vibrations and waves.* Murray, 1974.
BISHOP, R. E. D. *Vibration.* 2nd edn. Cambridge University Press, 1979.
CARNELL, R. C. *Physics principles at work; a resource book for teachers.* BP Educational, 1983.
CHAUNDY, D. C. F. Longman Physics Topics, *Waves.* Longman, 1972.
CRAC Hobsons Science Support Series, *Vibrations.* Hobsons Press, 1983.
CRAC Hobsons Science Support Series, *Waves and sound.* Hobsons Press, 1982.
CRAWFORD, F. S. *Berkeley Physics Course* Volume 3, *Waves.* McGraw-Hill, 1968.
FEATHER, N. Introduction to Physics, *Volume 1 Mass, length and time.* Penguin, 1961. (First published by Edinburgh University Press, 1959.)
GOULD, R. T. *John Harrison and his timekeepers.* 4th edn. National Maritime Museum, 1978.
GRIFFIN, D. R. Science Study Series No 4, *Echoes of bats and men.* Heinemann, 1962. (Out of print.)
HOWSE, D. *Greenwich time and the discovery of the longitude.* Oxford University Press, 1980.
LAITHWAITE, E. R. *Propulsion without wheels.* Hodder and Stoughton, 1965.
OPEN UNIVERSITY *Science foundation course S101.* Unit 4, 'Earthquake waves and the Earth's interior'. Open University Press, 1979.
PSSC *Physics.* 5th edn. Heath, 1981.
SCSST Physics Plus, *Sonar.* Hobsons Press, 1985.
SHIVE, J. N. and WEBER, R. L. *Similarities in physics.* Hilger, 1982.
TABOR, D. *Gases, liquids and solids.* Cambridge University Press, 1980.
TRICKER, R. A. R. *Bores, breakers, waves, and wakes.* Mills and Boon, 1965. (Out of print.)
WALKER, J. R. *The flying circus of physics.* Wiley, 1978.
WARREN, J. W. *Understanding force.* Murray, 1979.

Scientific American references

Note: Although *Scientific American* Offprints can no longer be purchased in Europe, the titles listed here may still be available in many schools and colleges.

BASCOM, W. 'Ocean waves'. *Scientific American.* Volume **201**(2), Aug. 1959. (Offprint No. 828.)
BERNSTEIN, J. 'Tsunamis'. *Scientific American.* Volume **191**(2), Aug. 1954. (Offprint No 829.)
BULLEN, K. E. 'The interior of the Earth'. *Scientific American.* Volume **193**(3), Sept. 1955. (Offprint No 804.)
GRIFFIN, D. R. 'More about bat radar'. *Scientific American.* Volume **199**(1), July, 1958. (Offprint No 1121.)
LYONS, H. 'Atomic clocks'. *Scientific American.* Volume **196**(2), Feb. 1967. (Offprint No 225.)
OLIVER, J. 'Long earthquake waves'. *Scientific American.* Volume **200**(3), Mar. 1959. (Offprint No 227.)

Unit E Field and potential

ARONS, A. B. *Development of concepts of physics.* Addison-Wesley, 1965. (Out of print.)

ASIMOV, I. *The collapsing Universe: the story of black holes.* Corgi, 1978.

BRONOWSKI, J. *The ascent of Man.* Futura, 1981. (First published by B.B.C., 1973.)

CALDER, N. *The key to the Universe.* B.B.C., 1977. (Out of print.)

COHEN, I. B. 'Newton's discovery of gravity'. *Scientific American.* Volume **244**(3), March 1981, p. 122.

DAVIES, P. C. W. *The forces of nature.* Cambridge University Press, 1979.

HALLIDAY, D. and RESNICK, R. *Physics Part 1.* 3rd edn. Wiley, 1977.

HALLIDAY, D. and RESNICK, R. *Physics Part 2.* 3rd edn. Wiley, 1978.

KOESTLER, A. *The sleepwalkers.* Penguin, 1970. (First published by Hutchinson, 1968.)

NEWTON, I. (MOTTE, A. trans.) *Principia.* University of California Press, 1962.

PSSC *College physics.* 5th edn. Heath, 1981.

Unit F Radioactivity and the nuclear atom

Books for teachers

CONN, G. K. T. and TURNER, H. D. *The evolution of the nuclear atom.* Illiffe, 1965.

GRIFFITH, J. A. R. and TEBBUTT, M. J. *Notes for guidance for the nuclear physics option.* Joint Matriculation Board, 1981.

LITTLEFIELD, T. A. and THORLEY, N. *Atomic and nuclear physics.* 3rd edn. van Nostrand Reinhold Co., 1979.

SEMAT, H. and ALBRIGHT, J. R. *Introduction to atomic and nuclear physics.* 5th edn. Chapman and Hall, 1973.

WARREN, J. W. *Understanding force.* Murray, 1979.

WEHR, M. R., RICHARDS, J. A., and ADAIR, T. W. *Physics of the atom.* 4th edn. Addison-Wesley, 1984.

WEIDNER, R. T. and SELLS, R. L. *Elementary modern physics.* 3rd edn. Allyn and Bacon, 1980.

Data books

NUFFIELD REVISED ADVANCED SCIENCE *Book of data.* 2nd edn. Longman, 1984.

TENNANT, R. M. (Ed.) *Science data book.* Oliver and Boyd, 1971.

Other books

BENNET, G. A. G. *Electricity and modern physics.* 2nd edn. Arnold, 1974.

CARO, D. E., McDONELL, J. A., and SPICER, B. M. *Modern physics.* 3rd edn. Arnold, 1978.

DAVIES, P. C. W. *The forces of nature.* Cambridge University Press, 1979.

LEWIS, J. L. Longman Physics Topics, *Electrons and atoms.* Longman, 1972.

LEWIS, J. L. and WENHAM, E. J. Longman Physics Topics, *Radioactivity.* Longman, 1970.

NUFFIELD REVISED ADVANCED CHEMISTRY *Students' book I* and *Teachers' guide I.* Longman, 1984.

PROJECT PHYSICS Text Unit 6, *The nucleus.* Holt, Rinehart, and Winston, 1981.

TABOR, D. *Gases, liquids and solids.* Cambridge University Press, 1980.

WRIGHT, S. (Ed.) *Classical scientific papers – physics.* Mills and Boon, 1964. (Out of print.)

Journal reference

PEACOCK, T. A. H. 'Nuclear chemistry: some applications of radioisotope techniques in chemistry and biology'. *School Science Review* **45**, 157, 1964.

Rutherford Appleton Laboratory Monographs

CLARK, D. H. *The Universe and Man.* 1981.

CLOSE, F. E. *Atoms, particles, leptons and quarks.* 1980.

DAMERELL, C. J. S. *Experimental particle physics.* 1981.

These can be obtained free from:
The Library
Rutherford and Appleton Laboratories
Chilton
Didcot
Oxfordshire OX11 0QX

Unit G Energy sources

BASSETT, C. R. and PRITCHARD, M. D. W. *Heating*. Longman, 1969.
CHAPMAN, P. *Fuel's paradise*. Penguin, 1980.
CRAWLEY, G. M. *Energy*. Collier Macmillan, 1975.
ELKINGTON, J. *Sun traps*. Penguin, 1984.
FOLEY, G. *The energy question*. 2nd edn. Penguin, 1981.
GRIFFITH, J. A. R. and TEBBUTT, M. J. *Notes for guidance for the nuclear physics option*. Joint Matriculation Board, 1981.
*LEWIS, J. L. (Ed.) *Science in society*. Heinemann, 1981.
McMULLEN, J. T., MORGAN, R. and MURRAY, R. B. *Energy resources*. 2nd edn. Arnold, 1983.
RAMAGE, J. *Energy: a guide book*. Oxford University Press, 1983.
*SOLOMON, J. Science In A Social Context Series. Blackwell/ASE, 1983. (Several of the titles in this series will be relevant to this Unit.)
WARREN, J. W. *The teaching of physics* Butterworth, 1967.

*Though neither the *SISCON in schools* nor the *Science in Society* materials are specially intended for A-level physics students, several of the publications produced by these two projects are relevant to this Unit.

Journal references

OGBORN, J. M. 'Dialogue concerning two old sciences'. *Phys. Educ.* **11**, 272–6, 1976.
SCHMID, G. B. 'Energy and its carriers'. *Phys. Educ.* **17**, 212–218, 1982.
SUMMERS, M. K. 'Teaching heat: an analysis of misconceptions'. *School Science Review*. **64**, 670–6, 1983.
WARREN, J. W. 'The teaching of the concept of heat'. *Phys. Educ.* **7**, 41–44, 1972.
WARREN, J. W. 'The nature of energy'. *Eur. J. Sci. Educ.*, **4**(3), 295–297, 1982.
WARREN, J. W. 'Energy and its carriers: a critical analysis'. *Phys. Educ.* **18**, 209–212, 1983.

COMPUTER PROGRAMS AND VISUAL AIDS

Unit A Materials and mechanics

16 mm films

Films available from CFL Vision, Distribution Centre, Chalfont Grove, Gerrards Cross, Buckinghamshire, SL9 8TN. Tel: (02407) 4433. All except 'River to Cross' can be hired free of charge.

'Materials for the engineer', 15 minutes, 1972 (UK 2665).
'Let's make a model', 22 minutes, 1979 (UK 2874).
'The slender span', 12 minutes, 1966 (UK 2787).
'River to cross', 20 minutes, 1950 (UK 1253).
'Dynamic behaviour of tall buildings', 10 minutes, 1976 (UK 2794).
'We build for the world', 11 minutes, 1967 (UK 1878).

Television or video

'Electron diffraction', in the Granada TV series *Experiment: Physics*. Notes available from Granada Television Ltd, Manchester, M60 90EA, or Education Officer (Schools Information) of regional ITV companies.
'Understanding materials' (1982): a series of five video cassettes available on free loan from the Institution of Metallurgists, Northway House, High Road, Whetstone, London, N20 9LW, or from Department of Ceramics, Glasses and Polymers, Sheffield University, Elmfield, Northumberland Road, Sheffield, S10 2TZ.

Unit B Currents, circuits and charge

16 mm film

'Are there electrons? (The Millikan experiment)'. Colour, sound, 13 minutes. Viewtech, 161 Winchester Road, Brislington, Bristol BS4 3NJ. Tel: (0272) 773422.

Unit D Oscillations and waves

Film loops

Highly recommended:
'Tacoma Narrows Bridge collapse'. Ealing Scientific No. 80-2181/1 (super 8).
Also useful:
'Measurement of "G"'. Ealing Scientific No. 80-2124/1 (super 8).
'Soap film oscillations'. Ealing Scientific No. 80-2660/1 (super 8).
'Vibrations of a drum'. Ealing Scientific No. 80-3924/1 (super 8).
These are all available from: BFA Educational Films, 56 Copthorne Road, Leatherhead, Surrey. In due course they may also be available on videotape.
'Wind-induced oscillations'. Penguin. No. XX1671 (standard 8).
This is no longer available, but some schools may still have it.

Television or video

'Oscillations' – Physical Science Series. BBC TV.
This is no longer broadcast, but some schools may still have access to a recording.
'Vibrations of music' – Open University, Programme 9 for course S271 *Discovering physics.*
Open University programmes may be recorded by licence-holders. A school licence costs £40 and a college licence costs £80. Both may be obtained from: Guild Organization Ltd, Guild House, Oundle Road, Peterborough, PE2 9PZ.

Computer programs

'Longitudinal waves'. Heinemann Computers in Education.
'Transverse waves II'. Heinemann Computers in Education
'Waves' by R. G. Roscoe (obtainable through MUSE).
'SHM'. Longman Micro Software.
'Dynamic modelling system'. Longman Micro Software.

Unit E Field and potential

Computer programs

'Software for Nuffield Advanced Physics' and 'Dynamic modelling system'. Longman Micro Software.

Television or video

'The determination of the Newtonian constant of gravitation'. Programme 7 in the Granada TV series *Experiment: Physics.* Notes available from Granada Television Ltd, Manchester, M60 9EA, or Education Officer (Schools Information) of regional ITV companies.

Unit F Radioactivity and the nuclear atom

16 mm films

'The Rutherford model of the atom'. Colour, sound, 16 minutes. Viewtech, 161 Winchester Road, Brislington, Bristol BS4 3NJ. Tel: (0272) 773422.

Television or video

'The determination of a radioactive half-life', and 'Measurement of ionization potential' in the Granada TV series *Experiment: physics.* Notes available from Granada Televison Ltd, Manchester, M60 9EA, or Education Officer (Schools Information) of regional ITV companies.

Computer programs

'Dynamic modelling system'. Longman Micro Software.
BEARE, R. 'Alphafoil', Warwick Science Simulations Physics Pack 2, Longman Micro Software.
CAMPBELL, J. 'Simulation of alpha scattering' in *Physics software Pack 1*, Hutchinson Software, 1983.
HARRIS, J. 'Particle scattering' (Chelsea Science Simulations). Arnold, 1982.
SOFTWARE PRODUCTION ASSOCIATES
'ALPHA' is one of six programs in their *A-level Physics Suite.* From Software Production Associates, P.O. Box 59, Royal Leamington Spa, Warwickshire, CV31 3QA.

APPENDICES

APPENDIX I

Using the 'basic unit' electronics kit

A single 'basic unit' kit comprises the following:
3 basic units
2 indicator units
1 AND gate
1 switch module
1 bistable module
1 multivibrator module
1 beam splitter module

The three basic units may be used as NOR gates by considering only the two resistive inputs and the direct output. It might be helpful to affix some sort of label to make this clear to the students (see figure 1).

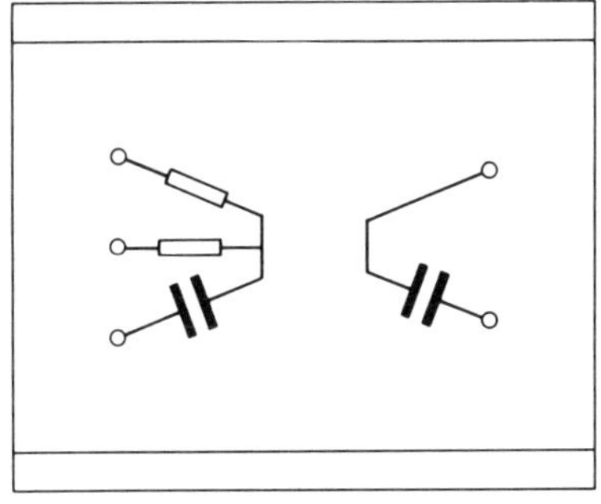

becomes

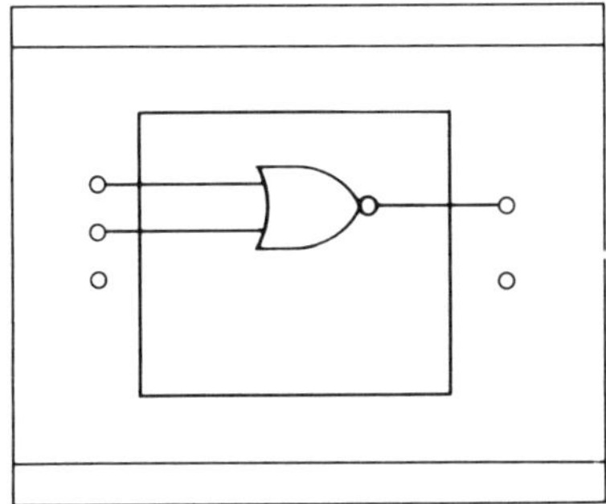

Figure 1

The indicator units and the AND gate may be used unmodified, as may the switch module, provided it has a bounce-free electronic switch circuit.

The bistable module may be used unmodified, but it differs in two ways from the specification of the new digital electronics kit. Firstly, the inputs and outputs are marked differently:

Old name		*New name*
input 1	becomes	C (clear)
input 2	becomes	S (set)
trigger	becomes	clock
output 1	becomes	Q
output 2	becomes	$\overline{\text{Q}}$

Again it may be helpful to affix some sort of label.

Secondly, the bistable module in the basic unit kit changes state when the trigger or clock input goes from a high state to a low state. Where this difference is important, appropriate instructions are given in *Teachers'* and *Students' guides.*

The multivibrator module may be used as a bistable module in the same way if the switches are in the 'R' position.

A high-frequency astable circuit may be made with the multivibrator with the switches in the 'C' position using the circuit in figure 2. Both variable resistors control the frequency which may vary from around 1 kHz down to a few hundred Hz.

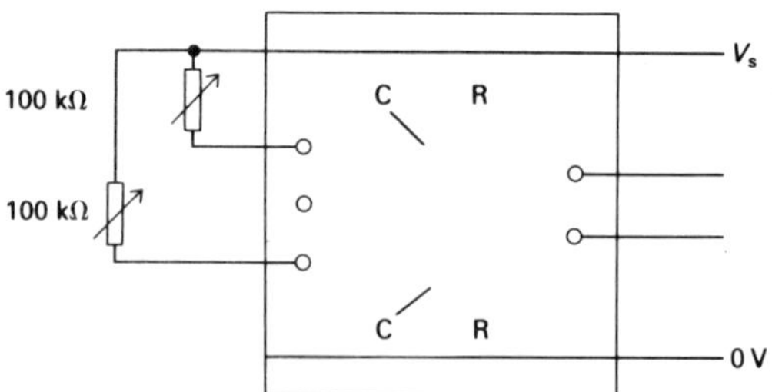

Figure 2

A low-frequency astable circuit may be made by connecting two basic units as shown in figure 3. Again the frequency is controlled by both variable resistors, this time from a few Hz down to perhaps 0.1 Hz.

The beam splitter module is not required for the revised course.

Although the quantity of gates that this provides is enough for many basic experiments to be performed by the class (with some sharing of modules for the more extensive experiments), it provides little introduction to NAND logic, except that which may be demonstrated by inverting the output from an AND gate with a NOR gate (basic unit). We would therefore recommend the addition of two further NAND gates for each student, or perhaps the addition of the quad NAND board, as described in the *Apparatus guide*, which may be used with the basic unit kit.

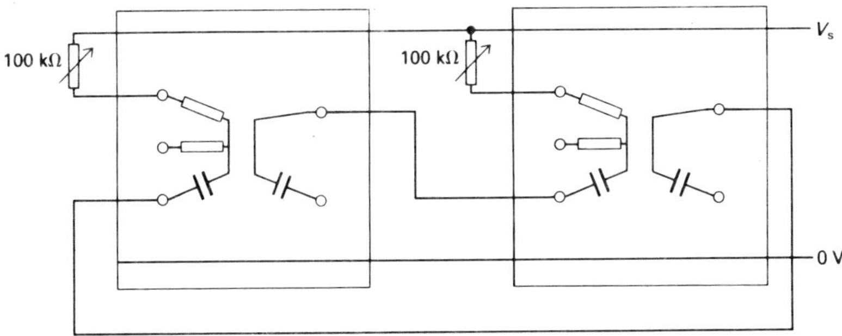

Figure 3
Low-frequency astable circuit from two basic units.

APPENDIX II

British Standard logic symbols

Teachers and students may come across the following British Standard symbols in addition to those used in the course. Note that for both systems, the presence of a circle at the output indicates an inverting gate.

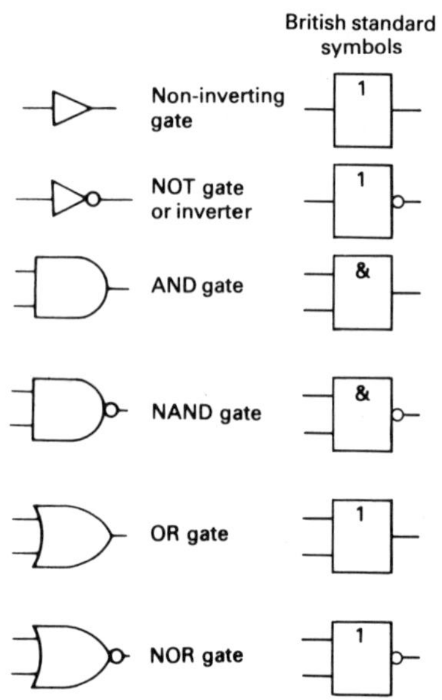

Figure 4

APPENDIX III

Television channels and nominal carrier frequencies (U.K. allocations)

Band IV

Channel	Carrier frequencies/MHz	
	Vision	Sound
21	471.25	477.25
22	479.25	485.25
23	487.25	493.25
24	495.25	501.25
25	503.25	509.25
26	511.25	517.25
27	519.25	525.25
28	527.25	533.25
29	535.25	541.25
30	543.25	549.25
31	551.25	557.25
32	559.25	565.25
33	567.25	573.25
34	575.25	581.25

Band V

Channel	Carrier frequencies/MHz	
	Vision	Sound
39	615.25	621.25
40	623.25	629.25
41	631.25	637.25
42	639.25	645.25
43	647.25	653.25
44	655.25	661.25
45	663.25	669.25
46	671.25	677.25
47	679.25	685.25
48	687.25	693.25
49	695.25	701.25
50	703.25	709.25
51	711.25	717.25
52	719.25	725.25
53	727.25	733.25
54	735.25	741.25
55	743.25	749.25
56	751.25	757.25
57	759.25	765.25
58	767.25	773.25
59	775.25	781.25
60	783.25	789.25
61	791.25	797.25
62	799.25	805.25
63	807.25	813.25
64	815.25	821.25
65	823.25	829.25
66	831.25	837.25
67	839.25	845.25
68	847.25	853.25

APPENDIX IV

Variable phase oscillator

This inexpensive and easily constructed module has been found invaluable in demonstrating the addition of waves of the same frequency and amplitude which differ in phase by any angle in the range 0–π radians.

It has been used in the lower school as a demonstration to explain superposition effects (especially the formation of Young's Fringes) and as a student experiment in the sixth form.

The phase shifter module is built around a 741 op-amp as shown in figure 5.

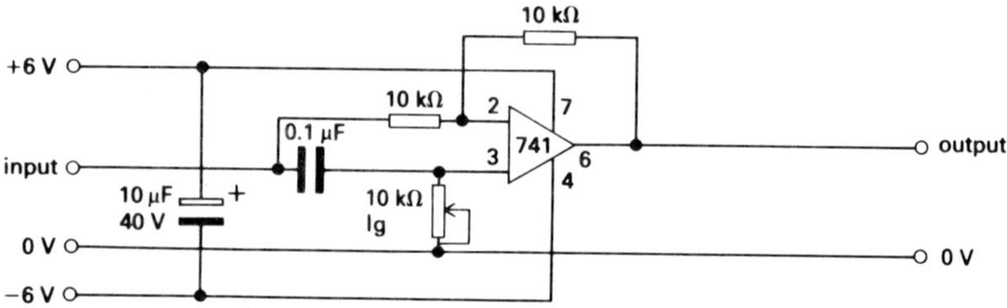

Figure 5

The cost is very low; most of it is the cost of the box and 4 mm sockets.

The module is connected in the circuit as shown in figure 6.

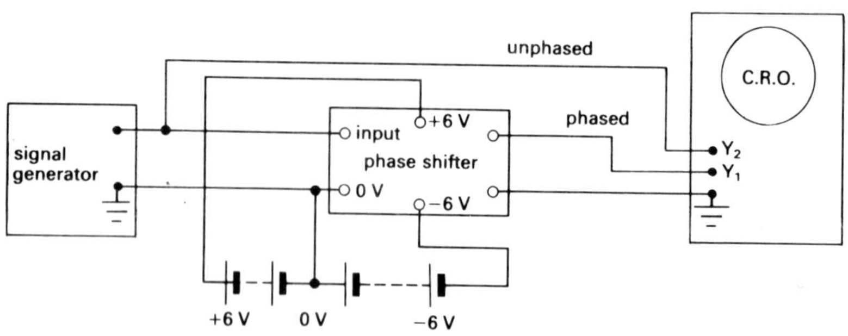

Figure 6

The output is shown on a dual beam oscilloscope; 'add' and 'X–Y' facilities are an advantage. Possible outputs are:
1 The two traces shown together, the shift in phase being immediately apparent.
2 The two traces added to show 'constructive interference' – two traces in phase; and 'destructive interference' as the two signals are brought into antiphase.
3 The Lissajou figures as the phase difference moves from 0 to $\pi/2$ to π radians.

With the component values given, the full phase shift is obtained in the range 1–10 kHz; best input frequency is obtained by tuning the signal generator input to obtain the full range in the Lissajou figure. A logarithmic potentiometer facilitates adjustment at the end of the range.

References

This module was developed by R. P. Ballingall.
The circuit shown was originally adapted from an article by Haglberg (*Phys. Teach.* **16** 58–59, 1978) and substantially the same circuit has since been published by Lopes, Melo, and Oliviera (*Phys. Educ.* **17** 238–240, 1982).

APPENDIX V

Lines of force

In earlier courses students may have used lines to sketch the shapes of magnetic fields. We do not stress the concept of lines of force in the course as it can lead to confusion and error. In figure 7(a), for instance, the field strength can be estimated at any point by reference to the spacing of equipotentials. If field lines are to be drawn so that the line density indicates the strength of the field then one must decide how many lines to draw and where, on the conductors, they should terminate. This is easier if the total charge and its distribution are known, but this is rarely the case.

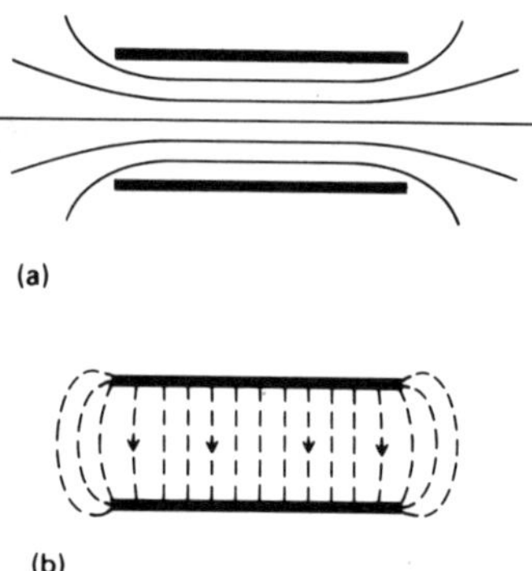

Figure 7
Two ways of representing a field.
(a) Equipotentials.
(b) Field lines.

Figure 7(b), for instance, suggests that charges only exist on the conductors where the field lines terminate. What happens between the lines – is there no field here? Of course there *is* a field here but students may easily fall into this confusion.

Also, in some circumstances, field lines cannot be correctly drawn. For example, beneath the surface of the Earth (figure 8) the gravitational field strength decreases with depth as shown by the increasing spacing of the equipotentials, yet the (erroneous) figure shows field lines getting closer together. The reason is that field lines will not be continuous through space which contains mass (remember Gauss's theorem) so it becomes very difficult to use them to represent the field.

However, the equipotentials still give a clear and true representation of the field at all points (even though the intervals between them must essentially be arbitrary).

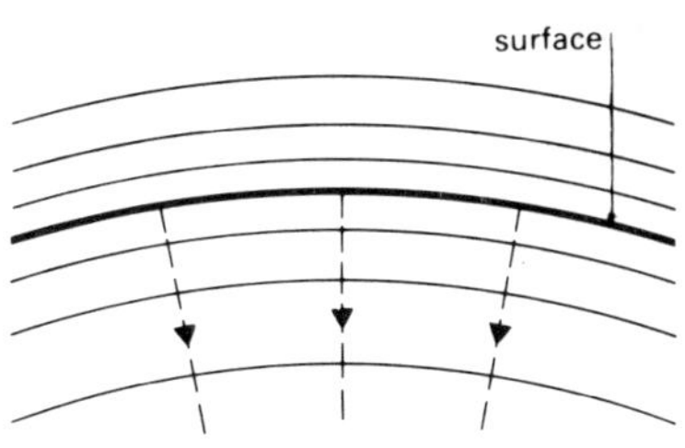

Figure 8
Equipotentials true, field lines false.

APPENDIX VI

Integration of field strength to obtain potential

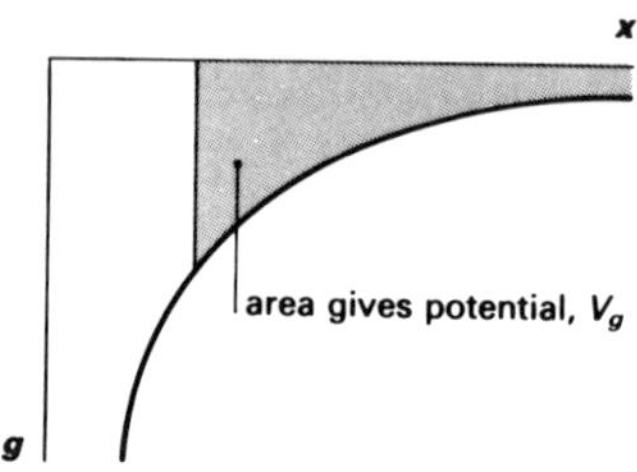

Figure 9
Area under a g against x graph.

The gravitational potential at a point is defined as the amount of energy required to bring unit mass from infinity (where its potential is zero) to that point. This may be found directly from a g against x graph by integration. Note, however, that the expression for g is negative ($g = -GM/x^2$), that the integration proceeds *from* infinity *to* the point, and that there is a negative sign in the relationship between field and potential gradient. This leads eventually to an expression for V_g which is also negative.

Since $g = -\dfrac{dV_g}{dx}$

then $$V_g = -\int_{\infty}^{r} g\,dx$$

$$= -\int_{\infty}^{r} -\frac{GM}{x^2}\,dx$$

$$= +GM\int_{\infty}^{r} \frac{dx}{x^2}$$

$$= GM\left[-\frac{1}{x}\right]_{\infty}^{r}$$

$$= GM\left[-\frac{1}{r} - \left(-\frac{1}{\infty}\right)\right]$$

$$= -\frac{GM}{r}$$

APPENDIX VII

The radial field and the uniform field

The link between the two shapes of field was discussed in the text in physical terms. Here we give a full integration of the effect of a sheet of charge which reveals the force constant k $\left(\text{in } F = k\dfrac{Q_1Q_2}{r^2}\right)$ as $\dfrac{1}{4\pi\varepsilon_0}$.

Integration of the field produced by a flat sheet of charge

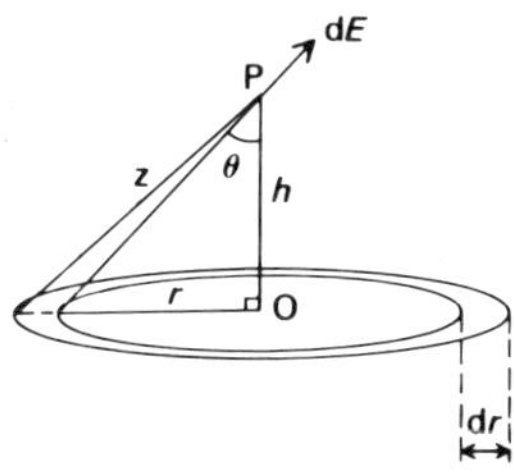

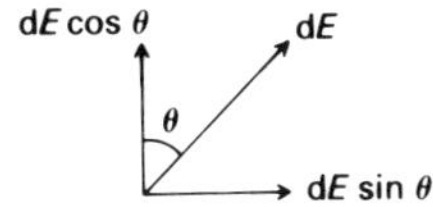

Figure 10
Field of one element of a ring of charge.

Consider a flat sheet with charge density σ. To calculate the total field at P, at height h above the sheet, consider a ring of charge centred on O of radius r and width dr.

The total area of this ring is $2\pi r\mathrm{d}r$

so it carries charge $2\pi\sigma r\mathrm{d}r$

A point charge of this value would make a contribution dE to the field at P of

$2\pi k\sigma r\mathrm{d}r/z^2$

By symmetry, the horizontal components of contributions from the ring cancel out leaving only the vertical component ($\mathrm{d}E\cos\theta$). To calculate the total field strength at P we must add up contributions from all rings from $\theta = 0$ to $\pi/2$ (*i.e.*, covering an infinite sheet).

$$z = \frac{h}{\cos\theta} = h\sec\theta$$

$$r = h\tan\theta \Rightarrow dr = h\sec^2\theta\, d\theta$$

$$E = 2\pi k\sigma \int_0^{\pi/2} \frac{r dr \cos\theta}{z^2}$$

$$= 2\pi k\sigma \int_0^{\pi/2} \frac{h\tan\theta\, h\sec^2\theta\, d\theta \cos\theta}{h^2\sec^2\theta}$$

$$= 2\pi k\sigma \int_0^{\pi/2} \sin\theta\, d\theta$$

$$= 2\pi k\sigma$$

Note that the field strength is independent of the height h: it is uniform.

Let us now consider two infinite parallel plates carrying opposite charges. Figure 11 shows the contributions to the field strength from each plate.

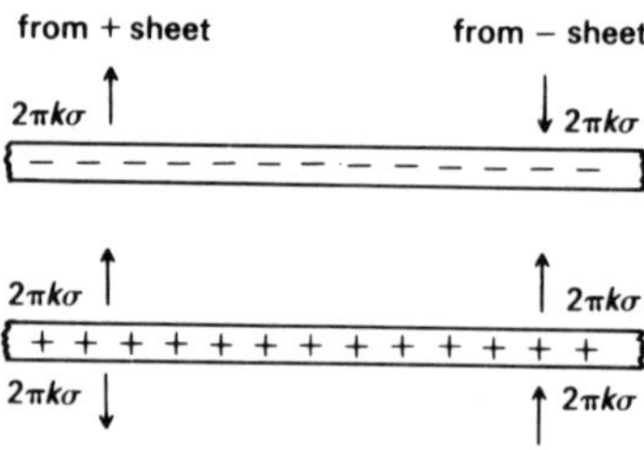

Figure 11
Adding up fields from 2 parallel sheets.

Adding the contributions from the two plates shows that there is no field outside the plates. The field between them is given by

$$E = 4\pi k\sigma$$

Comparing this with the expression we already have for parallel plates (page 312), that

$$E = \sigma/\varepsilon_0$$

we deduce that

$$k = \frac{1}{4\pi\varepsilon_0}$$

INDEX

E

F

G

H

I

K

L

M

N

O

P

T

U

V